PCR Protocols

Second Edition

METHODS IN MOLECULAR BIOLOGY™

John M. Walker, Series Editor

251. **HPLC of Peptides and Proteins:** *Methods and Protocols,* edited by *Marie-Isabel Aguilar, 2004*
250. **MAP Kinase Signaling Protocols,** edited by *Rony Seger, 2004*
249. **Cytokine Protocols,** edited by *Marc De Ley, 2004*
248. **Antibody Engineering:** *Methods and Protocols,* edited by *Benny K. C. Lo, 2004*
247. ***Drosophila* Cytogenetics Protocols,** edited by *Daryl S. Henderson, 2004*
246. **Gene Delivery to Mammalian Cells:** *Volume 2: Viral Gene Transfer Techniques,* edited by *William C. Heiser, 2004*
245. **Gene Delivery to Mammalian Cells:** *Volume 1: Nonviral Gene Transfer Techniques,* edited by *William C. Heiser, 2004*
244. **Protein Purification Protocols, Second Edition,** edited by *Paul Cutler, 2004*
243. **Chiral Separations:** *Methods and Protocols,* edited by *Gerald Gübitz and Martin G. Schmid, 2004*
242. **Atomic Force Microscopy:** *Methods and Applications,* edited by *Pier Carlo Braga and Davide Ricci, 2004*
241. **Cell Cycle Checkpoint Control Protocols,** edited by *Howard B. Lieberman, 2004*
240. **Mammalian Artificial Chromosomes:** *Methods and Protocols*, edited by *Vittorio Sgaramella and Sandro Eridani, 2003*
239. **Cell Migration in Inflammation and Immunity:** *Methods and Protocols,* edited by *Daniele D'Ambrosio and Francesco Sinigaglia, 2003*
238. **Biopolymer Methods in Tissue Engineering,** edited by *Anthony P. Hollander and Paul V. Hatton, 2003*
237. **G Protein Signaling:** *Methods and Protocols,* edited by *Alan V. Smrcka, 2003*
236. **Plant Functional Genomics**: *Methods and Protocols*, edited by *Erich Grotewold, 2003*
235. ***E. coli* Plasmid Vectors**: *Methods and Applications*, edited by *Nicola Casali and Andrew Preston, 2003*
234. **p53 Protocols,** edited by *Sumitra Deb and Swati Palit Deb, 2003*
233. **Protein Kinase C Protocols,** edited by *Alexandra C. Newton, 2003*
232. **Protein Misfolding and Disease:** *Principles and Protocols,* edited by *Peter Bross and Niels Gregersen, 2003*
231. **Directed Evolution Library Creation:** *Methods and Protocols*, edited by *Frances H. Arnold and George Georgiou, 2003*
230. **Directed Enzyme Evolution:** *Screening and Selection Methods,* edited by *Frances H. Arnold and George Georgiou, 2003*
229. **Lentivirus Gene Engineering Protocols,** edited by *Maurizio Federico, 2003*
228. **Membrane Protein Protocols:** *Expression, Purification, and Characterization,* edited by *Barry S. Selinsky, 2003*
227. **Membrane Transporters:** *Methods and Protocols,* edited by *Qing Yan, 2003*
226. **PCR Protocols,** *Second Edition,* edited by *John M. S. Bartlett and David Stirling, 2003*
225. **Inflammation Protocols,** edited by *Paul G. Winyard and Derek A. Willoughby, 2003*
224. **Functional Genomics:** *Methods and Protocols,* edited by *Michael J. Brownstein and Arkady Khodursky, 2003*
223. **Tumor Suppressor Genes:** *Volume 2: Regulation, Function, and Medicinal Applications,* edited by *Wafik S. El-Deiry, 2003*
222. **Tumor Suppressor Genes:** *Volume 1: Pathways and Isolation Strategies,* edited by *Wafik S. El-Deiry, 2003*
221. **Generation of cDNA Libraries:** *Methods and Protocols*, edited by *Shao-Yao Ying, 2003*
220. **Cancer Cytogenetics:** *Methods and Protocols,* edited by *John Swansbury, 2003*
219. **Cardiac Cell and Gene Transfer:** *Principles, Protocols, and Applications,* edited by *Joseph M. Metzger, 2003*
218. **Cancer Cell Signaling:** *Methods and Protocols,* edited by *David M. Terrian, 2003*
217. **Neurogenetics:** *Methods and Protocols,* edited by *Nicholas T. Potter, 2003*
216. **PCR Detection of Microbial Pathogens:** *Methods and Protocols,* edited by *Konrad Sachse and Joachim Frey, 2003*
215. **Cytokines and Colony Stimulating Factors:** *Methods and Protocols,* edited by *Dieter Körholz and Wieland Kiess, 2003*
214. **Superantigen Protocols,** edited by *Teresa Krakauer, 2003*
213. **Capillary Electrophoresis of Carbohydrates,** edited by *Pierre Thibault and Susumu Honda, 2003*
212. **Single Nucleotide Polymorphisms:** *Methods and Protocols,* edited by *Piu-Yan Kwok, 2003*
211. **Protein Sequencing Protocols, Second Edition,** edited by *Bryan John Smith, 2003*
210. **MHC Protocols,** edited by *Stephen H. Powis and Robert W. Vaughan, 2003*
209. **Transgenic Mouse Methods and Protocols,** edited by *Marten Hofker and Jan van Deursen, 2002*
208. **Peptide Nucleic Acids:** *Methods and Protocols,* edited by *Peter E. Nielsen, 2002*
207. **Recombinant Antibodies for Cancer Therapy:** *Methods and Protocols.* edited by *Martin Welschof and Jürgen Krauss, 2002*
206. **Endothelin Protocols,** edited by *Janet J. Maguire and Anthony P. Davenport, 2002*
205. ***E. coli* Gene Expression Protocols,** edited by *Peter E. Vaillancourt, 2002*
204. **Molecular Cytogenetics:** *Protocols and Applications,* edited by *Yao-Shan Fan, 2002*
203. ***In Situ* Detection of DNA Damage:** *Methods and Protocols,* edited by *Vladimir V. Didenko, 2002*
202. **Thyroid Hormone Receptors:** *Methods and Protocols,* edited by *Aria Baniahmad, 2002*
201. **Combinatorial Library Methods and Protocols,** edited by *Lisa B. English, 2002*
200. **DNA Methylation Protocols,** edited by *Ken I. Mills and Bernie H. Ramsahoye, 2002*
199. **Liposome Methods and Protocols,** edited by *Subhash C. Basu and Manju Basu, 2002*
198. **Neural Stem Cells:** *Methods and Protocols,* edited by *Tanja Zigova, Juan R. Sanchez-Ramos, and Paul R. Sanberg, 2002*
197. **Mitochondrial DNA:** *Methods and Protocols,* edited by *William C. Copeland, 2002*
196. **Oxidants and Antioxidants:** *Ultrastructure and Molecular Biology Protocols,* edited by *Donald Armstrong, 2002*
195. **Quantitative Trait Loci:** *Methods and Protocols*, edited by *Nicola J. Camp and Angela Cox, 2002*
194. **Posttranslational Modifications of Proteins:** *Tools for Functional Proteomics,* edited by *Christoph Kannicht, 2002*
193. **RT-PCR Protocols,** edited by *Joe O'Connell, 2002*
192. **PCR Cloning Protocols, Second Edition,** edited by *Bing-Yuan Chen and Harry W. Janes, 2002*
191. **Telomeres and Telomerase:** *Methods and Protocols,* edited by *John A. Double and Michael J. Thompson, 2002*
190. **High Throughput Screening:** *Methods and Protocols,* edited by *William P. Janzen, 2002*
189. **GTPase Protocols:** *The RAS Superfamily,* edited by *Edward J. Manser and Thomas Leung, 2002*
188. **Epithelial Cell Culture Protocols,** edited by *Clare Wise, 2002*
187. **PCR Mutation Detection Protocols,** edited by *Bimal D. M. Theophilus and Ralph Rapley, 2002*
186. **Oxidative Stress Biomarkers and Antioxidant Protocols,** edited by *Donald Armstrong, 2002*
185. **Embryonic Stem Cells:** *Methods and Protocols,* edited by *Kursad Turksen, 2002*
184. **Biostatistical Methods,** edited by *Stephen W. Looney, 2002*

METHODS IN MOLECULAR BIOLOGY™

PCR Protocols

Second Edition

Edited by

John M. S. Bartlett

Department of Surgery, Glasgow Royal Infirmary, Glasgow, UK

and

David Stirling

Department of Hematology, Royal Infirmary of Edinburgh, Edinburgh, UK

Humana Press **Totowa, New Jersey**

999 Riverview Drive, Suite 208
Totowa, New Jersey 07512

www.humanapress.com

This publication is printed on acid-free paper. ∞
ANSI Z39.48-1984 (American Standards Institute) Permanence of Paper for Printed Library Materials.

Production Editor: Robin B. Weisberg.

Cover illustration: Images showing colored bands from fluorescent genotyping. Artwork courtesy of David Stirling.

Cover design by Patricia F. Cleary.

For additional copies, pricing for bulk purchases, and/or information about other Humana titles, contact Humana at the above address or at any of the following numbers: Tel.: 973-256-1699; Fax: 973-256-8341; E-mail: humana@humanapr.com; or visit our Website: www.humanapress.com

Printed in the United States of America. 10 9 8 7 6 5 4 3 2 1

Library of Congress Cataloging in Publication Data

PCR protocols / edited by John M.S. Bartlett, David Stirling. --2nd ed.
p. cm. -- (Methods in molecular biology ; v. 226)
Includes bibliographical references and index.
ISBN 0-89603-642-1 (hardcover) (alk. paper); ISBN 0-89603-627-8 (combbound); 1-59259-384-4 (e-ISBN)
ISSN 1064-3745
1. Polymerase chain reaction—Laboratory manuals. I. Bartlett, John M. S. II. Stirling, David. III. Series.
QP606.D46P3595 2003
572'.43--dc21 2003042324

Preface

Since the discovery/invention of the polymerase chain reaction (PCR) in the mid-1980s, use of this technology has become so widespread that few researchers will progress far in their careers without exposure to PCR in one form or another. The robustness and exquisite sensitivity of PCR has led to the development of a multitude of protocols that exploit different aspects of this methodology. PCR has thus been used to analyze nucleic acids from all known phyla, and even some species extinct for hundreds of years. Disciplines as diverse as ecology, forensic medicine, food production, and archaeology employ PCR analyses as key components of population monitoring, genetic fingerprinting, quality assurance, and research. The place of PCR as a cornerstone of current and future scientific endeavor is beyond question.

Such has been the expansion in PCR methodologies that a revision of the previous edition of *PCR Protocols* dealing with PCR was considered essential. In undertaking this revision we were conscious that although much has changed, the fundamentals of PCR remain unaltered. Indeed in some areas, methods developed 5 or even 10 years ago remain as effective and useful as more recent developments. However, in other areas rapid progress has led to exciting new advances. We are also conscious of a "PCR generation gap," whereby many of the early developments in PCR have passed into folklore. The usefulness of this second edition of *PCR Protocols* to all researchers will, we trust, be enhanced by the provision of an introductory section dealing both with the history of PCR and the legal and practical issues surrounding the development of PCR-based methods today.

Such is the diversity of the PCR methodologies currently available that it would be impossible to provide an exhaustive PCR manual that covers all aspects of the technology. Many of the sections we have included can be, and in some cases have been, expanded into separate publications. Therefore in producing *PCR Protocols* we have sought to select those techniques that form a core manual for both the PCR novice and the adept. The methods we have selected represent worked examples from many fields that can be reproduced and adapted for use within the reader's laboratory. We have sought to provide both a primer to allow the reader to gain basic experience of different PCR techniques, and in-depth insight into a variety of the more complex applications of PCR. Each section has been introduced by an overview chapter whose aim is to provide researchers with an understanding of the applications and limitations of the techniques that follow. As with each volume in the *Methods in Molecular Biology* series, clear instructions for the performance of the various protocols is supplemented by additional technical notes that provide valuable insights into the working of the technique in question. Though often brief, these notes provide essential details that allow for a successful outcome.

We would like to express our gratitude to each of the contributors to *PCR Protocols*, 2nd edition, who have been patient over the period required to collate their contributions. Finally, thanks are owed to our wives, Dorothy and Muriel, for their patience during the long hours we have spent on this volume.

John M. S. Bartlett
David Stirling

Contents

PART IX. CLONING AND MUTAGENESIS

Contributors

Kenji Abe • *Department of Pathology, National Institute of Infectious Diseases, Tokyo, Japan*
Mitsuteru Akahoshi • *Genetic Polymorphism and Function Assay Group, Genetic Polymorphism Institute, RIKEN Yokohama Institute, Yokohama, Japan*
Yaqoub Ashhab • *Unidad de Inmunología, Hospital Universitario Germans Trias i Pujol, and Faculty of Medicine, Universitat Autònoma de Barcelona, Barcelona, Spain*
Michaela Aubele • *Institute of Pathology, GSF-National Research Center for Environment and Health, Neuherberg, Germany*
Sailen Barik • *Department of Biochemistry and Molecular Biology and Department of Microbiology and Immunology, University of South Alabama College of Medicine, Mobile, AL*
John M. S. Bartlett • *Endocrine Cancer Group, Division of Cancer and Molecular Pathology, University Department of Surgery, Glasgow Royal Infirmary, Glasgow, United Kingdom*
Peter B. Becker • *Adolf-Butenandt-Institut, Ludwig Maximilians Universität, Munchen, Germany*
Dean Befus • *Glaxo-Heritage Asthma Research Laboratory, Pulmonary Research Group, Division of Pulmonary Medicine, Department of Medicine, University of Alberta, Edmonton, Alberta, Canada*
Eva Belloso • *Unidad de Inmunología, Hospital Universitario Germans Trias i Pujol, and Faculty of Medicine, Universitat Autònoma de Barcelona, Barcelona, Spain*
William H. Benjamin, Jr. • *Departments of Pathology and Microbiology, University of Alabama at Birmingham, Birmingham, AL*
Edson Bertolini • *Department of Plant Protection and Biotechnology, Instituto Valenciano de Investigaciones Agrarias, Valencia, Spain*
Thomas C. G. Bosch • *Zoological Institute, University of Kiel, Kiel Germany*
David Boyle • *Sidney Kimmel Cancer Center, San Diego, CA*
Jens Bukh • *Hepatitis Viruses and Molecular Hepatitis Sections, Laboratory of Infectious Diseases, National Institute of Allergy and Infectious Diseases, National Institutes of Health, Bethesda, MD*
John H. Bull • *CMC International Ltd, Macclesfield, Cheshire, UK*
Emily Burke • *The Scripps Research Institute, La Jolla, CA*
Mariano Cambra • *Department of Plant Protection and Biotechnology, Instituto Valenciano de Investigaciones Agrarias, Valencia, Spain*
Peter Carroll • *Medlen & Carroll, LLP, Cambridge, MA*
Stephen Case-Green • *Department of Biochemistry, University of Oxford, Oxford, UK*
David Casimir • *Medlen & Carroll, LLP, Madison, WI*

JINGWEN CHEN • *Department of Technology Development, GlaxoSmithKline, Research Triangle Park, NC*
ALISON J. COFFEY • *The Wellcome Trust Sanger Institute, Cambridge, UK*
BIANCA CONTI-FINE • *Department of Biochemistry, University of Minnesota, St. Paul, MN*
FRIEDRICH W. CREMER • *Department of Internal Medicine, University of Heidelberg, Heidelberg, Germany*
RANIL S. DASSANAYAKE • *Faculty of Medicine, Department of Biochemistry and Molecular Biology, University of Colombo, Sri Lanka*
ORLANDO DOMINGUEZ • *Programa de Biotechnologia, Centro Nacional de Investigaciones Oncologicas, Madrid, Spain*
JEAN BAPTISTE DUMAS MILNE EDWARDS • *GENSET, Paris, France*
IAN DUNHAM • *The Wellcome Trust Sanger Institute, Cambridge, UK*
JOHN J. DUNN • *Biology Department, Brookhaven National Laboratory, Upton, NY*
JOANNE EDWARDS • *Endocrine Cancer Group, Division of Cancer and Molecular Pathology, University Department of Surgery, Glasgow Royal Infirmary, Glasgow, United Kingdom*
RUPERT EGENSPERGER • *Institute of Neuropathology, University of Münster, Münster, Germany*
MORTIMER M. ELKIND • *Department of Radiological Health Sciences, Colorado State University, Fort Collins, CO*
SUZANNE U. EMERSON • *Hepatitis Viruses and Molecular Hepatitis Sections, Laboratory of Infectious Diseases, National Institute of Allergy and Infectious Diseases, National Institutes of Health, Bethesda, MD*
OLGA ESTEBAN • *Department of Plant Protection and Biotechnology, Instituto Valenciano de Investigaciones Agrarias, Valencia, Spain*
MARK GILCHRIST • *Glaxo-Heritage Asthma Research Laboratory, Pulmonary Research Group, Division of Pulmonary Medicine, Department of Medicine, University of Alberta, Edmonton, Alberta, Canada*
JAMES J. GOING • *Department of Pathology, Glasgow Royal Infirmary, Glasgow, UK*
TORSTEN GOLDMANN • *Clinical and Experimental Pathology, Research Center Borstel, Borstel, Germany*
MANUEL B. GRAEBER • *Department of Neuropathology, Imperial College, London, UK*
HAIYING GRUNENWALD • *Epicentre Technologies Corp., Madison, WI*
MEI HAN • *Transplantation Immunology, Department of Internal Medicine, UT Southwestern Medical Center, Dallas, TX*
TAKUYA HIGASHIYAMA • *Molecular Biology Program, Sidney Kimmel Cancer Center, San Diego, CA*
ANTON H. N. HOPMAN • *Department of Molecular Cell Biology, Research Institute Growth & Development, University of Maastricht, The Netherlands*
ROBERT M. HORTON • *Attotron Biosensor Corp., Carson City, NV*
DAVID L. HYNDMAN • *Software Development, Automation Biology Drug Discovery Department, Ribapharm Inc., Costa Mesa, CA*
MARIE A. IANNONE • *Department of Gene Expression and Protein Biochemistry, GlaxoSmithKline, Research Triangle Park, NC*
CHRISTINE ICARD-LIEPKALNS • *GENSET, Paris, France*

DOUGLAS H. JONES • *Cedar Centre, Cedar Rapids, IA*
RON KERR • *Department of Haematology, Royal Infirmary of Edinburgh, Edinburgh, UK*
KOSTYA KHALTURIN • *Zoological Institute, University of Kiel, Kiel, Germany*
SIEGFRIED KÖSEL • *Labor Becker, Olgemöller & Kollegen, Munich, Germany*
SERGEJ KUZNETSOV • *Zoological Institute, University of Kiel, Kiel, Germany*
MAY-SUNG LI • *Bio-Informatics Group Inc., NC*
JINGMEI LIU • *Life Sciences Division, Los Alamos National Laboratory, Los Alamos, NM*
CHRISTOPH B. LÜCKING • *Department of Neurology, University of Munich, Munich, Germany*
JACQUES MALLET • *CNRS, Paris, France*
GEORGES-RAOUL MAZARS • *Université Paul Sabatier, Tailaise, France*
MARCIA A. MCALEER • *Department of Molecular Biology, Yamanouchi Research Institute, Oxford, UK*
MICHAEL MCCLELLAND • *Molecular Biology Program, Sidney Kimmel Cancer Center, San Diego, CA*
SUSAN MCDONAGH • *Department of Stomatology, University of California, San Francisco, CA*
MASATO MITSUHASHI • *Hitachi Chemical Research Center, Irvine, CA*
MARION MOOSE • *Department of Internal Medicine, University of Heidelberg, Heidelberg, Germany*
HITOSHI NAKASHIMA • *Medicine and Biosystemic Science, Kyushu University Graduate School of Medical Sciences, Fukuoka, Japan*
KARIN A. OIEN • *CRC Department of Medical Oncology, University of Glasgow, Glasgow, UK*
ANTONIO OLMOS • *Department of Plant Protection and Biotechnology, Instituto Valenciano de Investigaciones Agrarias, Valencia, Spain*
LYNN PASKINS • *a-1 biotech UK, Bedfordshire, UK*
HELEN PEARSON • *Molecular Endocrinology Unit, Western General Hospital, Edinburgh, UK*
GENEVIEVE PONT-KINGDON • *ARUP Institute for Clinical and Experimental Pathology, Salt Lake City, UT*
STEFFEN PORWOLLIK • *Molecular Biology Program, Sidney Kimmel Cancer Center, San Diego, CA*
GREGORY M. PRESTON • *Genomic and Proteomic Sciences, Pfizer Inc., Groton, CT*
CLARE PRITCHARD • *Harwell Microarray Facility, MRC Mammalian Genetics Unit, Harwell, Oxfordshire, UK*
RICARDO PUJOL-BORRELL • *Unidad de Inmunología, Hospital Universitario Germans Trias i Pujol, and Faculty of Medicine, Universitat Autònoma de Barcelona, Barcelona, Spain*
ROBERT H. PURCELL • *Hepatitis Viruses and Molecular Hepatitis Sections, Laboratory of Infectious Diseases, National Institute of Allergy and Infectious Diseases, National Institutes of Health, Bethesda, MD*
JEAN-PIERRE QUIVY • *Institut Curie, Section de Recherche UMR 218, Paris France*
RAGHAVANPILLAI RAJU • *Department of Biochemistry, University of Minnesota, St. Paul, MN*

Frans C. S. Ramaekers • *Department of Molecular Cell Biology, Research Institute Growth & Development, University of Maastricht, The Netherlands*
Philippe Ravassard • *GENSET, Paris, France*
Steven Ringquist • *Sidney Kimmel Cancer Center, San Diego, CA*
Rosa-Ana Risques • *Sidney Kimmel Cancer Center, San Diego, CA*
Kanokporn Rithidech • *Department of Pathology, State University of New York at Stony Brook, Stony Brook, NY*
Mary Ann Robinson • *Research Technologies Branch, National Institute of Allergy and Infectious Diseases, National Institutes of Health, Bethesda, MD*
Gaelle Rondeau • *Sidney Kimmel Cancer Center, San Diego, CA*
Lidia Sabater • *Unidad de Inmunología, Hospital Universitario Germans Trias i Pujol, and Faculty of Medicine, Universitat Autònoma de Barcelona, Barcelona, Spain*
Lakshman P. Samaranayake • *Oral Bio-Sciences, Faculty of Dentistry, The University of Hong Kong, Hong Kong, China*
Wera M. Schmerer • *Paleo DNA Laboratory, Department of Anthropology, Lakehead University, Thunder Bay, Ontario, Canada*
Zhiyuan Shen • *Life Sciences Division, Los Alamos National Laboratory, Los Alamos, NM*
Jan Smida • *Technische Universität München, Institute of Pathology, Munich, Germany*
Kim R. Smith • *Division of Infectious Diseases, Department of Medicine, University of Alabama at Birmingham, Birmingham, AL*
Edwin Southern • *Department of Biochemistry, University of Oxford, Oxford, UK*
Ernst J. M. Speel • *Research Institute Growth & Development, Department of Molecular Cell Biology, University of Maastricht, The Netherlands*
David Stirling • *Department of Hematology, Royal Infirmary of Edinburgh, Edinburgh, UK*
Anu Suomalainen • *Programme of Neurosciences and Department of Neurology, Biomedicum-Helsinki, Helsinki University, Helsinki, Finland*
Ann-Christine Syvänen • *Department of Medical Sciences, Uppsala University Hospital, Uppsala, Sweden*
Yosuke Tanaka • *Department of Internal Medicine, Sawara Hospital, Meinohama Nishi-ku, Japan*
J. David Taylor • *Department of Technology Development, GlaxoSmithKline, Research Triangle Park, NC*
Raymond Tellier • *Division of Microbiology, Department of Paediatric Laboratory Medicine, The Hospital for Sick Children, Toronto, Ontario, Canada*
Charles Theillet • *INSERM, Cancer Center of Montpellier, Montpellier, France*
Ken B. Waites, *Departments of Pathology and Microbiology, University of Alabama at Birmingham, Birmingham, AL*
Kai Wang • *PhenoGenomics Corp., Bothell, WA*
Xin Wang • *CELERA, an Applera Corporation Business, Rockville, MD*
Yi-Peng Wang • *Molecular Biology Program, Sidney Kimmel Cancer Center, San Diego, CA*

David Walsh • *Crop Science Department, Scottish Agricultural College, Edinburgh, UK*
Michael P. Weiner • *454 Corp., Branford, CT*
Robert L. Wells • *Radiological Health Sciences Department, Colorado State University, Fort Collins, CO*
John Welsh • *Molecular Biology Program, Sidney Kimmel Cancer Center, San Diego, CA*
Anne White • *DNASHEF Technologies, Royal Infirmary of Edinburgh, Edinburgh, UK*
Klaus Hermann Wiedorn • *Institut für Pathologie, Katharinenhospital Stuttgart, Stuttgart, Germany*
Stanley C. Winistorfer • *Department of Physiology, University of Iowa, Iowa City, IA*
Fei Ye • *IcAgen Inc., Durham, NC*
W. Scott Young III • *Laboratory of Cellular and Molecular Regulation, National Institute of Mental Health, National Institutes of Health, Bethesda, MD*

I

Introduction to PCR

1

A Short History of the Polymerase Chain Reaction

John M. S. Bartlett and David Stirling

The development of the polymerase chain reaction (PCR) has often been likened to the development of the Internet, and although this does risk overstating the impact of PCR outside the scientific community, the comparison works well on a number of levels. Both inventions have emerged in the last 20 years to the point where it is difficult to imagine life without them. Both have grown far beyond the confines of their original simple design and have created opportunities unimaginable before their invention. Both have also spawned a whole new vocabulary and professionals literate in that vocabulary. It is hard to believe that the technique that formed the cornerstone of the human genome project and is fundamental to many molecular biology laboratory protocols was discovered only 20 years ago. For many, the history and some of the enduring controversies are unknown yet, as with the discovery of the structure of DNA in the 1950s, the discovery of PCR is the subject of claim and counterclaim that has yet to be fully resolved. The key stages are reviewed here in brief for those for whom both the history and application of science holds interest.

The origins of PCR as we know it today sprang from key research performed in the early 1980s at Cetus Corporation in California. The story is that in the spring of 1983, Kary Mullis had the original idea for PCR while cruising in a Honda Civic on Highway 128 from San Francisco to Mendocino. This idea claimed to be the origin of the modern PCR technique used around the world today that forms the foundation of the key PCR patents. The results for Mullis were no less satisfying; after an initial $10,000 bonus from Cetus Corporation, he was awarded the 1993 Nobel Prize for chemistry.

The original concept for PCR, like many good ideas, was an amalgamation of several components that were already in existence: The synthesis of short lengths of single-stranded DNA (oligonucleotides) and the use of these to direct the target-specific synthesis of new DNA copies using DNA polymerases were already standard tools in the repertoire of the molecular biologists of the time. The novelty in Mullis's concept was using the juxtaposition of two oligonucleotides, complementary to opposite strands of the DNA, to specifically amplify the region between them and to achieve this in a repetitive manner so that the product of one round of polymerase activity was added to the pool of template for the next round, hence the chain reaction. In his *History of PCR (1)*, Paul Rabinow quotes Mullis as saying:

From: *Methods in Molecular Biology, Vol. 226: PCR Protocols, Second Edition*
Edited by: J. M. S. Bartlett and D. Stirling © Humana Press Inc., Totowa, NJ

> The thing that was the "Aha!" the "Eureka!" thing about PCR wasn't just putting those [things] together...the remarkable part is that you will pull out a little piece of DNA from its context, and that's what you will get amplified. That was the thing that said, "you could use this to isolate a fragment of DNA from a complex piece of DNA, from its context." That was what I think of as the genius thing....In a sense, I put together elements that were already there....You can't make up new elements, usually. The new element, if any, it was the combination, the way they were used....The fact that I would do it over and over again, and the fact that I would do it in just the way I did, that made it an invention...the legal wording is "presents an unanticipated solution to a long-standing problem," that's an invention and that was clearly PCR.

In fact, although Mullis is widely credited with the original invention of PCR, the successful application of PCR as we know it today required considerable further development by his colleagues at Cetus Corp, including colleagues in Henry Erlich's lab (2–4), and the timely isolation of a thermostable polymerase enzyme from a thermophilic bacterium isolated from thermal springs. Furthermore, challenges to the PCR patents held by Hoffman La Roche have claimed at least one incidence of "prior art," that is, that the original invention of PCR was known before Mullis's work in the mid-1980s. This challenge is based on early studies by Khorana et al. in the late 1960s and early 1970s (see chapter 2). Khorana's work used a method that he termed repair replication, and its similarity to PCR can be seen in the following steps: (1) annealing of primers to templates and template extension; (2) separation of the newly synthesized strand from the template; and (3) re-annealing of the primer and repetition of the cycle. Readers are referred to an extensive web-based literature on the patent challenges arising from this "prior art" and to chapter 2 herein for further details. Whatever the final outcome, it is clear that much of the work that has made PCR such a widely used methodology arose from the laboratories of Mullis and Erlich at Cetus in the mid-1980s.

The DNA polymerase originally used for the PCR was extracted from the bacterium *Escherichia coli*. Although this enzyme had been a valuable tool for a wide range of applications and had allowed the explosion in DNA sequencing technologies in the preceding decade, it had distinct disadvantages in PCR. For PCR, the reaction must be heated to denature the double-stranded DNA product after each round of synthesis. Unfortunately, heating also irreversibly inactivated the *E. coli* DNA polymerase, and therefore fresh aliquots of enzyme had to be added by hand at the start of each cycle. What was required was a DNA polymerase that remained stable during the DNA denaturation step performed at around 95°C. The solution was found when the bacterium *Thermophilus aquaticus* was isolated from hot springs, where it survived and proliferated at extremely high temperatures, and yielded a DNA polymerase that was not rapidly inactivated at high temperatures. Gelfand and his associates at Cetus purified and subsequently cloned this polymerase ***(5,6)***, allowing a complete PCR amplification to be created without opening the reaction tube. Furthermore, because the enzyme was isolated from a thermophilic organism, it functioned optimally at temperature of around 72°C, allowing the DNA synthesis step to be performed at higher temperatures than was possible with the *E. coli* enzyme, which ensured that the template DNA strand could be copied with higher fidelity as the result of a greater stringency of primer binding, eliminating the nonspecific products that had plagued earlier attempts at PCR amplification.

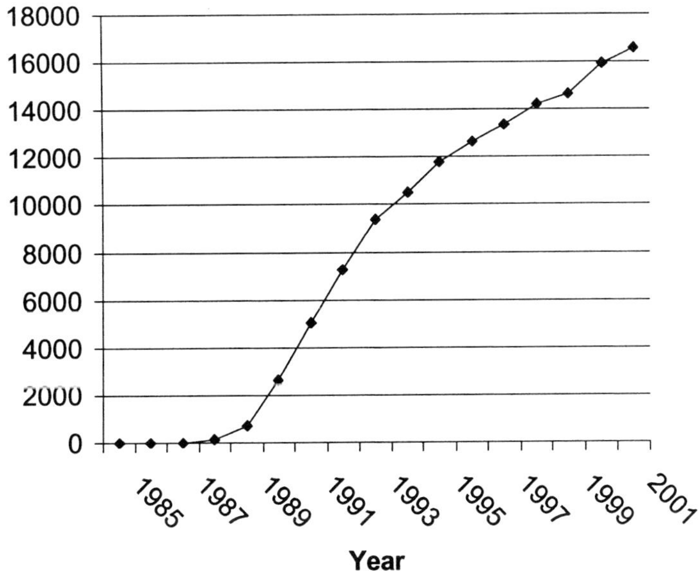

Fig. 1. Results of a PubMed search for articles containing the phrase "Polymerase Chain Reaction." Graph shows number of articles listed in each year.

However, even with this improvement, the PCR technique was laborious and slow, requiring manual transfer between water baths at different temperatures. The first thermocycling machine, "Mr Cycle," which replicated the temperature changes required for the PCR reaction without the need for manual transfer, was developed by Cetus to facilitate the addition of fresh thermolabile polymerases. After the purification of *Taq* polymerase, Cetus and Perkin–Elmer introduced the closed DNA thermal cyclers that are widely used today *(7)*.

That PCR has become one of the most widely used tools in molecular biology is clear from **Fig. 1**. What is not clear from this simplistic analysis of the literature is the huge range of questions that PCR is being used to answer. Another scientist at Cetus, Stephen Scharf, is quoted as stating that

> …the truly astonishing thing about PCR is precisely that it wasn't designed to solve a problem; once it existed, problems began to emerge to which it could be applied. One of PCR's distinctive characteristics is unquestionably its extraordinary versatility. That versatility is more than its 'applicability' to many different situations. PCR is a tool that has the power to create new situations for its use and those required to use it.

More than 3% of all PubMed citations now refer to PCR (**Fig. 2**). Techniques have been developed in areas as diverse as criminal forensic investigations, food science, ecological field studies, and diagnostic medicine. Just as diverse are the range of adaptations and variations on the original theme, some of which are exemplified in this volume. The enormous advances made in our understanding of the human genome (and that of many other species), would not have been possible, where it not for the remarkable simple and yet exquisitely adaptable technique which is PCR.

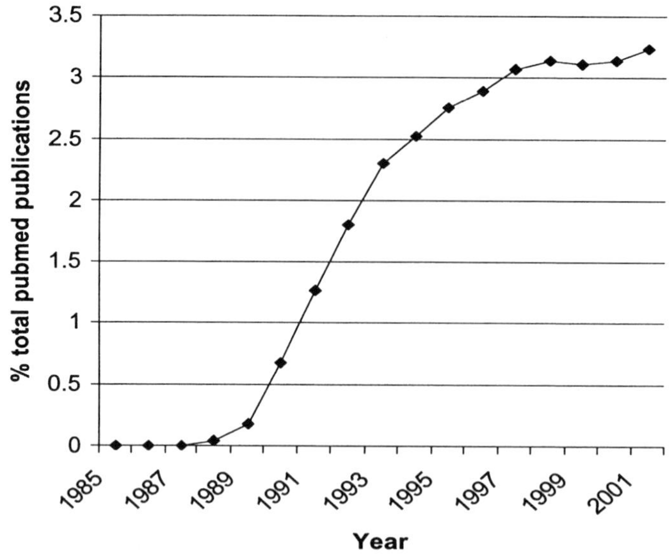

Fig. 2. Results of a PubMed search for articles containing the phrase "Polymerase Chain Reaction." Graph shows number of articles listed in each year expressed as a percentage of the total PubMed citations for each year.

References

1. Rabinow, P. (1996) *Making PCR: A Story of Biotechnology.* University of Chicago Press, Chicago.
2. Saiki, R., Scharf, S., Faloona, F., Mullis, K., Horn, G., and Erlich, H. (1985) Enzymatic amplification of beta-globin genomic sequences and restriction site analysis for diagnosis of sickle cell anemia. *Science* **230,** 1350–1354.
3. Mullis, K., Faloona, F., Scharf, S., Saiki, R., Horn, G., and Erlich, H. (1986) Specific enzymatic amplification of DNA in vitro: The polymerase chain reaction. *Cold Spring Harbor Symp. Quant. Biol.* **51,** 263–273.
4. Mullis, K. and Faloona, F. (1987) Specific synthesis of DNA in vitro via a polymerase-catalyzed chain reaction. *Methods Enzymol.* **155,** 335–350.
5. Saiki, R., Gelfand, D., Stoffel, S., Scharf, S., Higuchi, R., Horn, et al. (1988) Primer-directed enzymatic amplification of DNA with a thermostable DNA polymerase. *Science* **239,** 487–491.
6. Lawyer, F., Stoffer, S, Saiki, R., Chang, S., Landre, P., Abramson, R., et al. (1993) High-level expression, purification, and enzymatic characterization of full-length Thermus aquaticus DNA polymerase and a truncated form deficient in 5′ to 3′ exonuclease activity. *PCR Methods Appl.* **2,** 275–287.
7. http://www.si.edu/archives/ihd/videocatalog/9577.htm

2

PCR Patent Issues

Peter Carroll and David Casimir

1. Introduction

The science of the so-called polymerase chain reaction (PCR) is now well known. However, the legal story associated with PCR is, for the most part, not understood and constantly changing. This presents difficulties for scientists, whether in academia or industry, who wish to practice the PCR process. This chapter summarizes the major issues related to obtaining rights to practice PCR. The complexity of the patent system is explained with a few PCR-specific examples highlighted. The chapter also provides an overview of the exemption or exception from patent infringement associated with certain bona-fide researchers and discusses the status of certain high-profile patents covering aspects of the PCR process.

2. Intellectual Property Rights

Various aspects of the PCR process, including the method itself, are protected by patents in the United States and around the world. As a general rule, patents give the patent owner the exclusive right to make, use, and sell the compositions or process claimed by the patent. If someone makes, uses, or sells the patented invention in a country with an issued patent, the patent owner can invoke the legal system of that country to stop future infringing activities and possibly recover money from the infringer.

A patent owner has the right to allow, disallow, or set the terms under which people make, use, and sell the invention claimed in their patents. In an extreme situation, a patent owner can exclude everyone from making, using, and selling the invention, even under conditions where the patent owner does not produce the product themselves—effectively removing the invention from the public for the lifetime of the patent (typically 20 years from the filing date of the patent). If a patent owner chooses to allow others to make, use, or sell the invention, they can contractually control nearly every aspect of how the invention is disbursed to the public or to certain companies or individuals, so long as they are not unfairly controlling products not covered by the patent. For example, a patent owner can select or exclude certain fields of use for methods like PCR (e.g., research use, clinical use, etc.) while allowing others.

There are an extraordinary number of patents related to the PCR technology. For example, in the United States alone, there are more than 600 patents claiming aspects

From: *Methods in Molecular Biology, Vol. 226: PCR Protocols, Second Edition*
Edited by: J. M. S. Bartlett and D. Stirling © Humana Press Inc., Totowa, NJ

of PCR. Such patents cover the basic methods itself (originally owned by Cetus Corporation and now owned by Hoffmann-LaRoche), thermostable polymerases useful in PCR, as well as many non-PCR applications, (e.g., *Taq* polymerase, Tth polymerase, Pfu polymerase, KOD polymerase, Tne polymerase, Tma polymerase, modified polymerases, etc.), devices used in PCR (e.g., thermocyclers, sample tubes and vessels, solid supports, etc.), reagents (e.g., analyte-specific amplification primers, buffers, internal standards, etc.), and applications involving the PCR process (e.g., reverse-transcription PCR, nested PCR, multiplex PCR, nucleic acid sequencing, and detection of specific analytes). This collection of patents is owned by a wide variety of entities, including government agencies, corporations, individual inventors, and universities. However, the most significant patents (*see* **Table 1**), covering the basic PCR method, the most widely used polymerase (*Taq* polymerase), and thermocyclers, are assigned to Hoffmann-LaRoche and are controlled by Hoffmann-LaRoche or Applera Corporation (previously known as PE/Applied Biosystems) and are available to the public through an intricate web of licenses.

3. Navigating the PCR Patent Minefield

The following discussion focuses on issues related to the earliest and most basic PCR-related patents. A full analysis of the hundreds of PCR-related patents is not practical in an article this size, let alone a multivolume treatise. It is hoped that the following discussion will provide a preliminary framework for understanding the broad PCR patent landscape.

The early PCR patents now owned by Hoffmann-LaRoche have been aggressively enforced. In particular, the earliest patents intended to cover the basic PCR method and the *Taq* polymerase enzyme (U.S. Patent No. 4,683,202 to Kary Mullis, U.S. Patent No. 4,683,195 to Kary Mullis et al., U.S. Patent No. 4,889,818 to Gelfand et al. and foreign counterparts) have regularly been litigated and used to threaten litigation, even against academic researchers. This aggressive patent stance has created an environment of fear, confusion, and debate, particularly at universities and among academic researchers. Because of this aggressive patent enforcement, issues with respect to these patents are most relevant and are focused on herein.

3.1. Obtaining Rights to Practice PCR

In the case of the early PCR patents, Hoffmann-LaRoche, directly and through certain designated partners, has made PCR available to the public under specific conditions, depending on the intended use of the method (*see* <http://biochem.roche.com/PCRlicense.htm> for availability of licenses and current details). For example, for nonsequencing research use, PCR users have two options. They can individually negotiate a license from Applera (a proposition that is impractical for many researchers). Optionally, they can purchase "certain reagents" from a "licensed supplier" in conjunction with the use of "an authorized thermal cycler." This essentially means that the user must purchase thermostable enzymes and thermocyclers from suppliers licensed by Hoffmann-LaRoche or Applera. Not surprisingly, the price of these products from licensed suppliers greatly exceeds the price of equivalent products from nonlicensed suppliers. Indeed, thermostable enzymes from licensed suppliers may

Table 1
PCR Patents

U.S. patent number	Issue date	Expiration date	Related international patents	Claimed technology
4,683,195	07/28/87	03/28/05	Australia: 591104B Australia: 586233B Canada: 134012B Europe: 200362B Europc: 201184B Europe: 505012B Japan: 2546576B Japan: 2622327B Japan: 4067957B Japan: 4067960B	Amplification methods
4,683,202	07/28/87	03/28/05	Same as 4,683,195	Amplification methods
4,965,188	10/23/90	03/28/05	Australia: 586233B Australia: 591104B Australia: 594130B Australia: 632857B Canada: 1340121B Europe: 200362B Europe: 201184B Europe: 237362B Europe: 237362B Europe: 258017B Europe: 459532B Europe: 505012B Japan: 2502041B Japan: many others	Amplification methods using thermostable polymerases
4,889,818	12/26/86	12/26/06 (currently unenforceable)	Australia: 632857B Europe: 258017B Japan: 2502041B Japan: 2502042B Japan: 2719529B Japan: 3031434B Japan: 5074345B Japan: 8024570B	Purified *Taq* polymerase enzyme
5,079,352	01/07/92	01/07/09	Same as 4,889,818, plus Europe: 395736B Japan: 2511548B Japan: 2511548B	Recombinant *Taq* polymerase enzyme and fragments
5,038,852	8/13/91	08/13/08	Australia: 612316B Australia: 653932B Europe: 236069B Japan: 2613877B	Apparatus and method for performing automated amplification

cost more than twice as much as from nonlicensed suppliers ***(1)***. This elevated cost can place a substantial financial burden on researchers who require heavy PCR usage, particularly academic researchers on fixed and limited grant budgets. To the extent universities require their researchers to used licensed products, the aggregate cost increase for many large research universities is substantial. (For a list of *Taq* polymerase suppliers and prices, including licensed and unlicensed suppliers, *see* Constans, **ref.** ***2***).

3.2. Bona-Fide Researchers Are Not Infringers

As mentioned previously, Hoffmann-LaRoche has taken the position that academic researchers are infringers of their patents if they are not meeting the prescribed licensing requirements (e.g., not purchasing authorized reagents and equipment). At one point. several years ago, Hoffmann-LaRoche specifically named more than 40 American universities and government laboratories and more than 200 individual scientists as directly infringing certain patents through their basic research ***(3)***. Voicing the view of many researchers, Dr. Arthur Kornberg, professor emeritus at Stanford University and Nobel laureate, has stated that the actions by Hoffmann-LaRoche to restrain the use and extension of PCR technology by universities and nonprofit basic research institutions "violated practices and principles basic to the advancement of knowledge for the public welfare."

Fortunately for academic researchers, the laws of the United States and other jurisdictions agree with Dr. Kornberg. US patent law recognizes an exemption or exception from infringement associated with bona-fide research (i.e., not-for-profit activities). The experimental use exception to the patent infringement provisions of US law has its origins in the notion that "it could never have been the intention of the legislature to punish a man, who constructed…a [patented] machine merely for philosophical experiments…." ***(4)***. An authoritative discussion on the research use exception appears in the case *Roche Prods., Inc. v Bolar Pharmaceutical Co.* ***(5)***. Even though this case is generally considered to restrict the scope of the research use exemption (failing to find noninfringement where the defendant's acts were "solely for business reasons"), the case makes it clear that the exception is alive and well where the acts are "for amusement, to satisfy idle curiosity, or *for strictly philosophical inquiry*." Thus, to the extent that researchers' use of PCR is not applied to commercial applications or development (e.g., for-sale product development, for-profit diagnostic testing), the researchers cannot be considered infringers. For example, pure basic research, which describes most university research, cannot be considered commercial, and the researchers are not infringers. This applied to the PCR patents, as well as any other patent. Hoffmann-LaRoche has taken the position that "These researchers…are manifestly in the business of doing research in order to…attract private and government funding through the publication of their experiments in the scientific literature, create patentable inventions, and generate royalty income for themselves and their institutions through the licensing of such invention." However, current US law does not support this extraordinarily broad view of commercial activity, and Hoffmann-LaRoche seems to be alone in making such broad assertions.

Although the above discussion relates to the United States, researchers in other countries may or may not have the same exemption. The scope of this article does not permit a country-by-country analysis. However, it must be noted that many countries are in alignment with the position taken by US courts or provide an even broader exemption. For example, it is not considered an infringement in Canada to construct a patented article for the purpose of improving upon it or to ascertain whether a certain addition, subtraction, or improvement on it is workable. The Supreme Court of Canada spoke on this issue stating that "[N]o doubt if a man makes things merely by way of bona fide experiment, and not with the intention of selling and making use of the thing so made for the purpose of which a patent has been granted, but with the view of improving upon the invention the subject of the patent, or with the view of assessing whether an improvement can be made or not, that is not an invasion of the exclusive rights granted by the patent. Patent rights were never granted to prevent persons of ingenuity exercising their talents in a fair way." Likewise, UK law provides an exemption from infringement for acts that are performed privately and for purposes that are not commercial and for acts performed for experimental purposes relating to the subject matter of the invention. The experimental purposes may have a commercial end in view, but they are only exempt from infringement if they relate to the subject matter of the invention. For example, it has been held by the UK courts that trials conducted to discover something unknown or to test a hypothesis, to find out whether something which is known to work in specific conditions would work in different conditions, or even perhaps to see whether the experimenter could manufacture commercially in accordance with the patent can be regarded as experiments and exempted from infringement. Researchers in any particular country who wish to obtain current information about their ability to conduct research projects without incurring patent infringement liability should contact the patent office or an attorney in their respective countries. Unfortunately, there is very little literature addressing these issues, and because the law is constantly changing, older articles may not provide accurate information.

Even with uncertainties, it is clear that in many locations, researchers conducting basic research without a commercial end are free to practice in their field without fear or concern about the patent rights of others. Researchers at corporations likely cannot take advantage of such an infringement exemption. For researchers involved in work with a commercial link (e.g., researchers at private corporations, diagnostic laboratories reporting patient results for fees, academic research laboratories with private corporate collaborations, and the like), a license may be required. Unfortunately, each case needs to be evaluated on its own facts to determine whether a license is required and no general formula can be given. However, many corporations have personnel responsible for analyzing the need for, and acquisition of, patent rights. As such, bench scientists can generally go about their work without the burden of worrying about patent rights, or at a minimum, need only know the basic principles and issues so as to inform the appropriate personnel if potential patent issues arise.

3.3. Not Every Patent Is a Valid Patent

In addition to the experimental use exception, researchers, including commercial researchers, may obtain freedom from the early PCR patents because of problems with the patents themselves. Although issued patents are presumed valid and are enforceable until a court of law says otherwise, the early PCR patents have begun to fall under scrutiny and may not be upheld in the future such that the basic reagents and methods are no longer covered by patents. It must be emphasized that at this time most of the patents are still deemed valid and enforceable. However, researchers may wish to follow the events as they unfold with respect to the enforceability and validity of the PCR patents.

The first blow against the PCR patents was struck by Promega Corporation (Promega; Promega Corporation is a corporation headquartered in Madison, Wisconsin that produces for sale reagents and other products for the life science community.). HoffmannLaRoche filed an action against Promega on October 27, 1992 alleging breach of a contract for the sale of *Taq* DNA Polymerase (*Taq*), infringement of certain patents—the PCR Patents (United States Patent Nos. 4,683,195 and 4,683,202) and United States Patent No. 4,889,818—and related causes of action. At issue was United States Patent No. 4,889,818 (the '818 patent), entitled "Purified Thermostable Enzyme." Promega denied the allegations of the complaint and claimed, among other things, that the '818 patent was obtained by fraud and was therefore unenforceable. After a trial in 1999, a US court held that all of the claims of the '818 patent unenforceable for inequitable conduct or fraud. The unenforceable claims are provided below.

1. Purified thermostable Thermus aquaticus DNA polymerase that migrates on a denaturing polyacrylamide gel faster than phosphorylase B and more slowly than does bovine serum albumin and has an estimated molecular weight of 86,000 to 90,000 Dalton when compared with a phosphorylase B standard assigned a molecular weight of 92,500 Dalton.
2. The polymerase of claim 1 that is isolated from *Thermus aquaticus*.
3. The polymerase of claim 1 that is isolated from a recombinant organism transformed with a vector that codes for the expression of Thermis aquaticus DNA polymerase.

The court concluded that Promega had demonstrated by clear and convincing evidence that the applicants committed inequitable conduct by, among other things, withholding material information from the patent office; making misleading statements; making false claims; misrepresenting that experiments had been conducted when, in fact, they had not; and making deceptive, scientifically unwarranted comparisons. The court concluded that those misstatements or omissions were intentionally made to mislead the Patent Office. The court's decision has been appealed, and a decision from the Federal Circuit Court of Appeals is expected shortly. Pending the appeal court decision, the '818 patent is unenforceable.

Patents have also been invalidated in Australia and Europe. On November 12, 1997, the Australian Patent Office invalidated all claims concerning native *Taq* DNA polymerase and DNA polymerases from any other Thermus species, contained in a patent held by Hoffmann-La Roche (application no. 632857). The Australian Patent Office concluded that the enzyme had been previously purified in Moscow and published by Kaledin et al. ***(6)*** and that certain patent claims were unfairly broad. Although the case has been appealed, as of this writing, the *Taq* patent in Australia is unenforceable.

In Europe, on May 30, 2001, the opposition division of the European Patent Office held that claims in the thermostable enzyme patent EP 0258017B1 (a patent equivalent to the '818 patent in the United States) were unpatentable because they lacked an inventive step in view of previous publications to Kaledin et al. ***(6)*** and Chient et al. *(7)*, as well as knowledge generally know in the field at the time the patent application was filed.

Although it has not been determined yet whether the PCR method patents were procured with the same types of misleading and deceptive behavior, the PCR patents have been challenged based on an earlier invention by Dr. Gobind Khorana and coworkers in the late 1960s and early 1970s. Under US and many international patent laws, patent claims are not valid if they describe an invention that was used and/or disclosed by others prior to the filing date of the patent. The principle behind such rules is to prevent people from patenting, and thus removing from the public domain, things that the public already owns. Although the PCR patents make no mention of such work, DNA amplification and cycling reactions were conducted many years before the filing of the PCR patents in the laboratory of Dr. Khorana. Dr. Khorana's method, which he called "repair replication," involved the steps of the following: (1) extension from a primer annealed to a template; (2) separating strands; and (3) reannealing of primers to template to repeat the cycle. Dr. Khorana did not patent this work. Instead he dedicated it to the public. Unfortunately, at the time that Dr. Khorana discovered his amplification process, it was not practical to use the method for nucleic acid amplification, and the technique did not take off as a commercial method. At the time this work was disclosed, chemically synthesized DNA for use as primers was extremely expensive and cost-prohibitive for even limited use. Additionally, recombinantly produced enzymes were not available. Thus, not until the 1980s, when enzyme and oligonucleotide production became more routine, could one economically replicate Dr. Khorana's methods.

The validity of the PCR patents was challenged in 1990 by E.I. Dupont De Nemours & Co. (Dupont). Based on publicly available records, it appears that Dupont pointed to the work from Dr. Khorana's laboratory, arguing that all of the method steps required in the basic PCR method were taught by Dr. Khorana's publications and were in fact in the public domain. Hoffmann-LaRoche (who was positioned to acquire the technology) out-maneuvered Dupont by putting the Khorana papers in front of the United States Patent and Trademark Office in a reexamination procedure. Under reexamination, the patent holder has the ability to argue the patentability of an invention to the patent office without any input allowed by third parties, such as Dupont. As shown by publicly available records, during the reexamination procedure expert declarations were entered to raise doubt about the teaching of the Khorana references. As a result (not surprisingly), the Patent Office upheld the patents. Once a patent has issued in view of a reference, there is a strong presumption of validity that courts must acknowledge in any proceedings that later attempt to invalidate the patent in view of the reference.

In addition to the disadvantage caused by the reexamination procedure, publicly available records show that Dupont was not able to use several pieces of compelling evidence against the PCR patents. Dupont, although performing clever replication work to show the sufficiency of Dr. Khorana's disclosures (in direct contrast to the expert declarations submitted to the Patent Office during reexamination), did not submit the data in a timely manner in the proceedings. The judge ruled that the data should be excluded as untimely and prejudicial. Dupont also found additional references disclosing the earlier invention by Khorana, but did not provide them to the court in time and they were not considered. Thus, it seems that validity of the PCR patents was never truly tested in view of the work conducted by Dr. Khorana and his colleagues. Such a test, as well as others, may come in the near future as part of the Promega/HoffmannLaRoche litigation.

Should these or any additional patents be found invalid and unenforceable, the patent issues for researchers wishing to practice PCR will be greatly simplified. Interestingly, if it is found that one or more of the invalid or unenforceable patents were used to suppress competition in the market or to unfairly control the freedom of researchers, companies exerting such unfair market control may be subject to laws designed to prevent unfair and anticompetitive behavior. If a court were to rule that anticompetitive behavior was exercised, the violating patent owner may be forced to compensate those that were harmed. Although it is impossible to predict at this time the outcome of future court proceedings, researchers may wish to follow the progress of these cases. At a minimum, they offer perspective into the patent world and provide important subject matter for debate that is extremely relevant to shaping the future of patent public policy, an area that will increasingly play a role in the day-to-day lives of scientists.

References

1. Beck, S. (1998) Do you have a license? Products licensed for PCR in research applications. *The Scientist* **12,** 21.
2. Constans, J. (2001) Courts cast clouds over PCR pricing. *The Scientist* **15,** 1.
3. Finn, R. (1996) Ongoing patent dispute may have ramifications for academic researchers. *The Scientist* **10,** 1.
4. *Wittemore v Cutter*, 29 F. Cas. 1120 (C.C.D. Mass. 1813)(No. 17,600)(Story, J.).
5. *Roche Prods., Inc. v Bolar Pharmaceutical Co.*, 733 F.2d 858 (Fed. Cir. 1984).
6. Kaledin, A. S., Sliusarenko, A. G., and Gorodetskii, S. I. (1980) Isolation and properties of DNA polymerase from extreme thermophylic bacteria Thermus aquaticus YT-1. *Biokhimiia* **45,** 644–651. In Russian.
7. Chien, A., Edgar, D. B., and Trela, J. M. (1976) Deoxyribonucleic acid polymerase from the extreme thermophile Thermus Aquaticus. *J. Bacteriol.* **127,** 1550–1557.

3

Equipping and Establishing a PCR Laboratory

Susan McDonagh

1. Introduction

Polymerase chain reaction (PCR) is a very sensitive method of amplifying specific nucleic acid, but the system is susceptible to contamination from extraneous or previously amplified DNA strands *(1,2)*. Many specific copies of DNA are produced from each round of amplification *(3)* with a single aerosol containing up to 24,000 copies of amplified material *(4)*. The most important consideration when designing and equipping a laboratory for PCR is therefore to minimize the risk of contamination and ensure accurate results *(5,6)*. To do this, it is necessary to physically separate the different parts of the process and arrange them in a unidirectional workflow *(4)*. This avoids back flow of traffic and, along with restricted access, will reduce the risk of contamination and inaccurate results.

The way in which the workflow is arranged will depend on the amount of available space. If possible, different rooms should be used for reagent preparation, sample preparation, PCR (some also separate primary and secondary stages), and post-PCR processing (*see* **Fig. 1**). Each of these areas should contain dedicated equipment, protective clothing, and consumables *(1)*. Disposable gloves should be readily available for frequent changing to avoid cross contamination, and control material should be included in every run to monitor any contamination problems *(3)*.

2. Equipment

A list of basic equipment required for a PCR laboratory is given in **Table 1**.

2.1. Thermocyclers

This is obviously the most important piece of equipment in the laboratory, with many products available from different manufacturers. Thermocyclers can be supplied with a variety of reaction vessel formats, including 0.2- and 0.5-mL microtubes; strips of tubes; microtiter plates containing up to 384 wells; glass slides; and capillaries. Temperature ramp rates and uniform heat distribution across the block are important for consistent performance. These options, along with the consideration of laboratory requirements, are factors when purchasing a machine, and these specifications are obviously reflected in the cost. For example, if basic PCR is all that is required, equipment from the lower end of the range might suffice. These machines have programmable

From: *Methods in Molecular Biology, Vol. 226: PCR Protocols, Second Edition*
Edited by: J. M. S. Bartlett and D. Stirling © Humana Press Inc., Totowa, NJ

REAGENT PREPARATION
(Clean Room)

↓

SAMPLE PREPARATION

↓

1° PCR

↓

2° PCR

↓

POST-PCR

↓

DISCARD AREA

Fig. 1. Unidirectional flow in a PCR laboratory.

blocks, often with a heated lid, and a basic repertoire of cycling capabilities. If high throughput using many different protocols is required in a diagnostic setting, then a multiblock system with the advantage of adding satellite units may be appropriate. More specialized machines with gradient blocks suitable for rapid optimization studies or with specialized blocks for *in situ* PCR are also available.

Advances in technology have resulted in the development of real-time PCR systems, which allow rapid cycling (50 cycles in less than 30 min). These systems are expensive but provide benefits, including rapid throughput, efficient optimization, and further reducing the risk of contamination with reactions and product analysis occurring in a single tube.

2.2. Additional Equipment

Dedicated equipment for each area of the laboratory can be purchased from regular laboratory suppliers. Contamination can often arise from breaks and spills in equipment, such as centrifuges and waterbaths *(4)*; therefore, important considerations include the purchase of equipment that can be easily taken apart for decontamination (*see* **Note 1**).

All areas require dedicated pipets *(1)*. Plugged tips used with traditional pipets are generally cheaper and easier to use than positive displacement pipets (*see* **Note 2**). Storage space at 4°C and –20°C should be available in each area, along with access to –70°C freezer facilities.

Laminar hoods are not always recommended, except at the sample extraction stage, where they are required to protect the worker. Using individual workstations with

Table 1
Equipment Required

All	Reagent preparation	Sample preparation	1° PCR	2° PCR	Post-PCR
Pipets	Microfuge	Microfuge	Cyclers	Cyclers	Electrophoresis tanks
Refrigerator	Vortex	Vortex			Power packs
–20°C freezer	dH_2O source	Laminar cabinet			Microwave
Work stations	Ice machine				Gel viewing system
	Balance				Gel documentation system
	pH meter				

decontamination facilities reduces airflow throughout the laboratory and minimizes aerosol dispersal. These may simply consist of a disposable or wipeable tray on which the worker completes all operations before treating to remove any potential contaminating nucleic acids *(1,2)* (*see* **Note 3**). Some manufacturers produce purpose-built cabinets, which incorporate several decontamination and safety features.

An ice machine, distilled water supply, balance, and pH meter are required in the reagent preparation area, and a microwave is ideal for melting agarose for gel assembly in the post-PCR area.

2.3. Consumables

Disposable plastics rather than reusable glass should be used wherever possible, and high-quality consumables, for example, Rnase-free plasticware, should be used throughout the laboratory. It is also important to note that performance may be affected by different products from different suppliers, which was demonstrated in a study in which varying results were obtained when using microtubes supplied by a number of manufacturers *(2)*. Other factors have an inhibitory effect on PCR performance and should also be considered. Examples include methods, such as ultraviolet irradiation, which can affect reagents such as mineral oil *(7)*, therefore it is important to avoid exposure, and powder in gloves, which has been shown to inhibit PCR *(2,8)*; therefore, powder-free varieties are recommended (*see* **Note 4**).

3. Laboratory Layout

Work within the laboratory should be confined to the specific areas identified for that part of the procedure. Each of these areas is described below, but several points apply to all. These include removal of laboratory coat and gloves before moving into another part of the laboratory; provision of gloves for frequent change; avoidance of aerosols and drips; and decontamination of working area and equipment before and after use *(3)* (*see* **Note 3**). All reagents necessary for each process should be stored within the area in which the work is being performed *(3)*.

3.1. Reagent Preparation Laboratory

This area should be kept entirely free from samples and other potential sources of nucleic acid. Stock solutions and reagents should be made up, or diluted if purchased as concentrate, then dispensed in single use aliquots *(1,3,4)* or small volumes *(7)* and stored. This means that that they can be identified and discarded if contamination does arise *(9)*. Master mixes are made up here and added to reaction vessels before continuing onto the next stage of the process (*see* **Note 5**). If necessary, an oil overlay can also be added at this stage.

3.2. Sample Preparation Laboratory

Laminar flow cabinets are necessary for dealing with samples until they are inactivated and extracted, and these and other equipment should be decontaminated before and after each procedure (*see* **Note 3**). The equipment necessary will depend on the extraction methods used, but a microfuge, heating block, and vortex are minimal requirements.

3.3. PCR Laboratory

Primary and secondary PCR steps should be separated, preferably in different rooms, and certainly with separate thermocyclers; however, the layout of this area will depend on space and equipment available. Primary reactions containing master mix and nucleic acid should be assembled and placed on the appropriate thermocycler. After cycling, these are removed to the secondary PCR area, where reactions are assembled and placed on cyclers dedicated for this process. Other automated/integrated/single-round equipment should be positioned with secondary thermocyclers to reduce the risk of contamination (*see* **Note 6**).

3.4. Post-PCR Processing

All final amplified products should be dealt with in this area, which can be used for techniques, including electrophoresis, restriction fragment length polymorphism (RFLP), hybridization work, cloning, and sequencing. It is important that nothing from this area should go back through other areas involving preliminary steps but should be processed through a waste management or discard area.

4. Notes

1. For example, hot blocks are easier to decontaminate on a regular basis and are therefore a better option than water baths.
2. Normal tips can be used for post-PCR steps.
3. An ultraviolet irradiation source is valuable in reducing contamination; however, Cimino et al. *(10)* recommend caution when using this method alone. Otherwise, wash down all nonmetal surfaces with 0.1 *N* HCl, or 10% bleach, followed by water.
4. Nitrile gloves should be used for safety when handling ethidium bromide if used in gel electrophoresis.
5. As kit-based formats become available, reagent and master mix will be supplied, completely reducing the need for this area.
6. This setup will become more difficult as combined extraction/amplification and detection equipment become more available.

References

1. Wolcott, M. J. (1992) Nucleic acid-based detection methods. *Clin. Microbiol. Rev.* **5,** 370–386.
2. Wilson, I. G. (1997) Inhibition and facilitation of nucleic acid amplification. *Appl. Environ. Microbiol.* **63,** 3741–3751.

3. Sambrook, J., Fritsch, E. F., and Maniatis, T. (1989) In vitro application of DNA by the polymerase chain reaction, in *Molecular Cloning: A Laboratory Manual.* 2nd ed., Cold Spring Harbor Laboratory Press, New York, pp. 14.14.
4. Orrego, C. (1990) Organizing a laboratory for PCR work, in *PCR Protocols: A Guide to Methods and Applications* (Innes, M. L., Gelfand, D. H., Sninsky, J. J., White, T. J., eds.), Academic Press Inc., San Diego, pp. 447–454.
5. Baselski, V. S. (1996) The role of molecular diagnostics in the clinical microbiology laboratory. *Clin. Lab. Med.* **16,** 49–60.
6. Lisby, G. (1999) Application of nucleic acid amplification in clinical microbiology. *Mol. Biotechnol.* **12,** 75–79.
7. Hughes, M. S., Beck, L. A., and Skuse, R. A. (1994) Identification and elimination of DNA sequences in Taq DNA polymerase. *J. Clin. Microbiol.* **32,** 2007–2008.
8. De Lomas, J. G., Sunzeri, F. J., and Busch, M. P. (1992) False negative results by polymerase chain reaction due to contamination by glove powder. *Transfusion* **32,** 83–85.
9. Madej, R. and Scharf, S. (1990) Basic equipment and supplies, in *PCR protocols: A Guide to Methods and Applications* (Innes, M. L., Gelfand, D. H., Sninsky, J. J., White, T. J., eds.), Academic Press Inc., San Diego, pp. 455–459.
10. Cimino, G. D., Metchette, K., Isaacs, S. T., and Zhu, Y. S. (1990) More false positive problems. *Nature* **345,** 773–774.

4

Quality Control in PCR

David Stirling

1. Introduction

Polymerase chain reaction (PCR), like any laboratory procedure, can be subject to a range of experimental or procedural error. A clear consideration of where such potential errors may occur is essential to minimize their impact. Careful quality control of equipment and reagents is essential.

2. Equipment

The previous chapter dealt with the sort of equipment that is required to perform PCR. It is commonplace for an individual laboratory to contain many sets of equipment, each bought from different manufacturers, at different times, and subjected to various amounts of abuse from students who don't know any better and laboratory managers who do! In an ideal world, and any diagnostic or commercial laboratory, each piece of equipment should be serviced and calibrated on a regular basis, with careful records being kept of this maintenance. Unfortunately, not every laboratory has funds for full-service contracts on all equipment. There are a few fundamental procedures, however, which will reduce errors from equipment problems

- Be consistent in the equipment used for any given PCR. If it works on Monday but not Tuesday, this may simply be to the result of using a different PCR block. Even the most modern and expensive thermal cyclers deteriorate with age.
- Check pipetting devices on a regular basis (weekly is not excessive) to ensure they pipet the correct volume. This is easily performed by pipetting and weighing water. Most manufacturers produce inexpensive service packs for their pipettors.

3. Reagents

As with all laboratory procedures, it generally pays dividends to use high-quality reagents from reputable suppliers. You may well know someone who brews their own *Taq* polymerase in a vat in the garage, but do they control for batch-to-batch variability?

The design of optimum PCR primers will be discussed later. It is important to remember, however, that these are single-stranded DNA molecules and are therefore relatively labile. Repeated freeze/thawing will cause degradation to shorter products,

From: *Methods in Molecular Biology, Vol. 226: PCR Protocols, Second Edition*
Edited by: J. M. S. Bartlett and D. Stirling © Humana Press Inc., Totowa, NJ

which will either not anneal, or if the annealing temperature is low enough, will anneal promiscuously, yielding multiple products. Simple aliquoting primers into manageable volumes will reduce both the scope for contamination and degradation. This practice should also be adopted for dNTP stocks for the same reason.

4. Operator Errors

Anyone involved in teaching molecular techniques hears the same complaint again and again: "These reaction volumes are too small! . . . I can't see a microliter!" Even for those with many years of laboratory experience (perhaps especially for those), it can be difficult to adjust to dealing with small volume reactions. Although the obvious answer may be to increase the volume, this has both cost and efficiency implications.

- Use appropriate pipetting devices. A pipettor designed for the 20- to 200-µL range will not accurately dispense 10 µL.
- The use of master mixes not only reduces the dependence on accurately pipetting small volumes but also improves the control over reaction contents.
- Practice with the same reaction until consistent results are obtained.

5. PCR-Specific Difficulties

Although much of the above could apply to any analytical laboratory technique, PCR also is subject to the confounding problem of contamination. Cross contamination of samples is of concern in any discipline, and good laboratory practice, such as careful pipetting and the constant changing of disposable pipet tips, will minimize the opportunity of this occurring. Where PCR differs from most other procedure is in the production of vast quantities of the analyte during the procedure. The presence of billions of copies of potential template can create severe problems. These problems can be minimized by physically separating the pre- and postamplification processes (ideally in different rooms with different pipettors, etc.); however, they should constantly be monitored by the inclusion of appropriate controls.

6. Controls

- No DNA. Although it can seem extravagant to constantly set up reactions without template, this is the best way to monitor for contamination. A separate "no DNA" control should be set up for each master mix or each individual reaction. If contamination is discovered, the pipettor should be decontaminated (as per manufacturers guidelines), and the reagent aliquot should be rechecked or discarded.
- Positive control. PCR is often used simply to detect the presence of specific sequence. In such circumstances, it is essential to include at least one reaction with a template known to contain the sequence.
- Internal control. Even when master mixes have been used to ensure consistency of reaction components, and a positive control is used, there is the possibility that template may be omitted from individual tubes. This can be addressed by the inclusion within each reaction tube of primers, which will amplify a target known to consistently be present in the test DNA (*see* factor IX in Chapter 47 for example).

7. Regional Quality-Assurance Programs

In addition to the in-house precautions detailed above, there are a growing number of specialist quality-assurance programs that have been developed for most diagnostic PCR.

These programs distribute test material to laboratories, who then report their results centrally. Results from all participating centers are compared and confidential reports issued to each center. If such a program exists in your field, details will probably be available on the Internet: join it. If one doesn't already exist, consider starting one; it may generate enough revenue to pay for instrument service contracts!

II

Preparation of Nucleic Acid Templates

5

Extraction of Nucleic Acid Templates

John M. S. Bartlett

The following series of short technical descriptions covers the extraction of DNA and RNA from various starting materials. We have gathered these together at the beginning of this book to provide an easy reference. The polymerase chain reaction (PCR) techniques described in the rest of this book are, in the main, worked laboratory methods given detailed examples of the procedures used by the authors in their own research. However, the aim is to provide the reader with a method that may be translated into their own research; thus, although the description of ultrasensitive PCR focuses on viral genomes and cancer, this method may of course be equally applied to DNA from other sources. Rather than leaving the reader who is interested in applying this technique to ancient DNA or DNA from bone, etc., to search through the various chapters to find such a technique, we have collected these together for reference here.

PCR provides a simple method for the amplification and analysis of DNA; however, for most applications involving PCR, the DNA (or cDNA for RNA methods) must be in a reasonably pure state. Therefore, the first stage of any experimental procedure involving PCR based technologies is the provision of a pure suspension of nucleic acids, either RNA or DNA.

Extraction of nucleic acids is a fundamental precursor to almost all the techniques described within this volume. Isolation of RNA and DNA from blood and fresh tissues can be performed using a variety of techniques, which also form the basis of methods of extraction of these substrates from other sources. The sensitivity of PCR methods is now such that extraction of DNA and RNA from tissues fixed in formaldehyde and buffered formalin is considered routine, and we are now able to extract DNA from ancient tissues, feces, and many other sources. Indeed, in forensic science, DNA fingerprinting from sources as diverse as residual saliva on food and microscopic blood deposits is now possible! Indeed, the description of the extraction of DNA/RNA alone could probably fill several major chapters. It has, however, not proven desirable or feasible to be exhaustive in our approach to DNA/RNA extraction protocols, and we have therefore restricted these to major methods in use in many laboratories. Further references *(1–7)* that provide detailed reviews of methods for nucleic acid extraction and some recommended web sites are listed in the reference section.

From: *Methods in Molecular Biology, Vol. 226: PCR Protocols, Second Edition*
Edited by: J. M. S. Bartlett and D. Stirling © Humana Press Inc., Totowa, NJ

References

1. US Department of Commerce Molecular Biology Techniques Forum http://research.nwfsc.noaa.gov/protocols/methods/
2. http://www.stratagene.com/
3. http://www.promega.com/tbs/
4. http://www.highveld.com/protocols.html
5. http://www.dynal.no/
6. Sambrook, J., Fritsch, E. F., and Maniatis, T. (1989) *Molecular Cloning: A Laboratory Manual.* Cold Spring Harbor Laboratory Press, New York.
7. Bartlett, J. M. S., ed. (2000) *Ovarian Cancer: Methods & Protocols, Vol. 39, Methods in Molecular Biology*. Humana Press, Totowa, NJ.

6

Extraction of DNA from Whole Blood

John M. S. Bartlett and Anne White

1. Introduction

There are many differing protocols and a large number of commercially available kits used for the extraction of DNA from whole blood. This procedure is one we use routinely in both research and clinical service provision and is cheap and robust. It can also be applied to cell pellets from dispersed tissues or cell cultures (omitting the red blood lysis step.

2. Materials

This method uses standard chemicals that can be obtained from any major supplier; we use Sigma.

1. Waterbath set at 65°C.
2. Centrifuge tubes (15 mL; Falcon).
3. Microfuge (1.5 mL) tubes.
4. Tube roller/rotator.
5. Glass Pasteur pipets, heated to seal the end and curled to form a "loop" or "hook" for spooling DNA.
6. EDTA (0.5 *M*), pH 8.0: Add 146.1 g of anhydrous EDTA to 800 mL of distilled water. Adjust pH to 8.0 with NaOH pellets (this will require about 20 g). Make up to 1 L with distilled water. Autoclave at 15 p.s.i. for 15 min.
7. 1 *M* Tris-HCl, pH 7.6: Dissolve 121.1 g of Tris base in 800 mL of distilled water. Adjust pH with concentrated HCl (this requires about 60 mL). CAUTION: the addition of acid produces heat. Allow mixture to cool to room temperature before finally correcting pH. Make up to 1 L with distilled water. Autoclave at 15 p.s.i. for 15 min.
8. Reagent A: Red blood cell lysis: 0.01M Tris-HCl pH 7.4, 320 m*M* sucrose, 5 m*M* $MgCl_2$, 1% Triton X 100.
9. Add 10 mL of 1 *M* Tris, 109.54 g of sucrose, 0.47 g of $MgCl_2$, and 10 mL of Triton X-100 to 800 mL of distilled water. Adjust pH to 8.0, and make up to 1 L with distilled water. Autoclave at 10 p.s.i. for 10 min (*see* **Note 1**).
10. Reagent B: Cell lysis: 0.4 *M* Tris-HCl, 150 m*M* NaCl, 0.06 *M* EDTA, 1% sodium dodecyl sulphate, pH 8.0. Take 400 mL of 1 *M* Tris (pH 7.6), 120 mL of 0.5 *M* EDTA (pH 8.0), 8.76 g of NaCl, and adjust pH to 8.0. Make up to 1 L with distilled water. Autoclave 15 min at 15. p.s.i. After autoclaving, add 10 g of sodium dodecyl sulphate.

From: *Methods in Molecular Biology, Vol. 226: PCR Protocols, Second Edition*
Edited by: J. M. S. Bartlett and D. Stirling © Humana Press Inc., Totowa, NJ

11. 5 *M* sodium perchlorate: Dissolve 70 g of sodium perchlorate in 80 mL of distilled water. Make up to 100 mL.
12. TE Buffer, pH 7.6: Take 10 mL of 1 *M* Tris-HCl, pH 7.6, 2 mL of 0.5 *M* EDTA, and make up to 1 L with distilled water. Adjust pH to 7.6 and autoclave 15 min at 15. p.s.i.
13. Chloroform prechilled to 4°C.
14. Ethanol (100%) prechilled to 4°C.

3. Method

3.1. Blood Collection

1. Collect blood in either a heparin- or EDTA-containing Vacutainer by venipuncture (*see* **Note 2**). Store at room temperature and extract within the same working day.

3.2. DNA Extraction

To extract DNA from cell cultures or disaggregated tissues, omit **steps 1** through **3**.

1. Place 3 mL of whole blood in a 15-mL falcon tube.
2. Add 12 mL of reagent A.
3. Mix on a rolling or rotating blood mixer for 4 min at room temperature.
4. Centrifuge at 3000*g* for 5 min at room temperature.
5. Discard supernatant without disturbing cell pellet. Remove remaining moisture by inverting the tube and blotting onto tissue paper.
6. Add 1 mL of reagent B and vortex briefly to resuspend the cell pellet.
7. Add 250 µL of 5 *M* sodium perchlorate and mix by inverting tube several times.
8. Place tube in waterbath for 15 to 20 min at 65°C.
9. Allow to cool to room temperature.
10. Add 2 mL of ice-cold chloroform.
11. Mix on a rolling or rotating mixer for 30 to 60 min (*see* **Note 3**).
12. Centrifuge at 2400*g* for 2 min.
13. Transfer upper phase into a clean falcon tube using a sterile pipet.
14. Add 2 to 3 mL of ice-cold ethanol and invert gently to allow DNA to precipitate (*see* **Note 4**).
15. Using a freshly prepared flamed Pasteur pipet spool the DNA onto the hooked end (*see* **Note 5**).
16. Transfer to a 1.5-mL Eppendorf tube and allow to air dry (*see* **Note 6**).
17. Resuspend in 200 µL of TE buffer (*see* **Notes 7** and **8**).

4. Notes

1. Autoclaving sugars at high temperature can cause caramelization (browning), which degrades the sugars.
2. As will all body fluids, blood represents a potential biohazard. Care should be taken in all steps requiring handling of blood. If the subject is from a known high risk category (e.g., intravenous drug abusers) additional precautions may be required.
3. Rotation for less than 30 or over 60 min can reduce the DNA yield.
4. DNA should appear as a mucus-like strand in the solution phase.
5. Rotating the hooked end by rolling between thumb and forefinger usually works well. If the DNA adheres to the hook, break it off into the Eppendorf and resuspend the DNA before transferring to a fresh tube.
6. Ethanol will interfere with both measurements of DNA concentration and PCR reactions. However, overdrying the pellet will prolong the resuspension time.

7. The small amount of EDTA in TE will not affect PCR. We routinely use 1 μL per PCR reaction without adverse affects.
8. DNA can be quantified and diluted to a working concentration at this point or simply use 1 μL per PCR reaction; routinely, we expect 200 to 500 ng/μL DNA to be the yield of this procedure.

7

DNA Extraction from Tissue

Helen Pearson and David Stirling

1. Introduction

The following protocol is one of the longest-established methods of DNA extraction and works well with a wide range of solid tissues. Proteins are digested with proteinase K and extracted with phenol chloroform. DNA is then precipitated with ethanol. The resultant DNA (10–50 μg) is of high molecular weight and is a suitable template for long polymerase chain reaction (PCR).

2. Materials

1. Microfuge tubes (1.5 mL).
2. Shaking water bath or incubator with rotisserie.
3. Microfuge.
4. DNA digestion buffer: 50 m*M* Tris-HCl, 100 m*M* EDTA, 100 m*M* NaCl, 1% SDS, pH 8.0.
5. Proteinase K: 0.5 mg/mL in DNA digestion buffer.
6. Phenol/chloroform/isoamyl alcohol (25:24:1).
7. 100% EtOH.
8. 70% EtOH.
9. TE buffer: 10 m*M* Tris-HCl, 1 m*M* EDTA, pH 8.0.

3. Protocol

1. Place 0.1 to 0.5 g of tissue into polypropylene microfuge tube (*see* **Note 1**).
2. Add 0.5 mL of DNA digestion buffer with proteinase K (*see* **Note 2**).
3. Incubate overnight at 50 to 55°C with gentle shaking.
4. Spin tubes for 5 s at 500*g* to collect mix in bottom of tube.
5. Add 0.7 mL of phenol/chloroform/isoamyl alcohol (25:24:1).
6. Mix by inversion for 1 h (do not vortex).
7. Microfuge at 12,000*g* for 5 min and transfer 0.5 mL of the upper phase to new microfuge tube.
8. Add 1 mL of 100% ethanol at room temperature and gently invert until DNA precipitate forms (approx 1 min).
9. Microfuge at 12,000*g* for 5 min and discard supernatant.
10. Add 1 mL of 70% ethanol (–20°C) and invert several times. This ethanol wash removes excess salt, which may otherwise interfere with PCR.

From: *Methods in Molecular Biology, Vol. 226: PCR Protocols, Second Edition*
Edited by: J. M. S. Bartlett and D. Stirling © Humana Press Inc., Totowa, NJ

11. Microfuge at 12,000*g* for 5 min and discard supernatant.
12. Spin tubes for 5 s to collect any remaining ethanol in bottom of tube. Remove last drops of ethanol with fine pastette.
13. Air dry at room temperature for 10 to 15 min (any longer will render DNA difficult to redissolve).
14. Resuspend in 100 µL of TE and incubate at 65°C for 15 min to dissolve DNA (*see* **Note 3**).

4. Notes

1. Some tissues contain large amounts of connective tissue and are difficult to digest. These can be ground frozen in liquid Nitrogen and ground in a mortar and pestle before being digested with proteinase K.
2. Proteinase K solution can be kept for several days at 4°C.
3. Repeat pipetting through a narrow gauge tip can help this process.

8

Extraction of DNA from Microdissected Archival Tissues

James J. Going

1. Introduction

Many modern analytical methods require little material, and this has made feasible biochemical and molecular analyses of small tissue fragments, even individual cells, by microdissection of histological sections *(1,2)*. Polymerase chain reaction (PCR) can potentially be applied to the analysis of single DNA molecules, as in the analysis of single haploid cells, such as spermatozoa *(3)*. This sensitivity requires careful attention to technique and proper controls to avoid false-positive or other spurious results.

Microdissection techniques used by different research groups are diverse, and recent articles explore different techniques *(4–7)*. This chapter presents a technique of histological microdissection applicable to a variety of tissues.

Microdissection can be applied to paraffin or frozen sections of human and animal tissues, depending on availability, but in human studies, it may be necessary to work with formalin-fixed, paraffin-embedded archival tissues. Although fixed tissues have disadvantages, particularly the degradation of nucleic acids after fixation, which may make successful PCR amplification more difficult, better preservation of tissue morphology compensates. This may be important because one purpose of histological microdissection is to bring together molecular and morphological analysis of the same cells. Fixed tissues sections may be easier to handle than unfixed tissues during microdissection. This chapter concentrates on fixed tissues.

2. Materials

All reagents should be of molecular biology quality.

1. Proteinase K from *Tritirachium album* (Sigma), 20 mg/mL stock solution. Store 50-µL aliquots at –20°C, thaw, and dilute to 1 mL with digestion buffer containing 1% Tween to give working stock solution of proteinase K, 1 mg/mL for tissue digestion to release DNA.
2. Proteinase K digestion buffer, pH 8.3. (TRIS-HCl, 2.2 g/L; TRIS base 4.4 g/L; EDTA 0.37 g/L; separate batches of the buffer should be prepared detergent-free and containing 1% Tween).
3. Leica model M mechanical micromanipulator (other micromanipulators may be suitable).
4. Tungsten wire (0.5 mm in diameter) for dissection needles (or ready-made needles); bacteriological loop holders for mounting needles in micromanipulator.
5. Facility for electrolytic sharpening of tungsten needles (*see* **Subheading 3.5.**)

From: *Methods in Molecular Biology, Vol. 226: PCR Protocols, Second Edition*
Edited by: J. M. S. Bartlett and D. Stirling © Humana Press Inc., Totowa, NJ

3. Methods

3.1. Section Cutting

Careful clean techniques should be used when cutting sections for PCR analysis.

1. Use a new part of the microtome blade to cut each section to avoid possible the carryover of DNA from one tissue block to sections of the next.
2. Ribbons of sections should be floated out on a clean water bath and no buildup of section debris permitted.
3. Glass slides upon which sections are mounted should be scrupulously clean. Dry mounted sections at 56°C for 2 h (*see* **Note 1**).

3.2. Dewaxing and Staining Sections

Dewax sections completely before microdissection.

1. Immerse 6- to 7-µm sections in a slide rack for 10 min in xylene. Drain surplus xylene thoroughly from the sections to minimize the carryover of dissolved wax and then transfer to a second xylene bath for 2 min. Avoid breathing xylene vapor and be aware of the fire hazards of organic solvents.
2. Take sections through two baths of 99% industrial methylated spirits to remove xylene and a third bath of 95% industrial methylated spirits.
3. Transfer sections to distilled or deionized water of a satisfactory standard for preparing PCR reagents. Stain by immersion for 30 s in 0.05% w/v toluidine blue in distilled water (*see* **Note 2**).
4. Wash in distilled water (*see* **Note 3**).
5. After dissection, dehydrate and mount slides with a coverslip to provide a permanent record of the dissection (*see* **Note 4**).

3.3. Microdissection Tools

Successful dissection can be conducted freehand using sterile curved scalpel blades or sterile hypodermic needles, with or without a dissecting microscope.

1. Take a section for dissection from the water bath immediately before it is needed, drain it, and blot any surplus water from around the section with a disposable, lint-free tissue (do not touch the section).
2. Stroke the edge of the scalpel blade decisively across the section to remove a strip of tissue, the width of which will vary with the angle between blade and section (*see* **Note 5**).
3. For more precise dissection with the micromanipulator, an electrolytically sharpened tungsten needle is ideal (**Fig. 1**). This tool consists of 25 mm of tungsten wire, 0.5 mm in diameter, sharpened and polished electrolytically to a fine point (tip radius several microns).
4. Mount the needle in a collet-type bacteriological loop holder.
5. Mount the loop holder in the tool holder of the micromanipulator and angle it downward 25 to 30º.

3.4. Making and Maintaining Tungsten Microneedles

Tungsten needles can be obtained ready-made as ohmic probe needles but are easily fabricated from plain 0.5-mm tungsten wire (obtainable from suppliers of equipment for electron microscopy) This requires an electrolytic cell (*see* **Note 6**).

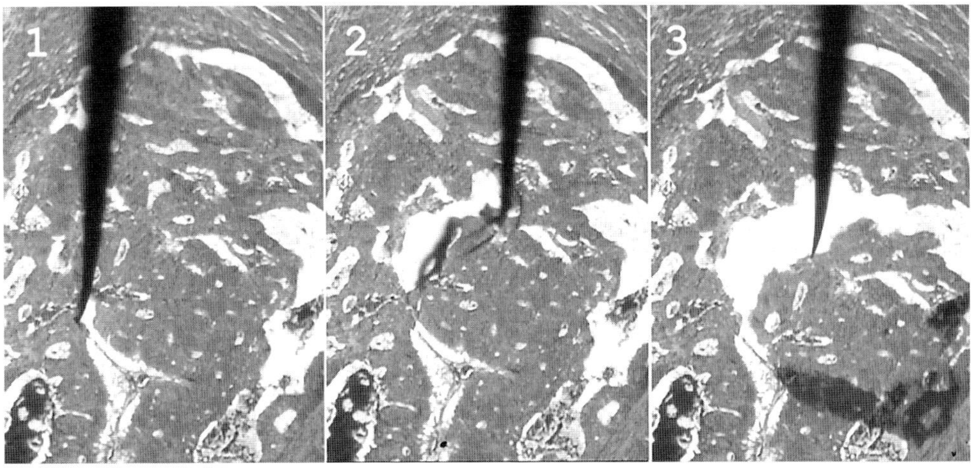

Fig. 1. Microdissection (stills from a video). Frame 1: the needle is placed at the periphery of an island of carcinoma cells (colonic carcinoma, paraffin section, toluidine blue stain). Frame 2: about one-quarter of the island of tumor cells has been separated from the glass slide. Frame 3: the whole island is completely detached, shown by slight rotation with respect to the position it occupied in the section. The sample is ready for collection, proteinase K digestion, and further analysis.

1. Obtain ~10 cm of platinum wire (e.g., from an old electrophoresis apparatus) and make a circular loop in it just small enough to fit inside a standard 20-mL universal container, with a "tail" 2 to 3 cm long.
2. Drill a 1-mm hole in the side of universal container near the base and thread the platinum wire tail through it, with the loop neatly placed at the base of the container. Seal the hole inside and out with an epoxy resin glue, for example, Araldite.
3. Almost fill the cell with 0.1 *M* potassium hydroxide (KOH; 1.2 g in 20 mL) in water (caution: KOH is caustic).
4. Connect the platinum wire cathode to the negative (–) terminal of a standard 9-V radio battery (dry cell), and make the tungsten wire, mounted in its bacteriological loop holder, the anode (+). Ensure that the cell cannot fall over and spill caustic electrolyte solution.
5. Complete the circuit by dipping 5 to 10 mm of the tungsten wire vertically into the KOH solution. Hydrogen bubbles will appear at the platinum cathode and nascent oxygen will remove tungsten from the anode, which sharpens to a fine, polished point. New needles are made this way and damaged needles refurbished. If sharpening does not occur, check the polarity.
6. Straighten bent needles by rolling firmly between glass slides, and then polish/resharpen.
7. Rinse needles with distilled water from a wash bottle after sharpening or resharpening to remove droplets of KOH electrolyte.
8. Store prepared needles in a covered Petri dish with the blunt end pressed lightly into a ring of modelling clay or similar. Handle with fine forceps.

3.5. Performing Microdissection

1. Retrieve the slide to be dissected from distilled water and dry the back of the slide and around the section using a clean disposable laboratory tissue.
2. Place the section on the microscope stage (*see* **Note 7**) and cover with a pool of proteinase K lysis buffer (without detergent) from a disposable sterile Pasteur pipet with a rubber-bulb

pipet filler. Spread the pool of buffer until it extends 3 to 4 mm beyond all edges of the section and is as deep as possible without spillage.

3. Center the area of cells in the section to be retrieved for subsequent analysis in the microscope field at an appropriate magnification.
4. Using the coarse motion controls of the micromanipulator, place the needle over the area to be dissected.
5. Lower the needle gently at the edge of the area to be dissected until its tip just touches the slide. Stop lowering the needle when a small lateral deflection of the needle tip occurs (*see* **Note 8**).
6. Microdissection techniques vary for different specimens. In general, attempt by blunt dissection to develop cleavage between groups of cells. Work around the area to be dissected, developing a split between the area to be kept and the area to be removed (**Fig. 1**). Then, use the point of the needle gradually to undermine the area to be recovered, pushing and pulling with the tip and side of the needle until the area to be retrieved has been peeled from the slide and floats freely in the buffer pool.
7. If it is not clear whether the fragment is still attached to the section, gently agitate the slide. Attached fragments do not move freely.
8. Sometimes a tissue fragment remains attached by a few strands of collagen; a second tungsten needle in a bacteriological loop holder, used freehand, will often detach it.

3.6. Retrieving Dissected Fragments

1. Bring the tip of the pipet close to the fragment to be retrieved and capture the fragment by suddenly releasing the pipet plunger, dragging fragment and a fixed volume of buffer into the pipet tip (*see* **Note 9**).
2. Expel the captured specimen into a labeled microcentrifuge tube, ready for further processing (*see* **Note 8**).
3. Check that the microdissected specimen is really in the tube. It helps if the fragment is easily seen with the naked eye. Use a magnifying lens or the microscope to make certain. Toluidine blue stained fragments are easy to see; unstained fragments may be practically invisible.

3.7. Extracting DNA: Proteinase K Digestion

In a fixed specimen, nucleic acids are present in a dense array of crosslinked proteins. Proteinase K digestion appears effectively to release them and make them available for subsequent PCR.

1. Add an equal volume of Proteinase K digestion buffer pH 8.3 containing 1 mg/mL of proteinase K and 1 mg/mL Proteinase K to each specimen tube.
2. Digest microdissected specimen in proteinase K (final concentration 500 μg/mL) at 37°C overnight in a water bath or incubator (digestion can continue over a weekend without detriment).
3. Heat specimens in a PCR block (95°C for 10), to inactivate PK.
4. Spin the specimen down by brief centrifugation.
5. Specimens are stable at room temperature for subsequent DNA PCR. Store for longer periods at 4°C or –20°C.
6. Accurate labeling is crucial (*see* **Note 10**).

3.8. DNA Purification after PK Digestion

The 25- or 50-μL sample remaining after PK digestion of a microdissected tissue fragment contains only small quantities of nucleic acids (DNA, RNA). One thousand cells contain about 6 ng of DNA and will contain at most 2000 copies of an amplifiable DNA sequence; only 1000 copies of each of two different alleles. An attempt to purify such quantities of DNA or RNA risks losing the specimen, although techniques have been described. Such purification does not obviously improve subsequent PCR amplification. Carefully optimize your PCR using the unpurified digest before you conclude that such purification is essential. Several published microdissection studies using the techniques described here have used 1-μL aliquots of sample as template ***(8–11)***. Strategies such as hot start, touchdown, nested, or real-time PCR may to help to obtain good PCR results by decreasing amplification of spurious products.

Techniques exist for whole-genome amplification of DNA from small samples, even single cells ***(12–14)***, to expand the template pool available for subsequent PCR analysis. The possibility of introducing artifacts should be considered, but with appropriate controls may be used for projects with scanty material.

3.9. Frozen vs Paraffin Sections

Frozen sections are less easy to microdissect than paraffin sections. Unfixed are less robust than fixed tissues and stand up less well to the manipulation necessary to detach the specimen.

3.10. Analysis of RNA from Microdissected Material

RNA from unfixed tissue may be more suitable for analysis than RNA from fixed tissue; however, RNA in unfixed tissue may be more susceptible to degradation, and RT-PCR of RNA from microdissected fixed tissue fragments can be achieved. Rather than performing the microdissection in a guanidinium-containing buffer, which removes toluidine blue from the section, performed microdissection in DEPC-treated distilled water and transfer fragments subsequently to RT-PCR buffers.

4. Notes

1. Drying influences the firmness with which sections adhere to the slide and ease of dissection. In general, slides coated with poly-L-lysine or treated with silanes should be avoided because it may be difficult to detach tissue from such slides. Conventional 3- to 4-μm histological sections are a compromise between ease of cutting, depth of staining, and visibility of cytological detail. For histological microdissection, slightly thicker sections (6–8 μm) contain more nucleic acid per unit area, and their slightly increased thickness does not usually cause interpretation problems. With sections over 10 μm, poorer visualization of cells is a disadvantage. Cut and dried sections should be stored in dust-free conditions. There seems to be no need to store them in a refrigerator or freezer. Disposable latex gloves should be worn for all manipulations to reduce contamination.
2. Staining reveals tissue structure, but unstained sections may be dissectible, especially if a serial hematoxylin and eosin section is available for reference. Staining often makes dissection easier. Toluidine blue staining is easy and does not seem to interfere with subsequent PCR, although this should be verified for particular applications.
3. Stained sections can be stored in distilled water until dissection. Refrigeration at 4°C in water overnight causes some destaining and sections may lift from the slide. Stained sections can be stored dry but it is best not to dewax and stain more sections than can be used in a single dissecting session.
4. A serial hematoxylin and eosin section shows what has been removed. A magnified photocopy or digital image of such a serial section can be annotated on hard copy or

digitally to record the dissection. Photocopying histological sections through an acetate sheet avoids scratching the photocopier glass or smearing it with mounting medium.

5. A scalpel blade vertically in contact with the section will remove a narrow strip of tissue. The blade at a flatter angle will remove a wider strip. Tissue so removed can often be rolled into a small ball or pill, picked up on the tip of the scalpel blade or sterile hypodermic needle and transferred to a microcentrifuge tube for further processing. Tissue is most easily handled when damp but without excess water, in which dispersed tissue fragments are hard to retrieve or too dry and brittle. Static electricity may be troublesome.
6. Bacteriological loop holders and tungsten wire are inexpensive. Either prepare enough needles in advance to use a new needle for each microdissection if necessary or briefly repolish the needle between samples, which exposes a new tungsten surface. The risk of DNA carryover on the needle from one specimen to the next appears largely theoretical. In a study of *ras* mutation in colorectal carcinomas ***(10)***, no false positives were detected in the analysis of several hundred separate microdissected normal and tumour tissue samples.
7. Any microscope can be used if there is enough space between objective and stage for access to the specimen. Ordinary or inverted standard microscopes can be used but a stereo dissecting microscope with relatively high maximum magnification (up to ×120) is ideal.
8. If the specimen adheres to the inside of the pipet tip, it can sometimes be dislodged by repeated in drawing and expulsion of buffer. It may be picked out with a needle, or the pipet tip may be cut off and placed with the tissue fragment in the tube for subsequent digestion. Sticking can usually be avoided by coating pipet tips before use with a silicone such as Sigmacote. Draw (e.g., 50 µL) of Sigmacote into the tip, expel it again, shake off any excess, and allow to dry. Coat tips an hour or two before you need them. Collect microdissected fragments for further processing in 500-µL tubes with screw-on lids sealed with rubber O rings. The O ring prevents loss of specimen volume by evaporation during subsequent processing and 500-µL tubes fit most PCR thermal cycling blocks for heating to inactivate Proteinase K after the digestion process is complete
9. Capture volumes of 12.5 or 25 µL work well. The pipet should be nearly vertical to minimize the risk of damaging the section. Use a Gilson-type pipet with a wide-bore polypropylene tip. Such tips are in many manufacturer's catalogs, or you can cut the end off an ordinary pipet tip. Steady your hand, holding the pipet, against your fist resting on the microscope stage. The fragment may be hard to retrieve if it lies flat on the section or slide. Briskly expelling buffer toward the fragment will usually lift the fragment to a level from which it may easily be recovered. Retrieval of dissected fragments without damage to the section is easier from a deep pool of buffer. The buffer should not contain detergent, such as Triton X-100 which, by reducing surface tension, prevents the formation of a deep standing pool of buffer.
10. Labeling must survive overnight incubation in the water bath and subsequent heat inactivation in the PCR thermal cycler block, the wells of which often contain oil traces which may dissolve even permanent marker pen inks. Write specimen numbers on the lid of the container as well. Typing correction fluid on the lid gives a white surface on which graphite pencil is permanent. Do not transpose lids. Do not open more than one specimen tube at a time, and replace the lid at once.

References

1. Schriever F., Freeman G., and Nadler L. M. (1991) Follicular dendritic cells contain a unique gene repertoire demonstrated by single-cell polymerase chain reaction. *Blood* **77,** 787–791.

2. Trumper, L. H., Brady, G., Bagg, A., Gray, D., Loke, S. L., Griesser, H., et al. (1993) Single-cell analysis of Hodgkin and Reed-Sternberg cells: Molecular heterogeneity of gene expression and p53 mutations. *Blood* **81,** 3097–3115.
3. Hubert, R., Weber, J. L., Schmitt, K., and Arnheim, N. (1992) A new source of polymorphic DNA markers for sperm typing: Analysis of microsatellite repeats in single cells. *Am. J. Hum. Genet.* **51,** 985–991.
4. Emmert-Buck, M. R., Bonner, R. F., Smith, P. D., Chuaqui, R. F., Zhuang, Z. P., Goldstein, S. R., et al. (1996) Laser capture microdissection. *Science* **274,** 998–1001.
5. Moskaluk, C. A. and Kern, S. E. (1997) Microdissection and polymerase chain reaction amplification of genomic DNA from histological tissue sections. *Am. J. Pathol.* **150,** 1547–1552.
6. Bohm, M., Wieland, I., Schutze, K., and Rubben, H. (1997) Microbeam MOMeNT—Non-contact laser microdissection of membrane-mounted native tissue. *Am. J. Pathol.* **151,** 63–67.
7. Fend, F., Emmert-Buck, M. R., Lee, J., Cole, K., Chaqui, R. F., Liotta, L. A., et al. (1998) Immuno-LCM: laser capture microdissection of immunostained frozen sections for RNA analysis. *Mod. Pathol.* **11,** A185.
8. Murphy, D. S., Hoare, S. F., Going, J. J., Mallon, E. A., George, W. D., Kaye, S. B., et al. (1995) Characterization of extensive genetic alterations in ductal carcinoma in situ by fluorescence in situ hybridization and molecular analysis. *J. Natl. Cancer Inst.* **87,** 1694–1704.
9. Going, J. J. and Lamb, R. F. (1996) Practical histological dissection for PCR analysis. *J. Pathol.* **179,** 121–124.
10. Lamb, R. F., Going, J. J., Pickford, I., and Birnie, G. D. (1996) Allelic imbalance at the NME1 locus in microdissected primary and metastatic human colorectal carcinomas is frequent, but not associated with metastasis to lymph nodes or liver. *Cancer Res.* **56,** 916–920.
11. Al Mulla, F., Going, J. J., Sowden, E. T. H. H., Winter, A., Pickford, I. R., and Birnie, G. D. (1998) Heterogeneity of mutant versus wild-type Ki-ras in primary and metastatic colorectal carcinomas, and association of codon-12 valine with early mortality. *J. Pathol.* **185,** 130–138.
12. Zhang, L., Cui, X., Schmitt, K., Hubert, R., Navidi, W., and Arnheim, N. (1992) Whole genome amplification from a single cell: implications for genetic analysis. *Proc. Natl. Acad. Sci. USA* **89,** 5847–5851.
13. Telenius, H., Carter, N. P., Bebb, C. E., Nordenskjold, M., Ponder, B. A. J., and Tunnacliffe, A. (1992) Degenerate oligonucleotide-primed PCR: General amplification of target DNA by a single degenerate primer. *Genomics* **13,** 718–725.
14. Grothues, D., Cantor, C. R., and Smith, C. L. (1993) PCR amplification of megabase DNA with tagged random primers (T-PCR). *Nucleic Acids Res.* **5,** 1321–1322.

9

RNA Extraction from Blood

Helen Pearson

1. Introduction

Based on the method of Chomczynski and Sacchi *(1)*, this is an extremely reliable method without the requirement for centrifugation over CsCl gradients. As with any RNA protocol, extreme care should be taken to exclude RNAse contamination, the greatest source of which will be the sample itself. All disposables and reagents should be RNAse free.

2. Materials

1. Microfuge tubes (1.5 mL).
2. Ice bucket.
3. Microfuge.
4. Red cell lysis buffer: 1.6 *M* sucrose, 5% Triton X-100, 25 m*M* $MgCl_2$, 60 m*M* Tris-HCl, pH 7.5; stored at 2–8°C and used cold.
5. Extraction Buffer: 5.25 *M* guanidinium thiocyanate, 50 m*M* Tris-Cl, pH. 6.4, 20 m*M* EDTA, 1% Triton X-100, 0.1 *M* β-mercaptoethanol (add immediately prior to use).
6. 2 *M* sodium acetate, pH 4.0.
7. Phenol (saturated with 1 *M* Tris HCl: 0.1 *M* EDTA, pH 8.0).
8. Chloroform:Iso-amyl alcohol (24:1).
9. Isopropyl alcohol.
10. 70% Ethanol.
11. RNAse-free distilled water.

3. Method

1. In a microfuge tube, mix 100 μL anticoagulated blood with 1 mL of red cell lysis buffer (*see* **Notes 1–3**).
2. Leave at room temperature with occasional shaking until the red cells have lysed and the solution translucent (usually within 5 min).
3. Microfuge for 30 s at 13,000*g* to pellet the white blood cells. Remove and discard supernatant.
4. Add 200 μL of extraction buffer and resuspend cell pellet by drawing through narrow gauge needle several times.
5. Add 20 μL of 2 *M* sodium acetate and mix gently by inversion.
6. Add 220 μL of phenol and mix gently by inversion.

From: *Methods in Molecular Biology, Vol. 226: PCR Protocols, Second Edition*
Edited by: J. M. S. Bartlett and D. Stirling © Humana Press Inc., Totowa, NJ

7. Add 60 µL of chloroform/isoamyl alcohol (24:1) and vortex vigorously.
8. Place on ice for 15 min.
9. Microfuge at 12,000*g* for 5 min and transfer the upper phase to new microfuge tube.
10. Add 200 µL of ice-cold isopropanol mix and store at –20°C for 30 min.
11. Microfuge at 12,000*g* for 15 min and discard supernatant.
12. Resuspend pellet in 200 µL of extraction buffer.
13. Repeat **steps 3** through **9**.
14. Wash pellet with 400 µL of cold 70% ethanol.
15. Microfuge at 12,000*g* for 5 min and discard supernatant.
16. Carefully remove last traces of ethanol from tube (folded sterile swab or kimwipe works well).
17. Resuspend in 100 µL of distilled water and incubate at 50°C for 15 min to dissolve RNA (*see* **Note 4**).

4. Notes

1. Blood stored at room temperature or 4°C should be mixed thoroughly prior to aliquots being removed.
2. Frozen blood samples should be allowed to thaw completely and mixed thoroughly before aliquots being removed. Although freezing lyses red blood cells, the red cell lysis step should still be performed to efficiently remove hemoglobin from the sample. Repeated freeze/thaw cycles should be avoided.
3. Buffy coat contains two to four times the amount of white blood cells per volume compared to fresh blood. Therefore, it is advisable to use only 50 µL of buffy coat diluted with 50 µL of phosphate-buffered saline as starting material for this protocol.
4. Repeat pipetting through a narrow gauge tip can help this process.

References

1. Chomczynski, P. and Sacchi. N. (1987) Single-step method of RNA isolation by acid guanidinium thiocyanate-phenol-chloroform extraction. *Anal. Biochem.* **162,** 156–159.

10

RNA Extraction from Frozen Tissue

John M. S. Bartlett

1. Introduction

RNA extraction is fundamental to all aspects of mRNA analysis. We include here a simple method that avoids the use of a mortar and pestle.

2. Materials

All chemicals, unless otherwise noted, were molecular biology grade and obtained from Sigma UK (Poole, Dorset). All glassware was pretreated with di-ethylpyrocarbonate (DEPC). All deionized distilled water was pretreated with DEPC and autoclaved (DEPC water). DEPC is a potent anti-RNAse agent.

2.1. DEPC Treatment of Glassware/Distilled Water

0.1% DEPC was added to distilled deionized water and glassware filled and left to stand overnight. The water was decanted and autoclaved (DEPC-treated water) and glassware sterilized at 220°C for 2 h (DEPC-treated glassware). DEPC is driven off by both procedures.

2.2. RNA Extraction

1. Braun Microdismembranator and Teflon vessels (Braun GmBH, Germany).
2. 3 *M* lithium chloride/6 *M* urea: Dissolve in 800 mL of DEPC water and make up to 1 L. The solution can be stored at 4°C for 3 to 6 mo.
3. 10 m*M* Tris-HCl, 0.5% sodium dodecyl sulphate (SDS) pH 7.5. Prepare stock solutions of 10% SDS and 0.5 *M* Tris-HCl (pH 7.5) in DEPC water. Stocks are stable at room temperature for up to 12 mo.
4. Proteinase K: Prepare 1 mg/mL w/v DEPC water stock and store at –20°C for up to 12 mo. Dilute in 10 m*M* Tris-HCl, 0.5% SDS as required, discard unused diluted enzyme.
5. Phenol:chloroform:isoamyl alcohol: phenol is presaturated with 10 m*M* Tris-HCl, pH 7.5. Prepare a mixture of 25:24:1 phenol:chloroform:isoamyl alcohol (v/v/v). Store at room temperature for up to 6 mo, shielding from light.

From: *Methods in Molecular Biology, Vol. 226: PCR Protocols, Second Edition*
Edited by: J. M. S. Bartlett and D. Stirling © Humana Press Inc., Totowa, NJ

3. Methods

3.1. RNA Extraction

1. Tissues should ideally be collected fresh and stored in liquid nitrogen. Routinely samples are collected on ice and transported for freezing within 30 to 60 min.
2. Tissues are disaggregated using a Braun-micro dismembranator. Teflon vessels and steel ball bearings are cooled in liquid nitrogen before use. Frozen tissue (50–500 mg) is placed in the vessel with a single ball bearing and agitated at 1000 cycles/second for 60 s. The vessel is then re-cooled in liquid nitrogen. This process is repeated until tissue is powdered (usually 2×; *see* **Note 1**).
3. Immediately after disaggregation of tissue, tissue material is resuspended while frozen by adding 1.5 mL of LiCl/Urea and transfer to a separate tube. The vessel is washed a further 2× with 1.5 mL of LiCl/Urea and the washing combined with the original sample. The resuspended medium is made up to 6 mL in LiCl/Urea and sonicated for 2× 30 s at maximum power using a probe sonicator. The sonicated samples are stored overnight at 4°C (*see* **Note 2**).
4. Centrifuge at 15,000*g*, 4°C for 30 min. The supernatant is discarded and the pellet washed with a further 6 mL of lithium chloride/urea, recentrifuged (15,000*g*, 4°C for 30 min) and the supernatant again discarded.
5. The pellet is resuspended in 6 mL of Tris-HCl/SDS with 50 µg/mL proteinase K (Boehringer Mannheim, UK) and incubated at 37°C for 20 min.
6. Samples are extracted with 100% phenol, followed by phenol:chloroform:isoamyl-alcohol. After each extraction, the sample is centrifuged at 2000*g* at room temperature for 10 min and the aqueous phase recovered.
7. After the final extraction, 300 µL of 8 *M* LiCl and 2.5 volumes absolute alcohol are added and samples stored at –20°C for 30 min overnight. RNA is pelleted by centrifugation at 4000*g*, 4°C for 45 min. The supernatant is discarded and the RNA pelleted dried and resuspended in DEPC-treated distilled water. Concentrations are estimated by optical density at 260/280 nm.

4. Notes

1. Disaggregation is critically dependent on tissue structure. Most tissues are readily disaggregated in two 60-s bursts. Other tissue types (e.g., fibrous tissues) may require longer periods to disrupt tissue. If a mechanical dismembranator is not available, other methods of tissue homogenization work equally well, either using a mortar and pestle or blade homogenizers.
2. Other methods can be used to lyse cells, such as passage through a syringe needle, etc. Extraction of RNA from solid tissues can be problematic because many of the commercial systems available for RNA extraction are validated for extraction of RNA from cell culture material or blood lymphocytes. These kits have often been less successful with tissue-derived material.

11

RNA Extraction from Tissue Sections

Helen Pearson

1. Introduction

There are two different methods of preparing tissue for histology: paraffin-embedding and freeze-embedding. Each has their advantages and drawbacks. Paraffin-embedded tissues (PET) produce optimum morphology but have comparatively poor molecular preservation and recovery. Although frozen sections have poorer histology, they allow excellent recovery of DNA and RNA for analysis.

Although fixation is performed to preserve the morphology of the living tissue, it does not necessarily have a beneficial effect on the DNA and RNA. Formalin, one of the most popular fixatives, crosslinks nucleic acids to protein, thus making the molecules rigid and susceptible to mechanical shearing. The duration of formalin fixation also seems to be important. Studies that have demonstrated DNA recovery around 200 base pairs recommend a period of fixation from 16 to 24 h but not any longer *(1)*.

RNA is a more labile species, and the paraffin-embedding process has been shown to greatly harm it. Many studies have shown that formalin fixation has the worst effects among commonly used fixatives and ethanol-based fixatives as having the best RNA preservation.

2. Materials

1. Microfuge tubes (1.5 mL).
2. Ice bucket.
3. Microfuge.
4. Extraction Buffer: 5.25 *M* guanidinium thiocyanate, 50 m*M* Tris-HCl, pH. 6.4, 20 m*M* EDTA, 1% Triton X-100, 0.1 *M* β-mercaptoethanol (add immediately before use.
5. Glycogen (10 mg/mL) in distilled water.
6. 2 *M* sodium acetate, pH 4.0.
7. Phenol (saturated with 1 *M* Tris-HCl, 0.1 *M* EDTA, pH 8.0).
8. Chloroform : iso-amyl alcohol (24 : 1).
9. Isopropyl alcohol.
10. 70% ethanol.
11. RNAse-free distilled water.

From: *Methods in Molecular Biology, Vol. 226: PCR Protocols, Second Edition*
Edited by: J. M. S. Bartlett and D. Stirling © Humana Press Inc., Totowa, NJ

3. Method

The method of Chomczynski and Sacchi *(2)* described in the previous protocol works well for fixed tissues with only two modifications.

1. The extraction is initiated by incubating tissue sections or microdissected cells in 500-µL extraction buffer for 5 min at room temperature with gentle agitation inverting several times.
2. Add 20 µL of 2 *M* sodium acetate and mix gently by inversion.
3. Add 220 µL of phenol and mix gently by inversion.
4. Add 60 µL of chloroform/isoamyl alcohol (24 : 1) and vortex vigorously.
5. Place on ice for 15 min.
6. Microfuge at 12,000*g* for 5 min and transfer the upper phase to new microfuge tube.
7. Add 1 to 2 µL of glycogen (10 mg/mL). Glycogen is a carrier that is used if RNA quantities are less than 1 µg. It also facilitates visualization of the pellet.
8. Add 200 µL of ice-cold isopropanol mix and store at –20°C for 30 min.
9. Microfuge at 12,000*g* for 15 min and discard supernatant.
10. Resuspend pellet in 200 µL of extraction buffer.
11. Repeat **steps 3** through **9**.
12. Wash pellet with 400 µL of cold 70% ethanol.
13. Microfuge at 12,000*g* for 5 min and discard supernatant.
14. Carefully remove last traces of ethanol from tube (folded sterile swab or kimwipe works well).
15. Resuspend in 100 µL of RNAse free distilled water and incubate at 50°C for 15 min to dissolve RNA (*see* **Note 1**).

4. Notes

1. Repeat pipetting through a narrow gauge tip can help this process.

References

1. Foss, R. D., Guha-Thakurta, N., Conran, R. M., and Gutman, P. (1994) Effects of fixative and fixation time on the extraction and polymerase chain reaction amplification of RNA from paraffin-embedded tissue. *Diagn. Mol. Pathol.* **3,** 148–155.
2. Chomczynski, P. and Sacchi, N. (1987) Single-step method of RNA isolation by acid guanidinium thiocyanate-phenol-chloroform extraction. *Anal. Biochem.* **162,** 156–159.

12

Dual DNA/RNA Extraction

David Stirling and John M. S. Bartlett

1. Introduction

It is sometimes desirable to extract both RNA and DNA from the same sample, especially when the sample is small. This can be achieved by isolating a total nucleic acid fraction that is then divided into two portions, which are treated differentially with either Dnase I (to remove DNA and recover RNA) or with RNase A (to selectively recover the DNA); however, this wastes half of the DNA and RNA. An alternative approach is to sequentially isolate the RNA and DNA fractions from the same sample. This protocol based on one reported by Chevillard *(1)*, begins by extracting RNA as in Chapter 9, but then re-extracts the DNA from the collected organic phases. The method described is for the extraction of both DNA/RNA from tissue but can be modified for either blood or cell lines (*see* **Notes 1** and **2**).

2. Materials

All chemicals, unless otherwise noted, are molecular biology grade and obtained from Sigma U.K. (Poole, Dorset). All glassware was pretreated with di-ethylpyrocarbonate (DEPC). All deionized distilled water was pretreated with DEPC and autoclaved (DEPC water). DEPC is a potent anti-RNAse agent.

2.1. DEPC Treatment of Glassware/Distilled Water

0.1% DEPC is added to distilled deionized water and glassware filled and left to stand overnight. The water was decanted and autoclaved (DEPC-treated water) and glassware sterilized at 220°C for 2 h (DEPC-treated glassware). DEPC is driven off by both procedures.

2.2. RNA Extraction

1. Braun Microdismembranator and Teflon vessels (Braun GmBH, Germany).
2. 3 *M* lithium chloride/6 *M* urea: Dissolve in 800 mL of DEPC water and make up to 1 L. This solution can be stored at 4°C for 3 to 6 mo.
3. 10 m*M* Tris-HCl, 0.5% sodium dodecyl sulphate (SDS), pH 7.5. Prepare stock solutions of 10% SDS, 0.5 *M* Tris-HCl (pH 7.5) in DEPC water. Stocks are stable at room temperature for up to 12 mo.

From: *Methods in Molecular Biology, Vol. 226: PCR Protocols, Second Edition*
Edited by: J. M. S. Bartlett and D. Stirling © Humana Press Inc., Totowa, NJ

4. Proteinase K: Prepare 1 mg/mL w/v DEPC water stock, which can be stored at –20°C for up to 12 mo. Dilute in 10 m*M* Tris-Cl/0.5% SDS as required, and discard unused diluted enzyme.
5. Phenol/chloroform/isoamyl alcohol: Phenol is presaturated with 10 m*M* Tris-HCl, pH 7.5. Prepare a mixture of 25:24:1 phenol:chloroform:isoamyl alcohol (v/v/v). Store at room temperature for up to 6 mo, shielded from light.
6. TE Buffer, pH 7.6: take 10 mL of 1 *M* Tris-HCl, pH 7.6, 2 mL of 0.5 *M* EDTA, and make up to 1 L with distilled water. Adjust pH to 7.6. Autoclave 15 min at 15. p.s.i.

2.3. DNA Extraction (as for RNA Plus)

1. Dual extraction buffer: 0.1 *M* NaCl, 10 m*M* Tris-HCl, pH 8.0, 1 m*M* EDTA, 1% SDS. Adjust to pH 12.0 with 5 *N* NaOH immediately before use.
2. 7.5 *M* ammonium acetate.

3. Methods

3.1. RNA Extraction

1. Tissues should ideally be collected fresh and stored in liquid nitrogen. Routinely samples are collected on ice and transported for freezing within 30 to 60 min.
2. Tissues are disaggregated using a Braun-micro dismembranator. Teflon vessels and steel ball bearings are cooled in liquid nitrogen before use. Frozen tissue (50 to 500 mg) is placed in the vessel with a single ball bearing and agitated at 1000 cycles/second for 60 s. The vessel is then recooled in liquid nitrogen. This process is repeated until tissue is powdered (usually twice; *see* **Note 3**).
3. Immediately after disaggregation of tissue, material is resuspended while frozen in 1.5 mL of LiCl/Urea and transferred to a separate tube. The vessel is washed a further 2× with 1.5 mL of LiCl/Urea and the washing combined with the original sample. The resuspended medium is made up to 6 mL in LiCl/Urea and sonicated for 2× 30 s at maximum power using a probe sonicator. The sonicated samples are stored overnight at 4°C (*see* **Note 4**).
4. Centrifuge at 15,000*g*, 4°C for 30 min. The supernatant is discarded and the pellet washed with a further 6 mL of lithium chloride/urea, recentrifuged (15,000*g* at 4°C for 30 min) and the supernatant again discarded.
5. The pellet is resuspended in 6 mL of Tris-HCl/SDS with 50 µg/mL proteinase K (Boehringer Mannheim, UK), and incubated at 37°C for 20 min.
6. Samples are mixed with an equal volume of phenol:chloroform:isoamyl-alcohol and mixed by inversion several times.
7. After mixing, the sample is centrifuged at 2000*g* at room temperature for 10 min and the aqueous phase recovered for RNA extraction, the organic phase is retained for DNA extraction.
8. Repeat **steps 6** and **7**.
9. After the final extraction, 300 µL of 8 *M* LiCl and 2.5 volumes absolute alcohol are added and samples stored at –20°C for 30 min overnight. RNA is pelleted by centrifugation at 4000*g*, 4°C for 45 min.
10. The supernatant is discarded and the RNA pelleted dried and resuspended in 50 µL of TE. Concentrations are estimated by optical density at 260/280 nm.

3.2. DNA Extraction

1. Combine the organic phases, including the interfaces from **step 7** above (both times), in a 15-mL polypropylene tube.
2. Add an equal volume of extraction buffer, vortex for 1 min, and place on ice for 10 min.

3. Centrifuge for 20 min at 10,000*g* 4°C.
4. Transfer aqueous phase to fresh tube and add 1/15 volume 7.5 *M* NH_4OAc and 2 volumes ice-cold EtOH. Incubate at –20°C for at least 1 h.
5. Centrifuge for 20 min at 10,000*g*, 4°C.
6. Carefully decant the supernatant and wash the pellet with 1 mL of 70% ethanol.
7. Centrifuge briefly to ensure that the pellet remains attached.
8. Carefully remove the supernatant and air dry pellet for 10 to 15 min.
9. Resuspend the DNA pellet in 50 µL of TE. Heat 5 min at 55°C then vortex thoroughly to dissolve the DNA.

4. Notes

1. For extraction from cell lines, scrape cells into 1.5 mL of lithium chloride/urea and then proceed from **step 3** of the RNA extraction above.
2. Extraction of DNA/RNA from blood can be achieved by collecting blood in either a heparin or EDTA containing Vacutainer by venipuncture. Store at room temperature and extract within the same working day. Whole blood (3 mL) is placed in a 15-mL polypropylene tube and mixed with 12 mL of red blood cell lysis solution (0.01 *M* Tris, pH 7.4, 320 m*M* sucrose, 5 m*M* $MgCl_2$, 1% Triton X 100). Blood is then mixed on a rolling or rotating blood mixer for 4 min at room temperature. Lymphocytes are recovered by centrifugation at 3000*g* for 5 min at room temperature. Then, proceed with RNA extract at **step 3** above.
3. Disaggregation is critically dependent on tissue structure. Most tissues are readily disaggregated in two 60-s bursts. Other tissue types (e.g., fibrous tissues) may require longer periods to disrupt tissue. If a mechanical dismembranator is not available, other methods of tissue homogenization work equally well, either using a mortar and pestle or blade homogenizers.
4. Other methods can be used to lyse cells, such as passage through a syringe needle, etc. Extraction of RNA from solid tissues can be problematic because many of the commercial systems available for RNA extraction are validated for extraction of RNA from cell culture material or blood lymphocytes. These kits have often been less successful with tissue-derived material.

References

1. Chevillard, S. (1993) A method for sequential extraction of RNA and DNA from the same sample, specially designed for a limited supply of biological material. *BioTechniques* 15, 22–24.

13

DNA Extraction from Fungi, Yeast, and Bacteria

David Stirling

1. Introduction

Although individual microorganisms may well require a unique DNA extraction procedure, here we include robust techniques for the preparation of DNA from fungi, yeast, and bacteria, which yield DNA suitable for a PCR template.

2. Materials

2.1. Fungal Extraction

1. CTAB extraction buffer: 0.1 *M* Tris HCl, pH 7.5, 1% CTAB (mixed hexadecyl trimethyl ammonium bromide), 0.7 *M* NaCl, 10 m*M* EDTA, 1% 2-mercaptoethanol. Add proteinase K to a final concentration of 0.3 mg/mL prior to use.
2. Chloroform : isoamyl alcohol (24 : 1).

2.2. Yeast Extraction

1. Yeast extraction buffer A: 2% Triton X-100, 1% sodium dodecyl sulphate, 100 m*M* NaCl, 10 m*M* Tris-HCl, pH 8.0, 1 m*M* EDTA, pH 8.0. Phenol : chloroform : isoamyl alcohol: Phenol is presaturated with 10 m*M* Tris-HCl, pH 7.5. Prepare a mixture of 25 : 24 : 1 phenol : chloroform : isoamyl alcohol (v/v/v). This solution can be stored at room temperature for up to 6 mo, shielded from light.
2. Glass beads. Diameter range 0.04–0.07 mm (Jencons Scientific Ltd, UK), suspended as 500 mg/mL slurry in distiller water.
3. Ammonium acetate (4 *M*).

2.3. Bacterial DNA Protocol

1. Lysozyme/RNase mixture: 10 mg/mL lysozyme, 1 mg/mL RNase, 50 m*M* Tris-HCl (pH 8.0). Store at –20°C in small aliquots. Do not refreeze after thawing.
2. STET: 8% sucrose, 5% Triton X-100, 50 m*M* Tris-HCl (pH 8.0), 50 m*M* EDTA, pH 8.0.
3. Filter sterilize and store at 4°C.

From: *Methods in Molecular Biology, Vol. 226: PCR Protocols, Second Edition*
Edited by: J. M. S. Bartlett and D. Stirling © Humana Press Inc., Totowa, NJ

3. Methods

3.1. Fungal Protocol

1. Grind 0.2 to 0.5 g (dry weight) of lyophilized mycellar pad in a mortar and pestle. Transfer to a 50-mL disposable centrifuge tube.
2. Add 10 mL (for a 0.5 g pad) of CTAB extraction buffer.
3. Gently mix to wet all the powdered pad.
4. Place in 65°C water bath for 30 min.
5. Cool and add an equal volume of chloroform/isoamyl alcohol (24:1).
6. Mix and centrifuge at 2000*g* for 10 min at room temperature.
7. Transfer aqueous supernatant to a new tube.
8. Add an equal volume of isopropanol.
9. High molecular weight DNA should precipitate upon mixing and can be spooled out with a glass rod or hook.
10. Rinse the spooled DNA with 70% ethanol.
11. Air dry, add 1 to 5 mL of TE containing 20 μg/ mL RNAse A. To resuspend the samples, place in 65°C bath or allow pellets to resuspend overnight at 4°C.

3.2. Yeast Protocol

1. Collect cells from fresh 5 mL culture by centrifugation at 2000*g* for 10 min and resuspend in 0.5 mL of water.
2. Transfer cells to 1.5-mL microfuge tube and collect by centrifugation at 15,000*g* for 10 min, pour off supernatant and resuspend in residual liquid.
3. Add 0.2 mL of buffer A, 200 μL of glass beads, and 0.2 mL of phenol:chloroform:isoamyl alcohol (25:24:1).
4. Vortex for 3 min and add 0.2 mL of TE.
5. Centrifuge at 15,000*g* for 5 min and then transfer aqueous to new tube.
6. Add 1 mL of 100% EtOH (room temperature), invert tube to mix, and centrifuge at 15,000*g* for 2 min.
7. Discard supernatant and resuspend pellet in 0.4 mL of TE (no need to dry pellet).
8. Add 10 μL of 4 *M* ammonium acetate, mix, and then add 1 mL of 100% EtOH and mix.
9. Centrifuge at 15,000*g* for 2 min and dry pellet. Resuspend in 50 μL of TE.

3.3. Bacterial DNA Protocol

1. Collect the bacteria from a 15-mL overnight culture into a 1.5-mL microfuge tube.
2. Resuspend pellet with 300 μL of STET buffer and add 30 μL of RNAse/lysozyme mixture.
3. Boil for 1 min 15 s.
4. Centrifuge at 15,000*g* for at least 15 min.
5. Take supernatant and phenol extract with 150 μL of STET-saturated phenol.
6. Spin and take supernatant. Add 1/10 volume 4 *M* lithium chloride (autoclaved). Let sit on ice for 5 to 10 min.
7. Spin and take supernatant. Add equal volume isopropanol at room temperature and incubate for 5 min.
8. Centrifuge at 15,000*g* for at least 15 min. No pellet will be visible.
9. IMPORTANT: wash with 80% ethanol (95% will cause the residual Triton to precipitate).
10. Resuspend pellet in 50 to 200 μL of TE.

14

Isolation of RNA Viruses from Biological Materials

Susan McDonagh

1. Introduction

The successful extraction of viral RNA from biological material requires rapid transport and adequate storage of samples because of the unstable nature of RNA. Samples should be received and processed within 6 h and the relevant fractions stored at –70°C until testing. Also, it is difficult to ascertain the efficiency of sample preparation methods; therefore, known standards should be processed alongside samples to assess the loss within the system *(1)*, particularly for quantitative applications.

2. Materials

Unless stated, all chemicals are supplied by Sigma, Poole, UK, or Merck. All stock solutions should be made using RNA-free water.

2.1. Sample Preparation

1. TNE-buffer: 0.11 *M* NaCl, 55 m*M* Tris (pH 8.0), 1.1 m*M* EDTA pH 8.0, 0.55% sodium dodecyl sulphate [SDS].
2. Poly-adenylic acid 2 mg/mL (Poly-A; Pharmacia, Upsala, Sweden).
3. Proteinase K, 10 mg/mL in water.
4. Phenol.
5. Phenol : chloroform (1 : 1).
6. Chloroform : isoamylalcohol, 50 : 1 (v/v).
7. 3 *M* sodium acetate, pH 5.2.
8. Ethanol 100% and 80%.
9. 10% SDS.

3. Methods

Because an ultrasensitive system is required, it is necessary to include an ultracentrifugation step where serum or plasma is concentrated by centrifugation at 27,000*g* for 1 h. Large volumes of 1 mL or greater can be used, and most of the supernatant (900 µL) can be removed before resuspending the pellet in the remaining 100 µL (*see* **Notes 1** and **2**).

1. Prepare a solution of TNE buffer, 0.5% SDS, 1 mg/mL proteinase K, and 40 µg/mL poly A. Pre-incubate the solution for 10 min at 37°C to inactivate RNases.

From: *Methods in Molecular Biology, Vol. 226: PCR Protocols, Second Edition*
Edited by: J. M. S. Bartlett and D. Stirling © Humana Press Inc., Totowa, NJ

2. Add 0.4 mL of TNE buffer with SDS, proteinase K, and poly A to the extraction tubes.
3. Add 100 µL of concentrated plasma or serum and mix immediately.
4. Incubate the lysates for 1.5 to 2 h in 37°C water bath.
5. Phenol extraction: Add 450 µL of phenol to the extraction tubes. Mix extensively and centrifuge at 13,000*g* for 10 min.
6. Phenol/chloroform extraction: Transfer the upper aqueous layer to a fresh tube and add 0.45 mL of phenol : chloroform (1 : 1). Vortex and centrifuge as above.
7. Chloroform/isoamylalcohol extraction: Transfer the upper aqueous layer to a fresh tube and add chloroform : isoamylalcohol (50 : 1). Vortex and centrifuge as above.
8. Ethanol precipitation: Transfer the aqueous layer to a fresh tube containing 40 µL of 3 *M* Na-acetate, pH 5.2, and 800 µL of ethanol. Mix and precipitate the nucleic acids at –20°C overnight or at –70°C for 30 min.
9. Collect the nucleic acid by centrifugation at 15,000*g* for 20 min at 0°C.
10. Discard the supernatant and wash the pellet with 1 mL 80% (v/v) ethanol.
11. Dry the precipitate on a dry block and dissolve in 25 µL of RNase-free water. These extracts can be stored at –70°C.

4. Notes

1. It is important to process the samples alongside any controls and standards to allow for the loss of RNA through sample preparation methods. This is especially important when preparing template for a limiting dilution curve or when using an external curve dilution series.
2. Although several commercial extraction systems have become available, we have not attained similar levels of sensitivity as with the protocol suggested here. However, extraction systems using guanidinium thiocyanate alongside phenol/chloroform or silica have been successfully used in ultrasensitive assays.

References

1. Clementi, M., Menzo, S., Bagnarelli, P., Manzin, A., Valenza, A., and Varaldo, P. E. (1993) Quantitative PCR and RT-PCR in virology. *PCR Meth. Appl.* **2,** 191–196.

15

Extraction of Ancient DNA

Wera M. Schmerer

1. Introduction

The DNA extraction process represents one of the critical stages in the analysis of degraded or ancient DNA. If polymerase chain reaction (PCR) amplification starts from a poor extract containing low template quantities, stochastic variation in the amplification of individual alleles may lead to allelic dropout *(1)*, resulting in a high risk of false-homozygous typing of a heterozygous sample *(2)*.

In analyzing repetitive sequences, such as STR loci, besides quantity, the quality of the extracted DNA used as template is of particular importance. PCR amplification of STR loci is generally accompanied by the generation of so called shadow bands, byproducts that are shortened in length by one repeat unit compared with the allelic product *(3)*. Because the accumulation of this artifact is increased with degradation of the DNA template *(4)* when amplifying ancient DNA, the intensity of a shadow band can exceed the intensity of the allelic product. As "artifact alleles" *(2,5)*, these products may be mistaken for a true allele of the sample, complicating the determination of a genotype or even resulting in false genotyping. Because a PCR may not be affected each time, independent amplifications of the same sample may result in differing genotyping results (e.g., *see* **ref.** ***6***) for the same sample *(2,7,8)*.

The degree at which these two artifacts occur is related to quantity and quality of extracted DNA amplified within a PCR *(1,4,8,9)*. Therefore, an optimized DNA extraction, to get the highest possible amount of target DNA with the best possible quality (which means with the highest possible reduction of additional degradation during the extraction procedure), is an essential precondition for optimal reproducibility of STR genotyping in the case of samples containing ancient DNA.

The protocols presented here are the results of a study with the aim to optimize the extraction of ancient DNA from historical skeletal material described by Schmerer et al. *(10,11)*. (For further detailed information on this study, also refer to **ref.** ***12***.)

The protocol described in detail represents a consensus protocol developed for the extraction of DNA with a wide range in degree of degradation (cf. **ref.** ***10***) and may be used regardless of the state of DNA preservation present in the skeletal material intended for analysis. Additional information on special protocols designed for three different degrees of DNA degradation is given in the table (*see* **Note 1**) and can be used to adapt the protocol in case of known DNA preservation.

From: *Methods in Molecular Biology, Vol. 226: PCR Protocols, Second Edition*
Edited by: J. M. S. Bartlett and D. Stirling © Humana Press Inc., Totowa, NJ

2. Materials

To apply the protocol described in the next section, the following chemicals and solutions are required.

1. 0.5 *M* EDTA solution (pH 8.3).
2. Sterile water (Ampuwa®, Fresenius).
3. Proteinase K-solution (approx 2 g/mL, e.g. Perkin–Elmer).
4. 70% phenol/chloroform/water (24:25:1, e.g., Rotipuran, Roth, or Perkin–Elmer) or, alternatively, a 5 *M* solution of sodium perchlorate in sterile water.
5. Chloroform (100%, e.g., Perkin–Elmer).
6. 2 *M* sodium acetate buffer (pH 4.5, Perkin–Elmer).
7. Isopropanol (abs., Merck).
8. Silica solution (Glasmilk™, Dianova).
9. Ethanol (abs.).

3. Methods

1. To prevent coprocessing of possible adhering contaminations, exposed surfaces of the bone sample are removed by the use of a scalpel. Subsequently, the material is exposed to ultraviolet light for 15 min on each of the surfaces.
2. According to the consistency of the material, samples are ground to a fine powder using a mixer mill (MM2, Retsch) or a mortar and pestle.
3. Bone powder (0.3 g) is mixed with 1.5 mL of 0.5 *M* EDTA-solution (pH 8.3) in a 2-mL reaction tube (Eppendorf), vortexed vigorously, and incubated for 96 h in a shaking waterbath at 20°C and constant shaking (for decalcification of the bone material).
4. Residues of the bone powder are pelleted by centrifugation at 3000*g* for 5 min in a 5415 C bench-top centrifuge (Eppendorf) or equivalent. The following steps are performed with the supernatants (approx 1.3 mL). These are transferred to an automated DNA extraction system (Gene Pure, Perkin–Elmer) or alternatively to a 15-mL tube (e.g., BlueMax™, Falcon).
5. The aqueous supernatants are mixed with 1.3 to 1.8 mL sterile water and incubated with 380 to 650 µL proteinase K-solution (approx 2 g/mL) at 60°C for 1.5 h.
6. Two volumes (2× supernatant volume) of 70% phenol/chloroform/water (24:25:1) are added to the solution. The suspension is constantly shaken for 6 min at room temperature. Alternatively, phenol can be replaced by 2 mL of a 5 *M* sodium perchlorate-solution (**ref. *13***, *see* **Notes 2** and **3**).
7. Using phenol—to facilitate the process—phase separation is performed at 60°C for 8 min and the phenolic layer is removed. Alternatively, the suspension can be separated by centrifugation at 4500*g* for 10 min. Using sodium perchlorate, no phase separation occurs. After incubation at 60°C for 8 min, the extraction mix is therefore processed according to **step 8**.
8. The aqueous phase is mixed with 4.0 to 5.3 mL of chloroform (100%) and the resulting suspension shaken for 6 min at room temperature.
9. Phase separation is performed at 60°C for 8 min and the chloroform phase is subsequently removed. Alternatively, the suspension can be separated by centrifugation at 4500*g* for 10 min.
10. DNA precipitation takes place in the presence of 64 to 120 µL of 2 *M* sodium acetate-buffer (pH 4.5) and 2.8 to 3.8 mL of isopropanol (abs.) at room temperature. After mixing these components for 1 min, the pH of the extraction mix should be determined. If necessary, the pH value should be adjusted to a maximum of 7.5 by further addition of sodium acetate buffer to ensure optimal precipitation of DNA (*see* **Note 4**). Subsequently 5 µL

Table 1
Extraction Parameters

	DNA preservation state*		
Parameters	High degradation	Intermediate degradation	Low degradation
Time of EDTA incubation	(96)–120 h	48 h	24 h
Temperature of EDTA incubation	20°C	(20)–30°C	(20)–30°C
Time of proteinase K incubation	(60)–90 min	90 min	60 min
Extraction reagent	Sodium perchlorate	Phenol (sodium perchlorate)	Phenol (sodium perchlorate)
Amount of sodium acetate solution (2 *M*)	64–117 µL	64–117 µL to (128–233 µL)	64–117 µL to (128–233 µL)
Additional purification	Wizard Prep	(Wizard Prep)	(Wizard Prep)

*DNA preservation state was determined by PCR amplification success and reproducibility of amplification results applying the basic protocol optimization was based upon ***(2,5,16)***.

The preservation state was defined as high degradation when specific products were detectable in up to 35% of the amplifications, displaying a low degree of reproducibility (12%) in typing results, and low degradation when specific products were detected in 70 to 90% of the amplificates, with a reproducibility of 86–91%.

of silica-solution (Glasmilk™, Dianova) is added and the mix is shaken again for 10 min at room temperature.

11. Using a DNA extraction device (Gene Pure, Perkin Elmer), the mix is removed by pressure and the DNA–silica complex is collected on a filter membrane. Alternatively, the mixture can be separated by centrifugation at 4500*g* for 10 min.
12. The DNA–silica complex is washed with 2.8 to 3.8 mL of ethanol (abs.) for 5 min at room temperature. Then, the alcohol is removed by renewed filtration or centrifugation. Co-extracted salts may be removed using ethane (80Y).
13. Using an automated DNA extractor, the silica-bound DNA is manually removed from the filter membrane with 500 µL of ethanol (abs.) and transferred to a 2-mL reaction tube for further processing.
14. The DNA-silica complex is pelleted by centrifugation for 4500*g* for 4 min (Centrifuge 5804, Eppendorf) and the ethanol is discarded.
15. The resulting pellet is air-dried for approx 30 min at room temperature, redissolved in 50 µL of sterile water (Ampuwa®, Fresenius), thermally eluted at 50°C for 5 min at constant shaking (thermal shaker 5437, Eppendorf), and stored at –20°C for further processing.

4. Notes

1. In case the state of DNA preservation of the material worked on is known, please change the parameters concerned according to **Table 1**.
2. The use of sodium perchlorate may result in an inhibition of DNA polymerase during the amplification process and therefore requires a further cleaning of the extracted DNA. For this purpose, the Wizard PCR Preps purification system (Promega) following a modified protocol ***(12,14)*** may be used.

3. The additional purification of the extract necessary when using sodium perchlorate results in loss of DNA. The use of sodium perchlorate, therefore, might be beneficial if brownish color of the bone meal or the extract indicates the presence of co-extracted polymerase-inhibiting substances like humic acids ***(15)***. In this case, an additional cleaning step would always be necessary and the use of sodium perchlorate instead of phenol ***(13)*** would not result in further loss of DNA.
4. Make sure to add the sodium-acetate buffer first, and to adjust the pH before adding isopropanol to avoid co-precipitation of salts.

5. Safety Note

Phenol and chloroform are aggressive organic solvents (please notice the safety data sheets provided by the manufacturers) and volatile even at room temperature, especially at the elevated temperature recommended to facilitate phase separation in **steps 7** and **9**. Avoid skin contact and inhalation. The use of a vented chemical hood is strongly recommended.

References

1. Kimpton, C., Fisher, D., Watson, S., Adams, M., Urquhard, A., Lygo, J., et al. (1994) Evaluation of an automated DNA profiling system employing multiplex amplification of four tetrameric STR loci. *Int. J. Leg. Med.* **106,** 302–311.
2. Schmerer, W. M., Hummel, S., and Herrmann, B. (1997) Reproduzierbarkeit von aDNA-typing. *Anthrop. Anz.* **55,** 199–206.
3. Weber, J. L. and May, P. E. (1989) Abundant class of human polymorphisms which can be typed using the polymerase chain reaction. *Am. J. Hum. Genet.* **44,** 388–396.
4. Murray, V., Monchawin, C., and England, P. R. (1993) The determination of the Sequences present in the shadow bands of a dinucleotide repeat PCR. *Nucleic Acids Res.* **21,** 2395–2398.
5. Schmerer, W. M. (1996) Reproduzierbarkeit von Mikrosatelliten-DNA-Amplifikation und Alleldetermination aus bodengelagertem Skelettmaterial. Unpublished diploma-thesis, Göttingen.
6. Ramos, M. D., Lalueza, C., Girbau, E., Perez-Perez, A., Quevedo, S., Turbon, D., et al. (1995) Amplifying dinucleotide microsatellite loci from bone and tooth samples of up to 5000 years of age: More inconsistency than usefulness. *Hum. Genet.* **96,** 205–212.
7. Schultes, T., Hummel, S., and Herrmann, B. (1997) Recognizing and overcoming inconsistencies in microsatellite typing of ancient DNA samples. *Ancient Biomol.* **1,** 227–233.
8. Ivanov, P. L. and Isaenko, M. V. (1999) Identification of human decomposed remains using the STR systems: effect on typing results. In: *Proceedings of the Second European Symposium on Human Identification 1998.* Promega Corporation, Innsbruck, Austria.
9. Lygo, J. E., Johnson, P. E., Holaway, D. J., Woodroffe, S., Whitaker, J. P., Clayton, T. M., et al. (1994) The validation of short tandem repeat (STR) loci for use in forensic casework. *Int. J. Leg. Med.* **107,** 77–89.
10. Schmerer, W. M., Hummel, S., and Herrmann, B. (1999) Optimized DNA extraction to improve reproducibility of short tandem repeat genotyping with highly degraded DNA as target. *Electrophoresis* **20,** 1712–1716.
11. Schmerer, W. M., Hummel, S., and Herrmann, B. (2000) STR-genotyping of archaeological human bone: Experimental design to improve reproducibility by optimisation of DNA extraction. *Anthrop. Anz.* **58,** 29–35.
12. Schmerer, W. M. (2000) *Optimierung der STR-Genotypenanalyse an Extrakten alter DNA aus bodengelagertem menschlichen Skelettmaterial.* Cuvillier Verlag, Göttingen.

13. Johns, M. B. and Paulus-Thomas, J. E. (1989) Purification of human genomic DNA from whole blood using sodium perchlorate in place of phenol. *Anal. Biochem.* **180,** 276–278.
14. Lassen, C. (1998) *Molekulare Geschlechtsdetermination der Traufkinder des Gräberfeldes Aegerten (Schweiz).* Cuvillier Verlag, Göttingen.
15. Cooper, A. (1992) Removal of colourings, inhibitors of PCR, and the carrier effect of PCR contamination from ancient DNA samples. *Ancient DNA Newsletter* **1,** 31–33.
16. Burger, J., Hummel, S., and Herrmann, B. (1997) Nachweis von DNA-Einzelkopiesequenzen aus prähistorischen Zähen. Liegemilieu als Faktor für den Erhalt von DNA. *Anthrop. Anz.* **55,** 193–198.

16

DNA Extraction from Plasma and Serum

David Stirling

1. Introduction

There are occasions where the only materiel available on a patient is stored plasma or serum samples. In normal individuals, the amount of DNA in these samples is very low but sufficient to serve as template for PCRs. Moreover, increased amounts of circulating DNA have been found in a variety of disorders, including cancer, autoimmune disease, and infection. Additionally, small amounts of fetal DNA have been detected in maternal plasma/serum during gestation. We have used the following protocol to successfully genotype archival plasma samples.

2. Materials

1. 10X SDS /Protein K: (Lauryl sulphate [SDS] 10 g/100 mL, Proteinase K 5 mg/mL).
2. TE (Tris EDTA) buffer: 10 m*M* Tris-HCl, 1 m*M* EDTA, pH 8.0.
3. Phenol : chloroform (1 : 1 v/v).
4. Glycogen (10 mg/mL).
5. 7.5 *M* Ammonium acetate.
6. 100% ethanol.
7. 70% ethanol.

3. Method

1. Place 1.5 mL of serum or plasma into a 15-mL centrifuge tube.
2. Add 1.5 mL of 1X SDS proteinase K solution in the tube containing the serum and mix well.
3. Digest overnight at 55°C in water bath.
4. Add 3 mL of phenol/chloroform solution.
5. Vortex 30 s and centrifuge for 10 min at 1000*g* using a swing-out rotor.
6. Transfer aqueous layer to fresh tube and repeat **steps 4** and **5**.
7. Transfer aqueous layer to fresh tube and add 5 µL of glycogen (10 mg/L), 1 mL of 7.5 *M* ammonium acetate, and 8 mL of 100% ethanol.
8. Mix by inverting and centrifuge at 2500*g* for 40 min.
9. Carefully remove supernatant and wash pellet in 10 mL of 70% ethanol.
10. Centrifuge at 2500*g* for 10 min. Carefully remove last traces of ethanol, and allow to air dry for 10 min before redissolving in 100 µL of TE.

From: *Methods in Molecular Biology, Vol. 226: PCR Protocols, Second Edition*
Edited by: J. M. S. Bartlett and D. Stirling © Humana Press Inc., Totowa, NJ

17

Technical Notes for the Detection of Nucleic Acids

John M. S. Bartlett

1. Introduction

In following any polymerase chain reaction (PCR)-based method, it is usual to identify the products of the reaction by some form of detection system. The majority of these still rely on size- and charge-based separation systems, although for some quantitative PCR applications, either direct measurement of fluorescence or indirect Enzyme-linked immunosorbent assay-based systems can be used. In this chapter, we summarize some of the most common methods for detection of nucleic acids as a handy reference for those seeking to validate their PCR reactions. Although there are many variants of these techniques, we have confined our reporting to methods of which we have direct experience and which offer broad applicability. Fluorescence detection of quantitative real-time PCR requires specialist equipment, and we have therefore omitted this from our discussions at present.

2. Gel Electrophoresis

By far the most common procedure for the analysis of nucleic acids is gel electrophoresis. This is a highly flexible approach that provides information on the size of the DNA molecule and under certain conditions can be used to discriminate different sequences of the same size. Electrophoresis can also be used to separate and purify nucleic acid fragments, to quantify allelic imbalances, etc. For DNA electrophoresis, the most common supports used are agarose and polyacrylamide. These are highly flexible because varying the concentration, and for polyacrylamide, the degree of crosslinking, can markedly alter the size range which can be discriminated. In general, polyacrylamide gels are more useful for separating smaller fragments of DNA (under 300–500 base pairs) and for applications where high resolution is required (such as analysis of microsatellites) because they are capable of resolving size differences of as little as 1 bp. Polyacrylamide gels can be run faster and at higher temperature than agarose gels. However, acrylamide is a neurotoxin and presents a safety hazard. Most laboratories now eschew the process of producing their own acrylamide solutions, relying instead on commercially prepared materials to circumvent this hazard. Polyacrylamide gels are also more difficult to pour and handle than agarose gels. Furthermore, using polyacrylamide gels requires, in general, the use of labeled

From: *Methods in Molecular Biology, Vol. 226: PCR Protocols, Second Edition*
Edited by: J. M. S. Bartlett and D. Stirling © Humana Press Inc., Totowa, NJ

Table 1
Recommended Agarose Concentrations for DNA Electrophoresis

Percentage agarose (w/v)	Molecular weight range
0.6%	1000–20,000 bp long PCR
1.0%	500–5000 bp
2%	100–2000 bp
3%	10–500 bp

nucleotides, although silver staining may be applied to these gels the fragility of the gel often precludes this.

Agarose gels, however, are more robust and easy to prepare. Although resolution is poorer, some modern forms of agarose claim separations that rival that of acrylamide gels. The major strength of agarose-based gels is the greater range of separation. Conventional agarose electrophoresis can separate DNAs from 200 to 50,000 bp which is more than adequate for PCR-based systems. Adaptations of agarose electrophoresis, for example, those using pulsed electric fields, can be used to separate DNA fragments of up to 10 Mbp.

2.1. Selecting Conditions for Agarose Gel Electrophoresis

Those interested in understanding the electrophysical properties governing migration of DNA in agarose supports can refer to a number of molecular biology texts that cover these areas or alternatively visit the web *(1)* for a useful guide to electrophoresis with agarose gels.

2.1.1. Agarose Concentrations

Table 1 outlines recommended agarose concentrations for gel electrophoresis of DNA. Note that modern specialist formulations of agarose may require alterations to this table (e.g., Nusieve, etc.).

2.1.2. Buffers

Two buffering systems are commonly used for agarose gel electrophoresis of DNA; of these, Tris acetate EDTA buffer (TAE) is more widely accepted because it facilitates recovery of material from agarose. However, it has a relatively low buffering capacity, and recirculation of buffer may be required over long electrophoresis runs (4–6 h). Tris borate EDTA (TBE) is preferred for small molecules and longer electrophoresis times because of its higher buffering capacity. We have used both with good results. Independent of the buffer selected, attention should be paid to the depth of buffer. In most systems 3 to 5 mm of buffer should cover the gel. Insufficient buffer may allow the gel to dry out during the run, whereas excessive buffer will reduce the current through the agarose support, promote heating and decrease DNA motility. Electrophoresis is performed applying a voltage of between 1 to 5 V/cm (where cm is the distance in centimeters between the electrodes in the gel tank).

2.1.3. Loading Dyes

Loading dyes serve two purposes: They are usually dense and promote the settling of the DNA to the base of the well. They usually contain dyes that migrate with the DNA and can be visualized to monitor the process of electrophoresis. Care should be taken, however, because the common dyes used (bromophenol blue and xylene cyanol) have different motilities in different agarose products, with differing buffers and differing gel percentages (*see* **ref. *3***).

2.1.4. DNA Dyes: Before or After?

Most dyes (ethidium bromide, SYBR green, etc.) that are used to stain DNA do so by intercalating into the DNA sequence. As such they, of necessity, alter both the structure and motility of the DNA. Although this can give spurious results, for many applications (such as confirmation of PCR products before cloning or sequencing) this is not a major issue. In these cases, the dye is often added to the loading buffer or gel before electrophoresis. Where an accurate determination of product size is required, such as in randomly amplified polymorphic DNA (RAPD) or allelotyping, products should be separated in the absence of dye and the gel stained thereafter.

2.1.5. Recovery of DNA from Agarose Gels

Having identified the product on agarose gels, often there is a requirement to sequence or otherwise analyze the product. Recovering DNA from agarose gels is a simple procedure that can be attempted in many ways ***(1)***. It is important, however, that the decision regarding recovery of DNA is taken before electrophoresis is attempted as the choice of agarose is probably the most crucial factor in determining DNA yield after extraction from agarose. Protocols for recovery of DNA from agarose can be accessed from the web (*see* **ref. *1*** for a good range of methods).

2.2. Polyacrylamide Gel Electrophoresis (PAGE) of DNA

Casting and running polyacrylamide gels is undoubtedly more complex and problematic than using agarose electrophoresis. Almost universally, polyacrylamide gels are supported between glass plates and must be polymerized by a chemically catalyzed reaction with the inherent problems caused by failed reagents from time to time. However, PAGE has distinct advantages in terms of resolution, capacity, and purity of DNA bands. The resolving power of PAGE is such that molecules of DNA with lengths differing by as little as 0.2% (2 bases/Kb) may be separated (although some novel agarose preparations can approach this for smaller fragments). PAGE has a higher capacity for DNA than agarose, with up to 10 µg of DNA per lane being applied without compromising results. DNA recovered from PAGE is extremely pure and has been used for microinjection into mouse embryos ***(2)***.

PAGE is equally applicable to separation of double-stranded or single-stranded DNA (under denaturing conditions); however, the size range of fragments that can be separated is more restrictive than for agarose because of the fragile nature of low percentage acrylamide gels (*see* **Table 2**).

PAGE gels are crosslinked using a varying ratio of *bis*-acrylamide in the monomeric solution; generally, the ratio of acrylamide to *bis*-acrylamide is around 30:1.

Table 2
Separation of Double-Stranded DNA: What Percentage of Acrylamide?

Acrylamide (% w/v)	DNA size range
3.5	1000–2000 bp
5.0	80–500 bp
8.0	60–400 bp
12.0	40–200 bp
15.0	25–150 bp
20.0	6–100 bp

2.2.1. Nondenaturing PAGE Conditions

Although both TAE and TBE can be used for PAGE, we recommend use of TBE because this generally provides sharper bands on low-percentage gels. Running gels in 1× TBE under low voltages (1–8 V/cm) will prevent denaturation of DNA caused by heating of the gel. Although the electrophoretic mobility of double-stranded DNA is inversely proportional to the log(fragment length), this relationship can be altered by both GC content and sequence. For this reason, PAGE is not a reliable means of sizing DNA fragments.

2.2.2. Denaturing PAGE

Denaturing page gels contain a denaturing agent (usually 6–8 *M* urea) that inhibits base pairing in nucleic acids. DNA fragments are loaded after a brief heat denaturation and electrophoresis at high voltages (ca 20 V/cm) ensures that high temperatures are maintained throughout the electrophoretic process. Denaturing gels are most commonly used to analyze sequencing reactions and to recover labeled DNA probes. As one would predict from their common use in sequence determination, the motility of DNA fragment under denaturing conditions is almost totally unaffected by sequence and base composition.

2.2.3. Recovery of DNA from Polyacrylamide Gels

In our experience, recovery of DNA from polyacrylamide gels is vastly simpler than recovery from agarose, largely because of the fact that much larger quantities of DNA are loaded onto the gel. Elution overnight by diffusion can recover 70 to 80% of the nucleic acid in Maxim–Gilbert's solution.

2.2.4. Radioisotopic Detection and Quantitation

One of the major advantages of polyacrylamide gels is the ability to dry them onto supports and produce autoradiographic images. These can either be analyzed by densitometry or used as a template to allow direct counting of the bands of interest. Although the latter approach is time consuming, it provides a highly quantitative analysis of DNA species in (for example) quantitative PCR ***(3)***.

2.2.5. Nonradioisotopic Detection and Quantitation

The increasing trend in modern molecular applications is away from the use of large amounts of radioactivity and to substitute biotinylated or digoxigenin-labeled DNA for 32-P- or 33-P-labeled nucleotides. More recently, directly fluorescently labeled nucleotides have been released which, with the appropriate systems, can also aid detection of nucleic acid. These approaches are largely applicable in northern and southern blotting procedures, and although these can be applied to sequencing (blotting of sequences) or other DNA gel procedures, they are time consuming and introduce an additional potential source of error. Other methods, such as silver staining, are more attractive because they do not require blotting, but they are generally less sensitive than radioisotopic procedures. With this proviso, they provide a useful and rapid means of detecting nucleic acids *in situ*, before proceeding with experiments.

2.3. Summary

It is of course impossible to detail all possible systems for detection of DNA in a simple chapter such as this one; however, we have included below examples of the major approaches to detection of nucleic acids that can be adapted to suit most needs. This is very much a general guide for broad application and other methods can be sourced from the references included below.

3. Materials

3.1. Agarose Gel Electrophoresis

1. Low melting point agarose (Sigma).
2. 50× TAE: 242 g of Trizma Base (Sigma) and 100 mL of 0.5 *M* EDTA (pH 8.0) are dissolved in 800 mL distilled water and autoclaved. Then, 57.1 mL of glacial acetic acid is added; make up to 1 L.
3. Gel loading buffer: 0.25% Bromophenol blue (w/v), 0.25% xylene cyanol FF (w/v), and 30% glycerol (v/v) in distilled water (*see* **Note 1**).
4. Molecular weight markers (e.g., 100-bp ladder; Gibco).
5. Gel apparatus and power pack.
6. Ethidium bromide (10 mg/mL) in water (*see* **Note 2**).
7. Ultraviolet light box and camera.
8. Activated charcoal.

3.1.1 Recovery of DNA from Agarose Gels

1. Scalpel blades.
2. β-agarose and buffer (Calbiochem, UK).
3. Phenol : chloroform 1 : 1 mixture (Sigma, UK).
4. Ammonium acetate (7.5 *M*).
5. 100% Ethanol.

3.1.2. PAGE

3.1.2.1. Nondenaturing PAGE

1. 40% Acrylamide (38 g of acrylamide, 2 g of *bis*-acrylamide, Bio–Rad, UK) in distilled water.

2. 5× TBE: 54 g of Tris-base, 27.5 g of boric acid, 20 mL of 0.5 *M* EDTA (pH 8.0) dissolved in 800 mL distilled water; make up 1 L.
3. 10% w/v ammonium persulphate (make fresh for each use).
4. TEMED (Bio–Rad).
5. Gel loading buffer: 0.25% Bromophenol blue (w/v), 0.25% xylene cyanol FF (w/v), and 30% glycerol (v/v) in distilled water (*see* **Note 1**).
6. Molecular weight markers (e.g., 100-bp ladder; Gibco).
7. Vertical gel electrophoresis system, spacers (1.0 mm; e.g., Bio–Rad Protean II).

3.1.2.2. Denaturing PAGE

1. 40% Acrylamide (38 g of acrylamide, 2 g of *bis*-acrylamide, Bio–Rad, UK) in distilled water.
2. 5× TBE: 54 g of Tris-base, 27.5 g of boric acid, 20 mL of 0.5 *M* EDTA (pH 8.0) dissolved in 800 mL of distilled water; make up to 1 L.
3. Urea (Sigma, UK).
4. 10% w/v ammonium persulphate (make fresh for each use).
5. Temed (Bio–Rad).
6. Radioactively labeled molecular weight markers: We have used *Hinf*1-digested plasmids labeled with Klenow enzyme (*see* **Note 3**).
7. Gel loading buffer: 0.25% Bromophenol blue (w/v), 0.25% xylene cyanol FF (w/v), and 30% glycerol (v/v) in distilled water (*see* **Note 1**).
8. Vertical gel electrophoresis system, spacers (0.6 mm).
9. Gel fixative: Prepare 2 L of 10% acetic acid and 10% methanol in distilled water.
10. Whatmann 3MM Paper.
11. Gel dryer.
12. Autoradiographic film and cassettes.
13. Developer and fixative for above.

3.1.3. Recovery of DNA from Acrylamide Gels

1. Maxam & Gilberts solution: 0.5 *M* ammonium acetate, 0.1% sodium dodecyl sulphate, and 1 m*M* EDTA; store at –20°C.
2. Transfer RNA (tRNA) 10 mg/mL and store at –20°C (optional).
3. Absolute ethanol (–20°C).
4. 70% ethanol (–20°C).

4. Methods

4.1. Agarose Gel Electrophoresis

1. Prepare a clean flat bed electrophoresis tray, with sealed ends (*see* **Note 4**) and a comb with the appropriate number of samples. Fill tank with 1× TAE.
2. Dissolve 3 g of low melting point agar in 100 mL of 1× TAE buffer.
3. Weigh the flask and note weight before microwaving.
4. Microwave until clear, take care not to over boil (*see* **Note 5**).
5. Allow to cool slightly, weigh, and add sufficient distilled water to make up to premicrowave weight.
6. Cool agarose to approx 50°C before pouring into gel support. Insert comb 0.5 to 1.0 mm above the base of the tank, add agarose and allow to set (30 to 60 min).
7. Remove buffer dams or tape and place gel in electrophoresis tank. Add sufficient buffer to cover gel to approx 1 mm.

8. Mix DNA with gel loading buffer by adding 1 : 1 DNA solution to gel loading buffer. Slowly add approx 5 to 10 µL sample per well. Add 1 µg of DNA molecular weight marker to 4 µL of water and 5 µL of gel loading buffer to one lane (*see* **Note 6**).
9. Close the lid of the tank to generate an electrical circuit (*see* **Note 7**) and run at a voltage of 1 to 5 volts/cm. Check bubbles are arising from the anode and that the dye is migrating into the gel.
10. When migration of markers is complete, remove gel and visualize under ultraviolet light and photograph.
11. Use intensity of bands to estimate DNA concentration; densitometric scanning of the photograph can also be used as appropriate.

4.1.1. Staining of Agarose or Acrylamide Gels

1. Remove the gel from the electrophoresis tank and stain in 0.5 to 1 µg/mL ethidium bromide in sufficient electrophoresis buffer to cover the gel for about 10 to 20 min at room temperature.
2. Destain in distilled water for 2 × 20 min.
3. Visualize gel (*see* **Note 7**).

4.1.2. Decontamination of Ethidium Bromide Solutions

Both electrophoresis buffers from ethidium bromide containing gels and staining/destaining of gels should be decontaminated before disposal (*see* **Note 2**).

1. Add 1 g/L activated charcoal.
2. Stir for 30 to 60 min at room temperature.
3. Remove charcoal by filtration and discard solution.
4. Place charcoal and agarose gel into solid waste for incineration (*see* **Note 8**).

4.1.3. Recovery of DNA from Agarose Gels

Recovery of DNA from agarose gels produces highly variable yields dependent on fragment length. Long fragments (5–10 kb and above) are prone to shearing if vortexed. We have described here the agarose method that is most widely applicable, in our hands; however, postrecovery purification with phenol chloroform greatly improves the purity of the DNA and is recommended where postrecovery enzyme modifications are planned. The use of TAE buffer is recommended and care should be taken not to overload the gel.

1. After running the gel, visualize the band of interest under ultraviolet light and excise using a scalpel (*see* **Note 9**). Remove the band as cleanly as possible; the lower the amount of agar present in the gel slice, the better the recovery.
2. Transfer up to 200 mg of agarose gel slice to a microcentrifuge tube and melt at 75°C (5–10 min. *see* **Note 10**) then cool to 45°C.
3. Estimate volume of melted gel and add 2% 50× agarose buffer and mix gently (*see* **Note 11**).
4. Add β-agarose (add 3–4 units per 100 mg of a 1% agarose gel), mix gently (*see* **Note 11**), and incubate at 45°C for 3 to 4 h (*see* **Note 12**).
5. Add an equal volume of phenol : chloroform and mix by gentle inversion 10 to 20 times.
6. Centrifuge at 3000*g* for 5 to 10 min.

7. Remove upper aqueous layer to a new microfuge tube and add 20 µL of 7.5 *M* ammonium acetate and 2 volumes of ice cold 100% ethanol. Mix by gentle inversion 10 to 20 times and let stand for 30 min (*see Note 13*).
8. Centrifuge at 15,000*g* for 10 min and discard supernatant.
9. Allow pellet to air dry and resuspend in distilled water.

4.2. PAGE

Many applications using PAGE are based on radiolabeling of DNA; however, we have successfully used this system also for silver-stained and ethidium bromide-stained gels.

4.2.1. Nondenaturing PAGE

1. Prepare a 6% (*see* **Note 14**) acrylamide gel by mixing 7.5 mL of 40% acrylamide, 10 mL of 5×TBE, and 32.5 mL of distilled water (50 mL total, sufficient for 2 × 20 cm protean II gels).
2. The protean II system is supplied with a gel-pouring apparatus, the vertical clamps seal the sides of the gel, and a rubber cushion seals the base. Add 25 µL of TEMED and 250 µL of fresh ammonium persulphate solution to the acrylamide and pour gels immediately (*see* **Note 15**). Insert combs and leave to polymerize for 30 to 60 min.
3. Remove combs and wash wells with 1× TBE (*see* **Note 16**). Prepare electrophoresis equipment and add 1× TBE to top and bottom buffer chambers. Cool electrophoresis equipment by passing water through central core (*see* **Note 17**).
4. Mix DNA with gel loading buffer by adding 1:1 DNA solution to gel loading buffer. Slowly add approx 5 to 10 µL of sample per well. Add 1 µg of DNA molecular weight marker to 4 µL of water and 5 µL of gel loading buffer to one lane (*see* **Note 6**).
5. Electrophorese at 30 mA for 2 to 3 h.
6. Carefully remove the top gel plate and thoroughly wet gloves and gel with 1× TBE. Gently remove gel onto a prewetted support prior to visualization under ultraviolet light (*see* **Note 18**).

4.2.2. Denaturing PAGE

Although nondenaturing PAGE is applicable to both unlabeled and labeled DNA, the fragile nature of the thin gels commonly used for denaturing PAGE makes radioactive labeling more common. We have, however, successfully used silver staining to visualize DNA on PAGE (*see* **ref. *1***). We have, however, described here conventional radioactive detection of single-stranded DNA.

1. To make a 6% denaturing gel: weigh 42 g of urea and add 15 mL of 40% acrylamide, 20 mL of 5× TBE, and 30 mL of distilled water. Stir at room temperature until urea dissolves completely.
2. Prepare vertical gel electrophoresis system for gel pouring, ensuring that the plates are clean (*see* **Note 19**), and seal with sleek tape.
2. Add 80 µL of TEMED and 800 µL of fresh ammonium persulphate solution to the acrylamide and pour gel immediately (*see* **Note 15**), tapping vigorously during pouring to free any air bubbles. Insert comb and leave to polymerize for 30 to 60 min.
3. Place the gel in a vertical electrophoresis tank and fill the upper and lower chambers with 1× TBE.
4. Mix DNA with gel loading buffer by adding 1:1 DNA solution to gel loading buffer. Slowly add approx 1 to 3 µL of sample per well. Heat samples to 85°C for 5 min and cool

on ice before loading. Add 1 µg of radioactively labeled DNA molecular weight marker to 1 µL of gel loading buffer in one lane.

5. Electrophorese at 30 mA for 2 to 3 h or at constant temperature (*see* **Note 20**).
6. Remove gel from the electrophoresis tank and discard radioactive buffer.
7. Place gel plates on tissue paper and insert a scalpel between the back plate and spacers. Remove the backplate trying to ensure that the gel remains on the front plate (*see* **Note 21**).
8. Immerse gel, on plate, in gel fixation solution and fix for 15 to 30 min. Carefully remove the gel, on the plate, from the fixative, drain, and cover with a dry piece of Whatman 3MM filter paper.
9. Either invert the gel and carefully lift the glass plate away from the paper or lift the paper gently from the glass plate. The gel should stick to the paper.
10. Dry on a gel dryer for 60 min at 80°C under vacuum (*see* **Note 22**).
11. Cover gel with clingfilm and place autoradiography film directly over the gel and expose at –70°C (*see* **Note 23**).

4.2.3. Recovery of DNA from Acrylamide Gels

Recovery of DNA from acrylamide is a relatively simple procedure because of the higher amounts of DNA that can be loaded on acrylamide gels, with yields of around 50 to 60% sufficient DNA recovered for most applications. We have only applied this technique to unfixed gels stained with ethidium bromide.

1. After running the gel, visualize the band of interest under ultraviolet light and excise using a scalpel (*see* **Note 9**). Remove the band as cleanly as possible; the lower the amount of acrylamide present in the gel slice, the better the recovery.
2. Transfer to a microcentrifuge tube and add 400 µL of Maxam & Gilberts solution.
2. Vortex and incubate overnight at 37°C.
3. Spin down the gel by pulse centrifugation for 30 s at 5000*g* in a bench-top centrifuge.
4. Remove supernatant to a separate Eppendorf tube (approx 350 µL).
5. Add 1 µL of 10 mg/mL tRNA (optional, omit if required).
6. Add 2.5 volumes of ice-cold 100% ethanol and mix gently.
7. Stand for 20 min and centrifuge for 10 min at 15,000*g*.
8. Decant supernatant and add 300 µL of 70% ethanol, mix gently.
9. Centrifuge for 10 min at 15,000*g*.
10. Decant supernatant and air dry pellet.
11. Resuspend in 20 µL of distilled water.

5. Notes

1. For some applications, lower concentrations of dyes can give better results, for example, when the DNA fragment runs close to one of the dyes. In this case, dilute loading buffer 1:5 in 30% glycerol.
2. Ethidium bromide is a carcinogen, teratogen, and mutagen. Handle with respect.
3. Digesting plasmids with enzymes such as *Hinf*I leaves 5′ overhangs that can be labeled with Klenow using the following reaction: 1 µg of digested plasmid in 5 µL of distilled water, 2 µL of 10 × repair buffer supplied with Klenow); 1 µL of a mixture containing 2 m*M* each of dGTP, dCTP, and dTTP; 2 µL of ^{35}S-dATP (approx 0.5 µCi), and 1 µL of Klenow fragment of DNA polymerase (5–10 units) in a total volume of 20 µL. Incubate for 30 min at room temperature and stop reaction by adding 2 µL of 0.25 *M* EDTA. Store at –20°C and use 0.5 to 1 µL per gel.

4. Many modern electrophoresis tanks are equipped with seals or with buffer dams to allow pouring of the gel or, alternatively, one can tape over the end of the tray with autoclave tape. Ensure gels are poured on a level surface.
5. Undissolved agarose appears as small lens-shaped flecks in the solution; reheat until the solution is clear. Take care not to superheat the solution because it may boil over.
6. Modern electrophoresis equipment will not allow circuits to be completed until the tank is sealed. If you have an older tank without this safety feature, discard it!
7. If the bands appear faint, with a dark background the gel is understained, return to buffer containing ethidium bromide and repeat. If the bands appear faint with a bright background the gel is overstained, destain in excess electrophoresis buffer (without ethidium bromide) for 10 to 20 min and revisualize.
8. Standard incineration is sufficient to decontaminate ethidium bromide.
9. Ultraviolet light produces strand breaks in DNA. Even exposure for short periods (30 s) can compromise DNA recovery. Ensure recovery of bands as quickly as possible. Where multiple bands are to be recovered, we have blocked the ultraviolet from parts of the gel using cardboard under the gel support. Some protocols recommend the addition of 1 m*M* guanosine or cytidine to the gel and electrophoresis buffers to protect DNA at the transillumination stage.
10. Ensure the agarose is completely melted to optimize digestion.
11. Vortexing can shear large DNA fragments, take care.
12. Although recommended incubation times are shorter, we have found extending the incubation time can improve the yield and ensure complete digestion of the agarose.
13. If the expected DNA concentration is less than 50 ng/μL, we recommend the addition of 5 μg of tRNA to enhance precipitation.
14. We have successfully worked with 4 to 20% acrylamide gels for various applications. Use of low concentration acrylamide is only possible in our experience where supports for the gels (Whatman paper) is provided and this limits the use to radioactive applications.
15. Polymerization will commence as soon as these catalysts are added and is temperature and time dependent, work quickly but carefully.
16. Wells will frequently contain small amounts of acrylamide solution that has notolymerisind because of contact with air (oxygen inhibits polymerization). Washing the wells prevents this from sinking to the base of the well and polymerizing giving uneven well depths.
17. The protean system is cooled by circulation of water through a central core. This is helpful when rapid results are required, but not essential.
18. Acrylamide gels, especially low percentage gels, are extremely fragile and tear easily, and handling them is an art learned through practice (and much frustration!). Siliconization of glass plates can aid removal of gels, but in our experience, providing the plates are clean separation is relatively easy. If your gel sticks to the glass, clean them thoroughly before use.
19. A common cause of tearing of gels and problems with pouring is dirty glass plates. Before use, scrub glass plates with concentrated detergent, rinse extensively with tap water and distilled water, and then with methylated spirits. Dry thoroughly before use. Clean spacers and combs by wiping with methylated spirits.
20. Modern powerpacks provide the option of running gels at constant temperature (usually 65°C), and we have found this to provide more consistent results.
21. The use of a silicon solution on the backplate after washing can facilitate separation. Be pragmatic: If the gel sticks to the back plate remove the front plate. When silver staining is to be used, stain the gel *in situ* on the plates.
22. If the gel cracks into a "crazy paving" type pattern while drying, this is indicative of poor fixation. Extend fixation period. This approach can be used with thicker gels (1–1.5 mm),

and higher percentage gels (up to 20%), longer fixation and drying times are needed. For high percentage gels adding 10% glycerol to the fixative helps preserve the gel.

23. If using 35-S, the use of enhancer screens is essential. Also, omit the cling film to ensure exposure. However, be careful to avoid moisture or the gel will stick to the film.

References

1. http://www.protocol-online.org
2. Sanbrook, J., Fritsch, E. F., and Maniatis, T. (1989) *Molecular Cloning: A Laboratory Manual.* 2nd ed. Cold Spring Harbor Laboratory, Cold Spring, New York.
3. Bartlett, J. M. S., Hulme, M. J., and Miller, W. R. (1996) Analysis of cAMP RI-alpha messenger-RNA expression in breast cancer—Evaluation of quantitative polymerase chain-reaction for routine use. *Br. J. Cancer* **73,** 1538–1544.

18

Technical Notes for the Recovery and Purification of PCR Products from Acrylamide Gels

David Stirling

1. Introduction

Although the best way of obtaining pure polymerase chain reaction (PCR) product will always be to optimize reaction conditions to yield only one product, there are still circumstances where DNA has to purified from gels. Several good commercial products exist for the recovery of DNA from agarose. Here, we present a reliable method of recovering DNA from polyacrylamide gels.

2. Materials

1. Sterile scalpel blade.
2. Microfuge tubes (0.5 and 1.5µL).
3. Sterile narrow-gauge needle (23-guage).
4. Tris-HCl EDTA buffer: 10 m*M* Tris-HCl, 10 m*M* EDTA, pH 8.0.
5. 3 *M* sodium acetate (pH 4.0).
6. 100% ethanol.
7. 70% ethanol.

3. Method

1. Stain the gel with ethidium bromide, visualize the band of interest, and dissect with scalpel blade.
2. Pierce hole in the bottom of a 0.5 ml Microfuge tube with narrow gauge needle.
3. Place 0.5 mL of TE in a 1.5-mL microfuge tube.
4. Place the acrylamide slice with the dissected DNA band into a 0.5-mL tube (with hole) and place inside the 1.5-mL tube (the lip will prevent it touching the bottom).
5. Microfuge the two-tube combination at 12,000*g* for 5 min. The acrylamide will be forced through the hole in the bottom of the 0.5-mL tube into the TE.
6. Discard the empty 0.5-mL tube.
7. Cap the 1.5-mL tube, vortex for 30 s, then incubate for 2 h overnight at 37°C.
8. Microfuge at 12,000*g* for 5 min.
9. Transfer supernatant to fresh tube.
10. Add 1/10 volume 3 *M* sodium acetate and 2 volumes 100% ethanol.

From: *Methods in Molecular Biology, Vol. 226: PCR Protocols, Second Edition*
Edited by: J. M. S. Bartlett and D. Stirling © Humana Press Inc., Totowa, NJ

11. Store at –20°C for at least 30 min.
12. Microfuge at 12,000*g* for 20 min at room temperature.
13. Discard supernatant and wash pellet in 70% ethanol.
14. Dry pellet and redissolve in 25 µL of TE.
15. If concentration of DNA is insufficient for further processing, re-amplify 1 µL in a second round of PCR.

III

Basic PCR Methods

19

PCR Primer Design

David L. Hyndman and Masato Mitsuhashi

1. Introduction

The selection of primers for a given polymerase chain reaction (PCR) can determine the efficiency and specificity of the PCR. Although in many cases successful PCR primers have been selected with little understanding of the principles involved, PCR can often only be achieved by using primers that are designed appropriately. Here, we give general recommendations for PCR primer selection and various aspects to be considered when designing primers.

2. General Primer Considerations

2.1. Location

The location of PCR primers is sometimes dictated by the purpose of the experiment. If the experiment is simply intended to identify the presence or absence of the sequence, then the location is of no consequence, so long as the amplification works well. If, however, the experiment is part of an assay for a particular allele of a gene, then the amplicon would be required to contain that region of interest.

2.2. Amplicon Size

In general, amplicons are from 100 to 1000 bp in length. The lower limit is caused by the typical need to be able to visualize the amplicon on an agarose gel. The upper limit of 1000 bp is the result of difficulties in amplifying large sequences. If an assay that does not require a minimum amplicon size is used, there is no theoretical minimum amplicon size.

2.3. Guanine/Cytosine (G/C) Content

Defined as the proportion of bases in the primer that are either G (guanine) or C (cytosine), good PCR primers are generally selected to have a G/C content between 40 and 60%. However, there is no well-defined reason for this, only that it has been considered preferable.

3. Considerations for Optimal PCR

The issues concerning PCR primer design can be divided into two categories: efficiency and specificity. Both of these are important to consider in most applications, but often the factors that promote one of these will adversely affect the other.

From: *Methods in Molecular Biology, Vol. 226: PCR Protocols, Second Edition*
Edited by: J. M. S. Bartlett and D. Stirling © Humana Press Inc., Totowa, NJ

3.1. Efficiency

Efficiency can be viewed as the proportion of templates that are used to synthesize new strands with each round of PCR, assuming the PCR primers are in a high abundance. A situation in which efficiency is the primary concern is the amplification of a purified template in which there is no chance of nonspecific PCR amplifications, so the main issue is that of primer binding to its target.

3.1.1. Melting Temperature

Melting temperature, or Tm, is defined for a given DNA duplex as the temperature at which half of the strands are hybridized and half of the strands are not hybridized. The original definition of T_m implied that the two complementary strands were in equal proportions. In cases where the two strands are not in equal proportions, such as a primer hybridizing to its target in PCR, the definition of Tm must be altered. Because in a PCR primer concentrations will be orders of magnitude higher than template concentrations, the interpretation is that the Tm is the temperature at which a primer is hybridized to half of the template strands. The T_m of a given primer/template combination depends on primer concentration, template concentration, and salt concentration. Generally, the template concentration is considered to be negligible compared with the primer concentration, so the formula for calculating Tm is simplified such that it does not contain a parameter for template concentration.

T_m can be expressed as follows:

$$T_m = \frac{\Delta H}{\Delta S + R\ln(c)} - 273.15 + 16.6 \log [Na^+] \; (\mathbf{1,2})$$

The primer concentration is c; the ΔH and ΔS refer to the total enthalpy and entropy of hybridization, respectively; and $[Na^+]$ is the sodium ion concentration but can refer to the total concentration of most monovalent cations (such as K^+). If there is Mg^{2+} or other divalent cations such as Mn^{2+}, the conversion is generally accepted as follows:

$$Na^+ = 4 \times [Mg^{2+}]^2 \; (3)$$

3.1.2. Calculating T_m with the Nearest Neighbor Model

The accurate calculation of T_m from a given sequence requires determining the ΔH and ΔS of the hybridization. The most successful method for this is through the use of the nearest neighbor model *(4–7)*. With the nearest neighbor model, every pair of adjacent base pairs makes a specific contribution to the overall ΔH and ΔS of the duplex. The total ΔH and ΔS are calculated by adding all of the component values plus a value for initiation of duplex formation. A table of ΔH and ΔS nearest neighbor values is shown in **Table 1** *(5)*.

3.1.3. Efficient PCR Primers and T_m

The most important issue for designing efficient PCR primers is that they must bind to the target site efficiently under the conditions of the PCR. This generally means not only that they bind at the annealing temperature but that if the annealing temperature

Table 1
Nearest Neighbor Thermodynamic Values for DNA Base Pairs

Base Pair	ΔH	ΔS	ΔG
aa/tt	–8.4	–23.6	–1.2
at/ta	–6.5	–18.8	–0.73
aa/at	–6.3	–18.5	–0.6
ca/gt	–7.4	–19.3	–1.38
gt/ca	–8.6	–23	–1.43
ct/ga	–6.1	–16.1	–1.16
ga/ct	–7.7	–20.3	–1.46
gc/cg	–11.1	–28.4	–2.28
gg/cc	–6.7	–15.6	–1.77

is so low that the thermostable DNA polymerase is not active, the primer must be bound at the temperature at which the polymerase becomes active in order to begin extension.

3.1.4. Typical Three-Step PCR

As an illustration, let us examine a typical PCR cycle, where there is a dissociation step of 30 s at 95°C, an annealing step of 1 min at 37°C, and an extension step of 3 min at 72°C. If the T_m of the primer, in the conditions of the reaction, is 65°C, then the following will happen. As the temperature decreases from 95°C to 37°C, at some point the primer will hybridize to the template. At this temperature, however, the thermostable DNA polymerase may be almost totally inactive. As the temperature rises towards 72°C, at some point the polymerase becomes active and will start to extend the primer along the template. As the temperature rises above 65°C, some of the primers will dissociate if they haven't been extended. Those that have extended sufficiently, however will form a more stable duplex as a result of their added base pairs from the extension, and they will not dissociate before reaching 72°C. During the extension at 72°C, the primers still hybridized will be fully extended to generate new strands.

Let's now imagine that with this scenario the T_m of the primer is 75°C. In this case, as the temperature rises from annealing to extension, even primers that have not extended will not dissociate. In this scenario, there will be almost total extension of every possible target strand. Therefore, this PCR would have a high degree of efficiency.

3.1.5. Two-Step PCR

PCR is sometimes performed in two steps without a discrete annealing step. In this case, the annealing takes place at the same temperature as the extension. This requires that the primers will hybridize to some degree at the extension temperature. If the extension temperature is 72°C, then a primer with a T_m of 72°C would be an efficient primer. Because the definition of T_m is the point at which half of the templates are in a duplex, a short time after reaching 72°C, half of the templates will be bound by a primer. Shortly after that, a large percentage of those templates will be extended by

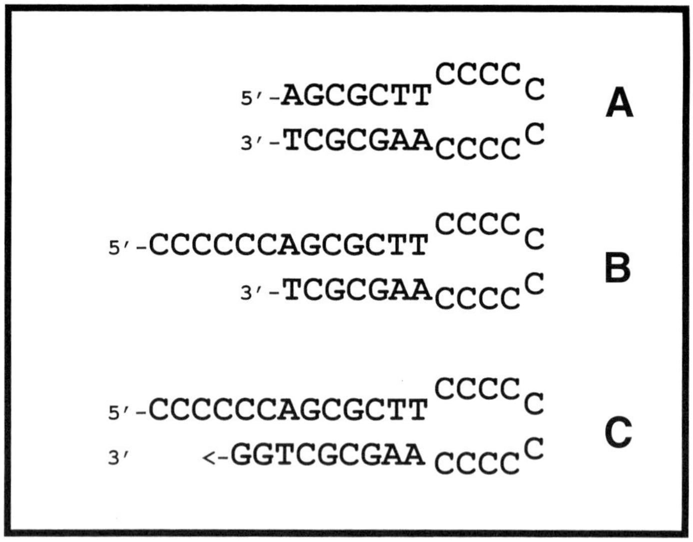

Fig. 1. Hairpin structures.

the polymerase and thereby taken out of the equilibrium between bound and unbound templates. Because the primer concentration will essentially be unchanged, half of the remaining templates will then be bound by primers and extended. In this way, most templates can be extended if the extension is done at the T_m of the primer.

If, however, the T_m of the primer is a few degrees below the extension temperature, only a small percentage of the primers will be hybridized, and the PCR will not be efficient. Therefore, very efficient and specific PCR can be performed with two-step cycling, but a sufficiently high primer T_m is very important.

3.1.6. Hairpins

A hairpin is a structure formed by a single DNA molecule in which a portion on one part of the DNA hybridizes to a complementary portion within the same DNA strand, forming a structure resembling a hairpin (**Fig. 1A**). When a PCR primer forms a hairpin, it adversely affects the primer's ability to bind and extend at the target site. In the worst case, the hairpin includes a base pair of the 3′-end and an overhang of the 5′-end (**Fig. 1B**). Such a structure allows the extension by DNA polymerase along the primer and will result in the formation of a primer that will not be complementary to the template and will not be extended if hybridized (**Fig. 1C**). In addition to removing primers from the mixture, this also will prevent native primers from binding as target sites that are bound by the extended primers. To avoid this, primers should be selected that do not have any possible hairpin structures if possible.

3.1.7. Primer-Dimer Formation

The hybridization of two primers together is referred to as a primer-dimer (**Fig. 2A**). There are two possibilities for these, homodimers and heterodimers. Homodimers are formed from the hybridization of the same species of primer together. Heterodimers are the duplex of two different primer sequences hybridizing together. The result of either of these is that the primers will not be as efficient in hybridizing to the target.

```
        5′-GCTTAGCTTTGCATG-3′          A
3′-TAGCTTTACTTCGAATCGA-5′

5′-TTGCATGGCTTAGCT-3′                  B
          3′-TCGAATCGATAGCTTTACT-5′

5′-TTGCATGGCTTAGCTATC->3′              C
          3′-TCGAATCGATAGCTTTACT-5′
```

Fig. 2. Dimer structures.

As with hairpins, the worst case is that in which the 3′-end of one of the primers is base paired and there is a 5′ overhang (**Fig. 2B**). In this case, the primer will extend, using the other primer as a template, rendering the extended primer unable to prime the desired template (**Fig. 2C**). Even worse than with hairpins, this situation leads to amplification of the primer dimers and rapid depletion of useable primers. To prevent this, primer pairs should be chosen such that primer-dimer formation is minimal.

3.1.8. 3′-Terminal Stability

3′-terminal stability can loosely be defined as the relative hybridization strength of the 3′-end of the primer. If the 3′-end of the primer has a low stability, it may not efficiently prime because of the transient fraying of the end of the duplex. Therefore, a higher 3′-terminal stability will improve priming efficiency. As will be mentioned later, however, this high stability can have an undesirable affect on specificity.

3.2. Specificity

Specificity can generally be defined as the tendency for a primer to hybridize to its intended target and not to other, nonspecific, targets. There are a few ways in which poor specificity can impair PCR. First, if primers are hybridizing to many locations nonspecifically, they will not be available to prime the target sequence. Second, if such nonspecific hybridization were to occur, priming could also occur at those nonspecific locations, which would effectively remove the primers from the reaction permanently. Finally, by priming nonspecifically, it can be possible to generate aberrant amplicons. This will not only obfuscate an assay for successful PCR, but will very rapidly consume the primers to remove them from the reaction for amplifying the intended target.

3.2.1. Specificity, T_m, and PCR Conditions

With respect to the annealing and extension temperatures chosen for the PCR reaction, there is a balance that must be reached between considerations of efficiency and specificity. As discussed previously, a more efficient PCR will result from having primer T_m equal to or above the extension temperature of the reaction. However, having a primer with a high T_m can often result in poor specificity. In such cases, partial hybridization of the primer may be likely and extension can occur from nonspecific sites. Such issues are less critical if highly specific primers can be selected as discussed here.

Probe		Tm	Sequence
XGENEA:P0006		65.05	cccccgattgggg

Found Match	Pos	Tm	Mismatches
HSU17969	273	56.82	g-----g------
HUMELA303	128	54.26	-----------tt
HUMPPE	212	54.26	-----------tt
HSU24498	35805	53.93	----a--------
HSDNASIAT	171	52.84	------gg-----
HSENO3	3296	52.84	------gg-----
HSU51021	2857	52.62	--------g---a
HS15D5R	108	52.46	t------g-----
HS121B1F	4	51.25	g------c-----
HS166B11R	17	51.25	g------c-----
HSV1077H7	4562	51.25	g------c-----
S69039	31	51.19	g--t---------
HSU52112	36943	50.77	agg----------
HSU71148	2762	50.77	gg----------c
HUMIGHVBL	317	50.77	ttg----------
HUMHTLVID	23	48.70	g-------g---a
HSU28838	1554	48.57	g------g----c
HUMCA2ATPY	1122	48.57	g------g----c
HUM1D9	11827	48.57	gg-----g-----
HUMNFAT3A	1306	48.54	g---g--------
HSU18018	160	48.46	------c-----c
HSEPB72E1	1230	48.31	a--t---------
HSSTOMAT01	476	48.31	a--t---------

Fig. 3. Hybridization simulation data.

3.2.2. Hybridization Simulation

The most precise way to view the specificity of a PCR primer is by hybridization simulation *(8)*. Hybridization simulation is the computer simulation of a hybridization of a primer with a specified database. This in silico analysis will identify all hybridization sites within the database for a candidate primer, allowing the user to select primers that will be the most specific.

It is important to realize that hybridization simulation is qualitatively different from a homology or similarity analysis *(9)*. Hybridization simulation uses a thermodynamic model with nearest neighbor values to calculate the mismatch Tm of hybridization for all hybridization sites. An example of hybridization simulation data is shown in **Fig. 3**.

Currently, hybridization simulation is only available from a single commercially available program, the HYBsimulator™. HYBsimulator allows for the screening of a large set of candidate primers and selection based on the hybridization simulation data.

3.2.3. Statistical Determination of Specificity

Various mathematical models exist in which the specificity of a given primer can be estimated based on the frequency of its constituent smaller sequences. One such method uses a table of frequencies of 6 mers found within a given genomic database

(10,11). The entire statistical frequency of the entire oligonucleotide is calculated based on the constituent 6-mers by starting with the 5′ terminal 6-mer frequency and multiplying the relative frequency of the 5-mer on the 3′-end of that 6 mer having the next nucleotide. This is repeated until the end of the oligonucleotide is reached.

For example, to calculate the frequency (f) of the 8-mer: CATAGCCT

$$f(CATAGCCT) =$$

$$f(CATAGC) \times \frac{4\,f(ATAGCC)}{f(ATAGCT) + f(ATAGCG) + f(ATAGCC) + f(ATAGCT)}$$

$$\times \frac{4\,f(TAGCCT)}{f(TAGCCA) + f(TAGCCG) + f(TAGCCC) + f(TAGCCT)}$$

where f(CATAGC) denotes the frequency of the 6-mer, CATAGC.

3.2.4. 3′-Terminal Effects

Partial hybridization of the primer at the 3′-terminus can permit extension by DNA polymerase. This could result in depletion of primers as well as possible nonspecific amplification; therefore, this type of partial hybridization should be minimized as much as possible.

There are two considerations for decreasing the chance of partial hybridization of the 3′-terminal: frequency and stability.

3.2.5. 3′-Terminal Frequency

If the 3′-terminal region has a sequence that has many occurrences in the DNA that will be in the reaction, then the likelihood of partial hybridization is greater. To minimize this, the primers can be selected such that the 3′-terminal region does not have a high frequency of occurrence within the genome of interest.

3.2.6. 3′-Terminal Stability

If the 3′-terminal region has a strong hybridization energy, then 3′-terminal partial hybridizations will be relatively more stable. More stable 3′-terminal hybridization will allow more false priming. Therefore if the primers are chosen such that the 3′-terminal has a low hybridization strength (also referred to as terminal stability), the primer is less likely to be priming as a result of such partial hybridization.

Note that in the efficiency section above, a stronger 3′-terminal stability is said to improve efficiency. Whether one should select primers with high or low terminal stability would depend on factors, such as the nature of the experiment (i.e., whether there will be a large amount of other DNA) and the nature of the gene (whether highly specific primers can be found).

3.2.7. Specificity Within the Target Sequence

If the primer is able to partially hybridize to a nonspecific region of the template, particularly undesirable effects can occur. If the nonspecific hybridization allows extension along the template such that a PCR product can be formed in conjunction with one of the primers binding at one of the actual binding sites, nonspecific amplicons can be generated. This problem is compounded if the nonspecific binding site is within the amplicon itself. For this reason, primers should be checked for nonspecific alternate hybridization sites within the target sequence.

4. Selecting Primers for Multiplex PCR

Multiplex PCR, in which several primer sets amplify several amplicons in the same reaction, add a degree of complexity to designing optimal primers. The additional issues to consider are those of possible heterodimer formation between all of the candidate primers and possible alternate hybridization sites within any of the target sequences. Some of the available primer design software provides functions for these types of designs.

5. Primer Design Software

Several programs are available for PCR primer design. As mentioned above, we think HYBsimulator is the most powerful such program and does provide PCR primer selection based on all criteria mentioned in this chapter. Other popular programs are Oligo™ and Primer Premier™, which provide a subset of these functions but are slightly easier to use.

References

1. Breslauer, K. J., Frank, R., Blocker, H., and Marky, L. (1986) Predicting DNA duplex stability from the base sequence. *Proc. Natl. Acad. Sci. USA* **83,** 3746–3750.
2. Freir, S. M., Kierzed, R., Jaeger, J. A., Sugimoto, N., Caruthers, M. H., Neilson, T., and Turner, D. H. (1986) Improved free-energy parameters for predictions of RNA duplex. *Biochemistry* **83,** 9373–9377.
3. Wetmer, J. (1991) DNA probes: Applications of the principles of nucleic acid hybridization. *Crit. Rev. Biochem. Mol. Biol.* **26,** 227–259.
4. Sugimoto, N., Nakano, S., Katoh, M., Matsumura, A., Nakamuta, H., Ohmichi, T., et al. (1995) Thermodynamic parameters to predict stability of RNA/DNA hybrid duplexes. *Biochemistry* **34,** 11,211–11,216.
5. SantaLucia, J., Allawi, H. T., and Seneviratne, P. A. (1996) Improved nearest-neighbor parameters for predicting DNA duplex stability. *Biochemistry* **35,** 3555–3562.
6. SantaLucia, J., Kierzed, R., and Turner, D. H. (1990) Effects of GA mismatches on the structure and thermodynamics of RNA internal loops. *Biochemistry* **29,** 8813–8819.
7. Sugimoto, N., Kierzed, R., Freier, S. M., and Turner, D. H. (1986) Energetics of internal GC mismatches in ribooligonulceotide helix. *Biochemistry* **25,** 5755–5759.
8. Hyndman, D., Cooper, A., Pruzinsky, S., Coad, D., and Mitsuhashi, M. (1996) Software to determine optimal oligonucleotide sequences based on hybridization simulation data. *BioTechniques* **20,** 1090–1096.
9. Altschul, S. F., Gish, W., Miller, W., Myers, E. W., and Lipman, D. J. (1990) Basic local alignment search tool. *J. Mol. Biol.* **215,** 403–410.
10. Han, J., Hsu, C., Zhu, Z., Longshore, J., and Finley, H. (1994) Over-representation of the disease associated (CAG) and (CGG) repeats in the human genome. *Nucleic Acids Res.* **22,** 1735–1740.
11. Han, J., Zhu, Z., Hsu, C., and Finley, W. (1994) Selection of antisense oligonucleotides on the basis of genomic frequency of the target sequence. *Antisense Res. Devel.* **4,** 53–65.

20

Optimization of Polymerase Chain Reactions

Haiying Grunenwald

1. Introduction

The polymerase chain reaction (PCR) is a powerful method for fast in vitro enzymatic amplifications of specific DNA sequences. PCR amplifications can be grouped into three different categories: standard PCR, long PCR, and multiplex PCR. Standard PCR involves amplification of a single DNA sequence that is less than 5 kb in length and is useful for a variety of applications, such as cycle sequencing, cloning, mutation detection, etc. Long PCR is used for the amplification of a single sequence that is longer than 5 kb and up to 40 kb in length. Its applications include long-range sequencing; amplification of complete genes; PCR-based detection and diagnosis of medically important large-gene insertions or deletions; molecular cloning; and assembly and production of larger recombinant constructions for PCR-based mutagenesis *(1,2)*. The third category, multiplex PCR, is used for the amplification of multiple sequences that are less than 5 kb in length. Its applications include forensic studies; pathogen identification; linkage analysis; template quantitation, genetic disease diagnosis; and population genetics *(3–5)*. Unfortunately, there is no single set of conditions that is optimal for all PCR. Therefore, each PCR is likely to require specific optimization for the template/primer pairs chosen. Lack of optimization often results in problems, such as no detectable PCR product or low efficiency amplification of the chosen template; the presence of nonspecific bands or smeary background; the formation of "primer-dimers" that compete with the chosen template/primer set for amplification; or mutations caused by errors in nucleotide incorporation. It is particularly important to optimize PCR that will be used for repetitive diagnostic or analytical procedures where optimal amplification is required. The objective of this chapter is to discuss the parameters that may affect the specificity, fidelity, and efficiency of PCR, as well as approaches that can be taken to achieve optimal PCR amplifications.

Optimization of a particular PCR can be time consuming and complicated because of the various parameters that are involved. These parameters include the following: (1) quality and concentration of DNA template; (2) design and concentration of primers; (3) concentration of magnesium ions; (4) concentration of the four deoxynucleotides (dNTPs); (5) PCR buffer systems; (6) selection and concentration of DNA polymerase; (7) PCR thermal cycling conditions; (8) addition and concentrations of PCR additives/cosolvents; and (9) use of the "hot start" technique. Optimization of PCR

From: *Methods in Molecular Biology, Vol. 226: PCR Protocols, Second Edition*
Edited by: J. M. S. Bartlett and D. Stirling © Humana Press Inc., Totowa, NJ

may be affected by each of these parameters individually, as well as the combined interdependent effects of any of these parameters.

2. Materials

1. Template DNA (e.g., plasmid DNA, genomic DNA).
2. Forward and reverse PCR primers.
3. $MgCl_2$ (25 m*M*).
4. dNTPs (a mixture of 2.5 m*M* dATP, dCTP, dGTP, and dTTP).
5. 10× PCR buffer: 500 m*M* KCl, 100 m*M* Tris-HCl, pH 8.3, 25°C.
6. Thermal stable DNA polymerase (e.g., *Taq* DNA polymerase).
7. PCR additives/cosolvents (optional; e.g., betaine, glycerol, DMSO, formamide, bovine serum albumin, ammonium sulfate, polyethylene glycol, gelatin, Tween-20, Triton X-100, β-mercaptoethanol, or tetramethylammonium chloride).

3. Methods

3.1. Setting Up PCR

The common volume of a PCR is 10, 25, 50, or 100 µL. Although larger volumes are easier to pipet, they also use up a larger amount of reagents, which is less economical. All of the reaction components can be mixed in together in a 0.5-mL PCR tube in any sequence except for the DNA polymerase, which should be added last. It is recommended to mix all the components right before PCR cycling. Although it is not necessary to set up the PCR on ice, some published protocols recommend it.

For each PCR, the following components are mixed together:

1. Template DNA (1–500 ng).
2. Primers (0.05–1.0 µ*M*).
3. Mg^{2+} (0.5–5 m*M*).
4. dNTP (20–200 µ*M* each).
5. 1× PCR buffer: 1 m*M* Tris-HCl and 5 m*M* KCl.
6. DNA polymerase (0.5–2.5 U for each 50 µL of PCR).

As a real-life example, the following PCR was set up to amplify the *cII* gene from bacteriophage lambda DNA (total volume = 50 µL):

1. 1 µL of 1 ng/µL lambda DNA (final amount = 1 ng).
2. 1 µL of 50 µ*M* forward PCR primer (final concentration = 1 µ*M*).
3. 1 µL of 50 µ*M* reverse PCR primer (final concentration = 1 µ*M*).
4. 5 µL of 25 m*M* $MgCl_2$ (final concentration = 2.5 m*M*).
5. 4 µL of 2.5 m*M* dNTPs (final concentration = 200 µ*M*).
6. 5 µL of 10× PCR buffer (final concentration = 1×).
7. 0.25 µL of 5 U/µL *Taq* DNA polymerase (final amount = 1.25 U).

3.2. PCR Cycling

A common PCR cycling program usually starts with an initial dissociation step at 92 to 95°C for 2 to 5 min to ensure the complete separation of the DNA strands. Most PCR will reach sufficient amplification after 20 to 40 cycles of strand denaturation at 90 to 98°C for 10 s to 1 min, primer annealing at 55 to 70°C for 30 s to 1 min, and primer extension at 72 to 74°C for 1 min per kilobase of expected PCR product. It is suggested that a final extension step of 5 to 10 min at 72°C will ensure that all amplicons are fully extended, although no solid evidence proves that this step is necessary. For example, the cycling program used to amplify the previously described lambda *cII* gene is as follows: initial denaturation for 4 min at 94°C, followed by 30 cycles of 1 min at 94°C, 1 min at 55°C, and 1 min at 72°C, then held at 4°C.

3.3. Verifying PCR Amplification

To measure the success of a PCR amplification, 5 to 10 µL of the final PCR product is run on a 1 or 2% agarose gel and visualized by staining with ethidium bromide. The critical questions are as follows: (1) Is there a band on the gel? (2) Is the band at the expected size? (3) Are there any nonspecific bands beside the expected PCR band on the gel? (4) Is there smear on the gel? A successful PCR amplification should display a single band with the expected size without nonspecific bands and smear.

4. Notes

1. The quality and concentration of DNA templates can directly affect the outcome of PCR amplifications. To achieve satisfactory amplification, certain baseline conditions may be used as a starting point for optimizing a PCR amplification. For a typical PCR, 10^4 to 10^7 molecules of template DNA is recommended. For long PCR (>5 kb), 10^7 to 10^8 molecules of a high copy number template DNA (e.g., 1–10 ng of lambda DNA) is recommended. For amplification from genomic DNA, use 100 to 500 ng of template DNA. In multiplex PCR, two- to fivefold more DNA template than what is needed for a typical PCR should be used.

   There are several methods for purifying DNA for PCR amplification, including commercially available kits, as well as standard methods ***(6)***. Long PCR amplification has the most stringent requirement for the quality of template DNA. Care should be taken to prevent template DNA damages from nicking, shearing, and depurination ***(7)***. Analysis by pulsed-field agarose gel electrophoresis is typically recommended for template DNA used in long PCR to assure its purity and integrity.
2. Appropriate primer design, as well as use of the proper primer concentration, is critical for successful PCR amplification. There are a variety of computer programs available for designing primers and they vary significantly in selection criteria, comprehensiveness, and user-friendliness ***(8–11)***. The purpose of primer design is to achieve a balance between the specificity and efficiency of an amplification. Specificity defines how frequently mispriming occurs, whereas efficiency represents the increase of the amount of PCR product over a given number of cycles. The following guidelines should be considered when designing primers. The optimal primer size is usually between 18 and 28 bases. Shorter primers are generally less specific but may result in more efficient PCR, whereas longer primers improve specificity yet can be less efficient. Primers from both directions should have melting temperatures (T_m, defined as the dissociation temperature of the primer/template duplex) that are within 2 to 5°C of each other. This will assure that the proper annealing temperature for both primers is achieved. For primers shorter than 20 bases, an estimate of T_m can be calculated as $T_m = 4 (G + C) + 2 (A + T)$ ***(12)***. However, for longer primers, correct estimation of T_m requires a "nearest-neighbor" calculation, which takes into account thermodynamic parameters of a chosen primer and is used by most of the available computer programs for primer design ***(13,14)***. Avoid complementary sequences within a primer or between the two primers. This will reduce formation of primer-dimers that can compete with the amplification of the desired PCR product, as well as the formation of secondary structures within a primer. Primers with T_m higher than

50°C will generally provide specific and efficient amplifications. For long PCR, a T_m of 62 to 70°C is recommended. In addition, a GC content of 40 to 60% is desirable for primers because this assures a higher T_m and therefore increases specificity. If AT content is high, use primers of 28 to 35 bases in length for long PCR. Also, avoid continuous stretches of purines or pyrimidine, as well as multiple repeats of thymidine residues at the 3′ end of the primer. It is extremely critical when designing primers for multiplex PCR to make sure that all primers have similar melting temperatures and do not contain sequences complementary to each other. Concentrations of primers used in PCR will also influence amplification specificity and efficiency. In general, primer concentrations between 0.05 to 1.0 μ*M* (or 0.5–100 pmol) are routinely used in 100 μL of PCR depending on the specific application. Higher primer concentrations can result in nonspecific priming and formation of primer-dimers, whereas lower primer concentrations may adversely affect PCR efficiency. Use the primers at a 1:1 concentration ratio to assure specificity and efficiency of amplifications. Lower concentrations of primers are more desirable for multiplex PCR because of the number of primer sets present in the reaction.

3. Magnesium concentration is critical to the success of PCR amplification because it may affect DNA polymerase activity and fidelity, DNA strand denaturation temperatures of both template and PCR product, primer annealing, PCR specificity, and primer-dimer formation. Excess magnesium results in accumulation of nonspecific amplification products seen as multiple bands on an agarose gel, whereas insufficient magnesium results in reduced yield of the desired PCR product. Common magnesium concentrations used in PCR are between 0.5 to 5 m*M*. It is important to optimize the magnesium concentration used for each individual PCR because DNA polymerases require free magnesium for their activity in addition to that bound by template DNA, primers, and dNTPs. Also, trace amounts of EDTA or other chelators may be present in primer stock solutions or template DNA. Therefore, each PCR should contain 0.5 to 2.5 m*M* magnesium over the total dNTP concentrations ***(15)***. A simple way to optimize magnesium concentration is to first perform a series of reactions in which the magnesium concentration is varied between 0.5 and 5 m*M* in 0.5-m*M* increments. After the concentration range is narrowed, perform a second round of reactions and vary the magnesium concentration in 0.2- to 0.3-m*M* increments.
4. Concentration of dNTPs can affect the yield, specificity, and fidelity of a PCR amplification. Concentrations of 20 to 200 μ*M* of each dNTP has been used to obtain successful PCR amplifications. Stock solutions of each dNTP are adjusted to pH 7.0 and diluted to a 10 m*M* final concentration. Commercially available premixed dNTP solutions with concentrations of 2.5 m*M* or individual dNTP stock solutions of 10 m*M* may be used. Lower concentrations of dNTPs minimize mispriming and reduce the likelihood of extending misincorporated nucleotides, which in turn increase specificity and fidelity of PCR amplifications ***(16)***. Because dNTPs are typically added in excess to a PCR, one should determine the lowest dNTPs concentration appropriate for the length and composition of the target sequence. Although 250 μ*M* of each of the dNTPs appears to be sufficient for long PCR ***(17)***, amplifications of sequences longer than 20 kb may require dNTP concentrations as high as 400 to 500 μ*M* each in a given 50-μL reaction. It is critical not to use a large excess of dNTP because higher dNTP concentrations increase the error rate of DNA polymerases. In fact, millimolar concentrations of dNTPs actually inhibit *Taq* DNA polymerase ***(18)***.
5. Standard PCR amplifications using *Taq* DNA polymerase are performed in 10 m*M* Tris-HCl (pH 8.3–8.4 at 20–25°C) and 50 m*M* KCl. For Tth and Tfl DNA polymerases, and DNA polymerases with proofreading activity (for example, *Pwo*, *Pfu*, *Tli*, and Vent DNA polymerases (New England Biolabs), a buffer system of 50 m*M* Tris-HCl (pH 9.0 at 25°C) and 20 m*M* $(NH_4)_2SO_4$ is normally used. These standard buffer systems are available

commercially and have been shown to produce satisfactory PCR amplifications in most cases. However, long PCR requires a different buffer system. For example, 20 to 25 m*M* Tricine (pH 8.7 at 25°C) and 80 to 85 m*M* potassium acetate (pH 8.3–8.7 at 25°C) ***(19)***, as well as 25 m*M* Tris-HCl (pH 8.9 at 25°C) and 100 m*M* KCl ***(2)***, have been used successfully in long PCR amplifications when used in conjunction with *rTth* DNA polymerase (Perkin–Elmer). The use of less temperature-sensitive buffers, such as Tricine, may enhance the ability to obtain long PCR amplifications.

6. The most common enzyme used for PCR amplification is *Taq* DNA polymerase because of its thermostability and processivity (i.e., the number of nucleotides replicated before the enzyme dissociates from the DNA template). It was originally purified from the gram-negative thermophilic bacterium *Thermus aquaticus* ***(20)***. Highly purified *Taq* DNA polymerase exhibits a temperature optimum of 75 to 80°C ***(21)***. The half-life of *Taq* DNA polymerase is 40 min at 95°C, which is sufficient to remain active over 30 or more cycles, during which the enzyme is transiently exposed to extremely high denaturation temperatures. *Taq* DNA polymerase has an extension rate of 35 to 100 nucleotides per second at 72°C ***(16)***, which is the most common extension temperature for PCR amplifications. A recommended concentration range for *Taq* DNA polymerase is between 1 and 2.5 units per 100 µL of PCR. However, different concentrations of *Taq* DNA polymerase may be required with respect to individual target template sequences or primers. Increasing the amount of *Taq* DNA polymerase beyond the 2.5 units/reaction can in some cases increase PCR efficiency. However, adding more *Taq* DNA polymerase can sometimes increase the yield of nonspecific PCR products at the expense of the desired product. When optimizing the *Taq* DNA polymerase concentration for a particular PCR, testing a range of 0.5 to 5 units per 100 µL of reaction in 0.5-unit increments is recommended, followed by analysis of the PCR products by gel electrophoresis to determine the amplification specificity and efficiency.

   The other important property of *Taq* DNA polymerase is its fidelity, which is measured as error rate. The error rate for *Taq* DNA polymerase, which lacks proofreading $3' \rightarrow 5'$ exonuclease activity, is estimated at approx 1 to 2×10^{-5} errors (or mutation frequency) per nucleotide per duplication ***(22–24)***. For many applications, this does not present any problems. However, for some sequencing, cloning, and long PCR applications, it is essential to have few, or no incorporation errors. In situations where "high fidelity" is required, DNA polymerases with $3' \rightarrow 5'$ proofreading activity (for example, *Pfu* or *Pwo* DNA polymerases) are recommended. The estimated error rates for these proofreading enzymes is approx 1 to 2×10^{-6} errors per nucleotide per duplication ***(23,24)***, representing a 10-fold improvement over standard *Taq* DNA polymerase. It is important to note that these proofreading enzymes with lower error rates also have lower extension rates, resulting in lower PCR efficiency. Therefore, more amplification cycles are required to achieve adequate amount of amplified DNA.

   DNA polymerase characteristics, such as extension rate, processivity, fidelity, thermostability, and thermal activity profile, are important in long PCR. PerkinElmer's *rTth* DNA polymerase (the recombinant form of the DNA polymerase from *T. thermophilus*) has been shown to perform consistent long PCR at 0.5 to 2.5 units per 50 µL of reaction ***(17)***. Use of a mixture of *Taq* DNA polymerase with a proofreading enzyme, such as *Pfu* DNA polymerase, at a 20:1 ratio (2.5 units per 50 µL of reaction) will also enhance the reliability of long PCR amplifications.

7. Optimization of PCR thermal cycling conditions includes determination of cycle number, the temperature and incubation time period for template denaturation, primer annealing, and primer extension. The optimum number of cycles depends mainly on the starting concentration of template DNA. Because many PCRs start with very limiting amount

of template DNA, a sufficient number of cycles are required to achieve satisfactory amplifications. However, PCR amplification is not an unlimited process. A common mistake is to execute too many cycles. The exponential amplification of PCR will continue up until the point when the product reaches about 10^{-8} *M* (about 10^{12} molecule in a 100-µL reaction). The reaction enters a linear phase where exponential accumulation of the product is attenuated. This is termed the plateau effect ***(25)***. In most PCRs, amplifications plateau after about 20 to 40 cycles. The other major limitation of standard PCR is the amount of DNA polymerase included in the reaction. The combination of thermal inactivation of the DNA polymerase after each denaturation step, reduction in denaturation efficiency, and the reduced efficiency of primer annealing (caused by increasing competition from the template), will cause the reaction to terminate. Although too few cycles of PCR result in low product yield, most PCR amplifications are performed for no more than 20 to 40 cycles.

It is often helpful to precede the first cycling denaturation step with an initial dissociation step at 92 to 95°C for 2 to 5 min to ensure the complete separation of the DNA strands. Template denaturation temperatures range 90 to 98°C. The duration of denaturation ranges from 10 s to 1 min. Although it only takes a few seconds to denature DNA at its strand-separation temperature, it is appropriate to use a higher denaturation temperature and a longer incubation time for some templates, such as those templates with high GC content, to achieve complete denaturation. Although a higher temperature and a longer incubation period result in a more complete denaturation of the DNA template, it can also cause depurination of the DNA template, which in turn reduces amplification efficiency. It is also important to note that higher denaturation temperatures will reduce the amount of active DNA polymerase available for amplification. The half-life of *Taq* DNA polymerase activity is more than 2 h at 92.5°C, 40 min at 95°C, and 5 min at 97.5°C.

The optimal primer annealing temperature for a particular PCR amplification depends on the base composition, nucleotide sequence, length, and concentration of the primers ***(26)***. A typical primer annealing temperature is 5°C below the calculated Tm of the primers. Annealing temperatures from 55 to 70°C generally yield the best results. Increasing the annealing temperature enhances discrimination against incorrectly annealed primers and reduces mis-extension of incorrect nucleotides at the 3′ end of primers. Therefore, a higher annealing temperature increases amplification specificity. For these reasons, when using primers with higher T_ms such as for long PCR, a higher annealing temperature (i.e., 60–70°C) should be used. It is also important to note that at typical primer concentrations of 200 µ*M* each, annealing requires only a few seconds. However, incubation times from 30 s to 1 min are generally recommended to assure successful primer annealing.

Primer extension time depends on the length and concentration of the target sequence, as well as the extension temperature. *Taq* DNA polymerase extends at a rate of 0.25 nucleotides per second at 22°C, 1.5 nucleotides per second at 37°C, 24 nucleotides per second at 55°C, greater than 60 nucleotides per second at 70°C, and 150 nucleotides per second at 75 to 80°C ***(18)***. Therefore, at the commonly chosen extension temperature of 72°C, *Taq* DNA polymerase is expected to extend at the rate of greater than 3500 nucleotides per minute. Thus, as a general rule, an extension time of 1 min per kilobase is more than sufficient to generate the expected PCR product. For PCR products up to 2 kb in length, an extension time of 1 min at 72°C is sufficient. A final extension step of 5 to 10 min at 72°C may be added in order to ensure that all amplicons are fully extended.

In addition to the conventional three-step cycling programs, two-step cycling programs that combine primer annealing and extension in one step are also widely used. Annealing and extension can be combined because most thermostable DNA polymerases can actively extend off the primers over the entire range of commonly chosen annealing and extension temperatures. Two-step cycling programs are generally applied when a high annealing temperature is used, such as 65 to 70°C. Because a higher annealing temperature improves amplification specificity, it is argued by some investigators that better PCR results may be obtained using a two-step cycling program *(27)*. Here is an example of a typical two-step cycling program: initial denaturation at 94°C for 2 min, followed by 30 cycles of 1 min at 94°C and 1 min at 68°C, then held at 4°C.

It is reasonable to follow the above rationale in defining optimal cycling conditions for long PCR. However, there are a few rules that must be considered when performing long PCR. Use moderate denaturation temperatures and short incubation time periods to maintain the integrity of the long DNA templates. Choose primers with a high Tm so that a higher annealing temperature can be used to increase specificity. Two-step cycling programs are more frequently used than three-step cycling programs. For example, denaturation at 92 to 95°C for 10 to 30 s, followed by annealing and extension at 65 to 68°C for 1 min per kilobase will increase the probability of obtaining the desired product. Programming an increase in extension time automatically in later cycles may also improve the yields of the amplification. Because DNA polymerases extend primers discontinuously through a succession of reactions, increasing the extension time in each of the later PCR cycles could increase the likelihood of synthesizing long PCR products. For example, perform the extension at 1 min per kilobase for the first 10 cycles, then lengthen the extension time 10 to 20 s for each of the next 20 cycles. Here is an example of a typical cycling program used to amplify a 10-kb PCR product: initial denaturation at 94°C for 2 min, followed by 16 cycles of 30 s at 94°C and 10 min at 68°C, then 12 cycles of 30 s at 94°C and 10 mins at 68°C with a 15-s extension per cycle, then held at 4°C.

8. The majority of PCR amplifications can be successfully performed after optimizing the above parameters. However, there are some PCR amplifications, for example, those using templates with high guanine/cytosine content or stable secondary structures, that may still amplify inefficiently, resulting in little or no desired product and/or nonspecific products. Incorporation of the nucleotide analog 7-deaza-2′-deoxyguanosine triphosphate (c7dGTP) in addition to deoxyguanosine triphosphate (dGTP) helps destabilize secondary structures of DNA and reduces the formation of nonspecific products *(28)*. However, the most effective and frequently used strategy is addition of various organic additives or cosolvents. The most commonly used cosolvents and their concentration ranges are: dimethyl sulfoxide (DMSO; 1–10%), glycerol (5–20%), formamide (1.25–10%), bovine serum albumin (10–100 μg/mL), ammonium sulfate $(NH_4)_2SO_4$; 15–30 m*M*), polyethylene glycol (5–15%), gelatin (0.01%), non-ionic detergents (such as Tween 20 and Triton X-100; 0.05–0.1%), β-mercaptoethanol, tetramethylammonium chloride(TMAC), and *N,N,N*-trimethyglycine (betaine) (1–3 *M*) ***(15,17,29–34)***.

The mechanisms underlying enhancement of PCR by many of these cosolvents are not well defined. It is suggested that some cosolvents, such as DMSO, formamide, glycerol, and polyethylene glycol, may affect the Tm of the primers, the thermal activity profile of *Taq* DNA polymerase, as well as the degree of product strand separation ***(18)***. Gelatin, bovine serum albumin, and nonionic detergents, such as Tween-20 and Triton X-100, are thought to stabilize DNA polymerases ***(18)***. TMAC is used to eliminate nonspecific priming ***(34)***. $(NH_4)_2SO_4$ may increase the ionic strength of the reaction mixture, altering the denaturation and annealing temperatures of DNA, and may affect polymerase activity. Betaine has been shown to increase the thermostability of DNA polymerases, as well as to alter DNA stability such that GC-rich regions melt at temperatures more similar to AT-rich regions ***(29)***.

It is important to carefully choose the appropriate cosolvents and correct concentrations to effectively improve PCR amplifications. Cosolvent concentrations should be no greater than absolutely necessary for optimal amplification, as they may reduce DNA polymerase activity. For example, DMSO at a final concentration of 10% can reduce *Taq* DNA polymerase activity by up to 50% ***(15)***. In addition to improving standard PCR, multiplex PCR performance has also been shown to improve when using DMSO ***(34)***, Tween-20 and Triton X-100 ***(32)***, β-mercaptoethanol ***(33)***, TMAC ***(34)***, and betaine ***(35)***. Long PCR has been enhanced using glycerol, gelatin ***(17)***, DMSO ***(19)***, and betaine ***(36)***.

9. The "hot start" technique enhances PCR specificity by eliminating the production of nonspecific products and primer-dimers during the initial steps of PCR ***(36)***. This is because even a brief incubation of a PCR mix at temperatures significantly below the Tm can result in primer-dimer formation and nonspecific priming. The purpose of a hot start is to withhold one of the critical components from the reaction until the temperature in the first cycle rises above the annealing temperature. There are various methods of performing a hot start. Manual hot start is performed by withholding one of the reaction components, such as the DNA polymerase or magnesium, and adding it only after the reaction temperature rises above 80°C during the first denaturation step. Wax-mediated hot start involves addition of a wax layer separating the component being withheld from the remainder of the reaction mix. During the temperature increase in the first denaturation step, the wax melts and the withheld component is mixed with the rest of the reaction components, starting the amplification reaction. The beads for wax layer can be made in the laboratory ***(37,38)*** or purchased commercially (Ampliwax™ PCR Gems, PerkinElmer). Hot start *Taq* DNA polymerase is constructed through the addition of an anti-*Taq* DNA polymerase antibody (*Taq*Start™ Antibody, Clontech). The antibody will prevent the DNA polymerase activity until the temperature rises during the initial denaturation step. The increased temperature dissociates and degrades the bound antibody, initiating PCR amplification. Hot start is commonly used for multiplex and long PCR amplifications.
10. In addition to all of the PCR optimization strategies discussed above, there are also commercially available buffer systems for fast and easy PCR optimization. Companies, such as Boehringer Mannheim, Stratagene, Invitrogen, and Epicentre Technologies, offer various buffer systems for PCR optimization. These buffer systems can be divided into two categories. One category (e.g., PCR optimization kit from Boehringer Mannheim) contains a set of 16 buffers that combine different pH (8.3, 8.6, 8.9, and 9.2) and various concentrations of magnesium (1.0, 1.5, 2.0, and 3.5 m*M*). There are also four different cosolvents, DMSO, glycerol, gelatin, and $(NH_4)_2SO_4$, provided separately for additional optimization. Because the cosolvents are not premixed in the buffers, inclusion of these cosolvents will require a second set of optimization reactions. The other category of buffer system (e.g., PCR optimization kits from Epicentre Technologies) contains variable concentrations of magnesium (1.5, 2.5, and 3.5 m*M*) and a betaine-containing enhancer. Because all necessary reaction components, including the cosolvent (betaine), are premixed in this buffer system, only one set of optimization reactions are performed.
11. The following conditions may lead to less than optimal PCR amplifications. The possible solutions for each condition are discussed.

    If little or not desired PCR product is detected:

    a. Too little DNA template is present in the reaction. Increase the amount of template DNA.
    b. The template DNA is damaged or degraded. Assure the purity and integrity of the DNA template by minimizing damage from nicking and shearing.
    c. Insufficient DNA polymerase is present in the reaction. Increase the DNA polymerase concentration in increments of 0.5 units per 100 µL of reaction.
    d. Insufficient number of cycles was performed. Increase cycle number by 5 to 10 cycles.
    e. Check for inhibitor(s) during template DNA preparation. Repurification of the DNA template may remove some inhibitors of PCR.

f. Magnesium concentration is too low. Increase magnesium concentration in increments of 0.1 m*M*.
g. The denaturation time is too long or too short. Adjust denaturation time in increments of 5 s.
h. Add cosolvents that enhance PCR amplification.
i. The denaturation temperature is too high or too low. Change denaturation temperature in increments of 1°C.
j. The primer annealing temperature is too high. Lower annealing temperature in increments of 2°C.
k. The primer extension period is too short. Increase extension time in increments of 1 minute.
l. Re-amplify dilutions (1:10 to 1:1000) of the first round of PCR amplification using nested primers.
m. Perform hot start.
n. Perform Touchdown (TD)/Stepdown (SD) PCR cycling program. TD or SD PCR uses a temperature cycling protocol that is performed at decreasing annealing temperatures. The cycling program begins at an annealing temperature a few degrees above the calculated Tm of the primers. This ensures that the first primer-template hybridization events involve only those sequences with the greatest specificity. The annealing temperature is decreased 1 to 4°C every other cycle to approx 10°C below the calculated Tm to permit exponential amplification ***(39)***. Here is an example of a typical TD/SD cycling program: initial denaturation at 94°C for 1 mi followed by 20 cycles of 10 s at 92°C and 20 s at 70°C with an 0.5°C decrease of temperature per cycle, then another 20 cycles of 10 s at 92°C and 30 s at 60°C with a 1-s extension per cycle and hold at 4°C. Hot start must be used with these cycling programs.
o. Review primer design and composition. Design new primers and try PCR again.

If multiple product bands or smear is detected:

a. Too much DNA template is present in the reactions. Decrease the amount of DNA template in the reaction mix.
b. Annealing temperature is too low. Increase annealing temperature in increments of 2°C.
c. DNA polymerase concentration is too high. Decrease enzyme concentration in increments of 0.5 units per 100-µL reaction.
d. Magnesium concentration is too high. Decrease the magnesium concentration in increments of 0.1 m*M*.
e. Denaturation time is too short. Increase the denaturation time in increments of 5 s.
f. Denaturation temperature is too low. Increase the denaturation time in increments of 1°C.
g. Cycle number is too high. Reduce the cycle number by 5 to 10 cycle.
h. Perform hot start.
i. Alter concentrations of cosolvents.
j. Perform TD/SD PCR.
k. Extension time is too long. Reduce the extension time in increments of 1 min.
l. Check for carry-over contamination. Set up PCR in a different area.
m. Review primer design and composition. Design new primers and try PCR again.

References

1. Higuchi, R. (1989) Using PCR to engineer DNA, in *PCR Technology: Principles and Applications for DNA Amplifications* (Erlich, H. A., ed.), Stockton Press, Inc., New York, pp. 61–70.

2. Foord, O. S. and Rose, E. A. (1995) Long-distance PCR, in *PCR Primer* (Dieffenback, C. W. and Dveksler, G. S., ed.), Cold Spring Harbor Laboratory Press, Cold Spring, NY, pp. 63–77.
3. Edwards, A., Civitello, A., Hammond, H. A., and Caskey, C. T. (1991) DNA typing and genetic mapping with trimeric and tetrameric tandem repeats. *Am. J. Hum. Genet.* **49,** 746–756.
4. Edwards, A., Hammond, H. A., Jin, L., Caskey, C. T., and Chakroborty, R. (1992) Genetic-variation at five trimeric and tetrameric tandem repeat loci in four human population groups. *Genomics* **12,** 241–253.
5. Klimpton, C. P., Gill, P., Walton, A., Urquhart, A., Millican, E. S., and Adams, M. (1993) Automated DNA profiling employing multiplex amplification of short tandem repeat loci. *PCR Methods Appl.* **3,** 13–21.
6. Sambrook, J., Fritsch, E. F., and Maniatis, T. (1989) *Molecular Cloning: A laboratory Manual.* Cold Spring Harbor Laboratory Press, Cold Spring Harbor, New York.
7. Lindahl, T. and Nyberg, B. (1972) Rate of depurination of native deoxyribonucleic acid. *Biochemistry* **11,** 3610–3618.
8. Rychlik, W. and Rhoads, R. E. (1989) A computer program for choosing optimal oligonucleotides for filter hybridization, sequencing and in vitro amplification of DNA. *Nucleic Acids Res.* **17,** 8543–8551.
9. Lowe, T. M. J., Sharefkin, J., Yang, S. Q., and Dieffenback, C. W. (1990) A computer program for selection of oligonucleotide primers for polymerase chain reaction. *Nucleic Acids Res.* **18,** 1757–1761.
10. O'Hara, P. J. and Venezia, D. (1991) PRIMGEN, a tool for designing primers from multiple alignments. *CABIOS* **7,** 533–534.
11. Montpetit, M. L., Cassol, S., Salas, T., and O'Shaughnessy, M. V. (1992) OLIGOSCAN: A computer program to assist in the design of PCR primers homologous to multiple DNA sequences. *J. Virol. Methods* **36,** 119–128.
12. Suggs, S. V., Hirose, T., Myake, D. H., Kawashima, M. J., Johnson, K. I., and Wallace, R. B. (1981) Using purified genes, ICN-UCLA Symp. *Mol. Cell. Biol.* **23,** 683–693.
13. Breslauer, K. J., Ronald, F., Blocker, H., and Marky, L. A. (1986) Predicting DNA duplex stability from the base sequence. *Proc. Natl. Acad. Sci. USA* **83,** 3746–3750.
14. Freier, S. M., Kierzek, R., Jaeger, J. A., Sugimoto, N., Caruthers, M. H., Neilson, T., et al. (1986) Improved free-energy parameters for predictions of RNA duplex stability. *Proc. Natl. Acad. Sci. USA* **83,** 9373–9377.
15. Innis, M. A. and Gelfand, D. H. (1990) Optimization of PCRs, in *PCR Protocols: A Guide to Methods and Applications* (Gelfand, D. H., Sninsky, J. J., Innis, M. A., and White, H., eds.), Academic Press, San Diego, CA, pp. 3–12.
16. Innis, M. A., Myambo, K. B., Gelfand, D. H., and Brow, M. A. D. (1988) DNA sequencing with Thermus aquaticus DNA polymerase and direct sequencing of polymerase chain reaction-amplified DNA. *Proc. Natl. Acad. Sci. USA* **85,** 9436–9440.
17. Ohler, L. and Rose, E. A. (1992) Optimization of long distance PCR using a transposon-based model system. *PCR Methods Appl.* **2,** 51–59.
18. Gelfand, D. H. (1989) *Taq* DNA polymerase, in *PCR Technology: Principles and Applications for DNA Amplification* (Erlich, H. A., ed.), Stockton Press, New York, pp. 17–22.
19. Cheng, S. (1995) Longer PCR amplifications, in *PCR Strategies* (Innis, M. A., Gelfand, D. H., and Sninsky, J. J., eds.) Academic Press, San Diego, CA, pp. 313–324.
20. Brock, T. D. and Freeze, H. (1969) *Thermus aquaticus* gene, a non-sporulating extreme thermophile. *J. Bacteriol.* **98,** 289–297.
21. Giebel, L. B. and Spritz, R. A. (1990) Site-directed mutagenesis using the double-stranded DNA fragment as a PCR primer. *Nucleic Acids Res.* **18,** 4947.

22. Saike, R. K., Gelfand, D. H., Stoffel, S., Scharf, S. J., Higuchi, R. Horu, G. T., Mullis, K. B., and Erlich, H. A. (1988) Primer-directed enzymatic amplification of DNA with a thermostable DNA polymerase. *Science* **239,** 487–491.
23. Flaman, J.-M., Frebourg, T., Moreau, V., Charbonnier, R., Martin, C., Ishioka, C., et al. (1994). A rapid PCR fidelity assay. *Nucleic Acids Res.* **22,** 3259–3260.
24. Cline, J., Braman, J. C., and Hogrefe, H. H. (1996) PCR fidelity of *Pfu* DNA polymerase and other thermostable DNA polymerases. *Nucleic Acids Res.* **24,** 3546–3551.
25. Sardelli, A. D. (1993) Plateau effect-understanding PCR limitations, in *Amplifications: A Forum for PCR Users.* Perkin–Elmer Corp., Norwald, CT, pp. 1.
26. Saiki, R. K. (1989) The design and optimization of the PCR, in *PCR Technology* (Erlich, H. A., ed.), Stockton Press, New York, pp. 7–16.
27. Kim, H. S. and Smithies, O. (1988) Recombinant fragment assay for gene targeting based on the polymerase chain reaction. *Nucleic Acids Res.* **16,** 8887–8903.
28. McConlogue, L., Brow, M. D., and Innis, M. A. (1988) Structure-independent DNA amplification by PCR using 7-deaza-2′-deoxyguanosine. *Nucleic Acids Res.* **16,** 9869.
29. Mytelka, D. S. and Chamberlin, M. J. (1996) Analysis and suppression of DNA polymerase pauses associated with a trinucleotide consensus. *Nucleic Acids Res.* **24,** 2774–2781.
30. Pomp, D. and Medrano, J. F. (1991) Organic solvents as facilitators of polymerase chain reaction. *BioTechniques* **10,** 58–59.
31. Newton, C. R. and Graham, A. (1994) *PCR, Bios Scientific*, Publishers Ltd., Oxford.
32. Levinson, G., Fields, R. A., Harton, G. L., Palmer, F. T., Maddleena, A., Fugger, E. F., et al. (1992) Reliable gender screening for human preimplantation embryos, using multiple DNA target-sequences. *Hum. Reprod.* **7,** 1304–1313.
33. Chamberlain, J. S., Gibbs, R. A., Ranier, J. E., Nguyen, P. N., and Caskey, C. T. (1988) Deletion screening of the Duchenne muscular dystrophy locus via multiplex DNA amplification. *Nucleic Acids Res.* **16,** 11,141–11,156.
34. Uggozoli, L. and Wallace, B. (1992) Application of an allele-specific polymerase chain reaction to the direct determination of ABO blood group genotypes. *Genomics* **12,** 670–674.
35. Henke, W., Herdel, K., Jung, K., Schnorr, D., and Loening, S. A. (1997) Betaine improves the PCR amplification of GC-rich DNA sequences. *Nucleic Acids Res.* **25,** 3957–3958.
36. Hengen, P. N. (1997) Optimizing multiplex and LA-PCR with betaine. *Trends Biochem. Sci.* **22,** 225–226.
37. Bassam, B. J. and Caetano-Anolles, G. (1993) Automated "hot start" PCR using mineral oil and paraffin wax. *BioTechniques* **14,** 30–34.
38. Wainwright, L. A. and Seifert, H. S. (1993) Paraffin beads can replace mineral oil as an evaporation barrier in PCR. *BioTechniques* **14,** 34–36.
39. Rous, K. H. (1995) Optimization and troubleshooting in PCR, in *PCR Primer* (Diefferbach, C. W. and Dveksler, G. S. eds.), CSH Press, New York, pp. 53–62.

21

Subcycling PCR for Long-Distance Amplifications of Regions with High and Low Guanine–Cystine Content

Amplification of the Intron 22 Inversion of the FVIII *Gene*

David Stirling

1. Introduction

Hemophilia A is an X-linked disorder caused by mutations in the *factor VIII* gene. Around 50% of all patients with severe hemophilia A share a common mutation. This intron 22 inversion results from homologous recombination of a sequence within intron 22 of the *factor VIII* gene and identical sequence around 500 kb telomeric to the gene. Although this inversion could be detected by Southern blotting, the development of a long polymerase chain reaction (PCR) assay, and its subsequent improvement by subcycling PCR (S-PCR), greatly facilitated the provision of diagnostic services ***(1,2)***.

In S-PCR, the annealing/elongation step is composed of subcycles of shuttling between a low and a high temperature. S-PCR produces consistent amplification of the various segments produced by wild-type, mutant, and carrier individuals. S-PCR is a robust variant of PCR, which may be of use in amplification of long segments in which the guanine–cytosine (GC) content varies among the segments, multiplex amplification of long segments, and multiplex amplification of short segments in which the GC content varies among the segments. We have also found it greatly improves the success of amplification from partially degraded templates.

2. Materials

Except where otherwise stated, all reagents are from Sigma Chemical Company, Poole, UK.

1. Thermal cycler.
2. Expand Long Template PCR System (Roche).
3. Dimethyl sulfoxide.
4. DNTP (Promega) Supplied separately (dATP, dCTP, dGTP, dTTP) at concentrations of 100 m*M*. Use at 10 m*M*. Aliquot 10 µL of stock dNTP into labeled tubes containing 90 µL of sterile distilled water. Store at –20°C.
5. 7-Deaza GTP (Roche).

From: *Methods in Molecular Biology, Vol. 226: PCR Protocols, Second Edition*
Edited by: J. M. S. Bartlett and D. Stirling © Humana Press Inc., Totowa, NJ

Table 1
Thermal Cycler Program for Subcycling PCR

Step	Temp/command	Time (mm:ss)/repetitions
1	95	2:00
2	94	00:10
3	63	00:05
4	68	5:00
5	Go to step 3	4 items
6	Go to step 2	15 times
7	94	00:10
8	63	00:10
9	68	6:00 (+ 10s per cycle)
10	Go to step 8	4 times
11	Go to step 7	15 times
12	68	30:00
13	End	

6. Oligonucleotides: INT22-P (5′-GCC CTG CCT GTC CAT TAC ACT GAT GAC ATT ATG CTG AC-3′); INT22-Q (5′-GGC CCT ACA ACC ATT CTG CCT TTC ACT TTC AGT GCA ATA-3′); INT22-A (5′-CAC AAG GGG GAA GAG TGT GAG GGT GTG GGA TAA GAA-3′); and INT22-B (5′-CCC CAA ACT ATA ACC AGC ACC TTG AAC TTC CCC TCT CAT A-3′). The oligonucleotides are diluted to 100 pmol/μL for storage. Subaliquot to minimize freeze-thaw cycles and store at –20°C.
7. PCR Mastermix 1. For each reaction to be run, the following are included: 10× buffer 2 (2.5 μmL); TAQ mix (0.94 μL); and water (5.69 μL).
8. PCR Mastermix 2. For each reaction to be run, the following are included: 10 m*M* aATP (1.25 μL); 10 m*M* aTTP (1.25 μL); 10 m*M* aACP (1.25 μL); 10 m*M* aGTP (0.625 μL); 10 m*M* deaza aGTP (0.625 μL); P/Q primer mix (5 μL); A (or B) primer mix (5 μL); 100% DMSO (1.875 μL).

3. Procedure

1. Remove reagents from freezer and allow to thaw fully then mix thoroughly and briefly centrifuge before pipetting. Reagents to be thawed: buffer 2 from Expand Long Template PCR Kit, dNTPs, and deaza dGTP.
2. Label 0.2 mL of flat-cap PCR tubes with the DNA number and primer grouping, if applicable (e.g., APQ).
3. Using sterile pipet tips, pipet 0.5 μL of each DNA sample into the appropriately labeled tube. Ensure that the DNA is pipetted directly into the bottom the tube. If concentration of DNA is very low, 1 μL of DNA should be added. Place tubes on ice.
4. Prepare 1 in 50 dilution of A oligo, 1 in 50 dilution of B oligo, and 1 in 25 dilution of P and Q oligo pairing. The number of samples being tested will determine the actual quantities required.
5. Prepare PCR mastermix 1 and 2 for appropriate number of tests: Place both mastermixes on ice!
6. For each reaction, aliquot 14.9 μL of Master Mix 2 into the labeled PCR tubes (which already contain the DNA template). Keep on ice.
7. Immediately before amplification, add 9.2 μL of Master Mix 1. Spin briefly and proceed to PCR step immediately.

8. Place tubes in thermal cycler and start the program (*see* **Table 1**).
9. When cycle is complete, remove PCR products and keep in fridge prior to electrophoresis.
10. Dilute 4 μL of each PCR product into 6 μL of sterile distilled water. This initial dilution may need adjusting depending on electrophoresis results picture.
11. Electrophorese 1 μL of this diluted DNA on an 0.8% agarose gel.
12. The inversion is associated with a band of 11 kb; the normal allele is associated with a band of 12 kb. An upper 12-kb band only indicates the inversion is not present; a lower 11-kb band only indicates the inversion is present in an affected male; and bands at both 11 kb and 12 kb indicate a female carrier of the inversion.

References

1. Liu, et al. (1998) Single tube polymerase chain reaction for rapid diagnosis of the inversion hotspot mutation in haemophilia A (letter). *Blood* **92,** 1458–1459.
2. Liu and Sommer (1998) Sub-cycling PCR for multiplex long distance amplifications of regions with High and Low GC content: Application to the inversion hotspot in the FVIII gene. *BioTechniques* **25,** 1022–1028.

22

Rapid Amplification of cDNA Ends

Xin Wang and W. Scott Young III

1. Introduction

The identification and isolation of full-length cDNAs can be a frustrating and time-consuming experience, especially for genes with a low abundance of expression or with large transcripts. Traditionally, full-length cDNAs are obtained from cDNA libraries by hybridization with radioisotope-labeled probes. This labor-intensive and tedious procedure often produces incomplete sequences and sometimes includes intronic sequence. To obtain full-length cDNAs, investigators had to rescreen libraries with larger numbers of clones (or upstream probes), not always successfully. The combination of rapid amplification of cDNA ends (RACE) and long-distance polymerase chain reaction (PCR) with high fidelity makes it possible to obtain full-length cDNAs quickly without constructing or screening a cDNA library.

The principle of RACE is simple and elegant: An anchor sequence is added to the end of the cDNA to be used as PCR primer binding template. A universal primer complementary to the added anchor template is coupled with a gene-specific primer (based on a single short known sequence within the mRNA of interest) in a PCR to amplify regions with unknown sequence. Several strategies have been developed to isolate full-length cDNA using this anchored PCR technology, each using a unique way to add the anchor sequence to the end of the cDNA. In the first generation of RACE, homopolymeric tails (G or A) are added to 3′ end of cDNA to be used as an anchor sequence using the enzyme terminal deoxynucleotidyl transferase *(1)*. The second generation of RACE technique is based on the ability of T4 RNA ligase to ligate a single-stranded anchor sequence to the 3′ end of the first-strand cDNA *(2)*. Both methods are difficult to optimize because of inefficient enzymatic reactions. The third generation of RACE uses T4 DNA ligase to add a double-stranded anchor sequence to both ends of double-stranded cDNAs *(3)*, thus the resulting anchored cDNAs are suitable for both 5′- and 3′-RACE. A commercial kit called Marathon cDNA amplification kit has been built around this approach (Clontech Laboratories, Inc.).

In this chapter, the application of this third-generation RACE method to the isolation of several pineal-specific cDNAs ranging from 1.4 to 8.0 kb in size is outlined. In one case, a full-length 2.0-kb cDNA for a pineal-specific cDNA, PG25, was obtained by 5′-RACE using gene-specific primers based on 260 base pairs of known sequence located in the 3′ terminus of the mRNA *(4)*. In another case, two different versions of

From: *Methods in Molecular Biology, Vol. 226: PCR Protocols, Second Edition*
Edited by: J. M. S. Bartlett and D. Stirling © Humana Press Inc., Totowa, NJ

full-length cDNAs for a pineal-specific gene, *PG23*, were obtained in a single 5′-RACE reaction because the antisense gene-specific primer used was derived from the common 3′ portion of the mRNAs (277 bp known sequence from differential display PCR or DD-PCR, this chapter). This approach is suitable for cloning full-length cDNA quickly based on a short sequence information from the 3′ end of mRNA, such as that obtained by the DD-PCR technique ***(5)*** or expressed sequence tags ***(6)***. PG10.2 is a gene (8 kb mRNA) expressed only in the pineal gland and the outer nuclear layer of the retina ***(7)***. Two 4-kb cDNA fragments were obtained by 5′-RACE and 3′-RACE using primers derived from only a 145-bp known sequence. Thus, this approach is suitable for cloning full-length cDNA based on a short known sequence located anywhere on the mRNA, whether derived from arbitrarily primed PCR (AP-PCR RNA fingerprinting technique ***(8)*** or suppression subtractive hybridization (SSH) technology (Clontech, **ref. *9***). In the case of PG10.2, a traditional 3′-RACE protocol was used to obtain the 3′ portion of unknown sequence. Then, an anchored cDNA pool constructed using an antisense gene-specific primer for reverse transcription was used to obtain the 5′ portion of unknown sequence. This approach is especially reliable and useful for isolating full-length cDNAs for large transcripts (such as 8 kb for PG10.2) with low and restricted expression. It may take only a few weeks to go from identification of differentially expressed sequence tags (by DD-PCR, AP-PCR, or SSH technology) to full-length cDNA by this long-template PCR-based RACE. In the following sections, both 5′-RACE and 3′-RACE protocols using the above model systems will be described.

The efficiency of described RACE approaches are very satisfactory for obtaining full-length cDNAs quickly. The choice of method is dependent on the available resources and relevant experience. For some genes, the use of more than one method may be necessary. Another excellent approach for full-length cDNA cloning is using biotin-tailed oligonucleotide probes to capture full-length cDNA clones from a well-constructed cDNA library (Invitrogen). A commercial kit called GeneTrapper cDNA Positive Selection System has been built around this capture technology (Invitrogen).

2. Materials

1. RNA from source tissue (*see* **Note 1**): pineal gland and retina tissues were dissected from Sprague–Dawley rats (male, 200–250 g, Taconic Farms). Total cellular RNA was isolated using TRIzol Reagent (Invitrogen) according to the manufacturer's instructions. For preparation of double-stranded cDNA, poly (A)+ RNA was purified from 500 µg total RNA using a Poly (A) Quik mRNA isolation kit (Stratagene, La Jolla, CA).
2. dNTP mixture: an aqueous solution of each dNTP (dGTP, dATP, dTTP, and dCTP) from any reputable vendor of molecular biology reagents.
3. First-strand cDNA synthesis: SuperScript II (RNAse H-) reverse transcriptase (Invitrogen) was used with manufacturer supplied 5 X reverse transcription buffer (*see* **Note 2**).
4. cDNA synthesis primer: this may be a universal oligo-dT primer (5′-CACTATAGGC CATCGAGGCC$(T)_{20}$MN-3′) for 5′-RACE and/or 3′-RACE (*see* **Subheading 3.2.1.**) or an antisense gene-specific primer (5′-RACE only, *see* **Subheading 3.4.2.**) complementary to the rare mRNA of interest or located upstream on a large transcript (*see* **Note 3**).
5. Second-strand cDNA synthesis: a 20× second-strand enzyme mixture, 5× second-strand buffer and T4 DNA polymerase supplied in the Marathon cDNA amplification kit (Clontech Laboratories, Inc.). Combining the following enzymes makes extra second-strand enzyme mixture: *Escherichia coli* DNA polymerase I, *E. coli* DNA ligase, and *E. coli* RNAse H (*see* **Note 4**).

6. Adaptor (anchor sequence) ligation: A specially designed partially double-stranded adaptor supplied in the Marathon cDNA amplification kit (Clontech Laboratories, Inc.).
7. Oligonucleotide primers: gene-specific primers should be about 25 nucleotides long and around 50% guanine–cytosine. Primers with a melting temperature between 65 and 70°C give sufficient binding specificity (*see* **Note 5**). For some difficult genes, primers with 70°C melting temperature or higher should be used in a touchdown PCR (*see* **Note 6**). Anchor primers (AP) complementary to the adaptor sequence (for 5′-RACE or 3′-RACE) or to the tailing sequence on the cDNA synthesis oligo-dT primer are coupled with GSP to amplify the unknown sequences flanked by the paired primer set. AP-1 (5′-CCATCCTAATACGACT CACTATAGGGC-3′) and AP-2 (nested within the AP-1, 5′-ACTCACTATAGGGCTC GAGCGGC-3′) are supplied in Clontech's Marathon cDNA Amplification Kit.
8. PCR machine: all experimental data presented in this paper were carried out on a thermocycler from MJ Research, Inc (*see* **Note 7**).
9. DNA Polymerase: Expand PCR system (Boehringer-Mannheim). There are three buffer systems supplied by the manufacturer. Buffer 1 is sufficient for RACE PCR of expected product size of less than 10 kb.
10. PCR fragment purification: QIAEX II from QIAGEN Bioscience Corporation.
11. TA cloning: pGEM-T vector system (Promega Corporation).
12. Other general molecular laboratory equipment and reagents.
13. All reagents, including enzymes, should be mixed briefly immediately before use.

3. Methods

3.1. The Relative Abundance of the Transcript in the Source Tissue

1. A good understanding of expression profile is essential for the successful isolation of the full-length cDNA of interest. The efficiency of RACE PCR amplification largely depends on the relative abundance of the mRNA of interest in the poly (A)+ RNA sample extracted from the target tissue. RACE PCR should be performed on the tissue where the expression is most abundant. The higher the copy number of the mRNA in the cDNA pool, the better chance the full-length cDNA can be amplified to a critical mass visible on agarose gel (*see* **Note 8**).
2. PG23 is a pineal-specific gene identified using DD-PCR (unpublished data). In this case, we had no choice but to use pineal as our target tissue for 5′-RACE. Northern blot analysis revealed a mRNA doublet about 2.0 kb and 2.4 kb for PG23 (**Fig. 1A**, lane 1). The size of mRNA serves as a useful guide in identifying the correct RACE bands (**Fig. 1B**, lanes 1 and 2). In a similar fashion, another 2.0-kb full-length pineal-specific cDNA (PG25) was obtained using the RACE protocol outlined for PG23 (Fig. 3B in **ref. *4***).
3. PG10.2 is a gene (8 kb mRNA) expressed only in the pineal gland and the outer nuclear layer of the retina *(7)*. In this case, retina was used as target tissue source because PG10.2 has a much higher expression level in retina than in pineal even though it is only expressed in the outer nuclear layer of the retina (Fig. 3 in **ref. *7***). This report will describe the application of a traditional 3′-RACE (*see* **Subheading 3.3.**) and a gene-specific 5′-RACE (*see* **Subheading 3.4.**) to the isolation of the large transcript of PG10.2 (8.0 kb). A general strategy is schematically depicted in **Fig. 2**.

3.2. 5′-RACE Using a Universal Oligo-dT Primer for cDNA Synthesis (see Note 9)

1. First-strand cDNA synthesis: 1 µg of high-quality poly (A)+ mRNA isolated from fresh rat pineal was reverse-transcribed to first-strand cDNA in a 10-µL reaction containing 1 µL of 10× first-strand synthesis buffer, 1 µL of 0.1 *M* dithiothreitol, 1 µL of 10 m*M*

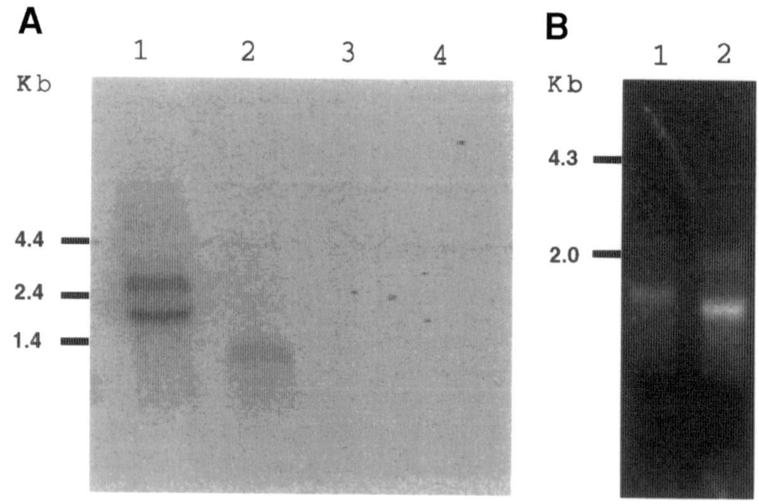

Fig. 1. **(A)** Northern blot analysis: 10 μg of total RNA were loaded in each lane; Pineal (lane 1), brain (lane 2), lung (lane 3), and kidney (lane 4). **(B)** RACE PCR: full-length cDNAs (1.4 kb and 2.0 kb) were obtained using a long template PCR-based RACE technique. Two primers, AGSP (5′- GGAACAGTCTGAGCTCTAAGCTAGG-3′, lane 1) and NAGSP (5′-CTAGAAGGATAAGTTCACGAGGGCC-3′, lane 2), were designed using a 277-bp sequence information from a sequenced DD-PCR product (**Fig. 2**) and coupled with AP-1 for 5′-RACE PCR.

dNTP, 1 μL of 10 μ*M* oligo-dT primer (5′-CACTATAGGCCATCGAGGCC(T)20MN-3′), and 1 μL of SuperScript II (RNase H-) reverse transcriptase (added last; see below). The reaction was performed in a thermocycler with the following parameters: 70°C for 10 min, 42°C for 50 min, and 50°C for 15 min. SuperScript II RNase H- reverse transcriptase (1 μL) was added to the reaction mixture after a 5-min incubation at 42°C. The reaction mixture was placed on ice before the next step.

2. Second-strand cDNA synthesis: The following components for second-strand cDNA synthesis were added to the above 10-μL first-strand reaction mixture: 48.4 μL of high-quality H_2O, 16 μL of 5× second-strand buffer, 1.6 μL of 10 m*M* dNTP mixture, and 4 μL of 20× second-strand enzyme cocktail. After mixing the contents briefly with gentle pipetting, the reaction was placed in the thermocycler for incubation at 16°C for 1.5 h. Then, 2 μL of T4 DNA polymerase was added and mixed well, and incubation continued at 16°C for another 45 min to create blunt end. Finally, 4 μL of the EDTA/Glycogen mixture (glycogen helps bring down the cDNA in a later precipitation) was added to stop the reaction.
3. Purification of blunt-ended double-stranded cDNA: For efficient recovery of the relatively small amount of the cDNA, the general phenol/chloroform extraction was performed twice using 100 μL of phenol:chloroform:isoamyl alcohol (25:24:1). The supernatant plus 0.5 volume of 4 *M* ammonium acetate and 2.5 volumes of 100% ethanol were placed at –20°C for at least 1 h and centrifuged for 10 min at 4°C. The recovered DNA pellet was washed in 1 mL of 75% ethanol and resuspended in 10-μL high-quality water after air-drying.
4. Adaptor ligation: A partially double-stranded adaptor was ligated to both ends of the above-purified double-stranded cDNA. Ten microliters of ligation mixture containing 5 μL of purified blunt-ended double-stranded cDNA, 2 μL of adaptor, 2 μL of 5× ligation

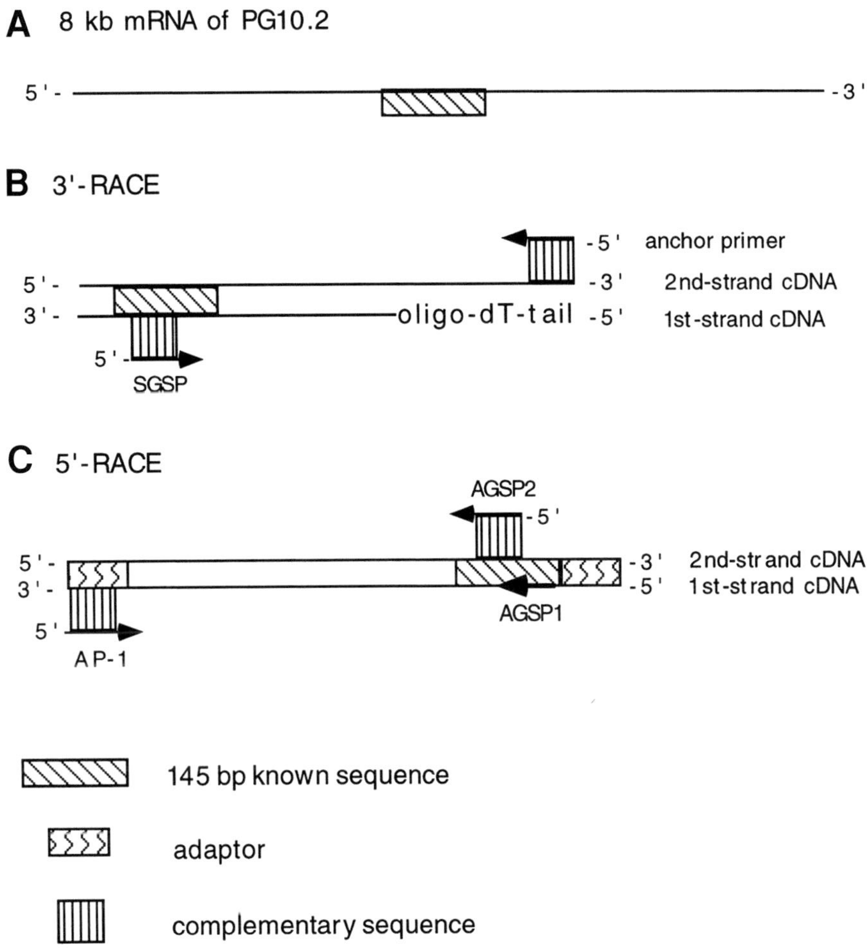

Fig. 2. Schematic representation of 5′- and 3′-RACE for the 8.0 kb cDNA of PG10.2. **(A)** The 145-bp sequence of PG10.2 is located in the middle of the 8-kb mRNA. **(B)** 3′-RACE. SGSP coupled with the anchor primer (complementary to the tail sequence of oligo-dT cDNA synthesis primer) for 3′-RACE. **(C)** 5′-RACE. AGSP2 coupled with the AP-1 anchor primer for 5′-RACE.

buffer, and 1 μL of T4 DNA ligase were incubated at 16°C overnight. The mixture was heated at 70°C to inactivate the ligase.

5. Primer design for *PG23* 5′-RACE: A short known sequence located at the 3′ end region of the gene is required for designing antisense gene-specific primers (AGSP) and a nested antisense gene-specific primer (NAGSP) for 5′-RACE. In the case of *PG23*, AGSP (5′- GGAACAGTCTGAGCTCTAAGCTAGG-3′) and NAGSP (5′-TCTAGAAGGATA-AGTTCACGAGGGCC-3′) were designed based on the 277-bp sequence information just upstream of the poly (A) tail as shown in **Fig. 3**.
6. PCR was performed in a 50-μL reaction containing the following components: 5 μL of 10× PCR buffer 1, 5 μL of the 1:100 diluted adaptor-ligated cDNAs, 1 μL of 10 m*M* dNTP, 1 μL of 10 μ*M* AP-1 linker primer, 1 μL of 10 μ*M* GSP, and 0.75 μL of DNA polymerase mixture (added last). The DNA polymerase mixture was added to the reaction after denaturation at 94°C for 1 min to reduce nonspecific amplification (*see* **Note 10**).

5'-GATCTGACTGCAGAGAACATTAAGTGGATCTGGACCTTTGCTGACACTGA-3'
3'-CTAGACTGACGTCTCTTGTAATTCACCTAGACCTGGAAACGACTGTGACT-5'

5'-GAATCTGGATCCACAAGAAGAAGCAAGCTGTTATAAATGCTACAGCCAGA-3'
3'-CTTAGACCTAGGTGTTCTTCTTCGTTCGACAATATTTACGATGTCGGTCT-5'

5'-AGAAAGGATGAGAATTCTGTTCTGCCCTGAGGGCCCTCGTGAACTTATCC-3'
3'-TCTTTCCTACTCTTAAGACAAGACGGGACTCCCGGGAGCACTTGAATAGG-5'
← NAGSP

5'-TTCTAGATTCCAGCCCTAGCTTAGAGCTCAGACTGTTCCTTACACAATGG-3'
3'-AAGATCTAAGGTCGGGATCGAATCTCGAGTCTGACAAGGAATGTGTTACC-5'
← AGSP

5'-TCTGGGCATAGCACCCCTTAAGCATGCTGGACTGACAGAACTAATATATA-3'
3'-AGACCCGTATCGTGGGGAATTCGTACGACCTGACTGTCTTGATTATATAT-5'

5'-TTTAAAGCGTGTCTGAAAAAAAAAAAAA-3'
3'-AAATTTCGCACAGACTTTTTTTTTTTTT-5'

Fig. 3. Expressed sequence tag of PG23. Sequence information of 277 bp was derived from a cloned DNA fragment using the DD-PCR method *(5)*. Double-strand sequences are shown for easy identification of primer's location (underlined). A 10-nucleotide primer (5′-GATCT-GACTGC-3′) and an olio-dT primer (5′-TTTTTTTTTTTTCA-3′) were used in DD-PCR for the original cloning of *PG23*. AGSP (5′- GGAACAGTCTGAGCTCTAAGCTAGG-3′) and NAGSP (5′-TCTAGAAGGATAAGTTCACGAGGGCC-3′) were designed for the 5′-RACE of full-length cDNA of PG23.

The sample was then subjected to 30 cycles of PCR on a thermocycler using the following parameters: 94°C for 30 s, 60°C for 30 s, and 68°C for 3 min. For RACE PCR of GC rich mRNAs, addition of some co-solvent may increase PCR efficiency (*see* **Note 11**).

7. Agarose gel electrophoresis of 5′-RACE PCR products: The above AGSP and NAGSP were used in two independent RACE PCR. Then, 20 µL was loaded on a 1% agarose gel and electrophoresed in 1× TAE buffer at 75 volts for 2 h. Each of these antisense primers produced bands of expected size on ethidium bromide stained agarose gels when coupled with the AP-1 primer (**Fig. 1B**, lanes 1 and 2). The band produced by the NAGSP (**Fig. 1B**, lane 2) is slightly smaller compared with the one generated by AGSP (**Fig. 1B**, lane 1), confirming that the bands are true RACE products. When only AGSP, NAGSP, or AP-1 were used in PCRs, these sharp bands disappeared (data not shown). The DNA bands of proper size were excised from agarose gel with a clean, sharp scalpel. DNA was extracted from the gel using the QIAEX II kit according to the manufacturer's instruction. A 15-µL elution was saved for later use.
8. The gel-purified DNA (5 µL) was used for cloning into the pGEM-T vector. Sequence information for PG23 mRNAs of 1.4 kb and 2.0 kb was subsequently determined (data not shown).
9. Nested second PCR: If the first PCR with AGSP and AP-1 is insufficient to produce a visible band(s) of expected size, a second round of PCR with NAGSP and AP-2 (nested within AP-1 primer) could be performed using 2.5 µL of 1:100 dilution of the first-PCR product as template. This second round of PCR usually yields defined fragments of expected size on ethidium bromide stained agarose gel. Sometimes no visible band

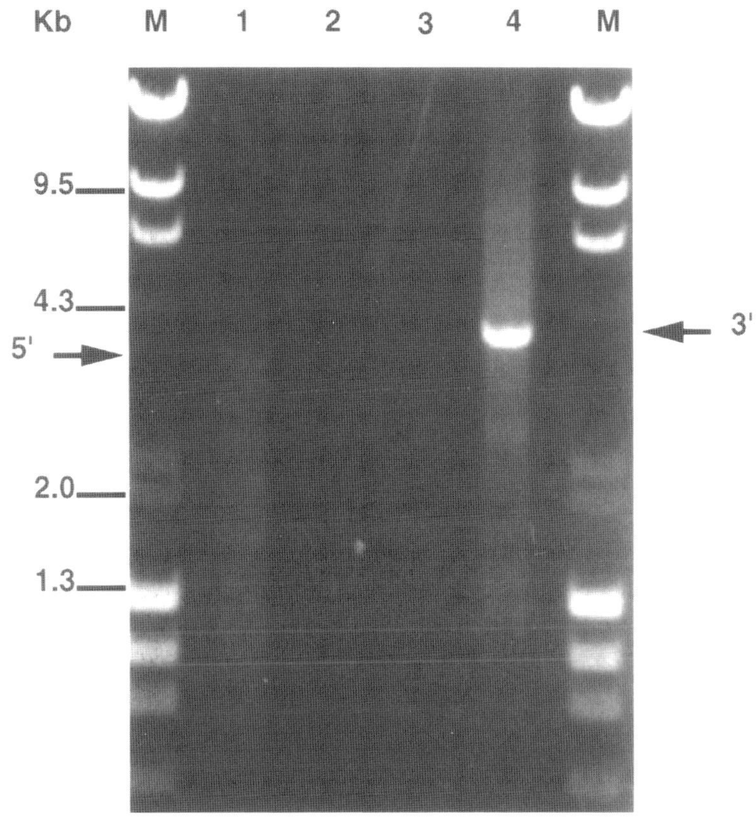

Fig. 4. 5′- and 3′-RACE of PG10.2. An antisense gene-specific primer 2 (AGSP2 of PG10.2, 5′-GGCAGTTCATCCACACTCAGGTACCCAG-3′) and AP-1 primer amplified a faint but sharp band of about 3.5 for 5′-RACE of PG10.2 (lane 1). Either AGSP2 (lane 2) or AP-1 (lane 3) alone failed to produce the band. A sense gene-specific primer (SGSP of PG10.2, 5′-GAGTGAGAAGCAAAGTGCAAATGCC-3′) and an anchor primer (5′-CCAAGCTATT-TAGGTGACACTATAGGCCATCGAGGCC-3′, priming at the end of the newly synthesized second-strand cDNA) amplified a band of 4 kb for the 3′-RACE (lane 4).

appeared even after a second round of PCR using nested primer set. This may due to non-optimized RACE PCR conditions or to an extremely low level of expression of the gene. Other methods could be used to identify the correct RACE clones (*see* **Note 12**).

3.3. 3′-RACE Using Tailed Oligo-dT Primer for cDNA Synthesis (see Note 9)

1. Northern analysis revealed a mRNA of 8.0 kb in size for PG10.2. Only 145 bp of sequence information for this DD-PCR fragment was available to be used as template for RACE primer design. After failing to find any positive cDNAs for PG10.2 in screens of two cDNA libraries, we suspected that this was probably caused by the relatively far upstream location of the 145 bp probe (Fig. 4 in **ref. 7**, underlined) and truncated clones in the cDNA libraries we examined. In order to obtain this potentially large 3′ portion of unknown sequence, we used a traditional 3′-RACE protocol outlined below to do 3′-RACE for PG10.2. In general, the extended sequence information would provide a much broader region for 5′-RACE primer design.

2. First-strand cDNA synthesis: same as above (*see* **Subheading 3.2.**).
3. Second-strand cDNA synthesis: same as above (*see* **Subheading 3.2.**). T4 DNA Polymerase was not used because blunt ending of cDNA was unnecessary in this case.
4. Purification of double-stranded cDNA: same as above (*see* **Subheading 3.2.**).
5. Primer design for PG10.2 3′-RACE: A sense gene-specific primer (SGSP of PG10.2, 5′-GAGTGAGAAGCAAAGTGCAAATGCC-3′) was designed based on the known 145 bp sequence information of PG10.2. An anchor primer (5′-CCAAGCTATTTAGGTGACAC-TATAGGCCATCGAGGCC-3′) was designed to prime at the end of the newly synthesized second-strand cDNA (**Fig. 2**). This end sequence on the newly synthesized second-strand cDNA is complementary to the 5′ tailing part of the oligo-dT primer used for first-strand cDNA synthesis.
6. PCR was performed as described (*see* **Subheading 3.2.**) with SGSP of PG10.2 and the above-mentioned anchor primer. The PCR product was electrophoresed on a 1% agarose gel. A 4-kb fragment 3′ to the 145 bp known sequence was amplified very efficiently by this 3′-RACE PCR (**Fig. 4**, lane 4).
7. The band was purified (as in **Subheading 3.2.**) and ligated into pGEM-T vector (as in **Subheading 3.2.**).

3.4. 5′-RACE Using Gene-Specific Primer for cDNA Synthesis (see Note 9)

1. To clone the full-length cDNA for the large transcript (8.0 kb mRNA) of PG10.2, a gene-specific RACE cDNA was generated and used as described below.
2. First-strand cDNA synthesis was performed essentially as described (*see* **Subheading 3.2.**) except that an antisense gene-specific primer (AGSP1 of PG10.2, 5′-TTCAAGGGCCAGT-CAGGCCGTAGGTCACAGACACTTTGAC-3′) based on 145-bp sequence information for PG10.2 was used for first-strand cDNA synthesis. The cDNA generated by this cDNA synthesis primer (GSP) is only suitable for 5′-RACE of this specific gene.
3. Second-strand cDNA synthesis: same as above (*see* **Subheading 3.2**).
4. Purification of double-stranded cDNA: same as above (*see* **Subheading 3.2**).
5. Adaptor ligation: same as above (*see* **Subheading 3.2.**).
6. Primer design for PG10.2 5′-RACE: An antisense gene-specific primer2 (AGSP2, 5′- GGCAGTTCATCCACACTCAGGTACCCAG-3′) based on 145 bp sequence information was designed for 5′-RACE of PG10.2. Notably, this primer was designed to be upstream of the AGSP1 used for cDNA synthesis (*see* **Subheading 3.4.**). In this design, AGSP2 is a nested primer relative to AGSP1 (**Fig. 2**).
7. PCR was performed as described (*see* **Subheading 3.2**). Agarose gel electrophoresis showed that the AGSP2 of PG10.2 coupled with AP1-linker primer (Clontech, complementary to the sequence of the ligated-adaptor) amplified a unique band of about 3.5 kb in size upstream of the 145-bp known sequence of the PG10.2 fragment (**Fig. 2**, lane1). When only the AGSP2 of PG10.2 or AP-1 was used in PCRs, this faint but sharp band disappeared (**Fig. 4**, lanes 2 and 3).
8. The band was purified (*see* **Subheading 3.2.**) and ligated into pGEM-T vector (as in **Subheading 3.2.**).
9. The 4368-bp sequence, including the entire 5′-RACE product and the 5′-portion of the 3′-RACE product, was determined (Fig. 4 in **ref.** *7*).

4. Notes

1. Template purity (free of DNA) and integrity (minimum degradation) are critical for effective isolation of full-length cDNA of interest. One should place the dissected tissue immediately on dry ice before isolating total and poly (A)+ RNAs. All standard

precautions for handling RNA should be followed carefully ***(10)***. TRIzol Reagent (Invitrogen) performs well with small quantities of tissue. Total RNA has been successfully extracted from punches of rat supraoptic and paraventricular nuclei and used to clone a gene preferentially expressed in the supraoptic and paraventricular nuclei of the brain by DD-PCR (unpublished data). High-quality poly (A)+ mRNA can be obtained from companies such as Clontech Laboratories, Inc.

2. Reverse transcriptase and related buffers supplied in the Marathon cDNA amplification (Clontech) may be used for the first-strand synthesis to meet most full-length cDNA cloning needs. SuperScript II (RNAse H-) reverse transcriptase (Invitrogen) is a preferred reverse transcriptase to generate longer first-strand cDNAs.
3. The oligo-dT primer (5′-TTCTAGAATTCAGCGGCCGC$(T)_{30}$MN-3′) supplied in the Marathon cDNA amplification kit (Clontech) is suitable as a universal cDNA synthesis primer. The two degenerate nucleotides (where M=A, G or C; N=A, G, T or C) place the oligo-dT primer at the beginning of the poly(A) tail and thus eliminate 3′-heterogeneity ***(11)***.
4. Combine 25 µL of *E. coli* DNA polymerase I (10.0 units/µL, Invitrogen), 3 µL of *E. coli* DNA ligase (10.0 units/µL, Invitrogen), 3 µL of *E. coli* RNase H (2.1 units/µL, Invitrogen), and 10 µL of high-quality H_2O to make 20× second-strand enzyme mixture. 5× second-strand buffer contains 500 m*M* KCL, 50 m*M* ammonium sulfate, 25 m*M* $MgCL_2$, 0.75 m*M* beta-NAD, 100 m*M* Tris (pH 7.5), and 0.25 mg/mL bovine serum albumin (Clontech's Marathon cDNA Amplification Kit).
5. The primer's melting temperature is an important parameter that greatly influences PCR specificity by reducing nonspecific priming events. High-melting point primers with melting temperatures between 65 and 70°C allow the use of higher annealing temperatures to enhance reaction specificity. The estimated melting temperature [(G+C) × 4 + (A+T) × 2] are not exact under PCR conditions but can be used as a starting point. Sense primers are used for 3′-RACE whereas anti-sense primers are used for 5′-RACE.
6. Touchdown PCR may help eliminate extraneous bands and increase yield. The first round of touchdown PCR has an annealing temperature 5 to 10°C higher than what is usually used. In each subsequent round, the annealing temperature is dropped a degree until the standard annealing temperature is reached.
7. GeneAmp PCR System 9600 (Applied Biosystems) gives more satisfactory amplification efficiency, especially for low abundant genes. Applied Biosystems GeneAmp 0.5-ml PCR tubes are preferred PCR tubes for critical PCR amplification.
8. Northern analysis is the most preferred method to provide an expression profile and size estimation of the mRNA. If northern data is not readily available, a quick reverse-transcription PCR survey using two GSPs (to produce a sizable PCR fragment) could be used to identify the best tissue source for the actual RACE PCR.
9. The 5′-RACE outlined in **Subheading 3.2.** is essentially as described in the Marathon cDNA amplification kit (Clontech Laboratories, Inc.) and the Expand PCR system (Boehringer Mannheim Biochemicals) except an oligo-dT primer (5′- CACTATAGGCCATC GAGGCC$(T)_{20}$MN-3′) was used for first-strand cDNA synthesis. The adaptor-ligated cDNA pool is essentially an uncloned cDNA library, which can be used to isolate full-length cDNAs for many different genes using gene-specific primers. These two-sided (5′ and 3′) anchored cDNAs permit 5′- and 3′-RACE PCR to be performed with the same cDNA pool. If the known sequence information is far upstream from the 3′ end of the mRNA (poly A tail), especially for the large transcript of the gene, the 3′-RACE protocol described in **Subheading 3.3.** is a preferred strategy to obtain this potentially long (often untranslated) region quickly and reliably. Although the ligation of the double-stranded adaptor using T4 DNA ligase is much more efficient compared to the inefficient tailing or the single-stranded anchor ligation, the adaptor actually tags only a portion of cDNAs. The

3′-RACE outlined in **Subheading 3.3.** is more efficient because it relies on the sequence introduced by the tail sequence 5′ of the oligo-dT sequence as the template for anchor primer binding. The cDNA pool always provides more template for this kind 3′-RACE than the one available from ligated adaptors, hence giving more robust 3′-RACE efficiency. In general, the newly extended sequence information provides a larger region for design of 5′-RACE primers. This is especially important if the available sequence information for primer design is very limited. The gene-specific 5′-RACE protocol outlined in **Subheading 3.4.** is useful for obtaining full-length cDNAs of large transcripts or those expressed at low levels. The cDNA produced by this protocol is only suitable for RACE PCR of unknown sequence upstream of the gene-specific primer that was used for first-strand cDNA synthesis.

10. Hot start PCR. The DNA polymerase is withheld from the reaction until the temperature of reaction tube is above the annealing temperature. This hot-start PCR is used to improve the specificity and sensitivity of the RACE PCR.
11. Co-solvent addition. For some primer/template systems such as GC-rich sequences, the addition of glycerol and/or DMSO to a final concentration of 5% has been found to enhance PCR yield and/or specificity.
12. The RACE PCR product may be cloned into pGEM-T vector even if agarose gel electrophoresis fails to show a visible band(s). One dozen to one hundred colonies could be screened for positive clones in colony hybridization using oligodeoxynucleotide probes downstream from the RACE primer used for RACE PCR.

Acknowledgments

All experimental data presented in this chapter were performed in the former Laboratory of Cell Biology, National Institute of Mental Health/National Institutes of Health. We would like to acknowledge Dr. Michael J. Brownstein for invaluable advice and generous support. We thank Dr. Herman E. Brockman for critical reading of the manuscript.

References

1. Frohman, M. A., Dush, M. K., and Martin, G. R. (1988) Rapid production of full-length cDNAs from rare transcripts: amplification using a single gene-specific oligonucleotide primer. *Proc. Natl. Acad. Sci. USA* **85,** 8998–9002.
2. Edwards, J. B. D. M., Delort, J., and Mallet, J. (1991) Oligodeoxyribonucleotide ligation to single-stranded cDNAs: A new tool for cloning 5′ ends of mRNAs and for constructing cDNA libraries by *in vitro* amplification. *Nucleic Acids Res.* **19,** 5227–5232.
3. Chenchik, A., Diatchenko, L., Moqadam, F., Tarabykin, V., Lukyanov, S., and Siebert, P. D. (1996) Full-length cDNA cloning and determination of mRNA 5′ and 3′ ends by amplification of adaptor-ligated cDNA. *BioTechniques* **21,** 526–532.
4. Wang, X., Brownstein, M. J., and Young, W.S., III (1997) PG25, a pineal-specific cDNA, cloned by differential display PCR (DDPCR) and rapid amplification of cDNA ends (RACE). *J. Neurosci. Methods* **73,** 187–191.
5. Liang, P. and Pardee, A. B. (1992) Differential display of eukaryotic messenger RNA by means of polymerase chain reaction. *Science* **257,** 967–971.
6. Sikela, J. M. and Auffray, C. (1993) Finding new genes faster than ever. *Nat. Genet.* **3,** 189–191.
7. Wang, X., Brownstein, M. J., and Young, W. S., III (1996) Sequence analysis of PG10.2, a gene expressed in the pineal gland and the outer nuclear layer of the retina. *Mol. Brain Res.* **41,** 269–278.

8. McClelland, M., Ralph, D., Cheng, R., and Welsh, J. (1994) Interactions among regulators of RNA abundance characterized using RNA fingerprinting by arbitrarily primed PCR. *Nucleic Acids Res.* **22,** 4419–4431.
9. Diatchenko, L., Lau, Y.-F. C., Campbell, A. P., Chenchik, A., Moqadam, F., Huang, B., et al. (1996) Suppression subtractive hybridization: A method for generating differentially regulated or tissue-specific cDNA probes and libraries. *Proc. Natl. Acad. Sci. USA* **93,** 6025–6030.
10. Sambrook, J., Fritsch, E. F., and Maniatis, T. (1989) *Molecular Cloning: A Laboratory Manual*, 2nd Ed. Cold Spring Harbor Laboratory, Cold Spring Harbor, New York.
11. Borson, N. D., Salo, W. L., and Drewes, L. R. (1992) A lock-docking oligo(dT) primer for 5′ and 3′ RACE PCR. *PCR Methods Appl.* **2,** 144–148.

23

Randomly Amplified Polymorphic DNA Fingerprinting

The Basics

Ranil S. Dassanayake and Lakshman P. Samaranayake

1. Introduction

The study of genetic polymorphism among diverse populations of organisms is a complex task. However, this can be accomplished by using newer tools, such as randomly amplified polymorphic DNA (RAPD). RAPD is a polymerase chain reaction (PCR) technique that relies on the generation of amplification products for a given nucleic acid using an amplification-based scanning technique driven by arbitrary priming oligonucleotides. The result is the generation of amplification products (amplicons) that represent a multiplicity of anonymous sites that are characteristic of the studied genome **(Fig. 1A,B)**.

In RAPD, the first few cycles are performed at a low stringency, which ensures the generation of products by priming with mismatches between the primer and the template. The subsequent PCR cycles are performed at a higher stringency (*see* **Note 1**), yielding products that have ends complementary to the primer. The amplified region consists of unstructured, hypervariable, mostly noncoding sequences that vary in length from one species to another. As the arbitrary priming depends upon the complimentary regions in the DNA template, differences in these regions lead to uniquely characteristic fingerprinting patterns *(1)*. These differences permit any organism to be characterized at the species or the strain level. However, it is noteworthy that the clarity of the species or strain discrimination, fingerprint complexity, and detection of DNA polymorphism are dependent on the primer that is selected for the RAPD assay.

A single primer approx 10 bp in size (40–70% guanine–cytosine content) is generally used in PCR fingerprinting. For the amplification of the target region, the distance between priming regions has to be not more than 3 to 4 Kb. Specificity in RAPD is defined as the ability to produce "consensus" fingerprints on multiple occasions as a result of a multiplicity of targeted arbitrary sites. To achieve such consensus and "stringent" fingerprints in arbitrary priming protocols, a decrease in primer length *(2)* or blocking of the interactions between amplicons by incorporating a mini-hairpin at the 5′ terminus of the oligonucleotide is required *(3)*.

There are a number of limitations in the routine use of RAPD profiles for detailed evaluation of the genomes of organisms. The necessarily short primers require stringent

From: *Methods in Molecular Biology, Vol. 226: PCR Protocols, Second Edition*
Edited by: J. M. S. Bartlett and D. Stirling © Humana Press Inc., Totowa, NJ

conditions for reproducible PCR *(4)*, because variable or absent PCR products may result depending on the purity, quantity, or the quality of the DNA templates *(5)*. The reproducibility of the RAPD technique can also be affected by minor changes in methodological aspects, such as the differences in the primer-to-template concentration ratio (*see* **Note 2**), variations of primer annealing temperatures, the cation concentration of PCR buffer, and magnesium ions in the reaction mixture (***6***, *see* **Note 3**). These parameters can dramatically affect the presence of low-intensity bands as well as the position and intensity of high-intensity bands. Moreover, some have reported that different lots of *Taq* polymerases (*see* **Note 4**) and the brand of the thermocycler used (*see* **Note 5**) can also affect the RAPD patterns, especially the low-intensity bands *(7)*. Furthermore, in some situations, the bands that show equal electrophoretic mobility may not be homologes, and missing bands may not necessarily reflect homology because they can be lost by nucleotide substitutions in either the PCR priming sites or by length mutations. These complications of identity can be resolved either by sequencing the homologous bands or using a band-specific probe. Such strategies have been used in several recent studies ***(8–10)***.

Our studies are mostly focused on the genomic analysis of the human fungal pathogen *Candida albicans*, and the following protocol is based on this experience. However, the principles guiding RAPD analyses are similar, and the following methods with minor variations would be applicable in general for many other organisms.

2. Materials

1. Thermocycler.
2. Agarose gel electrophoresis apparatus.
3. Ultraviolet transilluminator, Power supply (200 V and 150 mA).
4. Spectrophometer for determining DNA concentration.
5. Photographic unit that can capture ethidium bromide stained gel (e.g., Polaroid camera or digital image capture system, such as a CCD camera, a computer, and image analysis software for clear resolution of the gel image).
6. 10× PCR buffer: 100 m*M* Tris-HCl, 500 m*M* KCl, 15 m*M* $MgCl_2$, 0.01% (w/v) gelatin, pH 8.3.
7. 25 m*M* Magnesium chloride.
8. High-quality sterile, deionized water (more than 10 megaohm/cm) must be used for the preparation of all reagents and premixes.
9. Deoxynucleotide triphosphate stock solution (dNTP): 2 m*M* each of dGTP, dATP, dCTP, and dTTP. Ready-made dNTP (100 m*M*) solution (obtainable from Sigma, Pharmacia, Promega, etc.). Make aliquots, preferably 10 m*M*, and store in –20°C. If dNTP solutions are made from dry reagents, the pH of the solution should be adjusted to 7.5 with 0.1 *M* Tris or 0.1 *M* NaOH using a pH meter or strip of pH paper.
10. Primers: Lyophilized primers (5′GCGATCCCCA3) should be prepared at 100 μ*M* and 10 μ*M* concentration with deionized water; Store at –20°C.
11. *Taq* DNA polymerases: *Taq* DNA polymerase (5 U/μL; Sigma), Ampli *Taq* (5 U/μL; PerkinElmer), Stoffel fragment of *Taq* DNA polymerase (10 U/μL; Perkin–Elmer), or any other high-quality *Taq* polymerase is preferred.
12. Template DNA: 10 to 25 ng/μL stock solution containing good-quality, protein-free, nonsheared DNA; DNA can be resuspended in high-quality sterile, deionized water or TE (Tris-EDTA) pH 8.0. RNase (20 ng per 1 ng of DNA) treatment can increase amplification several-fold. RNase added should be DNase-free.
13. Sterile mineral oil.

14. GeneAmp® Thin-Walled Reaction Tubes (0.6 mL), designed for optimal fit in the DNA thermal cycler or DNA cycler 480 sample block.
15. Agarose (Sigma or similar product).
16. Ethidium bromide to visualize DNA
17. TBE buffer: 1 *M* Tris, 0.9 *M* boric acid, 0.01 *M* EDTA, pH 8.3.
18. Loading dye (6× concentration: 0.25% bromophenol blue, 0.25% xylene cyanol FF, and 40% (w/v) sucrose in water, keep at 4°C.
19. Electrophoretic size standards: PCR marker, 50–2000 bp, 123-bp DNA ladder, Supercoiled DNA ladder [2–16 kb], Sigma.

3. Methods

Programming of the thermal cycler for RAPD analysis is described first, followed by the PCR conditions used for the reaction.

3.1. RAPD PCR Programming

Program the thermal cycler (PerkinElmer model 480, which is a thermocycler with an average ramp speed) to denature for 1 min at 94°C, anneal for 2 min at 27°C (*see* **Note 1**), and allow primer extension for 2 min at 72°C for the first five cycles. Then, program for 45 cycles of 1 min denaturation at 94°C, 2 min of annealing at 32°C (*see* **Note 1**), and 2 min primer extension at 72°C. Set the final extension period for 15 min at 72°C. With a thermocycler with a faster ramp speed, like the PerkinElmer model 9600, a shorter protocol can be used, such as 45 cycles of 15 s at 94°C, 30 s at 27°C, and 1 min at 72°C.

3.2. Setting Up the PCR for RAPD Analysis

The following protocol is particularly suitable for RAPD analysis of *C. albicans*. Minor modifications of this method are useful for RAPD analysis of other *Candida* species. The components and procedure of the RAPD reaction mixture are listed as follows:

1. Thaw the PCR buffer, $MgCl_2$ solution, dNTPs, and primer solutions on ice and mix properly before use.
2. Prepare the PCR master mix in GeneAmp Thin-Walled Reaction Tubes containing approximately 50 ng of *C. albicans* genomic DNA (*see* **Note 6**), 5 µL of 10× PCR standard buffer, 200 µ*M* dNTPs, 5 µL of 25 m*M* $MgCl_2$ (*see* **Note 3**), 1 µ*M* of primer, 1.5 U *Taq* polymerase (Sigma, *see* **Note 4**), and make up to 50 µL of final reaction volume with double distilled deionized sterile water.
3. Mix these reagents well by vortexing and then overlay with mineral oil (40 µL) to prevent evaporation and internal condensation. Spin for a short period before amplification until a smooth interface appears between the aqueous and the mineral oil layer.
4. Preparation of the agarose gel is as follows: Separate the amplified products in 2% agarose gel because the molecular weight of the resultant amplicons are in the range of 0.3 to 4.2 kb (agarose has to be dissolved in the same electrophoresis running buffer, generally, 1× TBE. Include 0.5 µg/mL ethidium bromide during the preparation of the gel).
5. Sample preparation for gel loading is as follows: Retrieve amplified products by adding 150 µL of chloroform and pipetting out the aqueous droplet or by placing the reaction mixture over a piece of parafilm. Then, mix well the retrieved aqueous amplification mixture (12.5 µL) with the loading buffer (1.5 µL) and load into wells in an agarose gel together with an appropriate DNA size marker (ranging in size from 50 to 6 Kb) in a standard manner.

5. Electrophorese amplified products under 5 to 10 V/cm for 2 to 2.5 h. When the bromophenol blue travels three quarters of the length of the gel, visualize the gel under ultraviolet light and photograph.

4. Notes

1. Annealing temperature: This is an important parameter that needs optimization in RAPD and depends on primer length and sequence. The melting temperature (Tm) of a primer is proportional to both its length and the G + C content. Tm for primer template can be determined using the formula given below.

$$\text{Tm} = [(\text{number of A+T}) \times 2°\text{C} + (\text{number of G + C}) \times 4°\text{C}]$$

   However, optimal annealing temperature for a primer should be adjusted empirically. Generally, in RAPD, the first few cycles are performed at a low annealing temperature (`5°C below the calculated Tm) and subsequent cycles are performed at a high annealing temperature (`5°C above the calculated Tm).
2. Primer-template ratio: This is one of the important variables in the amplification reaction. Generally, moderate primer:template mass ratios ranging from 0.5 to 5000 are used.
3. Ionic composition: The concentration of ionic components is critical for RAPD. Of these, magnesium is important because different thermostable polymerases have different affinities for magnesium. Generally, the higher the concentration of the magnesium ions, the lower is the specificity, and vice versa. In our hands reproducible fingerprints for *C. albicans* isolates were obtained with magnesium ion concentrations of 2.5 m*M*. It is noteworthy that when DNA is dissolved in TE buffer, the magnesium ion concentration has to be increased (~3 m*M*) to obtain reproducible patterns. This is probably caused by the chelation of magnesium ions by EDTA, which lowers the effective ionic concentration in the reaction mixture. Further, Weaver et al. ***(11)*** reported that excess primer and template DNA could also modulate the activity of magnesium by sequestrating free magnesium ions and thus dampening the amplification reaction. The dNTP concentration also has a direct effect on the magnesium ion concentration in the reaction mixture as a result of the interaction between the mononucleotide and magnesium. Thus, a higher concentration of magnesium ions is necessary for amplifications with a higher concentration of dNTPs ***(12)***. On the contrary, high magnesium ion concentrations can lead to primer–template mismatching and thus decrease amplification stringency. Furthermore, magnesium ions can tightly bond with the sugar backbone of nucleotides and nucleic acids and therefore variation in the magnesium concentration has strong and complex effects on nucleic acid interactions.
4. DNA polymerase: The activity of polymerases is highly variable ***(13)*** and therefore, subtle differences in the specificity of these enzymes can influence the fingerprint profiles, and the multiplex ratio ***(14,15)***. The polymerase activity is regulated to a great degree by the buffer components and, thus, a recommended buffer has to be used for a particular polymerase. We have observed that the variations in the combinations of buffers and polymerases lead to inadequately resolved and incomplete fingerprints. Highly variable results are obtained in particular when different eubacterial DNA polymerases are used in the RAPD technique. On the contrary, *Thermal aquatics* Stoffel fragment is a truncated DNA polymerase that has wide magnesium tolerance and thermal stability and produces well-defined low molecular weight products (less than 500 bp) in general. RAPD with truncated DNA polymerases are known to produce good yields. Generally, *Taq* polymerase concentrations of 1 to 1.25 U/50 µL-reaction is used in RAPD. However, the reaction mixture of more than 2 U/µL can generate nonspecific products ***(16)***.
5. Thermal cycling parameters: These are of critical importance for optimization of the RAPD reaction ***(17)***. Thermal cycling parameters include template denaturation and

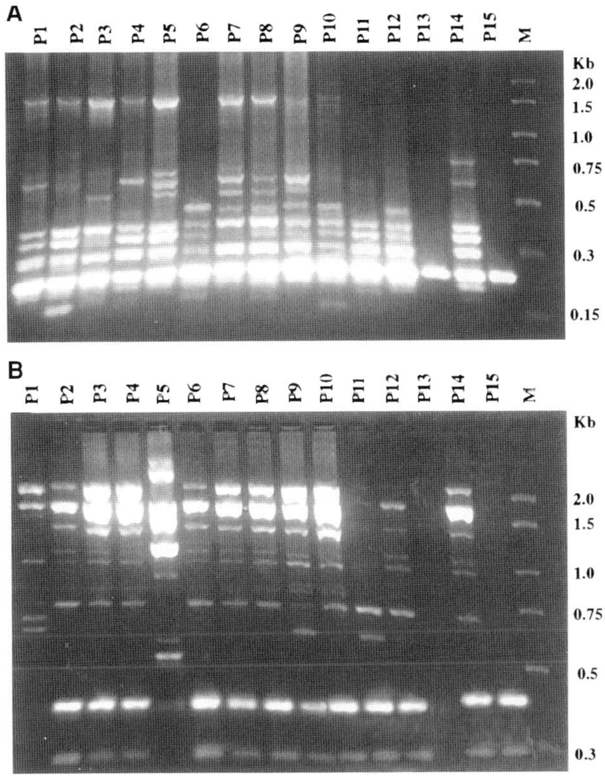

Fig. 1. RAPD fingerprinting of 15 clinical *Candida parapsilosis* isolates (P1 to P10 superficial and P11 to P15 systemic) with primers RSD12 (5′ GCA TAT CAA TAA GCG CAG GAA AAG 3′) (**A**) and RSD6 (5′ GCG ATC CCC A 3′) (**B**) obtained after electrophoretic separation on 1.2% agarose gel. M, PCR marker (Sigma). Sizes of bands indicate the number of base pairs ***(18)***.

annealing temperatures, cycle number and time duration of denaturation, annealing and primer extension period, and the type of thermal cycler used ***(4)***. Different varieties of thermocyclers, with intrinsic inhomogeneneities in rates of cooling and heating, can elicit incongruent fingerprints despite identical program settings and or reaction components. Similarly, subtle changes in the temperature within the same heat block can alter the mode of amplification. Annealing temperature also impacts the quality of the fingerprints produced and their reproducibility. It is noteworthy that the lifetime of a polymerizing agent can be extended considerably by reducing the duration of the denaturation temperature.

6. Template: It is well known that the purity of the template is critical for producing reproducible fingerprinting patterns; thus, DNA devoid of proteins and RNA must be used at all times. For instance, impurities, such as phenol remnants in DNA, can be removed by repeated washing in 70% ethanol. Also DNA isolation methods that lead to degradation of DNA and inhibit the activity of DNA polymerase should be avoided. Furthermore, the RAPD technique cannot be reproducibly used to amplify DNA beyond a minimal threshold (less than 5 ng) or at the other extreme, a very high concentration of template DNA (higher than 1 μg). Such attempts invariably lead to production of either "smears" or poor resolution of the amplicons. In general, 50 to 100 ng of DNA can be used for 50 μL of PCR reaction mixture ***(17)***.

References

1. Williams, J. G., Kubelik, A. R., Livak, K. J., Rafalski, J. A., and Tingey, S. V. (1990) DNA polymorphisms amplified by arbitrary primers are useful as genetic markers. *Nucleic Acids Res.* **18,** 6531–6535.
2. Caetano-Anolles, G., Bassam, B. J., and Gresshoff, P. M. (1992). Primer-template interactions during DNA amplification fingerprinting with single arbitrary oligonucleotides. *Mol. Gene Genet.* **235,** 157–165.
3. Caetano-Anolles, G. and Gresshoff, P. M. (1994). DNA amplification fingerprinting using arbitrary mini-hairpin oligonucleotide primers. *Biotechnology (NY)* **12,** 619–623.
4. Penner, G. A., Bush, A., Wise, R., Kim, W., Domier, L., Kasha, K., et al. (1993) Reproducibility of random amplified polymorphic DNA (RAPD) analysis among laboratories. *PCR Methods Appl.* **2,** 341–345.
5. Kubelik, A. R. and Szabo, L. J. (1995). High-GC primers are useful in RAPD analysis of fungi. *Curr. Genet.* **28,** 384–389.
6. Ellsworth, D. L., Rittenhouse, K. D., and Honeycutt, R. L. (1993) Artifactual variation in randomly amplified polymorphic DNA banding patterns. *BioTechniques* **14,** 214–217.
7. Meunier, J. R. and Grimont, P. A. (1993). Factors affecting reproducibility of random amplified polymorphic DNA fingerprinting. *Res. Microbiol.* **144,** 373–379.
8. Burt, A., Carter, D. A., Koenig, G. L., White, T. J., and Taylor, J. W. (1996) Molecular markers reveal cryptic sex in the human pathogen *Coccidioides immitis*. *Proc. Natl. Acad. Sci. USA* **93,** 770–773.
9. Graser, Y., Volovsek, M., Arrington, J., Schonian, G., Presber, W., Mitchell, T. G., et al. (1996) Molecular markers reveal that population structure of the human pathogen *Candida albicans* exhibits both clonality and recombination. *Proc. Natl. Acad. Sci. USA* **93,** 12,473–12,477.
10. Mondon, P., Brenier, M. P., Symoens, F., Rodriguez, E., Coursange, E., Chaib, F., et al. (1997) Molecular typing of *Aspergillus fumigatus* strains by sequence-specific DNA primer (SSDP) analysis. *FEMS Immunol. Med. Microbiol.* **17,** 95–102.
11. Weaver, K. R., Caetano-Anolles, G., Gresshoff, P. M., and Callahan, L. M. (1994) Isolation and cloning of DNA amplification products from silver-stained polyacrylamide gels. *BioTechniques* **16,** 226–227.
12. Blanchard, M. M., Taillon-Miller, P., Nowotny, P., and Nowotny, V. (1993) PCR buffer optimization with uniform temperature regimen to facilitate automation. *PCR Methods Appl.* **2,** 234–240.
13. Bej, A. K., Mahbubani, M. H., Boyce, M. J., and Atlas, R. M. (1994). Detection of Salmonella spp. in oysters by PCR. *Appl. Environ. Microbiol.* **60,** 368–373.
14. Bassam, B. J., Caetano-Anolles, G., and Gresshoff, P. M. (1992) DNA amplification fingerprinting of bacteria. *Appl. Microbiol. Biotechnol.* **38,** 70–76.
15. Schierwater, B. and Ender, A. (1993) Different thermostable DNA polymerases may amplify different RAPD products. *Nucleic Acids Res.* **21,** 4647–4648.
16. Saiki, R. K. (1989) The design and optimization of the PCR, in *PCR Technology: Principles and Applications for DNA Amplification.* (Erlich, H. A., ed.), Stockton Press, New York, pp. 7–16.
17. Yu, K. and Pauls, K. P. (1992) Optimization of the PCR program for RAPD analysis. *Nucleic Acids Res.* **20,** 2606.
18. Dassanayake, R. S. and Samaranayake, L. P. (2000). Characterization of the genetic diversity in superficial and systemic human isolates of *candida parapsilos* by randomly amplified polymorphic DNA (RAPD). *APMIS* **108,** 153–160.

24

Microsphere-Based Single Nucleotide Polymorphism Genotyping

Marie A. Iannone, J. David Taylor, Jingwen Chen, May-Sung Li, Fei Ye, and Michael P. Weiner

1. Introduction

Single nucleotide polymorphisms (SNPs) are single base differences in genomic DNA *(1)*. These single-base mutations, estimated to occur every 1000 bases, are thought to represent the most common form of genetic variation in the human genome *(2)*. Several million SNPs have been identified *(3)*. High-throughput analysis of these variations will be required to understand their contribution to disease. We outline a method that uses solution-based oligonucleotide ligation assay (OLA) *(4)* or single-base chain extension (SBCE) *(5,6)* for allele discrimination followed by hybridization to fluorescently encoded microspheres. Flow cytometric analysis of the microspheres' fluorescent profile yields rapid and accurate SNP genotyping *(7,8)*.

Allele discrimination by OLA or SBCE uses (1) polymerase chain reaction (PCR)-amplified genomic DNA that encompasses the SNP to be queried ('target' DNA); (2) a synthetic "capture" oligonucleotide probe; and (3) a fluorescent "reporter" (**Fig. 1**). In the allele discrimination reaction, an enzyme is used to covalently couple a reporter molecule to the capture probe in a target-dependent fashion. Each capture probe contains a sequence that is complementary to the target sequence and a ZipCode sequence that will associate the genotype result with a specific microsphere population. In OLA, a DNA ligase covalently couples the fluorescent reporter (a short oligonucleotide) to the capture probe if the capture and reporter probes correctly match the target DNA. In SBCE, a DNA polymerase adds a labeled dideoxynucleotide to the capture probe.

After thermal cycling to amplify the signal on the capture probes, the enzymatically reacted capture probes are incubated with a suspension of up to 100 populations of fluorescently encoded microspheres where each population is uniquely identified by its fluorescent profile *(9–11)*. Each microsphere population is covalently coupled to a different complementary ZipCode (cZipCode) oligonucleotide sequence that associates it with a capture probe and specific SNP allele. Flow cytometric analysis of the microspheres simultaneously identifies both the microsphere type and the fluorescent signal associated with the SNP genotype.

From: *Methods in Molecular Biology, Vol. 226: PCR Protocols, Second Edition*
Edited by: J. M. S. Bartlett and D. Stirling © Humana Press Inc., Totowa, NJ

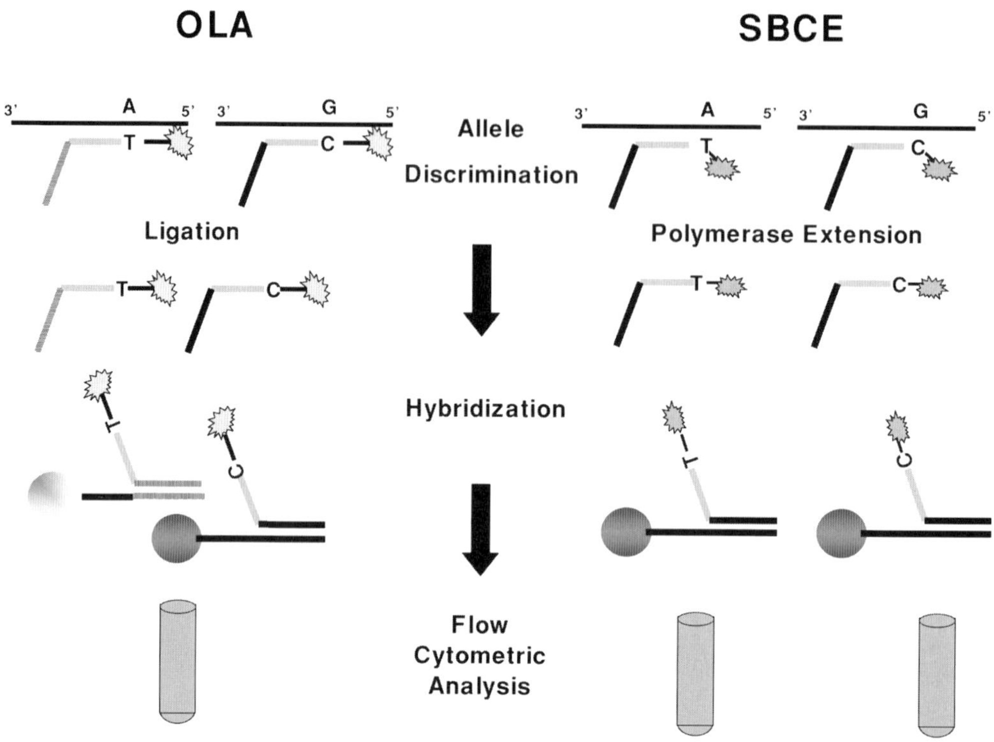

Fig. 1. Diagram of microsphere-based SNP genotyping by OLA or SBCE. Allele discrimination by OLA or SBCE uses PCR-amplified genomic DNA that encompasses the SNP (target DNA), synthetic capture oligonucleotide probe (capture probes extend up to the polymorphic base for OLA and end prior to the polymorphic base for SBCE), and a fluorescent reporter. Each capture probe contains both a sequence that is complementary to the target sequence and a unique 25-base ZipCode sequence that will link the genotyping result to a specific microsphere population. For OLA, the reporter is a short target-complementary oligonucleotide sequence that ends with a fluorescein molecule. If there is base pairing between the reporter and capture probe, DNA ligase will covalently couple the fluorescent reporter to the capture molecule. For the SBCE example shown above, the reporter is a fluorochrome or biotin-coupled ddTTP (or ddUTP) or ddCTP (each labeled ddNTP is used in a different tube). The DNA polymerase extends the capture probe by one base. In each case, the probes are thermally cycled to amplify the signal on the capture probes. A suspension of cZipCode-coupled microsphere populations is added and the capture probes are hybridized to the microspheres through their ZipCode tails. After washing, the fluorescent profiles of the microspheres are analyzed by flow cytometry. The OLA example shows a multiplexed reaction of a single SNP using two microsphere populations, one for each allele. More extensive multiplexing of SNPs can be conducted by combining more probes with complimentary microsphere populations. The SBCE reaction shows two uniplexed reactions (one for each allele). The SBCE reaction may be multiplexed by combining probes for different SNPs and assaying all A or G alleles in a single tube.

The three fluorescent colors associated with each microsphere (two to identify the microsphere population and one for the genotyping result) are determined by flow cytometry. We describe two different flow cytometric systems for SNP genotyping. The first uses a standard bench-top cytometer (FACSCalibur, BD Biosciences, San Jose, CA) with a 488-nm laser excitation source. Sixty-four individual populations of microspheres, manufactured by the Luminex Corporation (Austin, TX), are identified by their orange and red fluorescent profile. Reporter fluorescence is green. The second system uses a microsphere-dedicated flow cytometer, also manufactured by the Luminex Corp., called the LX-100. The two-laser system of the LX-100 uses a red laser (635 nm) to identify the microspheres (red and near infrared emission) and a green laser (532 nm) to excite the reporter fluorochrome (orange emission). One hundred individual microsphere populations are available for this system. The system is available with an XY platform for sampling directly from 96-well microtiter plates.

2. Materials

2.1. SNP Genotyping by OLA with Readout on a FACSCalibur Flow Cytometer

1. 2-[*N*-morpholino] ethenesulfonic acid (MES; Sigma, St. Louis, MO).
2. Microspheres, which are polystyrene beads with a carboxylated surface. Each population of microsphere has a unique profile of red and orange fluorescence (Luminex Corp., Austin, TX). Use 2.5×10^6 microspheres in 62 µL of 0.1 *M* MES (each population in a separate volume).
3. 1-ethyl-3-(3-dimethylaminopropyl) carbodiimide hydrochloride (EDC, Pierce, Rockford, IL).
4. Water containing 0.1% sodium dodecyl sulphate.
5. Water containing 0.02% Tween-20.
6. Tris [hydroxymethyl] aminomethane hydrochloride (10 m*M*)/ethylenediamine-tetraacetic acid (1 m*M*), pH 8.0 (TE).
7. 3.3× SSC: 0.5 *M* NaCl, 0.05 *M* Na Citrate, pH 7.0
8. Template DNA (human genomic DNA, 10–20 ng, lyophilized).
9. Target probe primers. These should be designed to yield 150- to 500-bp products (*see* **Table 1**).
10. AmpliTaq Gold (Applied Biosystems, Foster City, CA), 5 U/µL.
11. 10× PCR buffer I (Applied Biosystems, Foster City, CA). 10× buffer: 500 m*M* KCl, 100 m*M* Tris-HCl (pH 8.3), 15 m*M* $MgCl_2$, 0.01% gelatin (w/v).
12. dNTPs (Applied Biosystems, Foster City, CA); 10 m*M* total (2.5 m*M* of each dNTP).
13. *Taq* DNA ligase (40 U/µL) and 10× ligase buffer (New England BioLabs Inc., Beverly, MA). 10× buffer: 200 m*M* Tris HCl (pH 7.6), 250 m*M* K acetate, 100 m*M* Mg acetate, 100 m*M* dithiothreitol, 10 m*M* NAD, and 1% Triton X-100.
14. Oligonucleotides (*see* **Table 1** and **Notes** section): cZipCodes (Oligos etc, Bethel, ME); reporters (Oligos etc, Bethel, ME, or Biosource/Keystone, Camarillo, CA); captures (Biosource Keystone, Camarillo, CA); and fluorescein-labeled luciferase complement (Biosource Keystone, Camarillo, CA).
15. NaCl (5 *M*).
16. 12 × 75-mm polystyrene test tubes (Becton Dickinson Labware, Franklin Lakes, NJ).
17. FACSCalibur flow cytometer (BD Biosciences, San Jose, CA) equipped with Luminex Lab MAP hardware and software (Luminex Corp., Austin, TX).
18. FlowMetrix™ Calibration Microspheres (Luminex Corp., Austin, TX).
19. Quantum Fluorescence Kit for MESF units of FITC calibration particles (Sigma, St. Louis, MO). Add one drop from each of five separate populations of microspheres (each population has a different known amount of fluorescein molecules) to 1 mL of phosphate-buffered saline.
20. QuickCal software (Sigma, St. Louis, MO).

Table 1
Oligonucleotides

Description	Size (nt)	Modifications	Sequence*,†
SNP 7 amplicon	188	None	GTCCCAAGCT GCATGATTGC TCTTTCTCCT TCTTCCCTGA GTCTCTCTCC ATGCCCCTCA TCTCTTCCTT TTGCCCTCGC CTCTTCCATC CA**Y**GTCTTCC AAGGCCTGAT GCATTCATAA GTTGAAGCCC TCCCCAGATC CCCTTGGAGC CTCTGCCCTC CTCCAGCCCG GATGGCTCTC CTCCATTT
Forward PCR primer for SNP 7	20	None	GTC CCA AGC TGC ATG ATT GC
Reverse PCR primer for SNP 7	20	None	AAA TGG AGG AGA GCC ATC CG
cZipCode 14 with 20 nt **Luciferase tag**	45	5′ amino, C15 spacer	**CAG GCC AAG TAA CTT CTT CG**G GAT TGC ACC GTC AGC ACC ACC GAG
Complement to Luciferase tag	20	5′ biotin or FITC	CGA AGA AGT TAC TTG GCC TG
OLA reporter for SNP 7	14	5′ PO4, 3′ FITC	GTC TTC CAA GGC CT
OLA capture probe for C allele with **ZipCode 14**	50	None	**CTC GGT GGT GCT GAC GGT GCA ATC** CTT TTG CCC TCG CCT CTT CCA TCC Ac
OLA capture probe for T allele with **ZipCode 14**	50	None	**CTC GGT GGT GCT GAC GGT GCA ATC** CTT TTG CCC TCG CCT CTT CCA TCC At
SBCE capture probe with **ZipCode 14**	49	None	**CTC GGT GGT GCT GAC GGT GCA ATC** CTT TTG CCC TCG CCT CTT CCA TCC A

*All sequences are written 5′ to 3′.
†The polymorphic base at the 3′ end of the OLA capture probe sequence is shown in lower case.

2.2. SNP Genotyping by SBCE with Readout on the LX-100 Flow Cytometer

1. Microspheres (as described in **Subheading 2.1.**).
2. Template DNA (as described in **Subheading 2.1.**).
3. Target Probe Primers (as described in **Subheading 2.1.**).
4. Oligonucleotides (*see* **Table 1** and **Notes** section): cZipCodes (as described in **Subheading 2.1.**); captures (as described in **Subheading 2.1.**); and biotin-labeled luciferase complement: (Keystone Biosource, Camarillo, CA).
5. Shrimp alkaline phosphatase (2 U/μL) (Amersham Biosciences, Piscataway, NJ).
6. *Escherichia coli* Exonuclease I (10 U/μL) (Amersham Biosciences, Piscataway, NJ).
7. 2× SBCE reaction mix: (Use 10 μL per reaction) 160 m*M* Tris-HCl (pH 9.0), 4 m*M* $MgCl_2$, 50 n*M* of each capture probe, 2.4 units of AmpliTaq FS (Applied Biosystems, Foster City, CA), 2 μ*M* of the allele-specific biotin-labeled ddNTP (PerkinElmer Life Sciences, Inc., Boston, MA), and 2 μ*M* each of the other three unlabeled ddNTPs (Amersham Biosciences, Piscataway, NJ).
8. 1× SSC containing 0.02% Tween-20.
9. Streptavidin R-phycoerythrin conjugate, 0.1 mg/mL in phosphate-buffered saline, pH 7.2 (SA-PE, Molecular Probes, Eugene, OR).
10. LX-100 flow cytometer (Luminex Corp., Austin, TX), equipped with an XY plate sampler.
11. Instrument calibration particles for the LX-100 (CL1/CL2 and Reporter Calibrator Microspheres; Luminex Corp., Austin TX).

3. Methods

3.1. Coupling of cZipCodes to Microspheres (for Microspheres Analyzed on Either the FACSCalibur or the LX-100)

1. Combine 50 μL of microspheres (2.5×10^6 microspheres in 0.1 *M* MES) with 1 μL of amino-modified cZipCode oligonucleotide (1 m*M* in water).
2. At two separate times, add 10 μL of 30 mg/mL EDC in water to the microsphere mixture (at the beginning of the incubation and then after 30 min).
3. Incubate for 60 min at room temperature with occasional mixing and sonication to keep the microspheres unclumped and in suspension.
4. Add 200 μL of water containing 0.1% SDS. Vortex and centrifuge 5 min at 1100*g*. Carefully remove the supernatant. Add 200 μL of water containing 0.02% Tween 20. Vortex and centrifuge at 1100*g*.
5. Remove the supernatant and resuspend the microspheres in 200 μL of TE and store at 4°C (stable for 6 mo).

3.2. Determination of cZipCode Coupling Efficiency (for Microspheres Analyzed on Either the FACSCalibur or the LX-100)

1. Combine 10,000 coupled microspheres with 3 pmol of fluorescein-labeled luciferase complement (for microspheres to be run on a conventional cytometer) or biotin-labeled luciferase complement (for microspheres to be run on the LX-100) in 0.1 mL of 3.3× SSC. The different microsphere populations may be multiplexed at this point.
2. Heat the microsphere suspension for 2 min at 96°C to denature any secondary structure.
3. Incubate for 30 min at 45°C.
4. Add 200 μL of 1× SSC containing 0.02% Tween-20. Vortex and centrifuge 3 min at 1100*g*. Carefully remove the supernatant and resuspend in 300 μL of 1× SSC containing 0.02% Tween-20.

5. For microspheres to be analyzed on the LX-100, incubate an additional 30 min with 5 μL of SA-PE reagent at room temperature in the dark. Analyze orange fluorescence associated with the microspheres on the LX-100 without washing. Effective coupling reactions analyzed on our LX-100 yield 2000 to 4000 mean fluorescent intensity (MFI) units.
6. For analysis of green fluorescence associated with the microsphere populations, analyze on the FACSCalibur flow cytometer and convert MFI values to molecules equivalent soluble fluorochrome of fluorescein (MESF) (*see* **Subheading 3.7.**). Effective coupling reactions yield >100,000 MESF after background fluorescence contributed by the microspheres alone has been subtracted. The corrected MESF value will determine the number of molecules of cZipCode coupled per microsphere.

3.3. Generation of Target Probes by PCR Amplification (for Either OLA or SBCE)

1. In a Polyfiltronics 96-well plate, amplify 10 to 20 ng of genomic DNA per well in 15- to 30-μL reaction volumes. Each PCR reaction should contain 1.5 units of AmpliTaq Gold, 400 μ*M* dNTPs, 200 μ*M* forward PCR primer, and 200 μ*M* reverse PCR primer in 1× PCR buffer I (Applied Biosystems, Foster City, CA). PCR amplifications are uniplexed.
2. Set the thermal cycler to heat 10 min at 95°C min to activate the DNA polymerase, followed by 40 three-temperature amplification cycles holding at 94, 60, and 72°C for 30 s each and ending with an additional 5-min extension at 72°C.
3. Hold samples at 4°C following completion of the reaction.

3.4. OLA

1. Incubate the following in a total volume of 10 μL: 1× ligase buffer, 0.1 pmol of each capture oligonucleotide, 5 pmol of each reporter oligonucleotide, 3 to 20 ng of each dsDNA target probe (as determined by Picogreen™ staining, Molecular Probes, Eugene, OR), and 10 U *Taq* DNA ligase. Ligation reactions are multiplexed.
2. Incubate in a thermal cycler by heating to 96°C for 2 min, followed by 30 cycles of a two-step reaction (denaturation at 94°C for 15 s followed by ligation at 37°C for 1 min).
3. Hold samples at 4°C when the cycles are complete.

3.5. Hybridization of Capture Probes to Microspheres after OLA

1. Add cZipCode-coupled microsphere populations (5000 microspheres of each population) to the ligation reaction (microsphere populations are combined at this step).
2. Adjust the salt concentration to 500 m*M* NaCl by adding a small volume of 5 *M* NaCl.
3. Heat the mixture to 96°C for 2 min in a thermal cycler and incubate at 45°C from 2 h to overnight.
4. Wash microspheres with 200 μL of 1× SSC containing 0.02% Tween-20 by centrifuging at 1100*g* for 5 min.
5. Resuspend the microsphere suspensions in 300 μL of 1× SSC containing 0.02% Tween-20 just before flow cytometric analysis.

3.6. Flow Cytometric Analysis of Microspheres Hybridized to OLA Products on the FACSCalibur

1. Transfer the microsphere suspensions to 12 × 75-mm polystyrene test tubes.
2. Optimize the settings on the flow cytometer to analyze the microspheres using FlowMetrix™ Calibration Microspheres in conjunction with Luminex software.
3. Acquire a minimum of 100 microspheres from each population per tube.
4. Analyze a separate tube of calibration particles (Quantum Fluorescence Kit for MESF units of FITC) using the same instrument settings.

3.7. Analysis of FACSCalibur Data

1. Convert all green fluorescence measurements from MFI to MESF by either manually creating a calibration curve from the fluorescent intensities of the calibration particles or by using QuickCal software.
2. Adjust raw MESF values by subtracting the microsphere alone control MESF values to eliminate microsphere-contributed background fluorescence. This is necessary, because each microsphere population in the 64-microsphere set emits a small amount of fluorescence in the green reporter channel. Microsphere corrections range from 1000 to 30,000 MESF, depending on the microsphere population.
3. To adjust for tube-to-tube variability in the wash step, one may include a microsphere population with no cZipCode attached. Subtract the adjusted MESF (as described previously) of this negative control microsphere from the MESF of every microsphere type in that particular tube to normalize the data.
4. Merge the data from the two corresponding alleles and graph the results as x-y coordinates.

3.8. SBCE

1. PCR amplification of target probes is performed as described in **Subheading 3.3.**
2. To degrade the excess PCR primers and dNTPs before the SBCE assay, add 1 unit of shrimp alkaline phosphatase and 2 units of *E. coli* Exonuclease I directly to 10 µL of pooled PCR products (10 to 20 ng of each amplicon) and mix thoroughly.
3. Incubate at 37°C for 30 min and then for 15 min at 80°C to inactivate the enzymes.
4. For each SNP, set up similar reactions differing only by the choice of labeled ddNTP. Add 10 µL of pooled, treated PCR products to 10 µL of SBCE reaction mix (160 m*M* Tris-HCl, pH 9.0; 4 m*M* $MgCl_2$; 50 n*M* of each capture probe; 2.4 units of AmpliTaq FS; 2 µ*M* of the allele-specific biotin-labeled ddNTP; and 2 µ*M* each of the other three unlabeled ddNTPs).
5. Denature the reactions at 96°C for 2 min and follow with 30 amplification cycles at 94°C for 30 s, 55°C for 30 s, and 72°C for 30 s.
6. Hold reactions at 4°C.

3.9. Hybridization of Capture Probes to Microspheres after SBCE

1. To the wells of a standard 96-well microtiter plate in a total volume of 30 µL, add 1000 of each cZipCode-coupled microsphere population to 20 µL of SBCE reaction mixture. Adjust salt concentration to 500 m*M* NaCl and 13 m*M* EDTA.
2. Incubate the mixture at 40°C for 1 h.
3. Wash the microspheres with 150 µL of 1× SSC containing 0.02% Tween 20.
4. Centrifuge for 5 min at 1100*g* and remove the supernatants.
5. Resuspend microspheres in 60 µL of 1× SSC containing 0.02% Tween 20.
6. Add 5 µL of SA-PE to the microsphere-hybridized SBCE reaction products.
7. Incubate the mixture for 30 min at room temperature.

3.10. Flow Cytometric Analysis on the LX-100 and Data Analysis

1. Optimize the settings on the LX-100 to analyze the microspheres using Luminex calibration particles in conjunction with Luminex software.
2. For each microtiter well, analyze a minimum of 30 microspheres of each population.
3. Adjust the raw MFI values for microsphere background fluorescence by subtracting microsphere alone control MFI values from the MFI value of each corresponding microsphere sample. This is necessary because each microsphere population in the 100-microsphere set emits a small amount of fluorescence into the orange reporter channel. Microsphere corrections range from 1 to 100 MFI, depending upon the microsphere population.

4. To adjust for well-to-well variability in the wash step, each well may contain a microsphere population with no cZipCode attached. Subtract the adjusted MFI (as described above) of this negative control microsphere from the MFI of every microsphere type in that particular well.
5. Merge the data from the two corresponding alleles and graph the results as x-y coordinates.

4. Notes

1. Oligonucleotide probes for OLA: (1) cZipCodes. We have designed our 58 different cZipCode sequences to include: (a) 5′ amine group, an 18-atom spacer $(CH_3CH_2O)_6$ to minimize any potential interactions between the oligonucleotide sequence and the microsphere surface; (b) a common 20-base sequence from luciferase cDNA (5′-CAG GCC AAG TAA CTT CTT CG-3′) to test for oligonucleotide coupling efficiency; and (c) a 25-base, non-crossreacting cZipCode sequence derived from the *Mycobacterium tuberculosis* genome ***(7,8)*** to link each allele of SNP to a particular microsphere population. The cZipCode sequences have GC-contents between 56 and 72% and predicted Tm values of 61 to 68°C. Although this chapter does not outline the methodology, biotinylated cZipCodes may be coupled to Lumavidin-coated microspheres. (2) Reporters. These oligonucleotides are designed to hybridize to the target sequence immediately downstream of the capture probe. They are generally 8 to 20 bases in length and contain a 5′ phosphate group and a 3′ fluorescein modification. The phosphate group is required as a substrate for the ligase enzyme. Reporter probe Tm range from 36 to 40°C. (3) Captures. The 5′ end of each capture probe contains a 25-nucleotide ZipCode sequence and the 3′ end contains a 20 to 25 base target-specific sequence that extends to the polymorphic base. This orientation insures optimal ligase fidelity ***(12,13)***. Because each SNP has a minimum of two polymorphic bases, each SNP will require at least two different OLA capture probes. If the alleles are assayed in the same reaction volume, these capture probes will also require different ZipCode sequences. Target-specific capture sequences have a Tm of 51 to 56°C. (4) Fluorescein-labeled luciferase complement. A 5′ fluoresceinated oligo complementary to the 20 nucleotides of luciferase sequence in each cZipCode.
2. We have found that only 58 of the 64 red–orange populations Luminex provides could be run on our FACSCalibur with the Luminex software. The number of useable populations of microspheres may vary depending upon the cytometer model being used. We have used all 100 microsphere populations on the LX-100.
3. We have successfully multiplexed over 50 SBCE reactions.
4. Oligonucleotide probes for SBCE: (1) cZipCodes (same as OLA); (2) Captures. Similar to OLA captures with the exception that target-specific component of SBCE capture probes is designed to stop just short of the polymorphic base. For SBCE, each SNP will require only one capture probe. Different alleles are assayed in different reaction volumes. Each volume will have a different biotin-labeled ddNTP plus the other three unlabeled ddNTPs. (3) Biotin-labeled luciferase complement. A 5′ biotin-modified oligo complementary to the 20 nucleotides of luciferase sequence found in each cZipCode.
5. **Figure 2** shows representative SNP genotyping results from approx 100 patients for 6 A/G SNPs analyzed by SBCE on the LX-100. Homozygous and heterozygous clusters are readily discernible.

Fig. 2. *(see facing page)* SNP genotyping by SBCE with analysis on the LX-100. SNP genotyping results from approx 96 patients for 6 A/G SNPs. Biotinylated ddNTPs were used (one plate for G alleles, one for A alleles) followed by incubation with SA-PE. The results shown come from an experiment where 35 SNPs were genotyped in two microtiter plates. The points in the lower left corner represent either negative controls or failed PCR reactions.

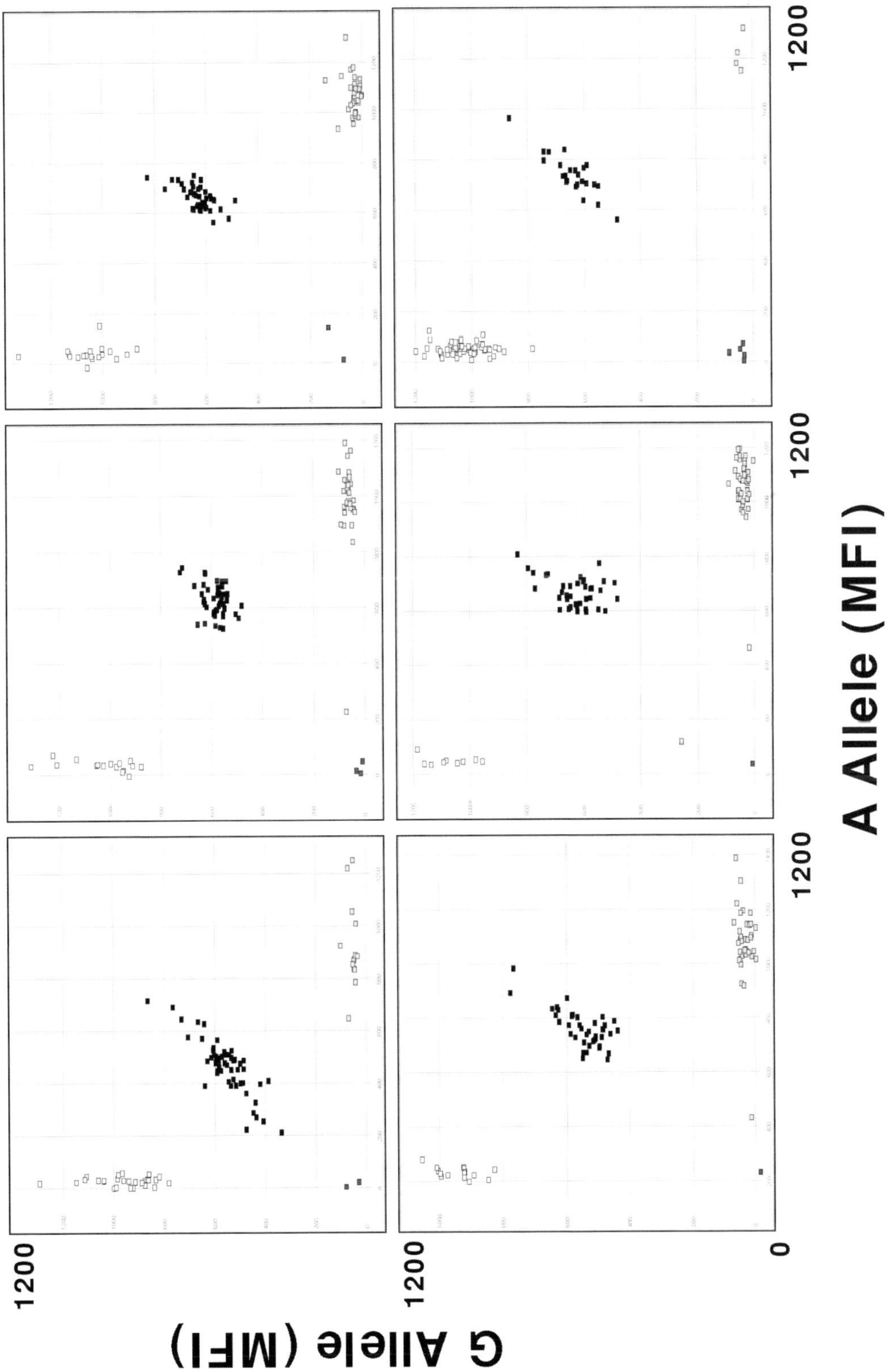
G Allele (MFI)
1200
1200
0
1200
1200
1200
A Allele (MFI)

6. Target concentration has a direct impact on signal intensity. Each panel in **Fig. 2** includes one negative control point. Additional points shown in the lower left are most likely the result of PCR failure.
7. High salt concentrations help hybridization efficiency. However, we have found that when analyzing samples on the LX-100, salt concentrations ≥400 m*M* result in optical disturbances. This may be caused by the different refractive indices of the high ionic strength core stream and the low ionic strength sheath fluid ***(14)***.
8. We have reserved one microsphere population and its associated cZipCode for use with an SBCE positive control. Each SBCE reaction includes a short (40mer) synthetic oligonucleotide target with a 4-fold degenerate position near its center and a complementary capture probe to insure incorporation of reporter signal for any nucleotide assayed. Use of this positive control in every SBCE assay well provides an excellent internal standard to assess well-to-well reaction variability and can be used to normalize signal intensities generated across a plate of 96 samples ***(14)***.
9. We have also used Rhodamine-6G (R6G)-labeled ddNTPs with the SBCE system on the LX-100. Although using this directly coupled fluorochrome saves an additional wash and incubation step, the fluorescent intensities were not as bright as PE. This is not unexpected since the quantum efficiency of PE is much better than R6G.
10. Genotyping by OLA permits allele multiplexing and has no requirement for shrimp alkaline phosphatase and Exonuclease I pretreatment of PCR target probes. The advantage of SBCE is that only one capture probe is required for each allele, thus saving on costs. The accuracy of genotyping using OLA or SBCE is >99% ***(7,8)***.
11. The inclusion of the detergent Tween 20 in the buffer helps reduce the loss of microspheres during wash steps.
12. Because the LX-100 uses a dual-laser system for microsphere identification and reporter measurement, spectral compensation is not required on the LX-100 as on the FACSCalibur.
13. Our current automated genotyping facility uses SBCE with 52 microsphere populations. Enhanced throughput may be realized by incorporating multiplexed PCR amplifications.
14. Conversion of MFI to MESF, although not required, offers several advantages. These advantages include (1) the use of a standard fluorescence unit; (2) the ability to compare data between experiments and instruments; and (3) normalization of signal variability in an instrument over time (caused by laser power shifts or PMT decline). Although not outlined in this chapter, data from the LX-100 may also be converted to MESF using calibration microspheres for PE. We have used QuantiBRITE PE beads from BD Biosciences for this purpose.
15. Although this chapter focuses on the use of Luminex microspheres for multiplexed analysis, there are other products available for developing multiplexed assays using a conventional bench-top flow cytometer such as the FACSCalibur (*see* http://www.spherotech.com, http://www.bangslabs.com). These products use one fluorescent parameter for the identification of the microsphere populations and have a more limited level of multiplexing. For manual acquisition of multiplexed assays on a conventional bench-top cytometer, (1) make sure microspheres can be seen in forward versus side scatter and gate to exclude doublets; (2) set the PMTs of fluorescent parameters to log acquisition; (3) adjust the PMT settings of fluorescent parameters so that unstained and brightly stained microspheres are not off-scale; and (4) adjust compensation settings to subtract reporter fluorescence from the other fluorescent channels. When properly compensated, microsphere populations with bright reporter staining will not move out of the regions that identify their population.
16. Microsphere-based SNP genotyping by OLA or SBCE is an accurate and rapid method for the analysis of SNPs.

References

1. Landegren, U., Nilsson, M., and Kwok, P. Y. (1998) Reading bits of genetic information: Methods for single-nucleotide polymorphism analysis. *Genome Res.* **8,** 769–776.
2. Cooper, D. N., Smith, B. A., Cooke, H. J., Niemann, S., and Schmidtke, J. (1985) An estimate of unique DNA sequence heterozygosity in the human genome. *Hum. Genet.* **69,** 201–205.
3. Marshall, E. (1999) Drug firms to create public database of genetic mutations. *Science* **284,** 406–407.
4. Abravaya, K., Carrino, J. J., Muldoon, S., and Lee, H. H. (1995) Detection of point mutations with a modified ligase chain reaction (Gap-LCR). *Nucleic Acids Res.* **23,** 675–682.
5. Syvanen, A. C. (1999) From gels to chips: "Minisequencing" primer extension for analysis of point mutations and single nucleotide polymorphisms. *Hum. Mutat.* **13,** 1–10.
6. Syvanen, A. C., Aalto-Setala, K., Harju. L., Kontula, K., and Soderlund, H. (1990) A primer-guided nucleotide incorporation assay in the genotyping of apolipoprotein E. *Genomics* **8,** 684–692.
7. Iannone, M. A., Taylor, J. D., Chen, J., Li, M.-S., Rivers, P., Slentz-Kesler, K. A., et al. (2000) Multiplexed single nucleotide polymorphism genotyping by oligonucleotide ligation and flow cytometry. *Cytometry* **39,** 131–140.
8. Chen, J., Iannone, M. A., Li, M-S., Taylor, J. D., Rivers, P., Nelson, A. J., et al. (2000) A micropshere-based assay for multiplexed single nucleotide polymorphism analysis using single base chain extension. *Genome Res.* **10,** 549–557.
9. Fulton, R. J., McDade, R. L., Smith, P. L., Kienker, J. J., and Kettman, J. R. (1997) Advanced multiplexed analysis with the FlowMetrix™ system. *Clin. Chem.* **43,** 1749–1756.
10. Kettman, J. R., Davies, T., Chandler, D., Oliver, K. G., and Fulton, R. J. (1998) Classification and properties of 64 multiplexed microsphere sets. *Cytometry* **33,** 234–243.
11. McDade, R. L. and Fulton, R. L. (1997) True multiplexed analysis by computer-enhanced flow cytometry. *Med. Dev. Diag. Indust.* **19,** 75–82.
12. Husain, I., Tomkinson, A. E., Burkhart, W. A., Moyer, M. B., Ramos, W., Mackey, Z. B., et al. (1995) Purification and characterization of DNA ligase III from bovine testes. Homology with DNA ligase II and vaccinia DNA ligase. *J. Biol. Chem.* **270,** 9683–9690.
13. Luo, J., Bergstrom, D. E., and Barany, F. (1996) Improving the fidelity of Thermus thermophilus DNA ligase. *Nucleic Acids Res.* **24,** 3071–3078.
14. Taylor, J. D., Briley, D., Nguyen, Q., Long K., Iannine, M. A., Li, M. S., et al. (2001) Flow cytometric platform for high-throughput single nucleotide polymorphism analysis. *Bioechniques* **30,** 661–669.

25

Ligase Chain Reaction

William H. Benjamin, Jr., Kim R. Smith, and Ken B. Waites

1. Introduction

The ligase chain reaction (LCR) is one of many techniques developed in recent years to detect specific nucleic acid sequences by amplification of nucleic acid targets. The LCR has been used for genotyping studies to detect tumors and identify the presence of specific genetic disorders such as sickle cell disease caused by known nucleotide changes that occur as a result of point mutations and has now become widely used in infectious disease detection, both in the diagnostic and research settings, primarily focusing on infections caused by microbes that have proven difficult to detect by traditional culture techniques. The LCR is now recognized as the method of choice for detection of urogenital infections due to *Chlamydia trachomatis* because of its greater sensitivity as compared to traditional cell culture or nonamplified DNA probes or antigen-detection assays. Other uses of the LCR have also been reported ***(1–8)***. When used for detection of infectious diseases, amplification tests such as the LCR have the additional advantages in that they do not require viable organisms in a specimen, a single specimen can be used to detect multiple different pathogens, provided suitable primers are available, and easily obtained specimens such as urine can be used for diagnostic purposes, making screening of large numbers of persons practical, as well as facilitating research to better understand the epidemiology of specific diseases.

Ligation of adjacent oligonucleotides while hybridized to a template was first investigated by Besmer et al. in 1972 ***(9)***. The first use of ligases to join oligonucleotides as a means to differentiate sequence variants was reported in 1988 ***(10,11)*** and the first "real" LCR utilizing cyclical denaturation hybridization and ligation of two pairs of oligomers was described in 1989 ***(12)***. Thermostable ligases that eliminate the necessity of adding ligase after each denaturation step were introduced by Barany in 1991 ***(1)***.

Simple LCR consists of two complementary oligonucleotide pairs (four oligonucleotides 20–35 nucleotides each) that are homologous to adjacent sequences on the target DNA, as opposed to two used in the polymerase chain reaction (PCR) assay. The adjacent pairs are ligated when they hybridize to the complementary sequence next to each other in a 3′ to 5′ orientation on the same strand of the target DNA. The 5′ nucleotide of the ends of the primers to be ligated must be phosphorylated. Newly ligated oligonucleotides become targets in subsequent cycles so logarithmic

From: *Methods in Molecular Biology, Vol. 226: PCR Protocols, Second Edition*
Edited by: J. M. S. Bartlett and D. Stirling © Humana Press Inc., Totowa, NJ

amplification occurs. The two complementary pairs can also, at low frequency, be blunt end ligated to each other and serve as template for amplification even though no target sequence was present in the original sample. Early work showed at least 10-fold greater efficiency of ligation of perfectly matched compared to a mismatched nucleotide at the ligation junction ***(1,8,12)***.

LCR-based systems have some advantages over the PCR-based amplification systems. Because there is no newly synthesized DNA, misincorporated nucleotides are not replicated in the product allowing amplification of a different sequence than that found in the target nucleic acid. The LCR reactions are also more specific for the 3′ nucleotide allowing for higher discriminatory power against mismatches at a single chosen site ***(1,3)***. Thus, LCR is very useful for determining the nucleotide at a specific site such as a single base change, e.g., single-nucleotide polymorphisms (SNPs) used in mapping complex genomes. The LCR cycle has only two short steps allowing for shorter amplification times. The usually small target of LCR, 36 to 60 nucleotides, does not require high-quality large fragment nucleic acids ***(13)***. The commercial LCR kit, the Abbott LCx System (Abbott Diagnostics, Abbott Park, IL, USA) is less affected by inhibitors in some specimens, such as fresh urines compared to PCR ***(14)***. However, the LCR is still subject to contamination that is at least as much of a problem as for the PCR.

There are several modifications to the simple LCR that have been used to overcome some of the problems that are inherent in this method. Owing to the specificity of the ligase reaction for perfect matches that is complicated by the blunt end ligation of double stranded probes PCR, the Ligase Detection Reaction (LDR) was developed ***(15)***. The amplification step for LDR is PCR using outside primers, then only one pair of adjacent primers is used to detect the proper sequence in the PCR product ***(1,7)***. Because there are no double stranded oligonucleotides to blunt end ligate, the background is very low and the linear detection of the PCR amplified sequence can be detected using ligation of oligonucleotides. Gap LCR utilizes a few base pair gaps between the two pair of oliogonucleotides on the primer, thus requiring a thermostable polymerase (*Taq* polymerase) to fill in the gap before ligation of adjacent primers can occur. The Gap LCR prevents amplification of the blunt end ligated primers because the oligonucleotide pairs cannot be amplified on blunt end ligated products ***(16,17)***.

1.1. Ligase

There are four thermostable ligases that can be used in LCR reactions. *Taq* ligase was the first described and is commercially available from New England Biolabs (www.uk.neb.com/neb/index.html) (Beverly, MA). Stratagene (www.stratagene.com) (La Jolla, CA) produces *Pfu* which is advertised as having higher ligation specificity and lower background than *Tth* that is marketed by Abbott Diagnostics (www.abbottdiagnostics.com) (Abbott Park, IL) and is used in their LCx diagnostic kit. Amplicase is sold by Epicenter Technologies (www.epicentre.com) (Madison, WI).

1.2. Probe Design

The most successful probes are 18 to 30 bases in length. Probes with high GC content or those that form stable secondary structures or dimers should be avoided if possible. It is difficult to predict activity, so empiric testing is still necessary ***(18)***. The 5′ nucleotide of the adjacent end must be phosphorylated to promote ligation.

The probe candidates must be tested for sensitivity by determining the lower level of detection of samples containing organisms of interest or their DNA. Specificity must be determined by testing related organisms, as well as other organisms likely to be found in specimens to be tested.

1.3. Target Sequence Selection

The sequence should be unique to the organism to be detected and multiple copies per organism increases sensitivity. If the ligase reaction is to be used to discriminate between single base pairs, the ultimate 3′ base is most sensitive, whereas the 5′ base is also relatively sensitive. Allele-specific PCRs used to detect single nucleotide polymorphisms are often not discriminating enough to differentiate between SNPs, whereas the ligase reactions are more discriminating against mismatches, especially on either side of the ligation site ***(1)***. All of the possible mismatches are discriminated against, but G-T and T-T less well, 1.5% as efficient as the matching nucleotide, while other mismatches were <0.4% using *Taq* ligase ***(1)***. All possible combinations of single base mismatches both on the 5′ and 3′ side of the ligation site have been tested for fidelity with *Tth* ***(3)***. T-G and G-T were less efficiently discriminated against (~2%) when on the 3′ side, whereas all perfect matches were >80%. When the mismatch was on the 5′ side the previous two and A-C, A-A, C-A, G-A and T-T, all were all less discriminated against than when on the 3′ side, same patterns as were found with T4 ligase. Intentionally introducing a mismatch in the third site from the 3′ end of the probes increased the discriminating power. Nucleotide analogs in the probes in the 2 and 3 location from the 3′ end also increased discriminatory power. Site-directed mutagenesis was used on *Tth* and mutants that increased discriminating power 4- and 11-fold were found by Luo et al. ***(3)***.

1.4. Detection of Amplification Products

Multiplex LCR using a mixture of probes differing by the 3′ nucleotide involved in the ligation, that are labeled by being one or two extra bases on the nonligating 5′ end, allows polyacrylamide gel electrophoresis (PAGE) differentiation of the one or two base changes by differences in migration in PAGE ***(1)***. The probes can also be labeled with several different, easily detected labels such as: 32P, fluorescent labels, immunologically detectable haptens (digoxigenen), (Roche Molecular Diagnostics, Indianapolis, IN) and after amplification and electrophoresis the signal is detected by Southern blot to determine if ligation has taken place. Ligated products can also be detected by having immuno-capture of one of the probe ends and after washing, detecting the second probe with an enzyme conjugate. Only ligated product will be captured and also have the end with the ligand for which the enzyme conjugated antibody is specific, IMx (Abbott) utilizes this method. The latter method is also employed in the commercially sold Abbott LCx for *C. trachomatis* and *N. gonorrhoeae*. The comparison of eight different nonradioactive methods of detecting the LCR products was reported by Winn-Deen ***(19)***.

1.5. Contamination Control

One problem with LCR is that the target is amplified, resulting in a contamination risk. The method commonly used to inactivate PCR products does not work because of the small size of the amplicon in LCR. The potential for contamination requires strict adherence to physical separation of setup and detection areas and other containment methods such as bleach treatment of lab benches, Ultraviolet irradiation of setup areas, unit premixes for setup, as well as use of aerosol barrier pipet tips. The commercial LCx instrument injects a binary inactivating agent that is capable of inactivating small amplicons by a factor of up to 109.

1.6. Inhibitors

LCR is less prone to problems with inhibitors from urogenital specimens than PCR for detection of *Chlamydia* ***(14,20)***. Freeze-thawing and dilution decreased the false-negative rate of PCR ***(21)***. Possible inhibitors of LCR were removed by a mildly acid wash to remove CaHPO4 from concentrated specimens being tested for the presence of acid fast bacteria ***(22)***. Potential effect of inhibitors on amplification results mandates rigorous quality control for all steps of the procedure as described in the technical procedures presented here.

1.7. LCR Applications

1.7.1. Noncommercialized Methods

Over the past few years, numerous publications have appeared that describe a number of potential applications of the LCR procedure, each with its own particular modifications of the basic procedure, using the varied available ligases, probe designs, relative positions, and detection methods. Table 1 shows references and lists the methods used for several diverse applications. A basic LCR procedure that can be adapted to detect nucleic acid of a variety of etiologic agents as well as eukaryotic polymorphisms, especially SNPs, is presented in further detail in **Subheading 2.**

1.7.2. Commercially Available LCR Kits*

As an artifact of the way the companies interested in molecular diagnostics have carved up the field, mostly driven by the ownership of rights to specific procedures, in the United States, LCR is primarily used to diagnose the sexually transmitted diseases caused by *C. trachomatis* and *N. gonorrhoeae.* Abbott Diagnostics markets a commercial LCR kit, the LCx, in which four oligonucleotide probes target and hybridize with a specific complementary single-stranded nucleotide sequence within the multicopy cryptic plasmid gene present in all serovars of *C. trachomatis* that is exposed during sample preparation in which the heating process causes the release of single-stranded DNA, leaving a gap of a few nucleotides between the probes ***(23)***. Polymerase then fills the gap with nucleotides in the LCR reaction mixture. Once the gap is filled, thermostable ligase covalently joins the pair of probes to form an amplification product that is complementary to the original target sequence that then serves as an additional target sequence for further rounds of amplification. Amplification occurs when the LCR reaction mixture and sample are incubated in a thermal cycler. During thermal cycling, the temperature is raised above the melting point of

*Abbott Diagnostics, which marketed the LCx Uriprobe that was FDA approved in 1994, is to be discontinued in June 2003 and after that time no commercial kits will use the LCR technology.

the hybridized amplification product causing it to dissociate from the original target sequence. Lowering the temperature allows more of the oligonucleotide probes to hybridize to the targets now available and to be ligated. The temperature continues to be cycled until sufficient numbers of target amplification product have accumulated. Ligated product is captured by antibody immobilized onto the surface of microparticles using a ligand attached to the end of one primer and then detected by an enzyme-conjugated antibody directed at a second reporter molecule at the distal end of the other primer. Only ligated product with both haptens covalently attached will generate a chemiluminescent signal. The amplification product accumulates exponentially and is detected by chemiluminescence on the automated Abbott LCx or IMx Analyzers. Specimens that can be used include cervical swabs and urethral swabs as well as urine from either men or women ***(24–26)***. The LCx kit also has the advantage of being multiplexed for the detection of *N. gonorrhoeae* in the same genitourinary specimen. The LCx target in *N. gonorrhoeae* is a 48 base pair DNA sequence in the multicopy *Opa* gene that is conserved in all strains studied to date and is specific to *N. gonorrhoeae.*

There are mixed results on the sensitivity of LCx compared to other methods of amplification for detection of urogenital infections. One report found PCR more sensitive than LCx. ***(27)***, whereas a second study found LCx more sensitive ***(28)***. Stary et al. compared the sensitivity and specificity of LCx and the Transcription Mediated Amplification (TMA) assay (GenProbe, Inc., San Diego, CA) using several different urogenital specimens. They found comparable results with endocervical and vulval swab samples, male urethral swabs and urine, but found a lower sensitivity with the TMA method for female first void urine ***(29)***. Carroll et al. found the overall sensitivity of LCx was better than GenProbe PACE 2 for chlamydia with very good specificity for both ***(30)***.

Abbott Diagnostics also makes a *Mycobacterium (M.) tuberculosis* LCx kit that is approved for diagnostic use in some European countries but not in the United States at present. This product is a Gap LCR assay with a few nucleotides that must be filled before ligation. It targets the protein antigen b genes of *M. tuberculosis*. This LCx kit has been evaluated in comparison to culture and/or other amplification methods such as the Roche PCR and the TMA with generally favorable results. Overall, most evaluations have found the LCx to perform better for detection of *M. tuberculosis* in smear-positive as opposed to smear-negative respiratory specimens ***(31–34)***. However, one investigation ***(35)*** found similar high sensitivity and specificity values for the LCx, exceeding 90% for smear-positive as well as smear-negative specimens, suggesting that the LCx assay has potential utility as a screening test for the rapid diagnosis of tuberculosis in high-risk patients.

2. Materials

2.1. Equipment

1. Microcentrifuge capable of speeds of ≥9000*g*.
2. Vortex mixer.
3. 20°C freezer for sample storage if not processed immediately.
4. LCx Analyzer or other detection system with appropriately labeled reagents specific for the system employed.
5. LCx Thermocycler.
6. LCx dry bath capable of heating from 60° to 100°C.

2.2. Supplies and Reagents

1. LCx kit (if commercial assay is being used, otherwise individual components as listed in **Table 1** and **Steps 2–9**).
2. *Taq* ligase or other type as described in **Subheading 1.1.** with appropriate buffers from same supplier.
3. Custom probes with primers chosen according to criteria outlined in **Subheadings 1.2.** and **1.3.**
4. Proteinase K (final concentration 200 μg/ml).
5. Mineral oil.
6. Specimen or analyte.
7. Specimen Collection and Transport System (Uriprobe) containing 0.5 ml transport buffer.
8. Sterile, preservative-free plastic screw-top containers for collection and transport of specimen (if using commercial LCx).
9. 100 μL aerosol barrier pipet tips and pipets.

3. Methods

3.1. Procedural Precautions

1. Work in laboratory using DNA amplification methods should always flow in a one-way direction beginning in the specimen preparation and processing area (Area 1), then moving to the amplification and detection area (Area 2). Do not bring any materials or equipment from Area 2 into Area 1.
2. Surface cleaning using a 1% (v/v) sodium hypochlorite solution followed by 70% (v/v) ethanol should be performed on bench tops and pipets prior to beginning the LCR Assay.
3. Chlorine solutions may pit equipment and metal. Use sufficient amounts or repeated applications of 70% ethanol until chlorine residue is no longer visible.

3.2. Prototype LCR Technique for Detection of Neisseria gonorrhoeae

The type of materials required for LCR assays varies greatly according to the ligase, amplification, and detection systems used, as well as the primers required that must be specific for the desired target. **Table 1** summarizes major types of LCR procedures and the various materials and equipment needed to perform the assays, including primers, ligases, buffers, templates, test conditions, targets, and detection systems. A more detailed description of the *Neisseria gonorrhoeae* detection LCR that was eventually modified and incorporated into the Abbott LCx system is described in **Subheadings 3.2.1.–3.3.9.** These methods were originally developed by Birkenmeyer and Armstrong and described in 1992 ***(36)***. Although this procedure was developed specifically for the detection of *N. gonorrhoeae* DNA, the basic principle can be adapted for the detection of other microbial agents, provided specific oligonucleotide primers are designed.

3.2.1. Primer Selection

1. Primers should be designed to be homologous to conserved sequences in *N. gonorrhoeae* that have mismatches with *N. meningitidis*, the species most homologous to *N. gonorrhoeae*. Because of the empiric nature of developing LCR, multiple primer sites should be developed. The site demonstrating the best sensitivity and specificity should be chosen and this must be validated as described previously in **Subheading 1.2.** In the example given, three sites were chosen corresponding to sequences immediately upstream of several of the *opa* gene family (*Opa-2* and *Opa-3*) and a site downstream of several of the *pil* gene family *(Pilin-2)* ***(36)***.

Table 1
Examples of Various Applications and Methods for the Ligase Chain Reaction

References	Barany (*1*) Wilson (*37*)	Jurinke (*2*)	Khanna (*5*) Luo (*3*) Zirvi (*6*)	Abravaya (*38*)	Day (*7*)	Reyes (*8*)
Ligase	*Taq* 15 U	*Pfu* 4 U	*Tth* 25 fmol	*Tth* 5000 U	*Tth* 10 U	Amplicase 1.5 U
Buffers	20 m*M*/7.6	20 m*M*/7.5	20 m*M*/7.6		10 m*M*/7.5	20 m*M*/8.3
Tris						
EPPS				50 m*M*/7.8		
KCl	100 m*M*	20 m*M*	100 m*M*	20 m*M*	50 m*M*	25 m*M*
$MgCl_2$	10 m*M*	10 m*M*	10 m*M*	30 m*M*	10 m*M*	10 m*M*
NaCl					400 m*M*	
EDTA	1 m*M*					
NAD+	10 m*M*		1 m*M*	10 µm		0.5 m*M*
DTT	10 m*M*	1 m*M*	10 m*M*			
NP-40		0.10%				
Triton X-100						0.01%
Primers	40 fmol	3.3 pmol	500 fmol (*38*)	30 nm	600 fmol	3.6 nm
Template	1 fmol (*1*); 6–60 µg (*37*)	0.74 fmol	500 fmol (*3*); 100 fmol (*5*); 50 fmol (*6*)	100 molecules in 500 ng genomic DNA	10–100 ng	250 ng
Conditions						
Denaturation	94°C, 1 min	92°C, 20 sec	94°C, 15 sec	85°C, 30 sec	95°C, 15 sec	91°C, 30 sec
Lgation	64°C, 4 min	60°C, 40 sec	65°C, 4 min	60°C, 30 sec	65°C, 4 min	55°C, 6 min
No. cycles	20–30	25	20	25	5 (LDR)[a]	27
Target	Sickle cell mutation (*1*); p53, Ha-ras (*37*)	*E. coli lac*I	Human eIF-4E (*3*); K-ras (*5*); human microsatellites (*6*)	HIV AZT-resistant mutants; *Chlamydia* cryptic plasmid	Human gene CYP21	Sickle cell mutation
Differential detection	Size-labeled primers (*1*); ^{32}P label (*37*)	Size-labeled primers; matrix-assisted laser desorption/ ionization time-of-flight-mass spectrometry	Fluorescent-labeled primers; microparticles coated with anti-hapten	Microparticles coated with anti-hapten; sandwich immunoassay performed using the Abbott IMx automated analyzer	Size-labeled primers; fluorescent labeled primers; detection in Perkin–Elmer GeneScan sequencing system	Arbitrary sequence on primer for capture hybridization; biotinylated primer streptavidin-alkaline phosphatase conjugate

[a]Ligase detection reaction uses ligase chain reaction to detect previously amplified product.

2. The primers for this technique are designed to have a gap of 4 to 5 Gs that would be filled by adding the single deoxynucleotide dGTP and *Taq* polymerase that would not add more nucleotides than are needed to fill the gap. The *Pilin-2* locus has 6 bp of nontargeted DNA added to one end that are complementary to the related primer.
4. The ends of the primers are labeled with fluorescein on one pair of homologous primers (one 5′ one 3′) and biotin on the other nonligating ends of the other two primers. The sequences of the primers are described in the original publication describing this technique ***(36)***.
5. The ligating 5′ ends are phosphorylated to facilitate ligation after the gap is filled.
6. All of the modifications of the oligonucleotides can be ordered from numerous companies that will custom synthesize oligonucleotides for specific purposes and targets.

3.2.2. Clinical Specimens

1. The type of specimen should be chosen according to the site that the organism is most likely to be detected in association with disease. For *N. gonorrhoeae,* urethral or endocervical swabs are collected and placed into 500 µl of specimen buffer containing 50 m*M* EPPS buffer [*N-(2-hydroxyethyl)piperazine-N′*-(3-propanesulfonic acid)] and 5 m*M* EDTA (pH 7.8). The tube can be maintained up to 24 to 48 h at room temperature if necessary for transport to the laboratory. The swabs are vigorously vortexed prior to removal and the sample is then frozen at -20°C until DNA extraction.
2. Organism lysis to release DNA is accomplished by adding proteinase K to a final concentration of 200 µg/ml and incubating for 1 h at 60°C.
3. Boiling for 10 min inactivates the proteinase K, further lyses the bacteria and denatures the DNA.
4. The cell debris is pelleted by centrifuging for 10 min at 4°C at 13,000*g*.
5. The supernatants can be stored at –20°C until used.

3.2.3. Amplification and Detection

1. Each 50 µl amplification mixture will contain 4 µl of sample, and a final concentration of: 20 m*M* Tris-HCl (pH 7.6 @25°C) 25 m*M* potassium acetate, 10 m*M* magnesium acetate, 10 m*M* DTT, and 1 m*M* NAD and 0.1% Triton X-100. This buffer supplied as 10X the concentrations needed can be purchased commercially (New England Biolabs, Beverly, MA) along with the *Taq* ligase.
2. Each of the four oligonucleotides within a set should be diluted in advance and the appropriate volumes of each added to obtain the 830 fmol of each of the four in a set in each reaction tube.
3. This mixture is overlaid with sterile mineral oil and heated for 3 min in a boiling water bath to ensure complete denaturation. After cooling, 1 unit of Amplitaq DNA polymerase (Perkin-Elmer Cetus, Norwalk, CT) and 15 units of *Thermus aquaticus (Taq)* DNA ligase (New England Biolabs) is added.
4. A positive control consisting of 270 cell equivalents of *N. gonorrhoeae* DNA in 80 ng/µl human placental DNA (Sigma Chemicals, Saint Louis, MO) and a negative control must be included in each run.
5. Gap LCR is performed in a thermocycler. The prototype technique for *N. gonorrhoeae* used 27 cycles *(Opa-2)*, 33 cycles *(Opa-3)* or 31 cycles *(Pilin-2)*. Each cycle consists of a denaturation step 85°C for 30 s and a hybridization, gap filling, ligation step of 60°C for *Opa-2* and *Pilin-2* and 53°C for *Opa-3*) for 1 min. When developing a new LCR procedure it is necessary to empirically test different numbers of cycles to determine the optimum for detection and specificity for each primer pair.
6. Detection is performed on 40 µl of each LCR reaction in the automated Abbott Imx, which uses anti-fluorescein-coated microparticles to capture the products of the gap LCR.

If ligation has taken place, the biotin on the other end of the product is detected by anti-biotin conjugated to alkaline phosphatase. Alkaline phosphatase is detected by adding methylumbelliferone phosphate that, when the phosphatase acts on the phosphate, yields a fluorescent product that is detected by the IMx. *(36)*. The amount of fluorescence produced is proportional to the ligated oligonucleotides.

7. Of the many methods used to detect ligation products, automated methods such as the IMx are the most labor efficient. Details and requirements regarding the actual use of automated systems for LCR detection are instrument-specific and must be performed with the manufacturers in order to ensure the validity and accuracy of the results.

3.3. Abbott LCx Commercial Automated Neisseria gonorrhoeae Detection System

The materials listed here and the procedures described in the subsequent section are based on what is needed for performance of the commercially sold Abbott LCx assay for detection of *N. gonorrhoeae* from urogenital swabs or voided urine from men or women. Due to the proprietary nature of the LCx technology, use of these specific materials, including specimen transport systems and oligonucleotide primers that must be purchased from the manufacturer in kit form is necessary for the successful performance of this type of LCR assay. Strict adherence to the manufacturer's instructions for sample collection, storage, and processing is necessary for satisfactory results. The LCx kits for *C. trachomatis* and *M. tuberculosis* follow the same general principles. Some modifications are necessary for the *M. tuberculosis* assay due to the nature of the different specimen types (respiratory secretions vs urogenital swabs or urine). It is possible to perform the LCR assay using other commercially available ligases and methods for detection of amplicons as described earlier, but the principles of the Abbott LCx and the notes regarding its performance are generally relevant for any type of LCR assay.

3.3.1. Specimen Collection

1. 15–20 mL of first-void urine should be collected into a sterile plastic, preservative-free container. It is desirable to obtain specimens from patients who have not urinated within 1 h prior to collection.
2. Swab specimens should be collected using the LCx swab collection and transport kit. For endocervical specimens in females, excess cervical mucus should be removed prior to sampling using the large-tipped cleaning swab provided in the collection system. When sampling the cervix, the small-tipped swab should be inserted into the endocervix and rotated for 15–30 s to ensure adequate sampling. In males, the small-tipped swab should be inserted 2–4 mm into the urethral meatus and rotated for 3 to 5 s. Swabs are then inoculated into the transport tube, broken off at the score line and then the cap is screwed securely onto the tube.

3.3.2. Specimen Storage

1. Time and temperature conditions must be adhered to for storage and transport of specimens. Swabs in the transport system can be stored at 2–30°C. If more than 24 h will elapse before processing, the swabs should ideally be refrigerated at 2–8°C or frozen at –20°C or below.
2. Urine specimens should be refrigerated immediately at 2–8°C and can be held at this temperature for up to 4 d before processing. If longer storage is necessary, swab or urine specimens can be frozen at –20°C or below for up to 60 d. Do not thaw urine until ready for testing.

3.3.3. Urine Specimen Preparation and Processing

1. Allow urine specimen to completely thaw if frozen. Mix urine in the urine collection cup by swirling to resuspend any settled material.
2. Using a pipet with aerosol barrier pipet tips, transfer 1 mL of mixed urine into the Urine Specimen Microfuge Tube from the Urine Specimen Preparation Kit.
3. Centrifuge at >9000*g* for 15 min in a microcentrifuge.
4. Using a fine-tipped, plastic disposable pipet, gently aspirate all of the urine supernatant. Be cautious not to contact or dislodge the pellet, which may be translucent. The time between centrifugation and removal of supernatant must not exceed 15 min.
5. Using a pipettor with aerosol barrier pipet tips, add 1 mL of LCx Urine Specimen Resuspension Buffer. Close lid of microfuge tube and resuspend the pellet by vortexing until the pellet is resuspended.
6. Secure tube closure with a cap lock until it clicks into place.
7. Insert specimen tubes in preheated dry bath wells. After the temperature of the heat block is stabilized at 97°C, heat specimens for 15 min.
8. Remove the specimen from the dry bath and allow to cool at room temperature for 15 min. Remove cap lock and discard.
9. Pulse-centrifuge the processed urine specimen in a microcentrifuge for a minimum of 10–15 s immediately before inoculating the LCx amplification vials.
10. The amplification reagent level in the LCx amplification vial should measure approx two-thirds of the conical part of the vial. If necessary, the vial may be pulse centrifuged in a microcentrifuge for 10–15 s.
11. Using a pipet with aerosol barrier pipet tips, add 100 µL of each processed urine specimen to the appropriately labeled LCx Amplification Vial, making sure each vial is securely closed and that only one at a time is open. The vial can now be transferred to Area 2 and placed immediately in the Thermal Cycler for amplification.

3.3.4. Swab Specimen Preparation

1. Allow specimen to completely thaw, if frozen.
2. Insert specimen tubes in preheated dry bath wells. After the temperature of the heat block is stabilized at 97°C, heat specimens for 15 min. Failure to reach 97 + 2°C could limit release of the DNA in the specimen causing false negative results.
3. Remove the specimen from the dry bath and allow to cool at room temperature for 15 min.
4. Unscrew cap and express swab along the side of the tube so that liquid drains back into the solution at the bottom of the tube. Discard swab and original closure, replacing with a new swab tube closure that is screwed on until it clicks into place.
5. The amplification reagent level in the LCx amplification vial should measure approx two-thirds of the conical part of the vial.
6. Using a pipet with aerosol barrier pipet tips, add 100 µL of each processed specimen to the appropriately labeled LCx Amplification Vial, making sure each vial is securely closed and that only one at a time is open. The vial can now be transferred to Area 2 and placed immediately in the Thermal Cycler for amplification.
7. The LCx negative control and calibrator must be prepared in conjunction with specimens to be tested and run in duplicate with each carousel of clinical specimens.

3.3.5. Quality Control Procedures

1. Negative control and calibrator preparations must take place in the dedicated Specimen Preparation Area (Area 1).

2. The LCx procedure requires that the negative control and the calibrator be run in duplicate with each carousel of specimen.
3. The negative control and calibrator are activated by the addition of 100 µL of LCx activation reagent. It is important to make sure correct volume is added or the run may be invalid. After addition, the contents of the bottles are then recapped and vortexed for 20 s. Each bottle of activated negative control or calibrator is designed to be used up to 48 h if stored at 2–8°C.
4. A positive control that monitors the entire assay procedure including the specimen processing step should be tested. A known positive urine specimen can be processed in parallel and tested with unknown specimens. The positive control should give a positive assay value (S/CO ratio >1.00). Each laboratory should establish a target value and limits from each control batch using statistical control rules. These target values and limits should be maintained throughout storage. Alternative choices for a positive control are type strains of the microorganism targeted in the assay, e.g., *N. gonorrhoeae.*

3.3.6. LCx Amplification Procedure

1. Turn the LCx Thermal Cycler on for at least 15 min prior to use.
2. Collect all LCx amplification vials containing samples, negative control and calibrator from Area 1 and transfer to Area 2 for thermal cycling.
3. LCx thermal cycling conditions should be edited to the following amplification parameters: 93°C for 1 s, 59°C for 1 s, 62°C for 1 min, 10 s for 40 cycles.
4. Place the amplification vials into the thermal cycler, and initiate run. After completion of the thermal cycler run, amplification product may remain at 15–30°C for up to 72 h prior to LCx detection.

3.3.7. Detection of Amplification Product

1. Refer to the LCx Analyzer Operations Manual for detailed instrument operation procedures. Before running the LCx Analyzer, check to see that LCx Inactivation Diluent (1) contains a minimum of 100 mL and the LCx System Diluent (2) contains a minimum of 250 mL. Remove the LCx Amplification Vials from the LCx Thermal Cycler.
2. Place LCx Reaction Cells into a Carousel; lock the carousel.
3. Pulse-centrifuge the LCx amplification vials in a microcentrifuge for 10–15 s before placing into the LCx reaction cells.
4. Place the amplification vials into the LCx reaction cells in the following order: negative controls in positions 1 and 2, calibrators in positions 3 and 4, and specimens in the remaining positions.
5. Place the carousel into the LCx Analyzer.
6. Lock the amplification vial Retainer.
7. Remove the LCx detection reagent Pack from 2–8°C storage, gently invert it five times, and open the reagent pack bottles in the numeric order: 1, 2, 3, 4.
8. Look for any film that may have formed over the opening of the reagent bottles. If present burst the bubble.
9. Place the LCx detection reagent pack into the LCx Analyzer.
10. Press Assay, then sample management to log in samples for the run. Press RUN on the LCx Analyzer control panel. Final assay results will be printed in approx 60 min.
11. Store the detection reagent pack at 2–8°C in original packaging, decontaminated with 1% v/v hypochlorite or separate from unopened LCx kits.
12. Remove the Carousel, individually remove the LCx reaction cells, and dispose appropriately. Rinse the carousel with water after each use.

3.3.8. Calculation of Results

1. *N. gonorrhoeae.* The LCx Assay uses MEIA detection on the LCx Analyzer to detect DNA. All calculations are performed automatically.
2. The presence or absence is determined by relating the LCx Assay results for the specimen to the Cutoff value. The Cutoff value is the mean RATE (c/s/s) of the LCx calibrator duplicates multiplied by 0.25.
3. Calculation of the Cutoff value:
   Cutoff value = 0.25 × (Mean of LCx Gonorrhea Calibrator RATES)
   The S/CO value is determined by calculating a ratio of the sample RATE to the Cutoff value

$$\frac{S}{CO} = \frac{\text{Sample RATE}}{\text{Cutoff Value}}$$

3.3.9. Interpretation of Results

N. gonorrhoeae S/CO Ratio	Result	Report
>1.20	LCx positive	*N. gonorrhoeae* DNA detected, and positive for *N. gonorrhoeae* by LCR amplification and MEIA detection.
<0.80	LCx negative	*N. gonorrhoeae* DNA not detected and presumed negative for *N. gonorrhoeae* by LCR amplification and MEIA detection. A presumed negative result may be caused by possible inhibition of the LCx method, collection variables or other factors
0.80–1.20	LCx equivocal	Repeat LCx test. If repeat test S/CO ratio is greater than or equal to 1.20, *N. gonorrhoeae* DNA detected, and positive for *N. gonorrhoeae* by LCR amplification and MEIA detection. If the repeat test is less than S/CO ratio 1.20, *N. gonorrhoeae* DNA not detected and presumed negative for *N. gonorrhoeae* by LCR amplification and MEIA detection.

4. Notes

1. Availability of commercial LCR kits for detection of *C. trachomatis* and *N. gonorrhoeae* that contain all of the major reagents needed to perform the assays, controls, and calibrators to confirm the validity of results obtained, and detailed operating instructions and guidelines for troubleshooting greatly facilitate performance of these types of assays. There are detailed description of error codes and a table for what to do for each message. However, these convenient commercially produced assays can only be used for these organisms, except for the availability of the LCx for *M. tuberculosis* in some countries. If other microorganisms are of interest, an in-house assay would need to be developed using LCR technology with the development and selection of appropriate primers, controls, and detection systems.
2. Contamination with DNA remains the major concern, despite physical separation of preparation and amplification procedures and attention to procedures such as frequent changing of gloves, especially between processing and amplification areas. Pipet tips should never enter any tube more than once. In the final step of the LCx analyzer, a chemical inactivation reaction takes place to destroy amplicons to reduce the likelihood

of contamination.

3. When using LCx tests for detection of microorganisms in urogenital specimens, the presence of excessive mucus or blood in the swab or urine specimen as well as use of feminine powder sprays can interfere with the assay and cause false negative results, thus proper specimen quality must be monitored.
4. In addition to the usual quality controls, reproducibility of LCx assays can be monitored by repeating random samples on a frequent basis.
5. Further quality control testing and recording is detailed for the LCx analyzer and thermocycler in the respective manuals for the equipment.
6. Failure to reach 97°C before amplification could limit release of the DNA in the specimen causing false negative results.
7. Test the processed urine specimen immediately, or store for up to 60 d at 2–8°C or –20°C prior to testing. If the processed urine specimen is stored frozen, it must be completely thawed prior to addition to the LCx Amplification Vial.
8. The LCx Analyzer first verifies that the assay results of the negative controls and calibrator are within the specified ranges of the LCx assay parameters by comparing the assay results of the negative control and calibrator to the values listed in the assay parameters. A run is valid when the individual and average results are within the values listed for CAL HIGH, CAL LOW, CAL AVE HIGH, CAL AVE LOW, NEG LOW, NEG HIGH, NEG AVE HIGH, and NEG AVE LOW parameters in the LCx assay parameters.
9. In the event of an invalid negative control or calibrator assay result, the assay results printout will identify the out-of-range result, the S/CO ratio of the specimen will not be calculated and a flag indicating an invalid result will occur in the NOTE column of the printout. Ensure the LCx Negative Controls and Calibrators are in the correct order on the MEIA carousel to avoid an invalid run.
10. Environmental quality control screening is also recommended. The laboratory should be monitored for the presence of amplification product by saturating one of the small-tipped swabs in transport buffer, then using it to wipe the desired area, including equipment, and processing according to the LCx procedures. If a positive reaction is detected, the decontamination steps using 1% sodium hypochlorite followed by 70% ethanol should be performed. The operation manual should be consulted for decontaminating equipment.

References

1. Barany, F. (1991) Genetic disease detection and DNA amplification using cloned thermostable ligase. *Proc. Natl. Acad. Sci. USA* **88,** 189–193.
2. Jurinke, C., van den Boom, D., Jacob, A., Tang, K., Worl, R., and Koster, H. (1996) Analysis of ligase chain reaction products via matrix-assisted laser desorption/ionization time-of-flight-mass spectrometry. *Anal. Biochem.* **237,** 174–181.
3. Luo, J., Bergstrom, D. E., and Barany, F. (1996) Improving the fidelity of *Thermus thermophilus* DNA ligase. *Nucleic Acids Res.* **24,** 3071–3078.
4. Khanna, M., Park, P., Zirvi, M., Cao, W., Picon, A., Day, J., et al. (1999) Multiplex PCR/LDR for detection of K-ras mutations in primary colon tumors. *Oncogene* **18,** 27–38.
5. Khanna, M., Cao, W., Zirvi, M., Paty, P., and Barany, F. (1999) Ligase detection reaction for identification of low abundance mutations. *Clin. Biochem.* **32,** 287–290.
6. Zirvi, M., Bergstrom, D. E., Saurage, A. S., Hammer, R. P., and Barany, F. (1999) Improved fidelity of thermostable ligases for detection of microsatellite repeat sequences using nucleoside analogs. *Nucleic Acids Res.* **27,** e41.
7. Day, D. J., Speiser, P. W., White, P. C., and Barany, F. (1995) Detection of steroid 21-hydroxylase alleles using gene-specific PCR and a multiplexed ligation detection reaction. *Genomics* **29,** 152–162.

8. Reyes, A. A., Carrera, P., Cardillo, E., Ugozzoli, L., Lowery, J. D., Lin, C. I., et al. (1997) Ligase chain reaction assay for human mutations: the Sickle Cell by LCR assay. *Clin. Chem.* **43,** 40–44.
9. Besmer, P., Miller, R. J., Caruthers, M. H., Kumar, A., Minamoto, K., Van de Sande, J. H., et al. (1972) Studies on polynucleotides. CXVII. Hybridization of polydeoxynucleotides with tyrosine transfer RNA sequences to the r- strand of phi80psu + 3 DNA. *J. Mol. Biol.* **72,** 503–522.
10. Landegren, U., Kaiser, R., Sanders, J., and Hood, L. (1988) A ligase-mediated gene detection technique. *Science* **241,** 1077–1080.
11. Alves, A. M. and Carr, F. J. (1988) Dot blot detection of point mutations with adjacently hybridising synthetic oligonucleotide probes. *Nucleic Acids Res.* **16,** 8723.
12. Wu, D. Y. and Wallace, R. B. (1989) The ligation amplification reaction (LAR)—amplification of specific DNA sequences using sequential rounds of template-dependent ligation. *Genomics* **4,** 560–569.
13. Morre, S. A., van Valkengoed, I. G., de Jong, A., Boeke, A. J., van Eijk, J. T., Meijer, C. J., and van den Brule, A. J. (1999) Mailed, home-obtained urine specimens: a reliable screening approach for detecting asymptomatic *Chlamydia trachomatis* infections. *J. Clin. Microbiol.* **37,** 976–980.
14. Chernesky, M. A., Jang, D., Sellors, J., Luinstra, K., Chong, S., Castriciano, S., and Mahony, J. B. (1997) Urinary inhibitors of polymerase chain reaction and ligase chain reaction and testing of multiple specimens may contribute to lower assay sensitivities for diagnosing *Chlamydia trachomatis* infected women. *Mol. Cell Probes* **11,** 243–249.
15. Prchal, J. T. and Guan, Y. L. (1993) A novel clonality assay based on transcriptional analysis of the active X chromosome. *Stem Cells* **11 (Suppl. 1),** 62–65.
16. Marshall, R. L., Laffler, T. G., Cerney, M. B., Sustachek, J. C., Kratochvil, J. D., and Morgan, R. L. (1994) Detection of HCV RNA by the asymmetric gap ligase chain reaction. *PCR Methods Appl.* **4,** 80–84.
17. Lee, H. H., Chernesky, M. A., Schachter, J., Burczak, J. D., Andrews, W. W., Muldoon, S., et al. (1995) Diagnosis of *Chlamydia trachomatis* genitourinary infection in women by ligase chain reaction assay of urine. *Lancet* **345,** 213–216.
18. Dille, B. J., Butzen, C. C., and Birkenmeyer, L. G. (1993) Amplification of *Chlamydia trachomatis* DNA by ligase chain reaction. *J. Clin. Microbiol.* **31,** 729–731.
19. Winn-Deen, E. S., Batt, C. A., and Wiedmann, M. (1993) Non-radioactive detection of *Mycobacterium tuberculosis* LCR products in a microtitre plate format. *Mol. Cell Probes* **7,** 179–186.
20. Gaydos, C. A., Howell, M. R., Quinn, T. C., Gaydos, J. C., and McKee, K. T., Jr. (1998) Use of ligase chain reaction with urine versus cervical culture for detection of *Chlamydia trachomatis* in an asymptomatic military population of pregnant and nonpregnant females attending Papanicolaou smear clinics. *J. Clin. Microbiol.* **36,** 1300–1304.
21. Berg, E. S., Anestad, G., Moi, H., Storvold, G., and Skaug, K. (1997) False-negative results of a ligase chain reaction assay to detect *Chlamydia trachomatis* due to inhibitors in urine. *Eur. J. Clin. Microbiol. Infect. Dis.* **16,** 727–731.
22. Leckie, G. W., Erickson, D. D., He, Q., Facey, I. E., Lin, B. C., Cao, J., and Halaka, F. G. (1998) Method for reduction of inhibition in a *Mycobacterium tuberculosis*-specific ligase chain reaction DNA amplification assay. *J. Clin. Microbiol.* **36,** 764–767.
23. Buimer, M., van Doornum, G. J., Ching, S., Peerbooms, P. G., Plier, P. K., Ram, D., and Lee, H. H. (1996) Detection of *Chlamydia trachomatis* and *Neisseria gonorrhoeae* by ligase chain reaction-based assays with clinical specimens from various sites: implications for diagnostic testing and screening. *J. Clin. Microbiol.* **34,** 2395–2400.

24. Stary, A. (1999) Correct samples for diagnostic tests in sexually transmitted diseases: which sample for which test? *FEMS Immunol Med Microbiol* **24,** 455–459.
25. Stary, A., Najim, B., and Lee, H. H. (1997) Vulval swabs as alternative specimens for ligase chain reaction detection of genital chlamydial infection in women. *J. Clin. Microbiol.* **35,** 836–838.
26. Schepetiuk, S., Kok, T., Martin, L., Waddell, R., and Higgins, G. (1997) Detection of *Chlamydia trachomatis* in urine samples by nucleic acid tests: comparison with culture and enzyme immunoassay of genital swab specimens. *J. Clin. Microbiol.* **35,** 3355–3357.
27. Dubuis, O., Gorgievski-Hrisoho, M., Germann, D., and Matter, L. (1997) Evaluation of 2-SP transport medium for detection of *Chlamydia trachomatis* and *Neisseria gonorrhoeae* by two automated amplification systems and culture for chlamydia. *J. Clin. Pathol.* **50,** 947–950.
28. Hook, E. W., III, Smith, K., Mullen, C., Stephens, J., Rinehardt, L., Pate, M. S., and Lee, H. H. (1997) Diagnosis of genitourinary *Chlamydia trachomatis* infections by using the ligase chain reaction on patient-obtained vaginal swabs. *J. Clin. Microbiol.* **35,** 2133–2135.
29. Stary, A., Schuh, E., Kerschbaumer, M., Gotz, B., and Lee, H. (1998) Performance of transcription-mediated amplification and ligase chain reaction assays for detection of chlamydial infection in urogenital samples obtained by invasive and noninvasive methods. *J. Clin. Microbiol.* **36,** 2666–2670.
30. Carroll, K. C., Aldeen, W. E., Morrison, M., Anderson, R., Lee, D., and Mottice, S. (1998) Evaluation of the Abbott LCx ligase chain reaction assay for detection of *Chlamydia trachomatis* and *Neisseria gonorrhoeae* in urine and genital swab specimens from a sexually transmitted disease clinic population. *J. Clin. Microbiol.* **36,** 1630–1633.
31. Garrino, M. G., Glupczynski, Y., Degraux, J., Nizet, H., and Delmee, M. (1999) Evaluation of the Abbott LCx *Mycobacterium tuberculosis* assay for direct detection of *Mycobacterium tuberculosis* complex in human samples. *J. Clin. Microbiol.* **37,** 229–232.
32. Moore, D. F. and Curry, J. I. (1998) Detection and identification of *Mycobacterium tuberculosis* directly from sputum sediments by ligase chain reaction. *J. Clin. Microbiol.* **36,** 1028–1031.
33. Ausina, V., Gamboa, F., Gazapo, E., Manterola, J. M., Lonca, J., Matas, L., et al. (1997) Evaluation of the semiautomated Abbott LCx *Mycobacterium tuberculosis* assay for direct detection of *Mycobacterium tuberculosis* in respiratory specimens. *J. Clin. Microbiol.* **35,** 1996–2002.
34. Lindbrathen, A., Gaustad, P., Hovig, B., and Tonjum, T. (1997) Direct detection of *Mycobacterium tuberculosis* complex in clinical samples from patients in Norway by ligase chain reaction. *J. Clin. Microbiol.* **35,** 3248–353.
35. Fadda, G., Ardito, F., Sanguinetti, M., Posteraro, B., Ortona, L., Chezzi, C., et al. (1998) Evaluation of the Abbott LCx *Mycobacterium tuberculosis* assay in comparison with culture methods in selected Italian patients. *New Microbiol.* **21,** 97–103.
36. Birkenmeyer, L. and Armstrong, A. S. (1992) Preliminary evaluation of the ligase chain reaction for specific detection of Neisseria gonorrhoeae. *J. Clin. Microbiol.* **30,** 3089–3094.
37. Wilson, V. L., Wei, Q., Wade, K. R., Chisa, M., Bailey, D., Kanstrup, C. M., et al. (1999) Needle-in-a-haystack detection and identification of base substitution mutations in human tissues. *Mutat. Res.* **406,** 79–100.
38. Abravaya, K., Carrino, J. J., Muldoon, S., and Lee, H. H. (1995) Detection of point mutations with a modified ligase chain reaction (Gap-LCR). *Nucleic Acids Res.* **23,** 675–682.

26

Nested RT-PCR in a Single Closed Tube

Antonio Olmos, Olga Esteban, Edson Bertolini, and Mariano Cambra

1. Introduction

There are several methods now in widespread use for detecting and characterizing specific RNA targets. These methods include *in situ* hybridization, Northern blotting, dot or slot blot, RNase protection assay, and reverse transcription coupled to polymerase chain reaction (RT-PCR). However, when the amount of RNA target is limited or restricted in its cellular or tissue distribution, the extreme sensitivity of the PCR allows the detection of minute quantities of RNA when coupled to an initial step that converts single-strand RNA to cDNA *(1–4)*. Nevertheless, when RT-PCR is applied for diagnostic purposes, the sensitivity usually afforded by this technique in routine detection tests is similar, or only slightly higher, to conventional enzyme-linked immunosorbent assay or hybridization techniques. This is frequently observed when dealing with poor-quality samples containing inhibitors of RT-PCR. The presence of different components of plant or animal origin, as well as specific RT-PCR conditions, may inhibit the reverse transcription and amplification by a number of mechanisms *(5)*.

Approaches have been developed to overcome these problems, including a previous immunocapture phase *(6–8)*, the immobilization of RNA targets on plastic surfaces *(9)* or on paper membranes by a direct printing or squashing of the sample *(10–12)*, preparation of crude extracts in simple buffers and subsequent dilution in sterile water, and other different protocols to prepare RNA targets free from interfering substances. Moreover, there are interesting alternatives to crude extracts or total nucleic acid preparations based on the use of commercially available resin- *(13)* or silica- *(14)* based kits.

Sensitivity and specificity problems associated with RT-PCR may be overcome by using nested RT-PCR. The process is based on two consecutive rounds of amplification *(15,16)*, the first round being an RT-PCR and the second a conventional PCR. The RT-PCR is performed using a pair of external primers. The 3′ end external primer (Pe1 in **Fig. 1**) is used for RT reaction and the 3′ and 5′ end (Pe2 in **Fig. 1**) primers are used to perform the first PCR. The resulting amplification product is transferred to another Eppendorf tube containing a second pair of nested primers (nested PCR) that are internal to the initial pair. Alternatively, one of the external primers and a single nested primer (heminested PCR) can also be used. The larger amplification fragment produced during the first reaction is used as a target for the second (nested or heminested) PCR. The concept is illustrated in **Fig. 1**.

From: *Methods in Molecular Biology, Vol. 226: PCR Protocols, Second Edition*
Edited by: J. M. S. Bartlett and D. Stirling © Humana Press Inc., Totowa, NJ

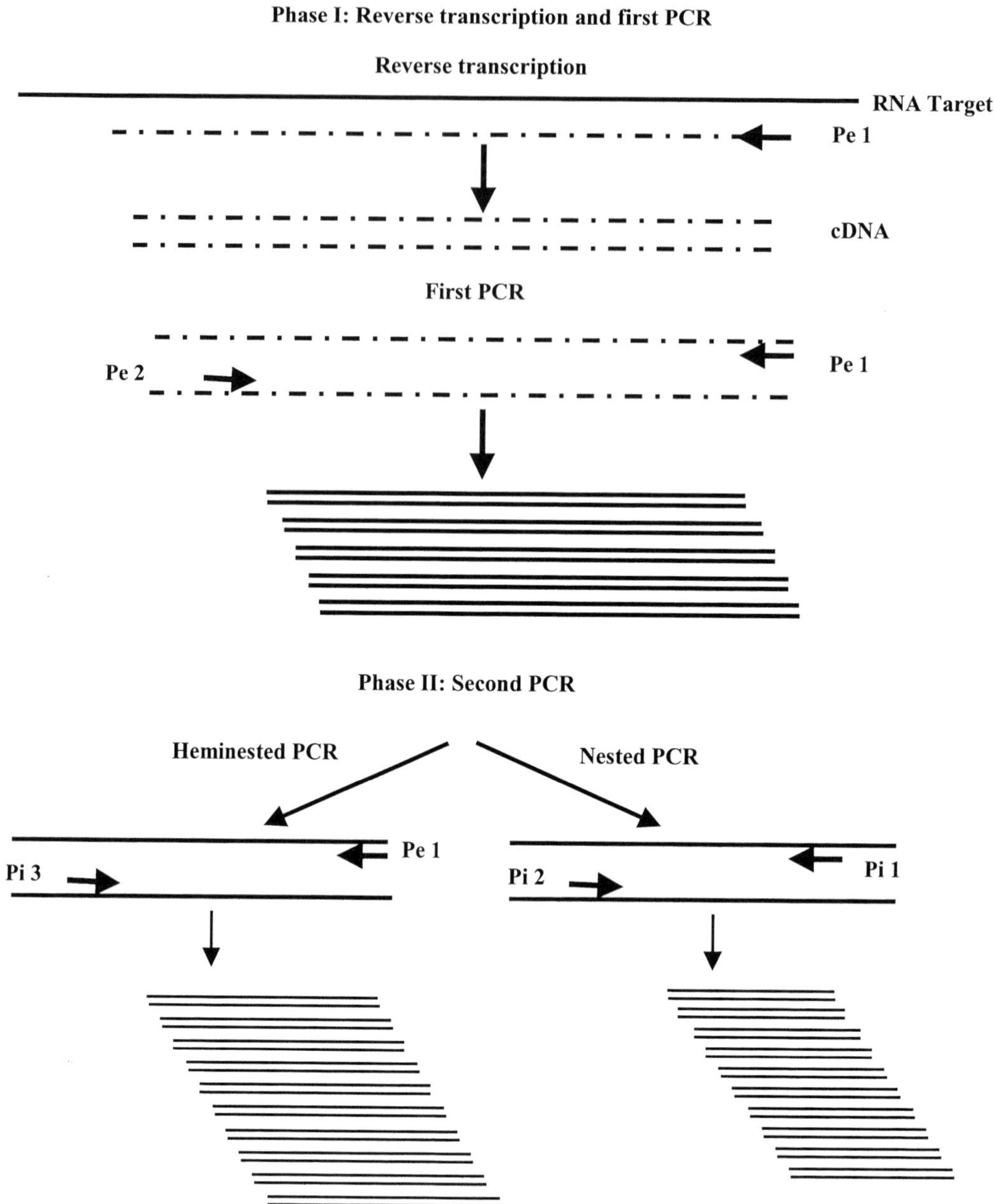

Fig. 1. RT-heminested/nested-PCR scheme. Phase I (reverse transcription and first PCR): RNA target is copied in a cDNA form by the reverse transcriptase that uses the 3′ end external oligo (Pe1) as the reaction primer. Subsequently, the first PCR is performed using Pe1 and the 5′ end external primer (Pe2) to produce as intermediate product that will act as the target in the second PCR amplification. Phase II (heminested or nested-PCR): a second PCR using a 5′ end internal primer (in the example Pi3) combined with one of the external primers (in the example Pe1; heminested-PCR) or two internal primers, the 3′ end internal primer (Pi1), and the 5′ end internal primer (Pi2; nested PCR). Note that the internal primers could be used for typing purposes of the larger fragment amplified in the first reaction.

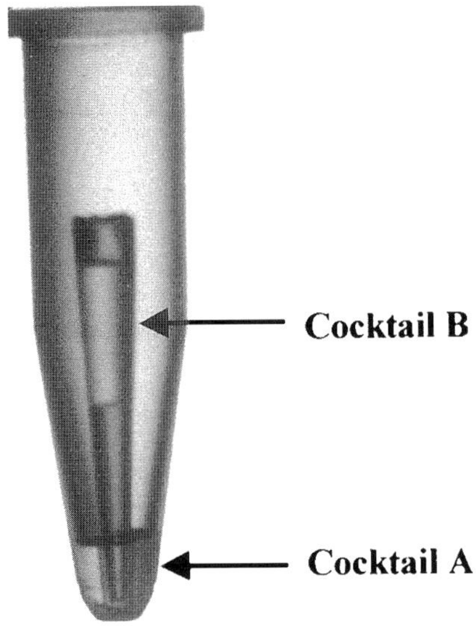

Fig. 2. Compartmentalized Eppendorf tube that allows capture (if necessary), reverse transcription, and nested PCR in a single closed tube. Cocktail A for the reverse transcription and first PCR mix is added to the bottom of the tube. Cocktail B for the second amplification mix is added into the pipet tip cone.

Several reports illustrate the potential of heminested and nested PCR in different fields *(11,12,17–19)*. However, the use of two rounds of amplification in different tubes enhances the risk of contamination, especially when the method is used on a large scale. To prevent this problem, some authors proposed single-tube nested PCR protocols *(20–22)*. However, a limitation to these approaches of nested RT-PCR is the need to accurately establish the ratio between an external and internal pair of primers and the use of limiting amounts of external primers to avoid its interference during the second amplification. Moreover, the external primers must be designed to anneal at a higher temperature than the internal primers. **Figure 2** illustrates a simple device based on the use of a compartmentalized Eppendorf tube (Spanish patent P9801642 of July 31, 1998) that allows RT reaction and nested PCR in a single closed tube in one manipulation. A small cone (the end of a standard 200-µL plastic pipet tip) is introduced into a 0.5-mL PCR tube allowing for the physical separation of the two different PCR cocktails in the same Eppendorf tube. The RT-PCR mix containing the external primers is added to the bottom of the tube and the PCR mix for the second amplification (heminested or nested) is added into the cone, where it remains as the result of capillary action (*see* **Note 1**). After RT-PCR, the Eppendorf tube is centrifuged to mix the products of the first reaction with the cocktail containing the internal primers. After this second round of PCR, the tube is finally opened to analyze the amplicons produced. The final result of this nested or heminested RT-PCR is a yield at least 100 times higher than a conventional RT-PCR.

The main advantages of this new approach of nested RT-PCR are the high sensitivity afforded without risk of contamination and the possibility of using external primers with the lowest annealing temperature and internal primers with the highest, in contrast to previously described protocols.

This nested RT-PCR protocol coupled with a preparation of squashed or printed samples on paper *(10)* would allow the detection of RNA targets from a number of viruses in individual insect vectors, as well as in plant materials, animal fluids, or tissues. The increased sensitivity provided with this method permits the amplification of RNA targets from individual viruliferous aphids carrying stylet-borne (nonpersistent) and semipersistent plant viruses *(11,12)* without the need of a preliminary purification of nucleic acids.

Conditions for multiplex nested RT-PCR can be easily established using the device described in **Fig. 2**. The simultaneous amplification of four RNA viruses and a bacterium from olive trees by multiplex nested RT-PCR without interference among primers has been successfully achieved in our laboratory *(23)*. This system can include the use, if necessary (*see* **Note 2**), of an immunocapture (IC) phase in the same tube (IC-nested RT-PCR) *(12)*. Recently, a new technique (called co-operational PCR/Co-PCR) for amplification of nucleic acids targets, based on a simple tetraprimer reaction has been described, with a sensitivity similiar to nested RT-PCR *(24)*.

2. Materials

1. Eppendorf tubes (*see* **Notes 2** and **3**) (Cultek, Thermowell tubes cat no: 6530).
2. Sterile 200-µL pipet tip cones (*see* **Fig. 2** and **Notes 1** and **4**)(Daslab, cat. no: 16-2001).
3. 10× RT-PCR buffer: 500 m*M* KCl, 100 m*M* Tris-HCl (pH 9.0 at 25°C), 1% Triton X-100 (supplied with *Taq* DNA polymerase).
4. Triton X-100 (Merck, Art. 8603).
5. DMSO (Sigma, cat. no. D8418).
6. $MgCl_2$ (25 m*M*; supplied with *Taq* DNA polymerase).
7. dNTPs (5 m*M*, Pharmacia, cat. no. 27-2035-02).
8. AMV reverse transcriptase (Pharmacia, cat. no. M5108).
9. *Taq* DNA polymerase (Promega, cat. no. M1865).
10. Primers for citrus tristeza virus (CTV) detection (*see* **Subheading 3.2., step 1a**):
    External primers (100 µ*M*)
    - PEX1 (5′ TAA ACA ACA CAC ACT CTA AGG 3′)
    - PEX2 (5′ CAT CTG ATT GAA GTG GAC 3′)

    Internal primers (100 µ*M*)
    - PIN1 (5′ GGT TCA CGC ATA CGT TAA GCC TCA CTT 3′)

    Primers for plum pox virus (PPV) detection (*see* **Subheading 3.2., step 1b**):
    External primers (100 µ*M*
    - P10 (5′ GAG AAA AGG ATG CTA ACA GGA 3′)
    - P20 (5′ AAA GCA TAC ATG CCA AGG TA 3′)

    Internal primers (100 µ*M*)
    - P1 (5′ ACC GAG ACC ACT ACA CTC CC 3′)
    - P2 (5′ CAG ACT ACA GCC TCG CCA GA 3′)
11. Micropipets (Gilson, pipetman P10, P20, P100, P200).
12. Microfuge (Heraeus, Biofuge PICO).
13. Thermal cycling machine with a heated lid (Techne, PHC-3).
14. Electrophoresis system (Bio–Rad, sub-cell gt/powerpac 300 power supply system Cat. No. 165-4350).
15. Ethidium bromide (10 µg/mL in water; AppliChem, A1152,0025).
16. Transilluminator (TDI, Sepctroline Model TC-312A 312 nm UV).

Table 1
Volume and Concentration of Reactives for CTV Detection by Nested RT-PCR

Ingredients	Cocktail A (RT-PCR)	Cocktail B (nested-PCR)
10X RT-PCR buffer	3.00 µl	1.00 µl
$MgCl_2$(25 mM)	3.60 µl	–
dNTPs (5mM each)	2.25 µl	–
Triton® X-100 (4%)	2.50 µl	–
3′ external primer (100 µM) (Pe1 in Fig. 1)	0.15 µl	3′ internal primer (100 µM) (Pi1 in Fig. 1) 0.80 µl
5′ external primer (100 µM) (Pe2 in Fig. 1)	0.15 µl	5′ internal primer (100 µM) (Pi2 in Fig. 1) 0.80 µl
DMSO (100%)	1.50 µl	–
H_2O	to 30.00 µl	to 10.00 µl
AMV (10 U/µl)	0.20 µl	–
Taq Pol (5 U/µl)	0.20 µl	–

3. Methods

3.1. Primer Design (see Note 5)

Sequenced regions of each RNA target can be recovered using the Nucleotide Sequence Search program located in the Entrez Browser program provided by the National Center for Biotechnology Information (NCBI) (http://www3.ncbi.nlm.nih.gov/Entrez; Bethesda, MD). Conserved regions for each target can be studied using the similarity search Advanced BLAST 2.0, with the blast program designed to support analysis of nucleotides (http://www3.ncbi.nlm.nih.gov/blast/blast.cgi?Jform=1) *(25)*. The alignment view could be performed as master-slave with identities to analyze significant nucleotide homologies in the molecular data retrieved from NCBI's integrated databases, GenBank, EMBL, and DDBJ. Specific nucleotide regions should be selected. Different specific primers with similar annealing temperature can be subsequently designed for the target of interest, based on Oligo software utility (http://www.lifescience-software.com/oligo.htm; LRS, Long Lake, MN). By using this methodology, appropriate PCR primers to different RNA targets can be easily and properly designed.

3.2. Nested PCR Product Generation

1. RT-PCR and nested PCR cocktail preparation (*see* **Fig. 2**).
   a. When the annealing temperatures between external and internal primers are very different (external 45°C and internal 60°C) and the external primers do not anneal at 60°C (for example, CTV detection), *see* **Table 1**.
   b. When the annealing temperatures of the external and internal primers are similar (external 50°C and internal 60°C) and the external primers anneal at 60°C (for example, PPV detection), *see* **Table 2**.
2. Add 30 µL of cocktail A for reverse transcription and external amplification to the bottom of an 0.5-mL Eppendorf tube.
3. Introduce the small plastic cone (*see* **Fig. 2** and **Notes 1** and **4**) and add 10 µL of cocktail B into the cone.

Table 2
Volume and Concentration of Reactives for PPV Detection by Nested RT-PCR

Ingredients	Cocktail A (RT-PCR)	Cocktail B (nested-PCR)
10X RT-PCR buffer	3.00 μl	1.00 μl
$MgCl_2$(25 mM)	3.60 μl	–
dNTPs (5 mM each)	2.25 μl	–
Triton® X-100 (4%)	2.50 μl	–
3′ external primer (1 μM) (Pe1 in Fig. 1)	3.00 μl3′ internal primer (100 μM) (Pi1 in Fig. 1)	0.80 μl
5′ external primer (1 μM) (Pe2 in Fig. 1)	3.00 μl5′ internal primer (100 μM) (Pi2 in Fig. 1)	0.80 μl
DMSO	1.50 μl	–
H_2O	to 30.00 μl	to 10.00 μl
AMV (10 U/μl)	0.20 μl	–
Taq Pol (5 U/μl)	0.20 μl	–

4. RT-PCR (*see* **Note 6**): 42°C for 45 min (reverse transcription), 94°C for 2 min (denaturation and reverse transcriptase inactivation), 20 to 25 cycles (92°C for 30 s [denaturation], 45 or 50°C (*see* **Subheading 3.2., step 1**) for 30 s [annealing], and 72°C for 1 min [extension], and final elongation at 72°C for 10 min.
5. Vortex the Eppendorf tube and centrifuge (pulse at 6000*g* for 2 s) to mix the second cocktail with the RT-PCR products.
6. Nested-PCR (*see* **Notes 7–9**): 35 to 40 cycles (92°C for 30 s [denaturation], 60°C [*see* **Subheading 3.2., step 1**] for 30 s [annealing], and 72°C for 1 min [extension] and final elongation at 72°C for 10 min.
7. Electrophoresis (*see* **Fig. 3**): Load 10 μL of amplification products onto a 3% agarose gel in 0.5× TAE and perform electrophoresis at 100 V for 30 min. Stain the agarose gel with ethidium bromide (0.5 μg/mL) for 15 min and visualize the amplicons under ultraviolet light.

4. Notes

1. The nested and heminested RT-PCR in a single closed tube described in this protocol is based on the use of a simple device (*see* **Fig. 2**). Other patented compartmentalized Eppendorf tubes with pockets could be commercialized by the industry in the near future, consequently avoiding the need to prepare cones. Accidental flow of the second PCR mix from the tip device might be caused by an incorrect manipulation of the device or by the use of pipet tips wider than standard. This can be solved by closing (heating) the tip end. In this case, after the RT-PCR it will be necessary to invert the tube, vortex to mix the reagents, and centrifuge to collect all components in the bottom of the tube.
2. A previous capture step (immunocapture or print capture) improves the detection of some RNA targets by RT-PCR. Conventional immunocapture is usually performed by pre-coating Eppendorf tubes with 100 μL of carbonate buffer (pH 9.6) containing immunoglobulins (2 μg/mL) of a known specificity ***(6–8)***. Normal immunoglobulins from nonimmunized rabbits, bovine serum albumin, or skimmed milk can be also successfully used in this capture phase. However, because of the high sensitivity of nested or heminested RT-PCR, this step can usually be omitted. In this case, the detection of RNA targets can be performed by direct incubation of 100 μL of the sample (crude extract or tissue print Triton X-100 extracts from

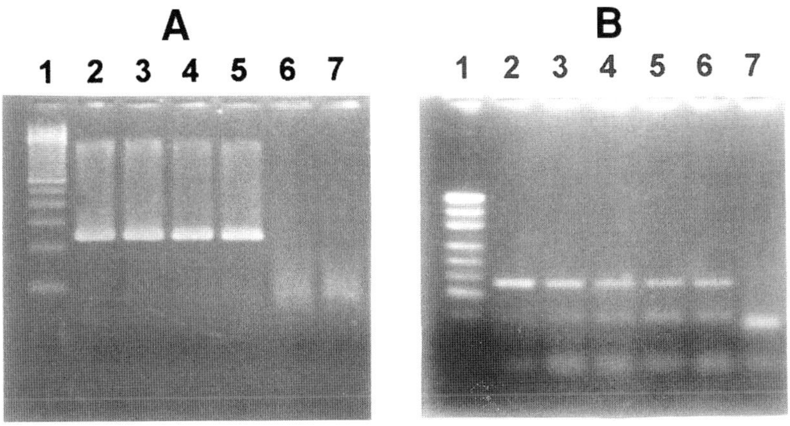

Fig. 3. Analysis of RT-nested-PCR products by agarose gel electrophoresis. **(A)** Detection of PPV targets from infected plant material. Lane 1: 100 bp molecular weight markers (Gibco BRL). Lanes 2 through 5: amplification products (243 bp) obtained from PPV-infected GF305 peach seedlings. Lanes 6 and 7: samples from healthy GF305 peach seedlings. **(B)** Detection of CTV targets from infected aphid vectors. Lane 1: pUC19DNA/*Msp*I(*Hpa*II) molecular weight markers 23 (MBI Fermentas). Lanes 2 through 6: amplification products (131 bp) obtained from individual *Aphis gossypii* fed on CTV-infected Washington Navel sweet orange. Lane 7: Sample from *Aphis gossypii* fed on healthy Washington Navel sweet orange.

paper; **ref. *10***) for 3 h at 37°C or overnight at 4°C in an uncoated Eppendorf tube. Before RT-PCR, the tube is washed twice with 150 µL of phosphate-buffered saline–0.05% Tween 20 (washing buffer). However, it may be useful to assay the ability of different plastic tubes to trap RNA targets in order to select the most appropriate for RT-PCR.

3. There are several commercially available thin-walled Eppendorf tubes for PCR. The protocol describes the use of 0.5-mL tubes because they are more easily compartmentalized with the end of a pipet tip (*see* **Fig. 2**).
4. The reaction must be performed in a thermal cycler with a heated lid to prevent the requirement for oil-overlay during the cycling events. The selected thermal cycler must allow the maintenance of cocktail B (*see* **Fig. 2**) in the cone during the cycling. If evaporation occurs close the tip device as in **Note 3** and/or adjust the lid temperature to 70 to 80°C during the first reaction (RT-PCR).
5. Primer design is critical to the success in amplifying RNA targets by conventional nested RT-PCR but it is even more critical when a multiplex nested strategy is used. The recommended primer design (*see* **Subheading 3.1.**) allows an optimal design of primers. The size of the amplified product should be small to ensure a good efficiency of the reaction and high sensitivity ***(26,27)***. Primers with a broad range of specificity must be designed from highly conserved genome sequences. Degenerate primers must be used for a universal detection of RNA targets belonging to a group. For characterization or typing of RNA targets, primers must be designed from discriminating or specific regions.
6. The goal of the first round of PCR after RT is to obtain a yield of the amplification product just enough to ensure adequate target for the second (nested or heminested) amplification. For this reason, the number of cycles can be as low as 20 to 25.
7. The goal of the second round of PCR (nested or heminested) is to achieve a high sensitivity and simultaneous specificity (if the method has been designed for typing RNA targets). The optimum number of cycles ranges from 35 to 40.

8. The sensitivity of nested RT-PCR in a single closed tube can be compared with the sensitivity obtained by conventional RT-PCR using internal or external primers.
9. The background of the nested RT-PCR must be reduced. Excessive cycling often results in the generation of multimeric PCR product that usually appears as a smear in the lane from the slot to the expected size of the amplified product in the agarose gel. Background also appears if the annealing temperature is not sufficient and/or the primer concentration is too high.

Acknowledgments

The authors wish to thank Dr. Mats Ohlin, University of Lund, Lund, Sweden, and Dr. Nuria Duran-Vila from IVIA, Valencia, Spain, for critical reading of the manuscript. The development of nested RT-PCR in a single closed tube has been funded by IVIA, INIA (SC98-060), CICYT (OLI96-2179), and EU-BIO96-0773 grants. O. Esteban and E. Bertolini were the recipients of PhD fellowships from Instituto Valenciano de Investigaciones Agarias and Agencia Española de Cooperación Internacional, respectively.

References

1. Hadidi, A., Levy, L., and Podleckis, E. V. (1995) Polymerase chain reaction technology in plant pathology, in *Molecular Methods in Plant Pathology* (Singh, R. P. and Singh, U. S., eds.), CRC/Lewis Press, Boca Raton, FL, pp. 167–187.
2. Candresse, T., Hammond, R. W., and Hadidi, A. (1996) Detection and identification of plant viruses and viroids using polymerase chain reaction (PCR), in *Control of Plant Virus Diseases*, APS Press, St. Paul, MN, pp. 399–416.
3. Singh, R. P. (1998) Reverse-transcription polymerase chain reaction for the detection of viruses from plants and aphids. *J. Virol. Methods* **74,** 125–138.
4. Hamoui, S., Benedetto, J. P., Garret, M., and Bonnet, J. (1994) Quantitation of mRNA species by RT-PCR on total mRNA population. *PCR Methods Appl.* **4,** 160–166.
5. Wilson, I. G. (1997) Inhibition and facilitation of nucleic acid amplification. *Appl. Environ. Microbiol.* **63,** 3741–3751.
6. Wetzel, T., Candresse, T., Macquaire, G., Ravelonandro, M., and Dunez, J. (1992) A highly sensitive immunocapture polymerase chain reaction method for plum pox potyvirus detection. *J. Virol. Methods* **39,** 27–37.
7. López-Moya, J. J., Cubero, J., López-Abella, D., and Díaz-Ruíz, J. R. (1992) Detection of cauliflower mosaic virus (CaMV) in single aphids by de polymerase chain reaction (PCR). *J. Virol. Methods* **37,** 129–138.
8. Nolasco, G., de Blas, C., Torres, V., and Ponz, F. (1993) A method combining immunocapture and PCR amplification in a microtiter plate for the detection of plant viruses and subviral pathogens. *J. Virol. Methods* **45,** 201–218.
9. Rowhani, A., Maningas, M. A., Lile, L. S., Daubert, S. D., and Golino, D. A. (1995) Development of a detection system for viruses of woody plants based on PCR analysis of immobilized virions. *Phytopathology* **85,** 347–352.
10. Olmos, A., Dasi, M. A., Candresse, T., and Cambra, M. (1996) Print-capture PCR: A simple and highly sensitive method for the detection of plum pox virus (PPV) in plant tissues. *Nucleic Acids Res.* **24,** 2192–2193.
11. Olmos, A., Cambra, M., Dasi, M. A., Candresse, T., Esteban, O., Gorris, M. T., and Asensio, M. (1997) Simultaneous detection and typing of plum pox potyvirus (PPV) isolates by Heminested-PCR and PCR-ELISA. *J. Virol. Methods* **68,** 127–137.

12. Olmos, A., Cambra, M., Esteban, O., Gorris, M. T., and Terrada, E. (1999) New device and method for capture, reverse transcription and nested-PCR in a single closed-tube. *Nucleic Acids Res.* **27,** 1564–1565.
13. Levy, L., Lee, I. M., and Hadidi, A. (1994) Simple and rapid preparation of infected plant tissue extracts for PCR amplification of virus, viroid and MLO nucleic acids. *J. Virol. Methods* **49,** 295–304.
14. Mackenzie, D. J., McLean, M. A., Mukerji, S., and Green, M. (1997) Improved RNA extraction from woody plants for the detection of viral pathogens by reverse transcription-polymerase chain reaction. *Plant Dis.* **81,** 222–226.
15. Simmonds, P., Balfe, P., Peutherer, J. F., Ludlam, C. A., Bishop, J. O., and Brown, A. J. (1990) Related articles human immunodeficiency virus-infected individuals contain provirus in small numbers of peripheral mononuclear cells and at low copy numbers. *J. Virol.* **64,** 864–872.
16. Porter-Jordan, K., Rosenberg, E. I., Keiser, J.F., Gross, J.D., Ross, A. M., Nasim, S., et al. (1990) Related articles nested polymerase chain reaction assay for the detection of cytomegalovirus overcomes false positives caused by contamination with fragmented DNA. *J. Med. Virol.* **30,** 85–91.
17. Arias, C. R., Garay, E., and Aznar, R. (1995) Nested PCR method for rapid and sensitive detection of Vibrio vulnificus in fish, sediments, and water. *Appl. Environ. Microbiol.* **61,** 3476–3478.
18. Green, J., Henshilwood, K., Gallimore, C. I., Brown, D. W. G., and Lees, D. N. (1998) A nested reverse transcriptase PCR assay for detection of small round-structured viruses in environmentally contaminated molluscan shellfish. *Appl. Environ. Microbiol.* **64,** 858–863.
19. Noyes, H., Reyburn, A. H., Bailey, J. W., and Smith, D. (1998) A nested-PCR-based schizodeme method for identifying Leishmania kinetoplast minicircle classes directly from clinical samples and its application to the study of the epidemiology of Leishmania tropica in Pakistan. *J. Clin. Microbiol.* **36,** 2877–2881.
20. Yourno, J. (1992) A method for nested PCR with single closed reaction tubes. *PCR Methods Appl.* **2,** 60–65.
21. Wolff, C., Hornschemeyer, D., Wolf, D., and Kleesiek, K. (1995) Single-tube nested PCR with room-temperature-stable reagents. *PCR Methods Appl.* **4,** 376–379.
22. Mutasa, E. S., Chwarszczynska, D. M., and Asher, M. J. C. (1996) Single-tube, nested PCR for the diagnosis of Polymyxa betae infection in sugar beet roots and colorimetric analysis of amplified products. *Phytopathology* **86,** 493–497.
23. Bertolini, E., Olmos, A., López, M. M., and Cambra, M. (2003) Multiplex nested RT-PCR in a single tube for sensitive and simultaneous detection of four RNA viruses and *Pseudomonas savastonoi pv savastanoi* in olive trees. *Phytopathology* (in press).
24. Olmos, A., Bertolini, E., and Cambra, M. (2002) Simultaneous and co-operational amplification (Co-PCR): a new concept for detection of plant viruses. *J. Virol. Methods* **106,** 51–59.
25. Altschul, S. F., Madden, T. L., Schäfer, A. A., Zhang, J., Zhang, Z., Miller, W., et al. (1997) Gapped BLAST and PSI-BLAST: a new generation of protein database search programs. *Nucleic Acids Res.* **25,** 3389–3402.
26. Rosner, A., Maslenin, L., and Spiegel, S. (1997) The use of short and long PCR products for improved detection of prunus necrotic ringspot virus in woody plants. *J. Virol. Methods* **67,** 135–141.
27. Singh, M. and Singh, R. P. (1997) Potato virus Y detection: sensitivity of RT-PCR depends on the size of fragment amplified. *Can. J. Plant Pathol.* **19,** 149–155.

27

Direct PCR from Serum

Application to Viral Genome Detection

Kenji Abe

1. Introduction

Nucleic acids used for polymerase chain reaction (PCR) assays usually are extracted by the phenol-chloroform method or an alternative rapid purification. The acid-guanidinium thiocyanate-phenol-chloroform method for RNA extraction and proteinase K digestion-phenol-chloroform method for DNA extraction from serum samples are used widely for PCR assays ***(1,2)***, but usually at least 3 h are needed for this step. Detection of RNA is more difficult and complex than that of DNA, mainly because of the contaminating RNase and the need to conduct an additional reverse transcription (RT) step. Additionally, another problem is the high cost of the reagents for RNA extraction. Recently, we reported that viral RNA and DNA are readily amplified directly from serum without purification of nucleic acids by direct (RT) PCR ***(3)***. The method is sensitive because as little as 1 µL of the initial serum produces a clearly visible amplified fragment, and there is no difference in the sensitivity and stability between the direct PCR and conventional PCR assay. Interestingly, the results of the direct PCR are much better when 1 to 2 µL is used rather than 3 to 5 µL of serum. This inability to amplify RNA/DNA from serum may be to the result of the masking of the target RNA/DNA by coagulated proteins during the initial heat denaturation or the presence in the serum of inhibitors (probably such as lipoprotein) of the enzyme reaction. The sensitivity of the direct PCR assay allows detection of as few as one copy of viral genome. This technique not only eliminates the risk of cross contamination during nucleic acid extraction but also is cost and time saving. In this chapter, we describe the method of the direct (RT) PCR assay and apply it for detection of hepatitis B virus (HBV) DNA and hepatitis C virus (HCV) RNA in serum specimens.

2. Materials

1. Thermal cycler.
2. AmpliTaq Gold DNA polymerase 5 U/µL (Perkin–Elmer).
3. 10× reaction buffer containing 15 m*M* $MgCl_2$ (supplied with AmpliTaq Gold DNA polymerase kit).
4. RNase inhibitor 40 U/µL (Promega).

From: *Methods in Molecular Biology, Vol. 226: PCR Protocols, Second Edition*
Edited by: J. M. S. Bartlett and D. Stirling © Humana Press Inc., Totowa, NJ

5. Moloney murine leukemia virus: 200 U(µL (M-MLV) reverse transcriptase (Gibco BRL).
6. Primer sequences used for detection of HBV DNA and HCV RNA by direct (RT) PCR are listed in **Table 1** (100 ng/µL).
7. Deoxy nucleotide triphosphate (dNTP) mixture: 10 m*M* (Pharmacia Biotech).
8. Phosphate-buffered saline (PBS): Dissolve 8 g of NaCl, 0.2 g of KCl, 1.44 g of Na_2HPO_4, and 0.24 g of KH_2PO_4 in 800 mL of distilled H_2O. Adjust the pH to 7.4 with HCl. Add H_2O to 1 L. Dispense the solution into aliquots and sterilize them by autoclaving for 20 min at 15 psi in. on liquid cycle. Store at room temperature.
9. RNase-free H_2O.
10. 50-bp DNA ladder (Pharmacia Biotech).

3. Methods

3.1 Pretreatment of Serum Samples (see Note 1)

1. Serum samples (4 µL) are diluted with PBS to the final volume of 20 µL (1:5 dilution, *see* **Note 1**).
2. The diluted serum samples are heated for 3 min at 95°C then cooled rapidly on ice for 3 to 5 min.
3. Five microliters of the heat-denatured diluted serum samples (= 1 µL of serum) are used as templates for the nested (RT) PCR.

3.2. PCR Reaction

1. Set up PCR mastermixes as follows.
   a. Reverse transcription and first PCR buffer (when performing DNA PCR omit RNase Inhibitor and M-MLV Reverse Transcriptase, *see* **step 4**).

Ingredients	for RNA µL	for DNA µL	Final concentration
10× AmpliTaq Gold buffer (containing 15 m*M* $MgCl_2$)	5	5	1× (1.5 m*M* $MgCl_2$)
10 m*M* dNTP mix	1	1	200 µ*M*
Sense primer (100 ng/µL)	1	1	100 ng
Antisense primer (100 ng/µL)	1	1	100 ng
AmpliTaq Gold DNA polymerase (5 U/µL)	0.4	0.4	2 U
RNase inhibitor (40 U/µL)	0.25	None	10 U
M-MLV reverse transcriptase (200 U/µL)	0.5	None	100 U
RNase-free H_2O	35.85	36.6	
Total		45 µL	45 µL

   b. Second PCR buffer

Ingredients	µL	Final concentration
10× AmpliTaq Gold buffer (containing 15 m*M* $MgCl_2$)	5	1× (1.5 m*M* $MgCl_2$)
10 m*M* dNTP mix	1	200 µ*M*
Sense primer (100 ng/µL)	1	100 ng
Antisense primer (100 ng/µL)	1	100 ng
AmpliTaq Gold DNA polymerase (5 U/µL)	0.4	2 U
RNase-free H_2O	39.1	
Total	47.5 µL	

Table 1
Primer Sequences Used for Nested PCR

Primer	Sequence	Primer pair	Product size
For HBV DNA (X region)			
MD24	5′-TGCCAACTGGATCCTTCGCGGGACGTCCTT-3′ (nt 1392-1421)	MD24/MD26 (1st)	233bp
MD26	5′-GTTCACGGTGGTATAAATG-3′ (nt 1625-1607)		
HBx1	5′-GTCCCCTTCTTCATCTGCCGT-3′ (nt 1487-1507)	HBx1/HBx2 (2nd)	117 bp
HBx2	5′-ACGTGCAGAGGTGAAGCGAAG-3′ (nt 1604-1584)		
For HCV RNA (5′-untranslated region)			
19	5′-GCGACACTCCACCATAGAT-3′ (nt 2-20)	19/20 (1st)	329 bp
20	5′-GCTCATGGTGCACGGTCTA-3′ (nt 330-312)		
21	5′-CTGTGAGGAACTACTGTCT-3′ (nt 28-46)	21/22 (2nd)	268 bp
22	5′-ACTCGCAAGCACCCTATCA-3′ (nt 295-277)		

Nucleotide positions deduced from HBVadr4 (see **ref.** ***4***) for HBV and HC-J1 for HCV (*see* **ref.** ***5***)

Table 2
Detection Rate of Hepatitis Virus Nucleic Acids in Serum. Comparison of Conventional PCR and Direct PCR

	HBV DNA	HCV RNA	HGV RNA
Conventional PCR Positive	21	56	32
Direct PCR-Positive	21	51	30
Agreement Rate	100%	91%	94%

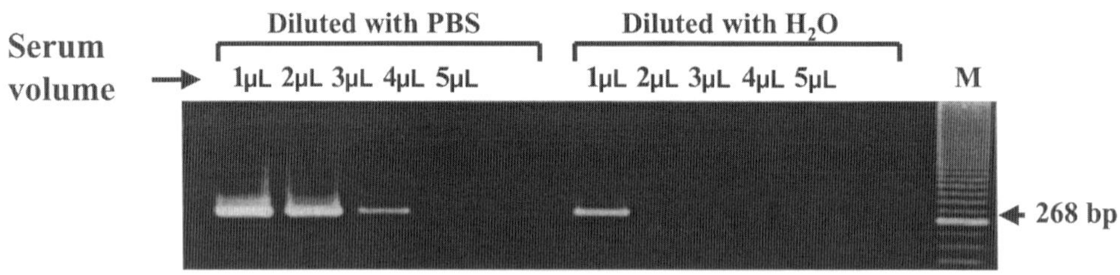

Fig. 1. Efficiency of serum volume and diluent as a factor in HCV-PCR reaction. A different amount of serum sample ranging in volume from 1 to 5 μL of serum was used. Each aliquot of serum, except for the 5 μL of volume, was diluted with PBS or H_2O to the final volume of 5 μL, respectively. M = 50-bp DNA ladder.

2. Perform the first PCR reaction by adding 45 μL of mastermix 1 to each 5-μL sample of heat-inactivated serum from **Subheading 3.1.** Then, program the thermocycler to incubate the samples for 50 min at 37°C for the initial RT step and then preheat at 95°C for 10 min to activate AmpliTaq Gold followed by 50 cycles consisting of 94°C for 20 s, 55°C for 20 s, and 72°C for 30 s using a Thermal Cycler. RT-PCR is performed with a one-step method combined with cDNA synthesis reaction, followed by the PCR reaction in a single tube (*see* **Note 2**). That is, for RNA of HCV, the first PCR is combined with the RT step in the same tube containing 50 μL of a reaction buffer as shown above. To obtain an automatic hot-start reaction, we use the AmpliTaq Gold DNA polymerase instead of regular thermostable DNA polymerase.
3. For the second reaction, 2.5 μL (1/20 volume) of the first PCR product is added to a tube containing second PCR buffer. The thermocycling for 50 cycles is performed as above, but omitting the initial 50 minute incubation at 37°C and the annealing temperature is raised to 60°C instead of 55°C for the second round of PCR.
4. The PCR products are electrophoresed on a 2% agarose gels staining with ethidium bromide and evaluated under ultraviolet light. The sizes of PCR products are estimated according to the migration pattern of 50-bp DNA ladder.

4. Notes

1. Examination of the optimal serum quantity reveals that 1 to 2 μL of serum give a readily amplifiable band of HCV RNA but 3 μL of serum give a faint band and 4 to 5 μL of serum do not yield any band (**Fig. 1**). This result is much better when PBS is used to dilute serum than that obtained using H_2O. It is similar even if this result is in the case of HBV.
2. The whole process of the direct RT-PCR can be completed within 6 h by combination with the one-step amplification method and the second round PCR.
3. Detection rate of HCV RNA and HBV DNA by direct PCR are consistent with the results obtained by conventional PCR (**Fig. 2** and **Table 2**).

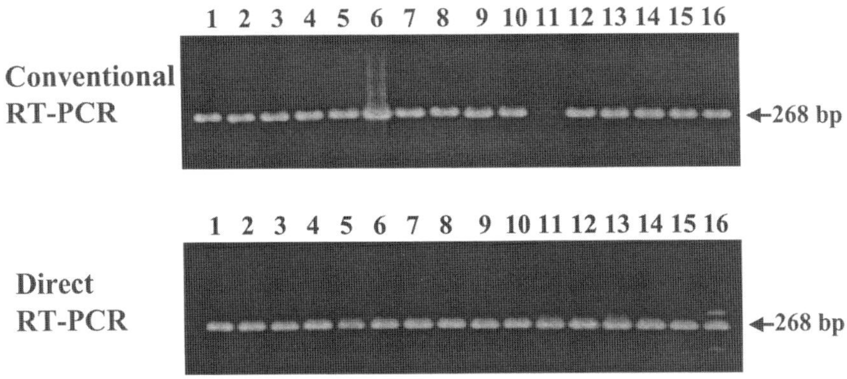

Fig. 2. Result of HCV-RNA detection in clinical serum samples. Comparison of conventional RT-PCR and direct RT-PCR.

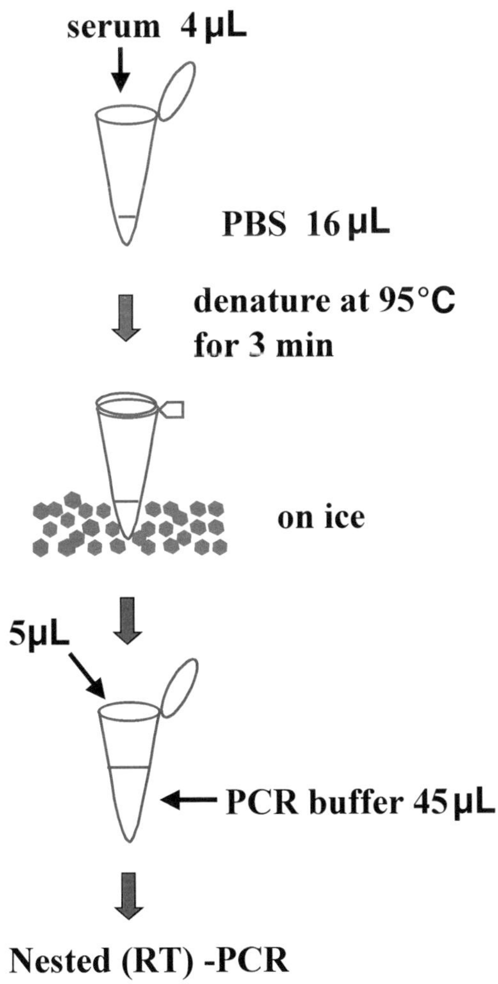

Fig. 3. An outline of direct (RT)-PCR.

4. For HBV, nested PCR using the same PCR buffer for RT-PCR, but without reverse transcriptase and omitting cDNA synthesis step is performed.
5. An outline of direct (RT)-PCR is shown in **Fig. 3.**

References

1. Chomczynski, P. and Sacchi, N. (1987) Single-step method of RNA isolation by acid guanidinium thiocyanate-phenol-chloroform extraction. *Anal. Biochem.* **162,** 156–159.
2. Kunkel, L. M., Smith, K. D., Boyer, S. H., Borgaonkar, D. S., Wachtel, S. S., Miller, O. J., et al. (1977) Analysis of human Y-chromosome-specific reiterated DNA in chromosome variants. *Proc. Natl. Acad. Sci. USA* **74,** 1245–1249.
3. Abe, K., Konomi, N., and Abdel-Hamid, M. (1998) Direct polymerase chain reaction for detection of hepatitis B, C and G virus genomes from serum without nucleic acid extraction—simple, rapid and highly sensitive method. *Hepatol. Res.* **13,** 62–70.
4. Fujiyama, A., Miyanchara, A., Nozaki, C., Yoneyama, T., Ohtoma, N., and Matsubara, K. (1983) Cloning and structural analyses of hepatitis B virus DNAs, subtype adr. *Nucleic Acids Res.* **11,** 4601–4610.
5. Okamoto, H., Okada, S., Sugiyama, Y., Yotsumoto, S., Tanaka, T., Yoshizawa, H., et al. (1990) the 5′-terminal sequence of the hepatitis C virus genome. *Jpn J. Exp. Med.* **60,** 167–177.

28

Long PCR Amplification of Large Fragments of Viral Genomes

A Technical Overview

Raymond Tellier, Jens Bukh, Suzanne U. Emerson, and Robert H. Purcell

1. Introduction

1.1. Long PCR

The polymerase chain reaction (PCR) has become an essential and ubiquitous tool for biological research and laboratory diagnostic applications. Until recently, reliable and sensitive amplification of large templates (several kb) was difficult to achieve. However, in 1994, an important breakthrough was reported by Barnes *(1)*. He hypothesized that a major obstacle to long PCR was the *Taq* DNA polymerase error rate, which causes mismatches that make elongation very inefficient. Many other thermostable DNA polymerases have a 3′ to 5′ exonuclease "proofreading" activity and a higher fidelity. However, the use of these polymerases alone does not reliably achieve long PCR, presumably because of excessive degradation of primers by the exonuclease activity *(1)*. The processivity of the enzyme may also be a factor. Of note, the 3′ to 5′ exonuclease activity alone is not a guarantee of high fidelity: Fidelity also depends on the degree of discrimination against misinsertion, the mismatch extension rate, and the rate of shuttling between polymerizing and proofreading modes *(2)*. The breakthrough reported by Barnes consisted in performing PCR with a mixture of two DNA polymerases: a major component consisting of a highly processive DNA polymerase and a minor component consisting of a DNA polymerase with a 3′ to 5′ exonuclease "proofreading" activity. With such enzyme mixes, reliable amplification of templates up to 35 kb in length was achieved *(1)*. The greater fidelity of long PCR enzyme mixes, relative to *Taq*, has been demonstrated *(1,2)*. Other modifications contribute to making long PCR possible, including optimization of the buffer and the thermal cycling conditions. It is especially important to address the drop in pH at high temperature observed with Tris-based buffers *(1,3)*. This is significant because DNA depurination is enhanced both by high temperature and by low pH. Because a larger DNA template is more likely than a small template to sustain depurination at sites within its boundaries, long PCR is inherently more vulnerable to depurination *(1,3)*. Interestingly, in contrast to 3′ to 5′ exonuclease-deficient DNA polymerases, the fidelity

From: *Methods in Molecular Biology, Vol. 226: PCR Protocols, Second Edition*
Edited by: J. M. S. Bartlett and D. Stirling © Humana Press Inc., Totowa, NJ

of Pfu (and therefore possibly of other proofreading DNA polymerases) actually increases with the pH *(2)*. In general, long PCR works better with rapid temperature transition, which is addressed by using smaller reaction and mineral oil volumes, thin-wall PCR tubes, and rapid thermocyclers *(1,3)*. Synthesis of amplicons of up to 42 kb has now been reported *(3)*.

There are currently several commercially available long PCR kits, made from different enzymes and requiring somewhat different protocols. It is expected that new thermostable enzymes will be introduced. More research will be needed to determine which enzyme mix is superior for which application.

1.2. Long RT-PCR

Long PCR from mRNAs or from the genomes of RNA viruses obviously necessitates the synthesis of a sufficient amount of long cDNA. In some systems, such as plant viruses, it is possible to start with a large amount of RNA *(4)*. For the more common scenario where the amount of RNA is limited, many have reported good results with the use of reverse-transcriptase enzymes that have been engineered to destroy the RNase H activity of the enzyme *(5–15)*. In principle, this would prevent the premature destruction of the RNA template before the synthesis of a large cDNA copy. A potential problem with the synthesis of long cDNA is the presence of strong secondary structures in some template RNAs. To some extent, this problem can be addressed by raising the temperature as high as enzyme stability permits during reverse transcription. For example, the RNase H-deficient reverse transcriptase Superscript II is active at temperatures up to 50°C (although its optimal temperature is between 42 and 45°C; **ref.** ***16***). Another approach is to use the thermostable DNA polymerase rTth which in the presence of Mn^{2+} has a reverse transcriptase activity at temperatures up to 70°C *(17,18)*. A fundamental limitation with this approach is that the RNA hydrolysis catalyzed by divalent cations is accelerated by high temperature *(17,19)*. Recent reports on the use of nucleic acid isostabilizers, such as betaine *(20)*, and the thermal stabilization of enzymes by trehalose *(18)* may lead to further refinements in this area.

2. Applications of Long PCR

2.1. General Applications

The availability of long PCR obviates some shortcomings of standard PCR: Not only does it speed up cloning or sequencing of genetic sequences of interest, but it extends the power of the PCR by enabling it to bridge large gaps of unknown or variable sequences between primers. As examples of the latter, there have been many reports of the use of long PCR to amplify introns of unknown sequence using primers targeting the adjacent exons (e.g., **refs.** ***21–24***), or to characterize transposons in *Drosophila* *(25)*. Long PCR has also been used to amplify the complete mitochondrial genome *(26)* and large genomic fragments for the determination of deletions in genetic diseases *(27,28)* or gene fusions in cancer *(29)*. Other applications in microbiology include bacterial typing *(30,31)* and amplification of *Plasmodium falciparum* DNA *(32)*.

2.2. Long PCR and Viruses

Long PCR has been applied with great success to viral genomes. For example, amplification of full-length, near full-length, or large fragments of the genome of many DNA viruses has been achieved, such as for the hepatitis B virus (HBV) *(8,33)*, the human papilloma virus *(34)*, SV-40 *(35)*, varicella-zoster virus *(36)*, the proviruses of the simian foamy virus of chimpanzees (SFVcpz) *(37)*, human T-cell leukemia virus 1 (HTLV-1) *(38)*, and human immunodeficiency virus type 1 (HIV-1) *(39–41)* and type 2 (HIV-2) *(42)*.

Long RT-PCR for RNA viruses has also been successful. Large fragments of the viral genome have been amplified for the tick-borne encephalitis virus *(6)*, the Norwalk-like viruses *(12)*, the hepatitis C virus (HCV) *(13)*, and the bovine torovirus *(43)*. The near full-length genome was amplified for HCV *(8,14)* and the full-length genome for the potato virus Y *(4)*, hepatitis A virus (HAV) *(7,9)*, hepatitis E virus *(44)*, poliovirus *(11)*, and coxsackie B2 virus *(10,15)*.

For HAV, we have shown that RNA transcribed directly from the full-length amplicon is infectious *(7,9)*. Furthermore, long PCR has greatly facilitated the construction of infectious clones or infectious cDNAs for HBV *(33)*, SV-40 *(35)*, SFVcpz *(37)*, HIV-1 *(41)*, tick borne encephalitis virus *(6)*, HAV *(9)*, coxsackie B2 *(15)*, coxsackie B6 *(45)*, HCV *(46,47)*, and potato virus Y *(4)*. These results confirm the high fidelity of long PCR.

In addition to streamlining the cloning and sequencing of viral genomes, long PCR presents other advantages for virology. For example, many viral sequences are toxic to *Escherichia coli*, leading to selection bias in cloning procedures *(48)*. This problem can be circumvented to a great extent by long PCR because large amplicons can be directly sequenced (or transcribed). Furthermore, direct sequencing of amplicons provides a certain protection against DNA polymerase mistakes during PCR: Unless these mistakes occur in the early cycles, at each position the majority of amplicons will have the correct nucleotide and the sequencing will provide the correct result, whereas cloning might select an amplicon with mistakes. However, one must remember that there are circumstances, such as when encountering homopolymeric stretches, tandem repeats, extreme AT or GC content, or strong secondary structures, in which polymerases can produce systematic mistakes *(49–52)*.

Long PCR is also of great interest for RNA viruses, which typically exist as "quasispecies," a mixed population containing several variants. Not only does direct sequencing of the amplicon yield immediately the consensus sequence, but it has been proposed that long PCR can be used to preserve and propagate a representative DNA version of such a "quasispecies" *(9,11)*. In fact, it has been shown that long PCR preserves the distribution of variants of poliovirus *(11)* and HIV-1 *(53)*. In the latter study, however, a relatively high rate of recombination between templates was observed after performing PCR *(53)*. Recombination was measured by cloning and sequencing, and thus this measurement is possibly affected by a selection bias from the cloning process. Nonetheless, because of this observation, and pending further research on this topic, the possibility of artifactual recombination should be kept in mind. Because the recombination is thought to occur as the result of incomplete transcripts acting as "mega primers" in subsequent PCR cycles *(53,54)*, careful optimization of the elongation time during PCR may minimize this problem. The use of polymerases with high processivity may also diminish the occurrence of recombination *(54)*.

References

1. Barnes, W. M. (1994) PCR amplification of up to 35-kb DNA with high fidelity and high yield from λ bacteriophage templates. *Proc. Natl. Acad. Sci. USA* **91,** 2216–2220.

2. Cline, J., Braman, J. C., and Hogrefe, H. H. (1996) PCR fidelity of Pfu DNA polymerase and other thermostable DNA polymerases. *Nucleic Acids Res.* **24,** 3546–3551.
3. Cheng, S., Fockle, C., Barnes, W. M., and Higuchi, R. (1994) Effective amplification of long targets from cloned inserts and human genomic DNA. *Proc. Natl. Acad Sci. USA* **91,** 5695–5699.
4. Fakhfakh, H., Vilaine, F., Makni, M., and Robaglia, C. (1996) Cell-free cloning and biolistic inoculation of an infectious cDNA of potato virus Y. *J. Gen. Virol.* **77,** 519–523.
5. Nathan, M., Mertz, L. M., and Fox, D. K. (1995) Optimizing long RT-PCR. *Focus* **17,** 78–80
6. Gritsun, T. S. and Gould, E. A. (1995) Infectious transcripts of tick-borne encephalitis virus, generated in days by RT-PCR. *Virology* **214,** 611–618.
7. Tellier, R., Bukh, J., Emerson, S. U., and Purcell, R. H. (1996) Amplification of the full-length hepatitis A virus genome by long reverse transcription-PCR and transcription of infectious RNA directly from the amplicon. *Proc. Natl. Acad. Sci. USA* **93,** 4370–4373. (Erratum published in *Proc. Natl. Acad. Sci. USA* 1997;94:10,485)
8. Tellier, R., Bukh, J., Emerson, S. U., Miller, R. H., and Purcell, R. H. (1996) Long PCR and its application to hepatitis viruses: amplification of hepatitis A, hepatitis B and hepatitis C virus genomes. *J. Clin. Microbiol.* **34,** 3085–3091. (Erratum published in *J. Clin. Microbiol.* 1997;35:2713).
9. Tellier, R., Bukh, J., Emerson, S. U., and Purcell, R. H. (1997) Amplification of the full length hepatitis A virus by long RT-PCR and generation of infectious RNA directly from the amplicon, in *Viral Hepatitis and Liver Disease Proceedings of the IX Triennial International Symposium on Viral Hepatitis and Liver Disease* (Rizzetto, M., Purcell, R. H., Gerin, J. L., and Verme G., eds.), Minerva Medica, Turin, pp. 48–50.
10. Gow, J. W., McGill, M. M., Behan, W. M. H., and Behan, P. O. (1996) Long RT-PCR amplification of full-length enterovirus genome. *BioTechniques* **20,** 582–584.
11. Chumakov, K. M. (1996) PCR engineering of viral quasispecies: A new method to preserve and manipulate genetic diversity of RNA virus populations *J. Virol.* **70,** 7331–7334.
12. Ando, T., Monroe, S. S., Noel, J. S., and Glass, R. I. (1997) A one tube method of reverse transcription-PCR to efficiently amplify a 3-kilobase region from the RNA polymerase gene to the poly(A) tail of small round-structured viruses (Norwalk-like viruses). *J. Clin. Microbiol.* **35,** 570–577.
13. Rispeter, K., Lu, M., Lechner, S., Zibert, A., and Roggendorf, M. (1997) Cloning and characterization of a complete open reading frame of the hepatitis C virus genome in only two cDNA fragments. *J. Gen. Virol.* **78,** 2751–2759.
14. Wang, L-F., Radkowski, M., Vargas, H., Rakela, J., and Laskus, T. (1997) Amplification and fusion of long fragments of hepatitis C virus genome. *J. Virol. Methods* **68,** 217–223.
15. Lindberg, A. M., Polacek, C., and Johansson, S. (1997) Amplification and cloning of complete enterovirus genomes by long distance PCR. *J. Virol. Methods* **65,** 191–199.
16. Gibco BRL (1994) Technical bulletin 18064-2. Gibco BRL, Gaithersburg, Md.
17. Myers, T. W. and Gelfand, D. H. (1991) Reverse transcription and DNA amplification by a Thermus thermophilus DNA polymerase. *Biochemistry* **30,** 7661–7666.
18. Carninci, P., Nishiyama, Y., Westover, A., Itoh, M., Nagaoka, S., Sasaki, N., et al. (1998) Thermostabilization and thermoactivation of thermolabile enzymes by trehalose and its application for the synthesis of full length cDNA. *Proc. Natl. Acad. Sci. USA* **95,** 520–524.
19. Lindahl, T. (1967) Irreversible heat inactivation of transfer ribonucleic acids. *J. Biol. Chem.* **242,** 1970–1973.
20. Henke, W., Herdel, K., Jung, K., Schnorr, D., and Loening, S. A. (1997) Betaine improves the PCR amplification of GC-rich DNA sequences. *Nucleic Acids Res.* **25,** 3957–3958.

21. Gaedigk, R., Karges, W., Hui, M. F., Scherer, S. W., and Dosch, H.-M. (1996) Genomic organization and transcript analysis of ICAp69, a target antigen in diabetic autoimmunity. *Genomics* **38,** 382–391.
22. Purroy, J., Bisceglia, L., Calonge, M. J., Zelante, L., Testar, X., Zorzano, A., et al. (1996) Genomic structure and organization of the human *rBAT* gene (SLC3A1). *Genomics* **37,** 249–252.
23. Nobile, C., Marchi, J., Nigro, V., Roberts, R. G., and Danieli, G. A. (1997) Exon-intron organization of the human dystrophin gene. *Genomics* **45,** 421–424.
24. Kelley, P. M., Weston, M. D., Chen, Z.-Y, Orten, D. J., Hasson, T., Overbeck, L. D., et al. (1997) The genomic structure of the gene defective in Usher syndrome type 1b (MYO7A). *Genomics* **40,** 73–79.
25. Gordadze, A. V. and Beněs, H. (1996) Long PCR-based technique for detection of transposon insertions in and around cloned genes of *Drosophila melanogaster. BioTechniques* **21,** 1062–1066.
26. Li, Y.-Y., Hengstenberg, C., and Maisch, B.(1995) Whole mitochondrial genome amplification reveals basal level multiple deletions in mtDNA of patients with dilated cardiomyopathy. *Biochem. Biophys. Res. Commun.* **10,** 211–218.
27. Bochmann, H., Gehrisch, S., and Jaross, W. (1996) Fast amplification of the low density lipoprotein receptor gene and detection of a large deletion by means of long polymerase chain reaction. *Eur. J. Clin. Chem. Clin. Biochem.* **34,** 955–959.
28. Hećimović, S., Barišić, I., Müller, A., Pertković, I., Barić, I., Ligutić, et al. (1997) Expand long PCR for fragile X mutation detection. *Clin. Genet.* **52,** 147–154.
29. Waggott, W., Lo, Y.-M. D., Bastard, C., Gatter, K. C., Leroux, D., Mason, D. Y., et al. (1995) Detection of NPM-ALK DNA rearrangement in CD30 positive anaplastic large-cell lymphoma. *Br. J. Haematol.* **89,** 905–907.
30. Smith-Vaughan, H. C., Sriprakash, K. S., Mathews, J. D., and Kemp, D. J. (1995) Long PCR-ribotyping of nontypeable Haemophilus influenzae. *J. Clin. Microbiol.* **33,** 1192–1195.
31. Hookey, J. V., Saunders, N. A., Clewley, J. P., Efstratiou, A., and George, R. C. (1996) Virulence regulon polymorphism in group A streptococci revealed by long PCR and implications for epidemiological and evolutionary studies. *J. Med. Microbiol.* **45,** 285–293.
32. Su, X.-Z., Wu, Y., Sifri, C. D., and Wellems, T. E. (1996) Reduced extension temperatures required for PCR amplification of extremely A+T– rich DNA. *Nucleic Acids Res.* **24,** 1574–1575.
33. Günther, S., Li, B.-C., Miska S., Krüger, D. H. Meisel, H., and Will, H. (1995) A novel method for efficient amplification of whole hepatitis B virus genomes permits rapid functional analysis and reveals deletion mutants in immunosuppressed patients. *J. Virol.* **69,** 5437–5444.
34. Stewart, A.-C. M., Gravitt, P. E., Cheng, S., and Wheeler, C. M. (1995) Generation of entire human papillomavirus genomes by long PCR: frequency of errors produced during amplification. *Genome Res.* **5,** 79–88.
35. Lednicky, J. A., Jafar, S., Wong, C., and Butel, J. S. (1997) High-fidelity PCR amplification of infectious copies of the complete simian virus 40 genome from plasmids and virus-infected cell lysates. *Gene* **184,** 189–195.
36. Takayama, M., Takayama, N., Inoue, N., and Kameoka, Y. (1996) Application of long PCR method to identification of variations in nucleotide sequences among Varicella-Zoster virus isolates. *J. Clin. Microbiol.* **34,** 2869–2874.
37. Herchenröder, O., Turek, R., Neumann-Haefelin, D., Rethwilm, A., and Schneider, J. (1995) Infectious proviral clones of chimpanzee foamy virus(SFVcpz) generated by long PCR reveal close functional relatedness to human foamy virus. *Virology* **214,** 685–689.

38. Tamiya, S., Matsuoka, M., Etoh, K.-I., Watanabe, T., Kamihira, S., Yamaguchi K., et al. (1996) Two types of defective human T-lymphotropic virus type 1 provirus in adult T-cell leukemia. *Blood* **88,** 3065–3073.
39. Salminen, M. O., Koch, C., Sanders-Buell, E., Ehrenberg, P. K., Michael, N. L., Carr, J. K., et al. (1995) Recovery of virtually full-length HIV-1 provirus of diverse subtypes from primary virus cultures using the polymerase chain reaction. *Virology* **213,** 80–86.
40. Salminen, M. O., Johansson, B., Sönnerborg, A., Ayehunie, S., Gotte, D., Leinikki, P., et al. (1996) Full-length sequence of an Ethiopian human immunodeficiency virus type 1 (HIV-1) isolate of genetic subtype C. *Aids Res. Hum. Retroviruses* **12,** 1329–1339.
41. Gao, F., Robertson, D. L., Carruthers, C. D., Morrison, S. G., Jian, B., Chen, Y., et al. (1998) A comprehensive panel of near-full-length clones and reference sequences for non-subtype B isolates of human immunodeficiency virus type 1. *J. Virol.* **72,** 5680–5698.
42. Damond, F., Loussert-Ajaka, I., Apetrei, C., Descamps, D., Souquière, S., Leprêtre, A., et al. (1998) Highly sensitive method for amplification of human immunodeficiency virus type 2 DNA. *J. Clin. Microbiol.* **36,** 809–811.
43. Duckmanton, L. M., Tellier, R., Liu, P., and Petric, M. (1998) Bovine torovirus: sequencing of the structural genes and expression of the nucleocapsid protein of Breda virus. *Virus Res.* **58,** 83–96.
44. Donati, M. C., Fagan, E. A., and Harrison, T. J. (1996) Sequence analysis of full length HEV clone derived directly from human liver in fulminant hepatitis E, in *Viral Hepatitis and Liver Disease, Proceedings of the IX Triennial International Symposium on Viral Hepatitis and Liver Disease* (Rizzetto, M., Purcell, R. H., Gerin, J., and Verme, G., eds.), Minerva Medica, Turin, pp. 313–316.
45. Martino, T. A., Tellier, R., Petric, M., Irwin, D. M., Afshar A., and Liu P. (1999) The complete consensus sequence of coxsackievirus B6 and generation of infectious clones by long RT-PCR. *Virus Res.* **64,** 77–86.
46. Yanagi, M., Purcell, R. H., Emerson S. U., and Bukh, J. (1997) Transcripts from a single full-length cDNA clone of hepatitis C virus are infectious when directly transfected into the liver of a chimpanzee. *Proc. Natl. Acad. Sci. USA* **94,** 8738–8743.
47. Yanagi, M., St. Claire, M., Shapiro, M., Emerson, S. U., Purcell., R. H., and Bukh, J. (1998) Transcripts of a chimeric cDNA clone of hepatitis C virus genotype 1b are infectious in vivo. *Virology* **244,** 161–172.
48. Forns, X., Bukh, J., Purcell, R. H., and Emerson, S. U. (1997) How *Escherichia coli* can bias the results of molecular cloning: preferential selection of defective genomes of hepatitis C virus during the cloning procedure. *Proc. Natl. Acad. Sci. USA* **94,** 13,909–13,914.
49. Loewen, P. C. and Switala, J. (1995) Template secondary structure can increase the error frequency of the DNA polymerase from Thermus aquaticus. *Gene* **164,** 59–63.
50. Wu, M. J., Chow, L. W., and Hsieh, M. (1998) Amplification of GAA/TTC triplet repeat in vitro: Preferential expansion of (TTC)n strand. *Biochem. Biophys. Acta* **1407,** 155–162.
51. Ji, J., Clegg, N. J., Peterson, K. R., Jackson, A. L., Laird, C. D., and Loeb, L. A. (1996) In vitro expansion of GGC:GCC repeats: identification of the preferred strand of expansion. *Nucleic Acids Res.* **24,** 2835–2840.
52. Kroutil, L. C., Register, K., Bebenek, K., and Kunkel, T. A. (1996) Exonucleolytic proofreading during replication of repetitive DNA. *Biochemistry* **35,** 1046–1053.
53. Yang, Y. L., Wang, G., Dorman, K., and Kaplan, A.H. (1996) Long polymerase chain reaction amplification of heterogeneous HIV type 1 templates produces recombination at a relatively high frequency. *Aids Res. and Hum. Retroviruses* **12,** 303–306.
54. Saiki, R. K., Gelfand, D. H., Stoffel, S., Scharf, S. J., Higuchi, R., Horn, G. T., et al. (1988) Primer directed enzymatic amplification of DNA with a thermostable DNA polymerase. *Science* **239,** 487–491.

29

Long PCR Methodology

Raymond Tellier, Jens Bukh, Suzanne U. Emerson, and Robert H. Purcell

1. Introduction

In this chapter, we detail protocols of long polymerase chain reaction (PCR) and long RT-PCR, which we have found to be versatile, sensitive, and straightforward to optimize. We have used these protocols with success on several different templates, including lambda phage DNA, HAV, HBV, HCV *(1)*, torovirus *(2)*, coxsackie B6 virus *(3)*, and human beta galactosidase mRNA (R. Tellier, unpublished data). The guanine–cytosine (GC) content of these templates varied from 37.8 to 58.8%. These protocols have also been used on genomic human DNA with success *(4)*.

The long PCR protocol we use is derived from the method described by Barnes *(5)* for KLA-16, a mixture of KlenTaq 1 and *Pfu*. We have replaced this mix by a commercially available mix, Advantage KlenTaq Polymerase mix (Clontech), a mix of KlenTaq 1, and Deep Vent. In our hands, this mix was found to produce more consistent results. KlenTaq 1 is an N-terminal deletion mutant of *Taq*, analogous to the Klenow fragment enzyme, and devoid of the 5′- 3′ exonuclease activity of *Taq* **(5)**. Unexpectedly, KlenTaq 1 was also found to be more thermostable and slightly more accurate than *Taq* ***(5,6)***, as well as less sensitive to variation in Mg^{2+} concentration and more processive than *Taq (7)*.

2. Materials

1. Water free of RNase, DNAse, and proteinase (ddH_2O). Aliquot and store at –20°C.
2. DTT (100 m*M*, Promega or Gibco BRL). Store at –20°C.
3. RNasin (20–40 U/µL; Promega). Store at –20°C.
4. Superscript II reverse transcriptase (Gibco BRL/Life Technologies). Store at –20°C.
5. 5× RT buffer (Gibco BRL). Store at –20°C.
6. RNase H (1–4 U/µL; Gibco BRL). Store at –20°C.
7. RNase T1 (900–3000 U/µL; Gibco BRL). Store at –20°C.
8. 50× Advantage KlenTaq Polymerase Mix (Clontech). Store at –20°C(*see* **Note 1**).
9. 10× KlenTaq PCR buffer (Clontech). Store at –20°C (*see* **Note 2**).
10. dNTP mix (10 m*M* each). Prepared in ddH_2O from 100 m*M* dNTP set (Pharmacia). Aliquot and store at –80°C.
11. Mineral oil (molecular biology grade, Sigma).
12. Primers. Aliquot at a concentration of 10 µ*M* in ddH_2O, store at –20°C (*see* **Note 3**).

From: *Methods in Molecular Biology, Vol. 226: PCR Protocols, Second Edition*
Edited by: J. M. S. Bartlett and D. Stirling © Humana Press Inc., Totowa, NJ

13. Thin-wall PCR tubes (Stratagene; *see* **Note 4**).
14. Robocycler 40 (Stratagene; *see* **Note 5**).

3. Methods: Long PCR

3.1. Long PCR of DNA

1. Components of the PCR are prepared in a master mix. For each reaction use the following: 10× KlenTaq PCR buffer (5 µL); 1.25 µL of dNTP mix (10 m*M* each) L; 1 µL of sense primer (10-µ*M* stock); 1 µL of antisense primer (10-µ*M* stock); 1 µL of KlenTaq Advantage; and 30.75 µL of ddH_2O.
2. Aliquot 40 µL in thin-wall 0.5-mL PCR tubes and keep at room temperature. Add 10 µL of the DNA template to the mix. Overlay with 40 µL of mineral oil.
3. Cycling parameters. We use a Robocycler 40; during each cycle of PCR we use the following parameters (*see* **Note 6**): denaturation: 99°C × 35 s; annealing: 67°C × 30 s; and elongation: 68°C × (optimal time for the targeted amplicon; *see* **Note 7**).

3.2. Nested Long PCR

1. We have shown that the strategy of nested PCR can be applied with success to long PCR. It requires a slightly modified master mix, including per reaction (*see* **Note 8**): 4.5 µL of 10× KlenTaq PCR buffer; 1.25 µL of dNTP mix (10 m*M* each); 1 µL of sense primer (10 µ*M*); 1 µL of antisense primer (10 µ*M*); 1 µL of KlenTaq Advantage; and 36.25 µL of ddH_2O.
2. Aliquot 45 µL of master mix in 0.5-mL thin-wall PCR tubes and overlay with 40 µL of mineral oil. Add 5 µL of the first-round PCR (*see* **Notes 9** and **10**).
3. Cycling parameters as in **Subheading 3.1.**

3.3. Long RT-PCR

3.3.1. Reverse Transcription

1. Components of the reverse transcription are prepared in a master mix. For each reaction use (*see* **Note 11**) the following: 4 µL of 5× RT buffer; 0.5 µL of RNasin; 1 µL of DTT (100 m*M*); 1 µL of dNTP mix (10 m*M* each); 2.5 µL of primer (10 µ*M*; *see* **Note 12**); and 1 µL of Superscript II.
2. Heat the 10 µL of RNA aliquot at 65°C for 2 min, and then put on ice.
3. Add 10 µL of master mix and incubate 1 h at 42°C (*see* **Note 13**).
4. Put on ice, add 1 µL of RNase H and 1 µL of RNase T1, and incubate 20 min at 37°C (*see* **Note 14**).
5. Keep on ice until used in long PCR or keep frozen for later use.

3.3.2. Long RT-PCR

Because of buffer incompatibility (the buffer for Superscript II contains KCl), not all of the RT reaction can be used in the PCR. We obtain good results by transferring a small amount of the RT reaction into a long PCR mix. The remainder of the RT reaction can then be frozen and used at a later time.

1. To amplify by long PCR the cDNA produced in the RT, we prepare a master mix as in **Subheading 3.1.** but corrected for adding the template in a volume of 2 µL. For each reaction, use the following: 5 µL of 10 × KlenTaq PCR buffer; 1.25 µL of dNTP mix (10 m*M* each); 1 µL of sense primer (10-µm stock); 1 µL of antisense primer (10-µm stock); 1 µL of KlenTaq Advantage; 38.75 µL of ddH_2O.

2. Aliquots (48 μL) of the master mix are placed in thin-wall 0.5-mL PCR tubes and overlaid with 40 μL of mineral oil.
3. Finally, 2 μL of the RT reaction is added. We neglect the contribution of the 2 μL of RT reaction to primer and dNTP concentrations, etc.

4. Notes

1. The KlenTaq Advantage polymerase mix also contains anti-*Taq* antibody, which ensures a "hot start" to the PCR. We have not found it necessary to include an initial prolonged denaturation step, so the way we are using the polymerase mix is in fact "time-release PCR."
2. The Clontech buffer is Tricine-based, and therefore is not subject to the same drop in pH at high temperatures that is seen with Tris-based buffers *(8)*. The buffer already contains Mg^{2+}; we have not found it necessary to modify the concentration of Mg^{2+} for any of the templates we have amplified. The buffer does contain KOAc instead of KCl, the latter having been shown to decrease the processivity of DNA polymerases *(8)*.
3. Primers: 10 pmol of each primer are used in the PCR. Because long PCR in general uses a higher annealing temperature than does standard PCR, a requirement of primer design is that the Tm must be high enough to hybridize at that temperature. Whereas 20 to 25 mers with a sufficiently high GC content will work, we have also often used primers of 30 to 40 bp without problems, and when incorporating promoters and cloning sites, primers of up to ~60 bp. Barnes *(5)* has also reported the use of amplicons of a few hundred bp as "mega primers." Primer design should include consideration of secondary structures, dimers, etc., but because of the high temperatures throughout the PCR cycle and the "hot start," there is in fact greater tolerance for weak secondary structures than with many "standard" PCR protocols. Because of the high annealing temperature, however, there is a low tolerance for mismatches between the primers and the template.
4. Long PCRs usually require rapid temperature transition (however, some templates are more tolerant than others), and several parameters are optimized toward this goal, including the small reaction volume (50 μL) and the small volume of oil. Because different volumes lead to different thermal capacity, great consistency is required for optimization. We use the Stratagene thin-wall tubes; we have found that some other brands may require, for some templates, different parameters: again, consistency is the essential element.
5. Long PCR requires a thermal cycler with fast temperature transitions; depending on the template and the size of the target for amplification, not all thermal cyclers will allow successful amplification *(5)*. In any case, cycling parameters must be determined for each different model of thermal cycler.
6. We usually use 35 cycles of PCR amplification. If greater yield or sensitivity is required, consider using nested PCR.
7. The elongation time is usually the only parameter to adjust in the protocol and depends on the template and the size of the targeted amplicon. Elongation times shorter or longer than the optimal time can result in sensitivity loss (or even negative results). Extension times that are too long are often associated with a "smearing" of the reaction product when it is electrophoresed on agarose gels. Optimal elongation time is best determined by using a serial dilution of the template (since suboptimal times may give positive reactions but with a loss of a few logs in sensitivity). The initial time should be chosen between 1 and 2 min per kb. Like other authors *(8)*, we have found that step-wise increase of the elongation time can be beneficial. Examples of elongation times for various templates can be found in Barnes *(5)* and Tellier et al. *(1)*. For example, to obtain amplicons from the tobacco mosaic virus (6.2 kb), HAV (7.5 kb), and HCV (9.25 kb), we found we could use the same elongation times: 9 min 45 s for the first 15 cycles, 11 min for the next 10 cycles, and

13 min for the last 10 cycles *(1)*. To minimize recombination events during PCR, one should probably err on the side of longer elongation times.

8. The master mix is merely corrected for a 5 μL volume of template, taking into account the amount of buffer carried from the first reaction.
9. For nested PCR, the primers of the second round must be internal to those of the first round (although they can overlap). If the same primers were to be used, artifactual bands produced by long PCR because of false priming would also get re-amplified. A second round of long PCR with the same primers can be performed, but this necessitates purification of the first round amplicon by agarose gel electrophoresis *(9)*.
10. The overall sensitivity that can be achieved can be close to standard PCR: for the lambda phage DNA, we have obtained an 11-kb amplicon from as few as 200 copies. When we used a nested long PCR protocol we could obtain, at the end of the whole process, a 5-kb amplicon starting from as few as 20 copies *(1)*.
11. This protocol assumes that the RNA is diluted in 10 m*M* DTT and 5% (vol/vol) RNasin (20–40 U/μL, Promega). If one uses RNA dissolved in RNase free water, increase the DTT in the RT master mix to 2 μL, and decrease the amount of primer to 1.5 μL.
12. The primer for the reverse transcription must be a template-specific primer and can be one of the primers used in the long PCR. Do not use random hexamers.
13. The optimal temperature for Superscript II is between 42 and 45°C *(10)*, but it is active up to 50°C. Although incubation at 50°C was less sensitive in our hands *(1)*, for some templates with strong secondary structures, it may be advantageous.
14. The treatment with RNase H and RNase T1 is not absolutely necessary, but we have found that it increases the sensitivity of the long RT-PCR *(1)*. Other authors have also reported benefits from the use of RNase H *(11–13)*. We have not tested separately the contributions of RNase H and RNase T1, but the benefits of RNase T1 are expected to vary depending on the amount of extraneous RNA present in the template.

References

1. Tellier, R., Bukh, J., Emerson, S. U., Miller, R. H., and Purcell, R. H. (1996) Long PCR and its application to hepatitis viruses: Amplification of hepatitis A, hepatitis B and hepatitis C virus genomes. *J. Clin. Microbiol.* **34,** 3085–3091. (Erratum published in *J. Clin. Microbiol.* 1997;35:2713).
2. Duckmanton, L. M., Tellier, R., Liu, P., and Petric, M. (1998) Bovine torovirus: Sequencing of the structural genes and expression of the nucleocapsid protein of Breda virus. *Virus Res.* **58,** 83–96.
3. Martino, T. A., Tellier, R., Petric, M., Irwin, D. M., Afshar, A., and Liu P. (1999) The complete consensus sequence of coxsackievirus B6 and generation of infectious clones by long RT-PCR. *Virus Res.* **64,** 77–86.
4. Chen, B., Rigat, B., Curry, C., and Mahuran, D. J. (1999) Structure of the GM2A gene: identification of an exon 2 nonsense mutation and a naturally occurring transcript with an in-frame deletion of exon 2. *Am. J. Hum. Genet.* **65,** 77–87.
5. Barnes, W. M. (1994) PCR amplification of up to 35-kb DNA with high fidelity and high yield from λ bacteriophage templates. *Proc. Natl. Acad. Sci. USA* **91,** 2216–2220.
6. Lawyer, F. C., Stoffel, S., Saiki, R. K., Chang, S. Y., Landre, P. A., Abramson, R. D., et al. (1993) High-level expression, purification, and enzymatic characterization of full-length *Thermus aquaticus* DNA polymerase and a truncated form deficient in 5′ to 3′ exonuclease activity. *Genome Res.* **2,** 275–287.
7. Clontech (1998) *User Manual for Advantage cDNA PCR kit and Advantage cDNA Polymerase Mix.*

8. Cheng, S., Fockle, C., Barnes, W. M., and Higuchi, R. (1994) Effective amplification of long targets from cloned inserts and human genomic DNA. *Proc. Natl. Acad Sci. USA* **91,** 5695–5699.
9. Tellier, R., Bukh, J., Emerson, S. U., and Purcell, R. H. (1997) Amplification of the full length hepatitis A virus by long RT-PCR and generation of infectious RNA directly from the amplicon in *Viral Hepatitis and Liver Disease, Proceedings of the IX Triennial International Symposium on Viral Hepatitis and Liver Disease* (Rizzetto M., Purcell R. H., Gerin J. L., and Verme, G., eds.), Minerva Medica, Turin, pp. 48–50.
10. Gibco BRL (1994) *Technical Bulletin 18064-2*. Gibco BRL, Gaithersburg, MD.
11. Nathan, M., Mertz, L. M., and Fox, D. K. (1995) Optimizing long RT-PCR. *Focus* **17,** 78–80.
12. Wang, L.-F., Radkowski, M., Vargas, H., Rakela, J., and Laskus, T. (1997) Amplification and fusion of long fragments of hepatitis C virus genome. *J. Virol. Methods* **68,** 217–223.
13. Lindberg, A. M., Polacek, C., and Johansson, S. (1997) Amplification and cloning of complete enterovirus genomes by long distance PCR. *J. Virol. Methods* **65,** 191–199.

IV

Ultrasensitive and Quantitative PCR

30

Qualitative and Quantitative PCR

A Technical Overview

David Stirling

1. Introduction

The nature of the polymerase chain reaction (PCR) process lends itself well to qualitative determinations. It transforms very small quantities of analyte into the realms of bucket chemistry, allowing specific gene portions to be directly visualized with ethidium bromide and ultraviolet light. It was these approaches that were first to be exploited in DNA analysis. The early PCR tests were for the presence or absence of a gene, transcript, or by inference whole organism (e.g., pathogen identification). These qualitative tests rapidly evolved to distinguish between related genes or organisms and accelerated the whole process of molecular taxonomy.

In contrast to this global utility and acceptance, the use of quantitative PCR has grown more slowly, for similar reasons. A chain reaction by definition is intuitively out of control. Surely the vagaries of chaos theory will obliterate any hope of quantitation; it matters not if the butterfly flaps once or one thousand times. Despite these early misgivings, a great many successful quantitative PCR protocols have been developed and gained general acceptance.

2. Qualitative PCR

There a number of considerations that should be taken into account when performing qualitative PCR. Careful thought at the outset of the design process can avoid a great deal of subsequent frustration. When the PCR product contains the expected amplicon (assuming all negative controls have been included and are negative), it is safe to assume the template was present in the starting sample. However, the converse is not true. There are a number of common causes for the failure of qualitative PCR other than the absence of template.

2.1. Template Concentration

Even the most basic qualitative PCR is dependent on quantitative changes in template concentration. The whole process is based on an ideally exponential amplification of starting material to a point where it can be readily seen and manipulated. Clearly, this ability will be influenced by both the starting concentration of template and the

From: *Methods in Molecular Biology, Vol. 226: PCR Protocols, Second Edition*
Edited by: J. M. S. Bartlett and D. Stirling © Humana Press Inc., Totowa, NJ

efficiency of the reaction. Although it is theoretically possible to amplify from a single copy of a sequence, if the reaction conditions are not optimized, such amplification may not reach a threshold for detection within the average 30-cycle PCR. If the object of the exercise is to determine the presence or absence of a specific sequence, elements of quantitation must be built into the reaction design to determine the required level of sensitivity. Where PCR is to be used to test for the presence of a pathogen for instance, consideration has to be given to the clinically significant levels. If a single organism is clinically significant, then clearly the test has to be capable of detecting a single copy. If, however, 1000 organisms are required for clinically significant event, the assay need not be as sensitive *(1)*.

2.2. Mixed Template Competition

If the template contains more than one species of DNA capable of being amplified by the primers, those templates that exist in greater initial concentrations or that amplify more efficiently may outcompete the remaining templates. For instance, PCR is used to amplify papilloma virus sequences in cervical cytology specimens. Consensus primers are used to amplify a broad spectrum of viral strains, which can then be characterized by restriction digest or probe hybridization. If the patient has a mixed infection, it is likely that the more abundant strain will be amplified preferentially, resulting in only one strain being represented in the PCR product *(2)*. This is perfectly adequate to determine which are the predominant strain(s) within a sample. However, if a complete description of the strains present is required, PCR should be performed to specifically amplify individual strains

3. Quantitative PCR: General Considerations

3.1. End-Point Analysis

Unfortunately, the cause of quantitative PCR has not promoted by the use of end-point analysis. This approach simply amplifies multiple templates under the same set of conditions and examines the amount of product at the end of the run. Because PCR exhibits a typical exponential amplification of product, complete with lag phase and plateau, it is easily possible for widely differing starting template concentrations to yield remarkably similar final product concentrations. This can be a useful technique but only if great care is taken to establish an appropriate end point. Reactions should be sampled after a wide range of cycle numbers, to delineate the exponential phase, and the subsequent plateau. Tests should then be designed to sample during the exponential phase of amplification, before primers, dNTPs, and or enzyme become limiting. As with all quantifications, it is greatly enhanced by the inclusion of a suitable internal control.

3.2. Limiting Dilution

PCR quantitation can be achieved using qualitative end points. For the PCR to proceed, there must me a theoretical minimum of one template per reaction. Thus, if a dilution series is performed on the template, a dilution can be reached where an aliquot of sample taken for PCR has a statistical probability of containing no template molecules. The higher the initial template concentration, the greater the dilution required needed to reach that point. This approach has been used with great success to quantify viral titers *(3)*.

3.3. Competitive Template PCR

If two different templates capable of being amplified by the same primers are present at the start of a PCR, one will tend to predominate by the end. This will be determined by a number of factors, such as relative starting concentrations, size, and internal sequence. If a competitive template is designed to be almost identical to the test sequence, the final dominance of test or competitor will entirely be the result of the initial concentration. Many groups, including myself, have used competitors identical to the test DNA apart from one base change (used to introduce a restriction site to distinguish the templates) in this way *(4)*. A dilution series of competitor is set up with a constant concentration of test sample and subjected to PCR. A plot of the relative product concentrations at the end of the amplification reveals a point where there is equivalent production from both template and competitor. This then yields the concentration of the test sample template. This is a very robust approach to quantitation, but is time consuming both to establish and to run, requiring many PCRs for each sample, and being difficult to automate.

3.4. Real-Time PCR

The development of fluorescent detection systems, capable of monitoring PCR product accumulation while it occurs has greatly improved the reliability of quantitative PCR. There is a well-established inverse correlation between the template concentration and the duration of the lag phase of the PCR. An entire generation of automated systems has been developed to exploit this. Using any of several different chemistries, the accumulation of PCR product is monitored until a predetermined threshold is reached. This lag time is then compared with a standard curve and a concentration calculated.

Lending itself very readily to automation, real-time PCR is becoming the method of choice for most quantitative PCR systems. Although the equipment for performing such analysis has been fairly expensive, second-generation systems are already on the market at much more reasonable costs.

One final note of caution: It is not uncommon for these systems to be based upon the detection of accumulated PCR product by the use of a fluorescent dye, such as SYBR green. Such systems can work very well when the PCRs are optimized to yield only the product to be quantified. If, however, a great many heterogeneous samples are to be analyzed, it is entirely possible that some will yield additional bands, confounding the results. The use of hybridization probe strategies will greatly reduce this possibility.

References

1. Tonjum, T., Klintz, L., Bergan, T., Baann, J., Furuberg, G., Cristea, M., et al. (1996) Direct detection of Mycobacterium tuberculosis in respiratory samples from patients in Scandinavia by polymerase chain reaction. *Clin. Microbiol. Infect.* **2,** 127–131.
2. O'Leary, J. J., Landers, R. J., Crowley. M., Healy, I., O'Donovan, M., Healy, V., et al. (1998) Human papillomavirus and mixed epithelial tumors of the endometrium. *Hum. Pathol.* **29,** 383–389.
3. Lee, T. H., Sunzeri, F. J., Tobler, L. H., Williams, B. G., Busch, M. P. Lee, T. H., et al. (1991) Quantitative assessment of HIV-1 DNA load by coamplification of HIV-1 gag and HLA-DQ-alpha genes. *AIDS* **5,** 683–691.
4. Stirling, D., Hannant, W. A., and Ludlam, C. A. (1998)Transcriptional activation of the factor VIII gene in liver cell lines by interleukin-6. *Thromb. Haemost.* **79,** 74–78.

31

Ultrasensitive PCR Detection of Tumor Cells in Myeloma

Friedrich W. Cremer and Marion Moos

1. Introduction

Chromosomal aberrations, such as translocations or inversions, described for a growing number of malignancies, are now widely used to detect tumor cells by polymerase chain reaction (PCR). However, in multiple myeloma (MM), no such ubiquitous PCR marker exists. Therefore, other means have been established to distinguish myeloma cells from normal cells. Because the plasma cells of a myeloma clone share an identical rearranged immunoglobulin gene sequence, it is possible to detect malignant cells with PCR primers specific for the VDJ rearrangement of the heavy chain of each myeloma clone. The sensitivity and specificity of this method, named allele-specific oligonucleotide (ASO) PCR, even with low proportions of malignant cells, has been proven *(1)*.

The heavy chains of the immunoglobulins have three highly variable regions near their amino-terminal end that mediate specific binding of the antigen. These regions are called complementarity determining regions (CDR1 to –3). The sequences of the heavy chain of the immunoglobulin genes are encoded on chromosome 14. In germline configuration, several hundred base pairs divide the approx 200 variable regions (V) from the 30 diversity regions (D). Several kilobase pairs downstream, 6 joining regions (J) are located. About 7000 base pairs to the 3′-end the constant regions of the heavy chain are encoded. During B-cell maturation, a random rearrangement of these regions occurs, in which one of the V-, one of the D-, and one of the J-regions are joined together, thus generating the VDJ-segment. Together with the sequence of the rearranged light chain, it determines the antigenic specificity of the immunoglobulin.

The rearrangement of V-, D-, and J-regions can generate over 35,000 different VDJ-sequences. The diversity of these sequences is further increased by three different mechanisms: (1) the recombination process is imprecise, thus nucleotides can be lost or stretches of bases that divide the different regions are not deleted; (2) nucleotides can be inserted without matrix at the junctions of V-, D-, and J-regions; and (3) somatic hypermutations, especially of the V-regions, further enhance the antigenic specificity of the immunoglobulin. Although CDR1- and CDR2-regions are entirely encoded by the V-region, the CDR3-region stretches over the 3′-end of the V-region, the D-region and the 5′-end of the J-region (*see* **Fig. 1**). Thus, the CDR3-region has the highest

From: *Methods in Molecular Biology, Vol. 226: PCR Protocols, Second Edition*
Edited by: J. M. S. Bartlett and D. Stirling © Humana Press Inc., Totowa, NJ

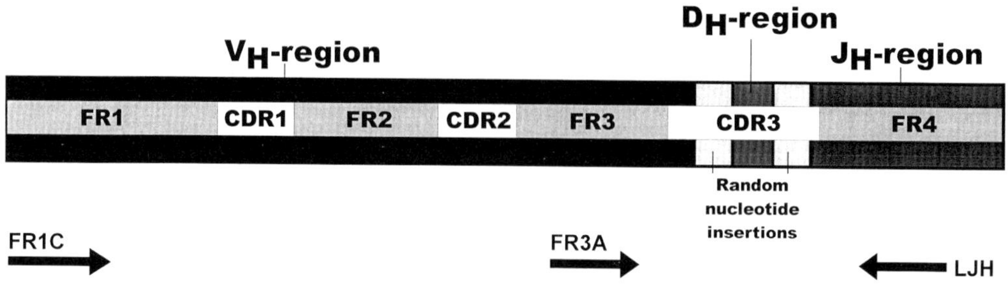

Fig. 1. Rearranged VDJ segment of a mature B-cell and strategies for consensus PCR to amplify the VDJ segment. The CDR regions of the heavy chain of the immunoglobulin gene are flanked by highly conserved segments, the framework (FR) regions 1 through 4. Consensus primers complementary to these FR-regions have been developed that can be used to amplify the enclosed highly variable CDR regions regardless of their sequence. In this chapter, two strategies are described using either FR1C or FR3A plus LJH.

diversity of all CDR-regions, and its sequence is a specific marker for each clone of B-cells *(2,3)*.

In MM, the CDR regions of the malignant cells are somatically hypermutated, and no further mutations occur, which would lead to an oligoclonal diversification *(4)*. This is a prerequisite for designing ASO primers complementary to the CDR regions of the malignant clone that allow the detection of myeloma cells by PCR.

For designing ASO primers, the sequence of the CDR regions of the malignant clone has to be determined. In a first step, the CDR regions of virtually all B-cells are amplified with consensus primers flanking the CDRs (*see* **Fig. 1**). Besides the strategies using FR1C *(5)* or FR3A *(6)* as sense primers and LJH as antisense primer, consensus PCR with family-specific primers complementary to the leader segment 5′-end to the V-region *(7)*, with a mixture of six FR1 family-specific primers *(8)*, with an FR2 consensus primer alone *(9)* or as a mixture with the FR1 family-specific primers VH5 and VH6 *(5)* and a mixture of JH1245, JH3, and JH6 antisense primers *(5)* has been described. After consensus PCR, the product of the malignant clone can be distinguished from normal ones clones by its predominant occurrence among the polyclonal CDR-regions. After sequencing (directly or after cloning), primers complementary to the CDR regions can be designed. These ASO primers have to be tested for specificity and sensitivity before they can be used for the PCR detection of cells of the malignant clone. PCR is quantified using the method of limiting dilutions.

2. Materials

2.1. Isolation of Nucleic Acids

1. Bone marrow (BM) sample with a high proportion of myeloma cells.
2. Ficoll separating solution (Biochrom, Berlin, Germany).
3. Phosphate-buffered saline (pH 7.4; Gibco BRL, Eggenstein, Germany).
4. TE buffer (10 m*M* Tris-HCl, pH 7.5, and 1 m*M* EDTA).
5. Reagents for RNA extraction, for example, Trizol Reagent (Gibco BRL).
6. Reagents for DNA extraction, for example, DNAzol (Gibco BRL).
7. Ethidium bromide-stained agarose gels (0.8 and 2%).

8. Denaturating agarose gels (1.2%) containing 2.2 m*M* formaldehyde for RNA.
9. PCR reagents and primers for amplifying reference genes like β-actin or GAPDH.

2.2. Consensus RT-PCR

1. GeneAmp RNA PCR Core Kit (Perkin–Elmer, Weiterstadt, Germany).
2. Amplitaq Gold DNA polymerase with GeneAmp 10× PCR buffer II (Perkin–Elmer) and $MgCl_2$ solution.
3. Mixture of dATP, dCTP, dGTP, and dTTP, concentration 2.5 m*M* each (stock solution 100 m*M*, Promega, Madison WI).
4. Consensus primers FR1C (5′-GGTGCAGCTGS(A/T)GSAGTC(A/G/T)GG-3′) ***(5)***, FR3A (5′-ACACGGCYSTGTATTACTGT-3′) ***(6)***, and LJH (5′-TGAGGAGACGGTGACC-3′) ***(6)***.
5. Ethidium bromide-stained agarose gels (2 and 5%, NuSieve 3:1 Agarose, FMC/Biozym, Oldendorf, Germany).

2.3. Identification of the Myeloma CDR Regions

1. Scalpel.
2. DNA purification kit (EasyPure DNA Purification kit, Biozym, Göttingen, Germany).
3. Cloning kit for PCR products (TOPO TA Cloning Kit, Invitrogen, De Schelp, The Netherlands).
4. LB medium, LB plates, with 50 µg/mL kanamycin (Boehringer Mannheim, Mannheim, Germany).
5. Lysis buffer (100 µg/mL of Proteinase K (Boehringer Mannheim) in 10 m*M* Tris-Cl, pH 7.5, and 1 m*M* EDTA).
6. PCR reagents (*see* **Subheading 2.2.**) and M13 universal (forward) and M13 reverse primers.
7. Sequitherm Cycle Sequencing kit (Biozym, Göttingen, Germany).

2.4. Quantitative PCR (qPCR) Using ASO Primers

1. DNA processing software, for example Gene Jockey II (Biosoft, Cambridge, United Kingdom).
2. PCR reagents, ASO primers as designed, LJH primer, ethidium bromide stained agarose gels (2% or 5%).
3. Buffy coat DNA from healthy donors.
4. Computer program for likelihood maximization and χ^2 minimization, for example, MAXLIKE. This program written in C (Watcom C/C++ version 10.6, Powersoft, Waterloo, Canada) running under DOS™ on an IBM™ compatible PC or the source code can be obtained for free according to the terms and conditions for copying, distribution, and modification of programs under the GNU general public license agreement from the authors (see contact address).
5. Myeloma or B-cell lines, for example U266 ***(10)*** or Daudi ***(11)***.

3. Methods

3.1. Isolation of Nucleic Acids

1. Collect bone marrow aspirate in two 10-mL syringes with heparin added for anticoagulation. This will be the starting material for both the identification of the VDJ sequence of the malignant clone and the testing of ASO primers.
2. Isolate mononucleated cells by centrifugation over Ficoll separating solution.
3. Wash cells twice in phosphate-buffered saline, pelleting between washes by centrifugation at 800*g* for 10 min.

4. Count cells and divide them into at least 1×10^7 cells for RNA extraction and the remaining cells for DNA extraction. If the cell pellet appears red, resuspend in TE buffer and vortex for 5 to 10 s to lyse remaining red blood cells, then pellet again quickly and remove TE buffer.
5. Add 250 µL of Trizol reagent per 1×10^7 cells then freeze cells at –20°C for later RNA extraction. Freeze the remaining cells immediately for later DNA extraction.
6. Thaw pelleted cells in Trizol reagent and extract RNA according to the manufacturers' instructions.
7. Thaw pelleted cells and isolate DNA, for example using the DNAzol reagent according to the manufacturers instructions or *see* Chapter 6 for DNA extraction protocol.
8. Determine concentration of RNA and DNA by optical density (OD) measurement at 260 and 280 nm. Adjust concentration of RNA to 500 ng/µL and of DNA to 100 ng/µL.
9. Confirm integrity of RNA by electrophoresis of 500 ng on a denaturing 1.2% agarose gel containing 2.2 m*M* formaldehyde.
10. Confirm integrity of DNA by electrophoresis of 100 ng on a 0.8% agarose gel.
11. Check quality of DNA or RNA by PCR amplifying a gene like β-actin in the case of DNA or a housekeeping gene like GAPDH in the case of RNA.

3.2. Consensus RT-PCR

1. Perform reverse transcription reaction in a total volume of 20 µL containing 2 µg of RNA using commercially available kits.
2. Prepare a PCR mixture containing 8 µL of 10× PCR buffer, 1 µL of 20 m*M* primer solutions FR1C or FR3A plus LJH-CL, and 2.5 U Amplitaq Gold DNA polymerase.
3. Add the total volume of the RT reaction mixture to the PCR mixture. The final volume should be 100 µL.
4. Amplify with a program consisting of 7 min of denaturation and enzyme activation at 94°C, followed by 40 cycles of denaturation for 1 min at 94°C and combined annealing and extension at 63°C (for primer FR3A) or 65°C (for primer FR1C) for 1 min, followed by a final extension step at 65°C for 5 min.
5. Resolve PCR products either on a 2% (PCR products are approx 350 bp in size if FR1C was used) or 5% (PCR products are approx 110 bp in size if FR3A was used) ethidium bromide stained agarose gels. The monoclonal CDR regions lead to a distinct band if the proportion of myeloma cells is high enough. This band can be distinguished from the surrounding smear of polyclonal CDR-regions. **Figure 2** shows an example of a consensus PCR (*see* **Note 1**).

3.3. Identification of the Myeloma CDR Regions

3.3.1. Cloning of Consensus PCR Products and Direct Lysis of Transformed Bacteria

1. Excise appropriately sized consensus PCR products from the agarose gel with a scalpel *(4)*.
2. Purify DNA from the agarose block using commercially available kits.
3. Clone-purify PCR products into plasmids and transform competent cells using kits featuring TA-cloning. If possible, use kanamycin rather than ampicillin for selection, as satellite colonies occur less frequently. Grow bacteria for 12 to 16 h at 37°C.
4. Pick single colonies and transfer to a replica plate. Spread on an area of approx 1 cm^2. Incubate for 12 to 16 h at 37°C. If a distinct band without surrounding smear was visible after consensus PCR, 10 to 20 colonies should be sufficient. If a band surrounded by a polyclonal smear was visible, 20 to 40 colonies should be picked.

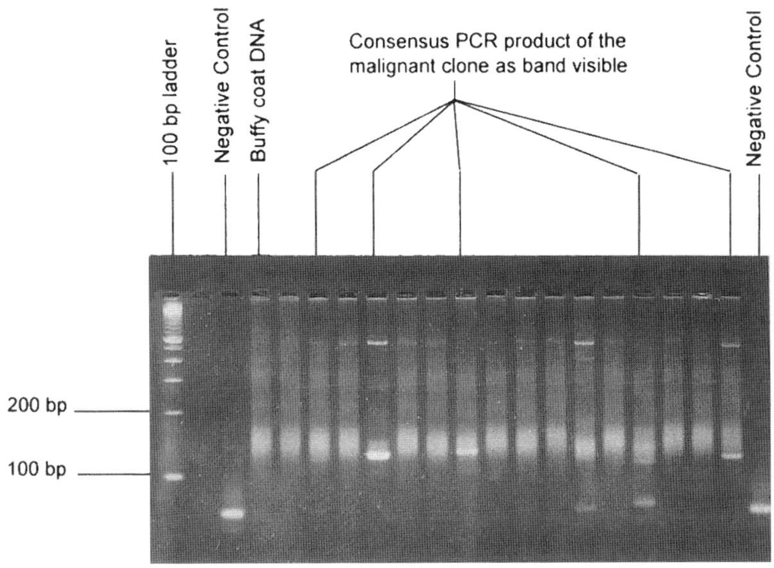

Fig. 2. Consensus PCR using primers FR3A and LJH. In 5 of 16 cases, the consensus PCR product of the malignant clone can be distinguished from the surrounding polyclonal smear. In all other cases, only a smear, like the one seen in buffy coat DNA, is visible. This figure also illustrates the low rate of distinct consensus PCR products obtained if using DNA instead of RNA. Detection rates can further be increased by using two different consensus strategies in parallel.

5. Scrape layer of bacteria from the plate and suspend in 100 µL of lysis buffer. Vortex vigorously.
6. Incubate the suspension at 65°C for 15 min.
7. Inactivate proteinase K at 95°C for 15 min.
8. Pellet debris by centrifugation at maximum speed and 4°C. Transfer the supernatant containing the plasmids to a fresh tube. Store at 4°C.
9. Analyze plasmids for inserts by PCR using primers complementary to the M13 sites of the vector. Set up a PCR containing 5 µL of 10× PCR buffer, 2 m*M* $MgCl_2$, each deoxynucleotide at 0.1 m*M*, primers M13 forward plus M13 reverse at 0.4 µ*M* each, 2.5 U *Taq*-DNA-polymerase, and 1 µL of plasmid solution. Amplify for 40 cycles of 1-min denaturation at 94°C, 1 minute annealing at 65°C, and extension at 72°C, followed by a final extension of 5 min at 72°C. Analyze on a 2% agarose gel.

3.3.2. Sequencing and Identification of the CDR Regions of the Myeloma Clone

1. Prepare sequencing reactions using commercially available kits according to the manufacturer's instructions with 4 to 8 µL of plasmid solution, and perform cycle sequencing.
2. Analyze on a sequencer, for example Alf-express (Pharmacia, Freiburg, Germany).
3. Analyze obtained sequences by aligning them to known CDR regions (examples for CDR3 regions are given in **Fig. 3**; *see* **Note 2**).
4. Identify the consensus PCR product of the malignant clone by its predominant occurrence among the polyclonal CDR-regions. B-cells of normal clones are not encountered more frequently than once in 20000 B-cells *(12)*. At least five identical clones should be typed to identify the myeloma clone.

```
Primer FR3A                                Hypervariable region                                                            Primer LJH
acacggcystgtattactgt acaaaagataagggtattt.....tgaatggtttg.....g.c.t.ac...t.ttgactcctggggccagggaaccct ggtcaccgtctcctca
acacggcystgtattactgt gca.cgcat.ggtctatgatagcatggattctttcgtctactacttctatggtctggacgtctggggccacgggacccc ggtcaccgtctcctca
acacggcystgtattactgt gtg.cg.ag.ggttta.g.tag.........ttcg..tggtac..caat..t.t.gactactggggccaaggaaccct ggtcaccgtctcctca
acacggcystgtattactgt gcga.g.a...ga......ccgg.gagc......cg..tggggt....a..ctttgactattggggccagggaaccct ggtcaccgtctcctca
acacggcystgtattactgt gcga.a.a...ga.....tcagg.gtgcat.tttcg..agggtgtcttcaccccttgactactggggccagggaaccct ggtcaccgtctcctca
acacggcystgtattactgt gcga.g.agcgggct.t.tacggagtggttattgcg.caattattactattacatggacctctggggcaaagggaccac ggtcaccgtctcctca
acacggcystgtattactgt gcga.a.a...gact.t.tcataacggggta.....ctactactacggtt...tggacgtctggggccaagggaccac ggtcaccgtctcctca
acacggcystgtattactgt gcgagg.ag..gtc.tagc.g.....a.ccctt...a.t..ta.cg.ctc....tggacgtctggggccaagggacctc ggtcaccgtctcctca
acacggcystgtattactgt gcgaga.at..tcc.cagt.gga..ta.caact...a.t..gg.ta.gcc....tgaattactggggccagggaaccct ggtcaccgtctcctca
acacggcystgtattactgt gcgaaagatcgagatctttccgggtgggtggtggtgactacgacgt.acgagt.ttgacaaatggggccagggaaccca ggtcaccgtctcctca

                                                  ggtgactacgacgt.acgag          Primer 1
                                             gggtggtggtgactacgacg               Primer 2
                                       ttccgggtgggtggtggtgac                    Primer 3
                                     tctttccgggtgggtggtgg                       Primer 4
```

Fig. 3. Alignment of sequences of CDR3 consensus PCR products of the malignant clones from 10 patients with MM and an example of four ASO primers designed for the last of the given sequences (testing of the primers is shown in **Fig. 4**). Aligning allows to distinguish variable and rather conserved segments. Primers are designed to be complementary to the highly variable region.

3.4. Quantitative PCR (qPCR) Using ASO Primers

To quantitate PCR results, essentially four different methods have been described. Although the measurement of the amount of PCR product, the coamplification of a control gene, and the competitive PCR rely on the quantitation of the generated product, the limiting dilution assay analyzes only positive and negative PCRs at different dilution levels ***(13)***. Amplification efficiencies and hence the amounts of PCR product can vary substantially from tube to tube despite identical and simultaneous processing of reaction mixtures ***(14)***, even if quantitation is performed in the exponential phase of amplification. Limiting dilutions are less prone to errors caused by these deviations. Standards can be used to control for these differences; however, in MM a system of standard and template would have to be established for each new patient. Therefore, we prefer the limiting dilutions assay, which analyzes only PCR positivity versus PCR negativity. A prerequisite of this method is that a single copy of target DNA can be detected.

*3.4.1. Designing Primers (see **Note 3**)*

1. ASO primers should be 18 to 23 bases long.
2. The initiation of the *Taq*-DNA-polymerase can be enhanced by 3′-ends with NS or SS as bases, but more than 2 G at the 3′-end should be avoided.
3. The 5′-end and 3′-end of primers should not be complementary to each other to avoid primer homodimerization.
4. The A/T to G/C ratio should be approx 50%, if possible.
5. There are several computer programs that can be used to check primers for hairpin formation and primer pairs for dimerization.
6. FR3-strategy: Identify hypervariable parts of the CDR3-region of the malignant clone by aligning the sequence to known CDR3-sequences. Avoid segments that are relatively conserved. Design ASO primer as forward primer according to above considerations. Use ASO primer together with LJH as antisense primer.

7. FR1-strategy: Distinguish FR regions and CDR regions by aligning sequence to known CDR sequences. Design the first ASO primer as the forward primer complementary to the CDR1 or CDR2 region (CDR1 regions are often rather conserved, thus CDR2-specific primers might be preferred). Design the second ASO primer as the antisense primer complementary to the CDR3 region. Use these primers as a pair for PCR.

3.4.2. Testing of ASO Primers

Because even carefully designed primers may not work as well as expected, only testing can distinguish between oligonucleotides that fulfill all requirements and those that are not specific or sensitive enough. CDR1/2- plus CDR3-specific ASO primer pairs do not necessarily give better results than CDR3-specific plus J-consensus primers. An example of a primer test is given in **Fig. 4**. Design several CDR3-specific or CDR1/CDR2-specific plus CDR3-specific (antisense) primers. An example for 4 ASO primers is given in **Fig. 3**.

1. Prepare PCRs containing 5 µL of 10× PCR buffer, 2 m*M* $MgCl_2$, each deoxynucleotide at 0.1 m*M*, the ASO primer pair or the CDR3-specific primer plus LJH-CL at 0.8 µ*M* each, and 2.5 U *Taq*-DNA-polymerase.
2. Add 500 ng of DNA from the initial BM sample to generate a positive control. Add 1000 ng of buffy coat DNA from healthy donors to generate a negative control that tests for specificity of the primers. Adjust with distilled water to a final volume of 50 µL. Repeat this for every primer combination to be tested.
3. Amplify with a program consisting of 7 min of preheating at 94°C, 60 cycles of 1 min of denaturation at 94°C, and 1 minute of annealing and extension at 63°C, ending with a final extension at 63°C for 5 min.
4. Analyze PCR products on a 2 (for FR1/2-strategy) or 5% (for FR3-strategy) ethidium bromide-stained agarose gel. Reactions containing DNA from the positive control should display a prominent band of the expected size (*see* **Note 4**). No such product should be visible in those with buffy coat DNA only. Avoid primers that generate a multitude of nonspecific products.

3.5. Quantitation by Limiting Dilutions

1. Isolate DNA from the sample to be assessed (*see* **Note 3**).
2. Determine the DNA concentration of the sample by OD measurement. Add distilled water to achieve a DNA concentration of 100 ng/µL.
3. Prepare a dilution series in 0.5 log steps (resulting in dilution levels of 1:3, 1:10, 1:33, etc.).
4. For each dilution level, set up a PCR containing 10 µL of the DNA solution. Use conditions as determined to be optimal for the ASO primers used (*see* **Subheading 3.7.**) and amplify.
5. Analyze PCR products on agarose gels. Determine the highest dilution that still generates a specific PCR product.
6. Set up five replicates of PCRs containing 10 µL of DNA solution from the highest dilution level that was PCR positive, five replicates with DNA from the lowest dilution level that was PCR negative, and five replicates with DNA from the next lowest dilution level that was PCR negative and amplify.
7. Analyze on an agarose gel.
8. Decrease or increase the dilution level analyzed in five replicates until the pattern of PCR results finally obtained shows a change from PCR positivity in all replicates (or the

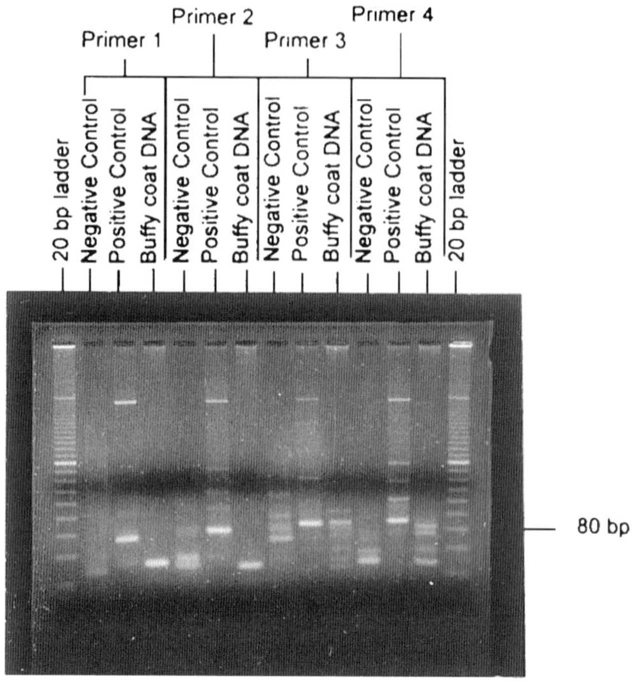

Fig. 4. Testing of primers (*see* **Fig. 3**) as described in **Subheading 3.4.2.** For each ASO primer, a negative control without added DNA, a positive control with DNA from the initial BM sample, and a specificity test with buffy coat DNA were amplified by PCR. Using primer 1, only the positive control leads to the PCR product of the expected size. Primers 2, 3, and 4 produce several nonspecific bands, even in the negative control. Primer 3 generates a false-positive PCR product of the same size with buffy coat DNA. Primer 1 was selected, and PCR conditions were further optimized for this oligonucleotide as described in **Subheading 3.7.**

undiluted level has been reached) to partial PCR positivity in the next higher dilution levels to PCR negativity in all replicates in the highest dilution levels (*see* **Fig. 5**).

9. To deduce the proportion of malignant cells in the undiluted sample from the pattern of PCR results, analyze by likelihood maximization and by χ^2-minimization *(15)*, for example, by using the MAXLIKE computer program. Ten microliters of the undiluted DNA solution (100 ng/µL) are equivalent to 165,000 cells, which is used as a starting point for the calculations. The program will return the most probable value for the initial proportion of malignant cells in the analyzed sample.
10. Compare the values obtained by likelihood maximization and by χ^2-minimization. If PCR results are plausible, both methods will yield consistent values (*see* **Note 5**).

3.6. Testing the Sensitivity of ASO Primers

A prerequisite for the analysis of the pattern of PCR results by χ^2-minimization or likelihood maximization is that a single copy of target DNA per PCR tube can be detected. It is advisable that at least the overall sensitivity of the assay should be tested, as described in **Subheading 3.6.1.**

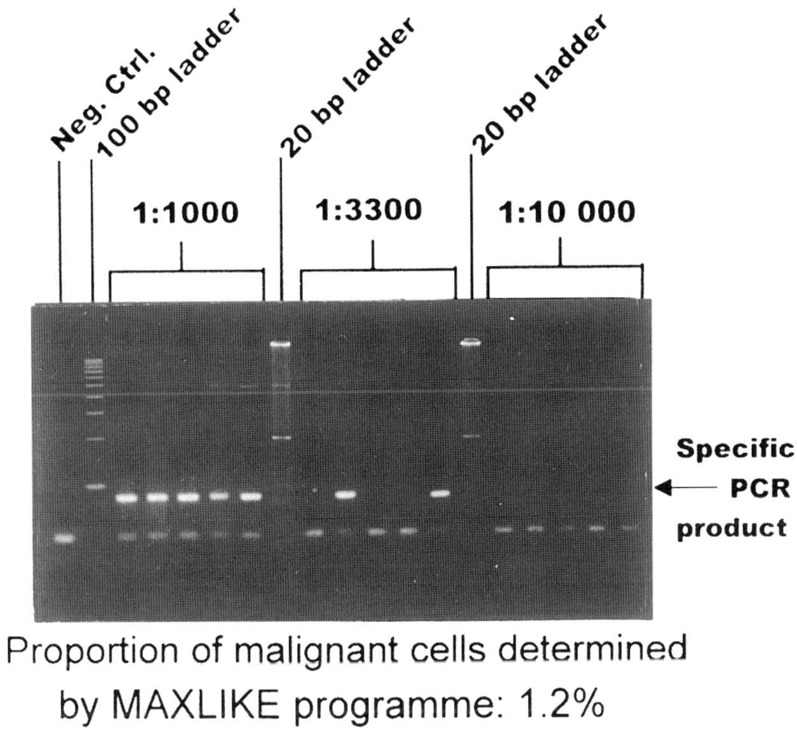

Fig. 5. Example for a qPCR of a PB sample from a patient with MM. The specific PCR product is 85 bp. The highest dilution level at which all five replicates are PCR positive is 1:1000, the lowest dilution level at which all five replicates are PCR negative is 1:10,000. Analysis of this pattern of PCR results by likelihood maximization and χ^2-minimization yielded a result of 1.2% of malignant cells.

3.6.1. Testing the Sensitivity of the Assay Using a Cell Line

1. Design primers for a B-cell or myeloma cell line and test for specificity.
2. Isolate DNA from cells of the cell line and from buffy coat of healthy donors and quantitate by OD measurement.
3. Mix DNA from the cell line with buffy coat DNA to simulate samples with different proportions of malignant cells (100%, 10%, 1%, 0.1%, 0.01%, and 0.001%).
4. Analyze all samples by qPCR (*see* **Chapter 6.3.**) using the appropriate ASO primers.
5. Compare the results of the qPCR with the simulated tumor loads.

3.6.2. Testing Primer Sensitivity by Comparison to Results of Flow Cytometry

1. Collect a BM sample with a proportion of malignant cells of greater than 2% from a patient in whom ASO primers have been devised.
2. Assess sample for CD38++, κ/γ-restricted cells by flow cytometry. This gives an approximate value for the proportion of malignant cells.
3. Isolate DNA from this sample (*see* **Subheading 3.1.**).
4. Quantitate the tumor load by qPCR as described in **Subheading 3.1.**

5. Compare the value determined by qPCR with the result of the analysis by flow cytometry. If the result obtained by qPCR is definitely lower than the one obtained by flow cytometry, then the ASO primer is most probably not able to detect a single copy of the template of the malignant clone.

3.7. Optimization of PCR Conditions for ASO Primers

1. As a starting point, use PCRs of 50 µL containing 5 µL of 10× PCR buffer, 2 m*M* $MgCl_2$, each deoxynucleotide at 0.1 m*M*, the ASO primer pair or the CDR3-specific primer plus LJH at 0.8 µ*M* each, 2.5 U *Taq*-DNA-polymerase, and a maximum of 1000 ng of DNA. Amplify with a program consisting of 7 min preheating at 94°C, 60 cycles of 1 min of denaturation at 94°C and 1 min of annealing and extension at 63°C, ending with a final extension at 63°C for 5 min.
2. If these amplification conditions lead to unsatisfactory results, for example, too many nonspecific products or low intensity of the specific product, optimize the reaction conditions. Always amplify a negative control without added DNA, a negative control with buffy coat DNA from healthy donors, and a positive control with DNA from a BM sample of the patient.
3. Vary the annealing temperature in steps of 1°C. Vary the $MgCl_2$ concentration in steps of 0.25 m*M*. These parameters work synergistically.
4. Vary the concentration of primers in 0.2-µ*M* steps in a range of 0.4 to 1.2 µ*M*. Higher concentrations of primers can lead to a higher sensitivity; however, this is hampered by the occurrence of more nonspecific products.
5. Increase the number of amplification cycles to improve sensitivity. Decrease the number of cycles to reduce nonspecific products. At least 50 cycles should be performed to allow detection of single copies of template, which is a prerequisite for the statistical analysis of limiting dilution series by likelihood maximization.

4. Notes

1. No distinct band or multiple bands are visible after consensus PCR: Some BM samples will not lead to a single distinct band of the expected size after consensus PCR. Either only a polyclonal smear is visible or, in rare cases, more than one band of the expected size. (1) Change the strategy used for consensus PCR (FR3 or FR1). Try DNA instead of RNA. Prepare a PCR mixture containing 10 µL of 10× PCR buffer, 2 m*M* $MgCl_2$, each deoxynucleotide at 0.05 m*M*, primers FR1C or FR3A plus LJH at 0.4 µ*M* each, and 2.5 U *Taq*-DNA-polymerase. Add 500 to 1000 ng of DNA and distilled water to a final volume of 100 µL. Amplify with a program consisting of 7 min denaturation and enzyme activation at 94°C, followed by 50 cycles of denaturation for 1 min at 94°C and combined annealing and extension at 63°C (for primer FR3A) or 65°C (for primer FR1C), followed by a final extension step at 65°C for 5 min. If possible, collect another BM aspirate and start again.
2. Sequence has no resemblance to known CDR regions: This is most probably caused by a nonspecific product of the consensus PCR. Perform consensus RT-PCR again.
3. Sequence of a CDR-region is unsuitable for primer design: Sometimes the sequence of a CDR region is too short, has too many base repeats, or has too many long stretches rich in G and C, which makes designing primers difficult. Because CDR1 and CDR2 region are shorter and less variable than the CDR3-region, one should never do without a CDR3-specific primer. If segments suitable for primer design are not long enough, identify at least a short segment that is apt. Devise primers complementary to this stretch, and elongate in 5′-end direction (never in 3′-direction, because this end is more important for specific annealing of the primer to the template). Use the CDR1/2- plus CDR3-specific primer strategy, if the CDR3 region is the one that is difficult for primer design.

4. The ASO primer does not generate specific PCR product or is not specific or sensitive enough: Optimize PCR conditions and test primer for specificity or sensitivity. Devise new primers. Change primer strategy (CDR3 or CDR1/ plus CDR3). If no specific product can be generated at all, it is most probable that the sequence of the CDR-regions is not that of the malignant clone.
5. qPCR results are implausible: This is most probably caused by a faulty dilution series. Prepare dilution series anew and reanalyze the sample by qPCR.

5. Discussion

The described qPCR assay with ASO primers complementary to CDR regions of the malignant clone can be used to quantitate the proportion of malignant cells in peripheral blood (PB) and BM, aliquots of leukapheresis products, sorted cell fractions, or in cultured cells. It is possible to use this method for quantitative follow up in the course of therapy ***(16)***, to determine the tumor load of different stem cell sources ***(17)***, to determine the effect of different mobilization regimens and the duration of collection on the number of malignant cells in leukapheresis products ***(18,19)*** or to assess the extent of involvement in the malignant process of different cell fractions, for example, in the CD19+ or CD20+ cells of PB ***(20,21)***. The qPCR assay has evolved into an accurate and very sensitive tool for the detection of malignant cells in MM.

References

1. Cremer, F. W., Kiel, K., Wallmeier, M., Goldschmidt, H., and Moos, M. (1997) A quantitative PCR assay for the detection of low amounts of malignant cells in multiple myeloma. *Ann. Oncol.* **8,** 633–636.
2. Tonegawa, S. (1983) Somatic generation of antibody diversity. *Nature* **302,** 575–581.
3. Walter, M. A., Surti, U., Hofker, M. H., and Cox, D. W. (1990) The physical organisation of the human immunoglobulin heavy chain gene complex. *Eur. Mol. Biol. Org. J.* **9,** 3303–3313.
4. Bakkus, M. H., Heirman, C., Van Riet, I., Van Camp, B., and Thielemans, K. (1992) Evidence that multiple myeloma Ig heavy chain VDJ genes contain somatic mutations but show no intraclonal variation. *Blood* **80,** 2326–2335.
5. Aubin, J., Davi, F., Nguyen-Salomon, F., Leboeuf, D., Debert, C., Taher, M., et al. (1995) Description of a novel FR1 IgH PCR strategy and its comparison with three other strategies for the detection of clonality in B cell malignancies. *Leukemia* **9,** 471–479.
6. Brisco, M. J., Tan, L. W., Orsborn, A. M., and Morley, A. A. (1990) Development of a highly sensitive assay, based on the polymerase chain reaction, for rare B-lymphocyte clones in a polyclonal population. *Br. J. Haematol.* **75,** 163–167.
7. Campbell, M. J., Zelenetz, A. D., Levy, S., Levy, R. (1992) Use of family specific leader region primers for PCR amplification of the human heavy chain variable region gene repertoire. *Mol. Immunol.* **29,** 193–203.
8. Deane, M., McCarthy, K. P., Wiedemann, I. M., and Norton, J. D. (1991) An improved method for detection of B-lymphoid clonality by polymerase chain reaction. *Br. J. Haematol.* **5,** 726–730.
9. Diss, T. C., Peng, H., Wotherspoon, A. C., Isaacson, P. G., and Pan, L. (1993) Detection of monoclonality in low-grade β-cell lymphoma using the polymerase chain reaction is dependent on primer selection and lymphoma type. *J. Pathol.* **169,** 291–295.
10. Moore, G. E. and Kitamura, H. (1968) Cell line derived from patient with myeloma. *N. Y. State J. Med.* **68,** 2054–2060.

11. Klein, E., Klein, G., Nadkarmi, J. F., Nadkarmi, J. J., Vigzell, H., and Clifford, P. (1968) Surface IgM-k specificity on a Burkitt lymphoma cell in vivo and in derived culture lines. *Cancer Res.* **28,** 1300–1310.
12. Yamada, M., Wasserman, R., Reichard, B. A., Shane, S., Caton, A. J., and Rovera, G. (1991) Preferential utilization of specific immunoglobulin heavy chain diversity and joining segments in adult human peripheral blood B lymphocytes. *J. Exp. Med.* **173,** 395–407.
13. Brisco, M. J., Condon, J., Sykes, P. J., Neoh, S. H., and Morley, A. A. (1991) Detection and quantitation of neoplastic cells in acute lymphoblastic leukaemia, by use of the polymerase chain reaction. *Br. J. Haematol.* **79,** 211–217.
14. Wiesner, R. J. (1992) Direct quantitation of picomolar concentrations of mRNAs by mathematical analysis of a reverse transcription/exponential polymerase chain reaction assay. *Nucleic Acids Res.* **51,** 5863–5864.
15. Taswell, C. (1981) Limiting dilution assays for the determination of immunocompetent cell frequencies. *J. Immunol.* **126,** 1614–1619.
16. Cremer, F. W., Dada, R., Kiel, K., Ehrbrecht, E., Goldschmidt, H., and Moos, M. (1998) Quantitative PCR monitoring of the effect of double high-dose therapy on the tumor load of patients with multiple myeloma. *Ann. Hematol.* **77 (Suppl. II),** 182 (abstract 722).
17. Vescio. R. A., Han, E. J., Schiller, G. J., Lee, J. C., Wu, C. H., Cao, J., et al. (1996) Quantitative comparison of multiple myeloma tumor contamination in bone marrow harvest and leukapheresis autografts. *Bone Marrow Transplant.* **18,** 103–110.
18. Cremer, F. W., Kiel, K., Wallmeier, M., Haas, R., Goldschmidt, H., and Moos, M. (1998) Leukapheresis products in multiple myeloma: lower tumor load after mobilization with cyclophosphamide plus granulocyte colony-stimulating factor (G-CSF) compared with G-CSF alone. *Exp. Hematol.* **26,** 969–975.
19. Kiel, K., Cremer, F. W., Ehrbrecht, E., Wallmeier, M., Hegenbart, U., Goldschmidt, H., aet al. (1998) First and second apheresis in patients with multiple myeloma: no differences in tumor load and hematopoietic stem cell yield. *Bone Marrow Transplant.* **21,** 1109–1115.
20. Kiel, K., Cremer, F. W., Rottenburger, C., Kallmeyer, C., Ehrbrecht, E., Atzberger, A., et al. (1998) Circulating tumor cells in patients with multiple myeloma in the course of high-dose therapy. *Ann. Hematol.* **77 (Suppl. II),** 182 (abstract 724).
21. Rottenburger, C., Kiel, K., Cremer, F. W., B″sing, T., Atzberger, A., Moldenhauer, G., et al. (1998) No differences in the tumor load in the CD19+ and CD20+ cell fractions of peripheral blood from patients with multiple myeloma post high-dose therapy and with peripheral blood stem cell support. *Ann. Hematol.* **77 (Suppl. II)**, 184 (abstract 731).

32

Ultrasensitive Quantitative PCR to Detect RNA Viruses

Susan McDonagh

1. Introduction

The use of quantitative polymerase chain reaction (qPCR) to detect RNA viruses has become increasingly important as a prognostic marker and in patient management, for example, in human immunodeficiency virus (HIV) and hepatitis C virus (HCV) infection. Drug therapies can be monitored by regularly checking viral load, indicating whether the regime is sufficient, or whether alternatives should be sought. It is therefore crucial that the systems used are ultrasensitive and give accurate and reproducible results.

There are several elements to consider in developing a reliable and accurate quantitative assay. First, rapid transport and storage of samples is important because of the unstable nature of RNA, as is the method of preparation. Samples should be received and processed within 6 h and the relevant fractions stored at –70°C until testing. Second, it is difficult to ascertain the efficiency of sample preparation methods; therefore, values obtained using qPCR may be several times lower than the actual copy number. Known standards should therefore be processed alongside samples to assess the loss within the system *(1)*. Third, the efficiency of the reverse transcriptase step has been assessed to be from 5 *(2)* to 10% *(3,4)*, and this must be taken into account when calculating viral load. Finally, there is a need to amplify all viral types equally, which can be a problem with the high mutation rate observed in RNA viruses. As a result, primers should be chosen from highly conserved noncoding regions. such as the 5′ noncoding region of HCV *(5)* and of enteroviruses *(6)*.

Several quantitative systems that use a variety of approaches have been developed and published. These include limiting dilution; the use of external standard curves; co-amplification of an internal reference template; and competitive PCR using an internal control. There are advantages and disadvantages with each of these systems, and the system of choice will depend on several factors. These methods have been used to detect as few as 40 copies of HCV *(7)*, 64 and 4 copies of HIV per ml of plasma or serum, respectively *(8,9)*, and less than 10 copies of enterovirus *(10)*.

1.1. Limiting Dilution

The linear relationship between the amount of template and product is the basis of the limiting dilution method. However, this only occurs over a limited range; therefore,

From: *Methods in Molecular Biology, Vol. 226: PCR Protocols, Second Edition*
Edited by: J. M. S. Bartlett and D. Stirling © Humana Press Inc., Totowa, NJ

the dilution method is a semiquantitative system. Another consideration is that several reactions have to be run if Poisson distribution analysis is to be conducted using the limiting dilution method. Perhaps the most important consideration is that tube-to-tube variation is not controlled using this method *(1)*.

1.2. External Standard Curves

This is based on the generation of an external standard curve where a dilution series of known amounts of template are carried out alongside each run.

1.3. Coamplification of an Internal Reference Template

The problem of tube-to-tube variation can be overcome by the coamplification of a single-copy cellular gene and the target sequence. This means that both templates should be affected by any variations of amplification efficiency. However, because different templates and reactions have different efficiencies, the relative amounts of product could vary. An added problem is that this system cannot be used to quantitate extracellular organisms and therefore is of no value in measuring viruses, including HIV and HCV, in plasma samples *(1)*. As a result, this method will not be considered further.

1.4. Competitive PCR Using an Internal Control

The use of a competitive internal control that is similar to the target apart from length or the existence of a restriction site may alleviate the problems encountered with other methods of qPCR, such as co-amplification of an internal reference template *(9)*. It will also act as a control for inhibition *(10)*. Both the target and control DNA should amplify with the same efficiency because they compete equally under given conditions and the ratio of products will therefore remain constant throughout amplification *(11)*. The point at which target and control DNA are equal can then be found by direct comparison using gel electrophoresis and densitometry or by enzyme immunoassay (EIA). This method has become the method of choice for quantitative assays, but it is expensive and time consuming because it requires several reactions per sample.

To conduct a competitive PCR, it is necessary to create a template that is virtually the same as the product to be amplified but which can be identified during the detection stage. Primer-directed mutagenesis can be used to create a deletion, insert, or substitution at any point in the fragment *(12,13)*. An outline of this procedure can be seen in **Fig. 1**. Two separate reactions are conducted, each using an external primer plus an internal primer containing the appropriate changes, for example, to create a novel restriction enzyme (RE) site. This can then be used as a competitive template, which can be differentiated from amplified wild-type product. The products of these reactions are purified, denatured, and renatured together then added to a PCR containing the outer primers. The resulting amplicon contains the mutated site and can be used directly or cloned to create as much template as required.

The analysis of products is also important. Agarose gel electrophoresis, sometimes with densitometry, is easy to perform and is suitable for differentiating wild type from mutated products after digestion with the appropriate enzyme.

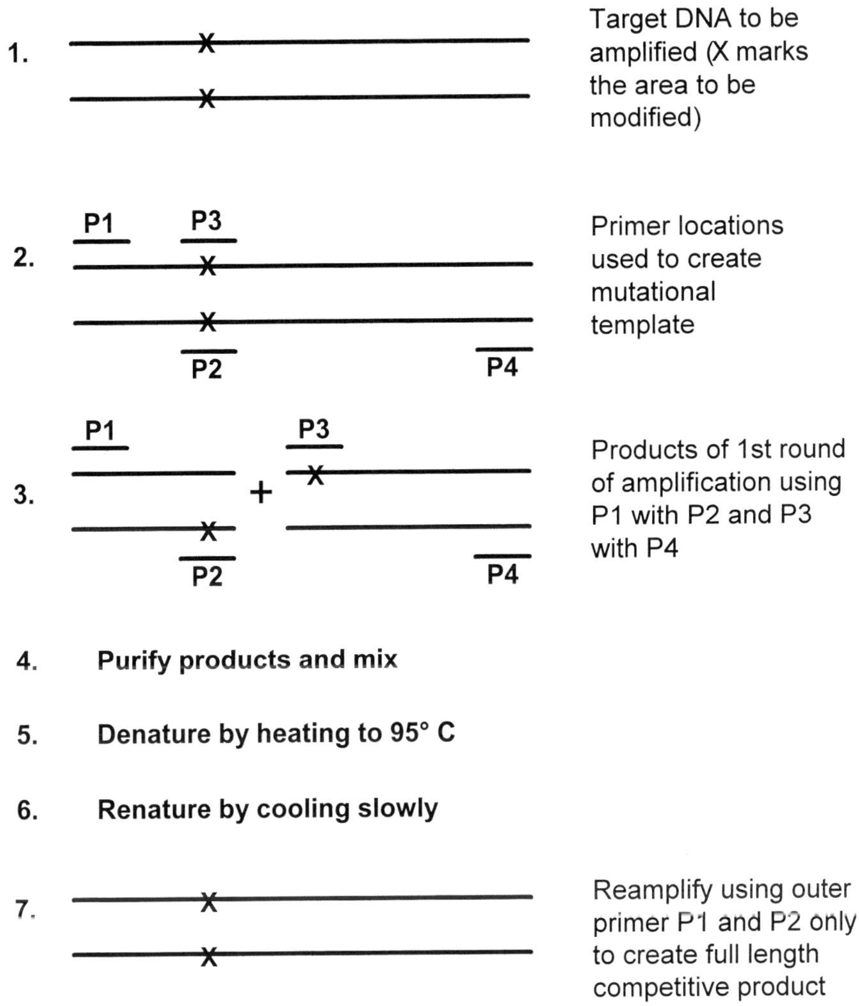

Fig. 1. Construction of a competitive PCR template. Figure modified from **ref.** ***12***.

2. Materials

Unless stated, all chemicals are supplied by Sigma (Poole, UK), or Merck. All stock solutions up to and including the PCR step should be made using RNA free water.

2.1. Reverse Transcription (see Note 1)

1. dNTPs (1 m*M*; Pharmacia).
2. Anti-Sense primer (100 ng/µL).
3. RNasin (20–40 units/µL; Promega, Southampton, UK).
4. Reverse transcriptase (RT) and 10× buffer, for example, AMV (Promega).
5. Dimethyl sulphoxide (DMSO).
6. Thermocycler (Hybaid, Teddington, UK).

2.2. Basic PCR

1. dNTPs (2 m*M*, 10× stock; Pharmacia; *see* **Note 2**).
2. *Taq* or *Pfu* DNA polymerase and 10× buffer (Sigma, Promega; *see* **Note 3**).
3. Forward and reverse primers (2 μ*M*; 10× stock).
4. Mineral oil (Sigma) or wax beads (Perkin–Elmer, Warrington, UK).
5. Thermocycler (Hybaid, Teddington, UK).

2.3. PCR to Create Mutated Competitor Fragments (e.g., Containing a Novel RE Site)

1. Forward and reverse primers used in conventional PCR (2 μ*M*).
2. Internal forward and reverse primers containing bases to create novel RE site (2 μ*M*).
3. DNA purification system (Geneclean, Bio 101, US).
4. PCR cloning system (TA cloning kit; Promega or Invitrogen; Leek, The Netherlands).
5. RE enzyme (to digest mutated product) or method of differentiating between wild-type and mutated product, such as a probe.

3. Methods

3.1. Reverse Transcription

1. Mix together the following reagents in an ice bath:
   7.5 μL of RNAse free water
   3 μL of DMSO
   100 ng of antisense virus-specific primer (in 1 μL)
   2 μL × 10 RT buffer
   3 μL of 1 m*M* dNTPs
   0.5 μL of RNAsin (20–40 units/μL)
   5 μL of RNA, prepared by extraction method
   1 μL of RT
2. Incubate for 30–60 min at 42°C followed by 5 min at 95°C and 5 min at 4°C.
3. Pulse spin tubes and add 5 μL to PCR reaction or store at –20°C until required (*see* **Note 4**).

3.2. PCR

3.2.1. Basic Reaction

1. Make up a master mix containing 5 μL of buffer (×10); 5 μL of dNTP (2 m*M*, final concentration 0.2 m*M*); 5 μL of each primer (2 μ*M*, final concentration 0.2 μ*M*); and 1 U DNA polymerase (*see* **Notes 2** and **3**).
2. Overlay with oil if not using a thermocycler with a heated lid, then add 5 μL of prepared cDNA through the oil.
3. A typical cycling protocol consists of denaturation at 95°C for 30 s, annealing at 5°C below the primer melting point for 30 s, and extension at 68 to 74°C for 45 s (*see* **Note 5**).
4. Carry out as few cycles as necessary because the reaction is only linear over a short range and this may distort the results.
5. To further increase sensitivity, it is beneficial to use nested PCR. As a general rule, the primary round should consist of 20 to 25 cycles and the secondary round up to 20 cycles. It may also be necessary to add diluted primary product (1/10) to the secondary round.

3.2.2. Limiting Dilution Method

A limiting dilution of cDNA is performed in 10-fold steps, and this is then added to the PCR. The sensitivity of the assay must be known, and the amount of cDNA in the samples can then be extrapolated by finding the detection end point. A specific volume of cDNA may also be added to a number of replicate reactions to give a Poisson distribution *(5,14,15)*.

3.2.3. The Use of External Standard Curves

A 10-fold dilution series covering from 1 to 10^5 molecules is used to produce a standard calibration curve. Amplified products can then be analyzed using the desired detection system, for example, by microwell capture so that optical density values can be compared with those of the standard curve *(4)*.

3.3. Competitive PCR Using an Internal Control

3.3.1. Creation of a Mutated Template Containing a Novel RE Site

1. Perform two PCR reactions, each containing an external primer plus the appropriate internal primer with mismatches to create the RE site. Approximately 20 to 30 cycles using a low annealing temperature will allow for mismatches (**Fig. 1**).
2. Analyze the products by agarose gel electrophoresis, excise the bands, and recover the DNA using Geneclean following manufacturer's protocols.
3. Heat the combined products to 94°C for 1 min, then cool and allow to renature.
4. Amplify the renatured products using the external primers and normal cycling conditions.
5. Check for the mutated site by RE digest (3 µL of product, 1 µL of 10× buffer, 5 µL of $dH2_2O$, 10 U/µL enzyme) at the appropriate temperature for 1 h. Analyze products by agarose gel electrophoresis. Alternatively, sequence the amplicon.
6. If required, the products may be cloned into a PCR cloning vector for further manipulation or simply used as PCR template to create as much competitor as necessary.

3.3.2. Quantitative PCR Using the Mutated Competitor

1. Perform a series of 10-fold titrations of mutated fragment with known copy number (1–10,000 copies; *see* **Note 6**).
2. Add each to a separate basic PCR along with the unknown quantity of DNA to be amplified and perform as few rounds as necessary (20–30 rounds).
3. Run products on 1 to 2% agarose and analyze by densitometry (gel or photograph; *see* **Fig. 2**).
4. The point at which wild-type and mutant intensity is equal identifies the amount of wild type DNA to within 1 log.

4. Notes

1. The RT and PCR stages may be performed in a single-tube format and also combined with hot start when wax beads are used to separate the stages *(7)*.
2. PCR DIG labeling mix must also be included if using EIA detection methods. Instead of adding combined dNTPs, add 200 µ*M* of dATP, dCTP, dGTP, 190 µ*M* of dTTP, and 10 µ*M* of Dig dUTP.
3. Pfu has been found to generate higher product yields *(4)*.
4. cDNA produced in this reaction appears to be slightly unstable. Long-term storage leads to significant reductions in the amount of target DNA.
5. Hot start PCR where a heating step at 95°C is required to activate the polymerase may increase amplification efficiency.
6. Further titrations can be performed, although this is expensive and time consuming.
7. The 118-bp fragment must also be taken into account when finding the point of equilibrium. Although it is possible to detect the end point of this reaction by eye (between lanes 6 and 7), it is more accurate to use densitometry.

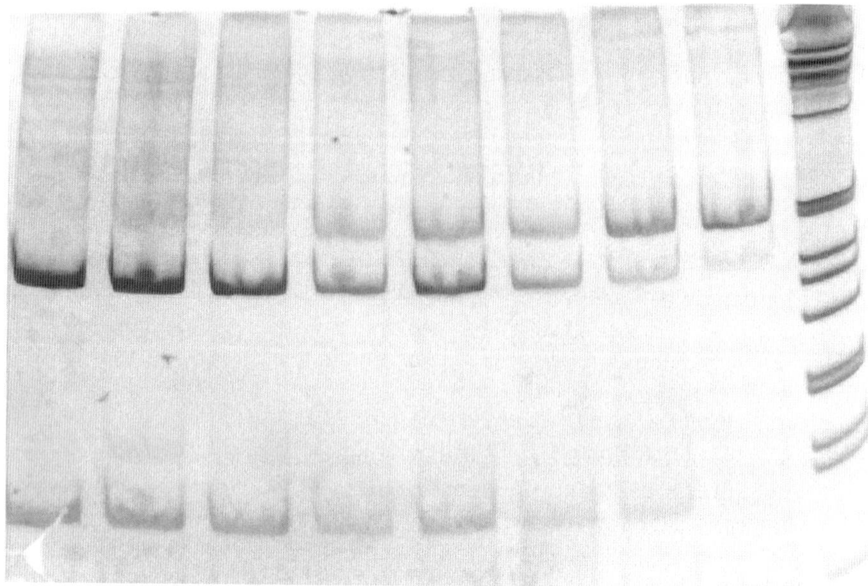

Fig. 2. Quantitative PCR products after RE digestion to reveal true products (522 bp) and mutated competitor products (404 and 118 bp). Lanes 1 through 8 contain 15 copies of template and decreasing amounts of mutated competitor (10,000, 2000, 1000, 200, 100, 20, 10, and 1 copy, respectively). The reaction was based on the amplification of a 522-bp product and a mutated competitor that was identical except for the creation of a *Sma*I site 118 bp into the fragment (*see* **Note 7**).

References

1. Clementi, M., Menzo, S., Bagnarelli, P., Manzin, A., Valenza, A., and Varaldo, P. E. (1993) Quantitative PCR and RT-PCR in virology. *PCR Methods Appl.* **2,** 191–196.
2. Zhang, L. Q., Simmonds, P., Ludlam, C. A., and Leigh Brown, A. J. (1991) Detection, quantification and sequencing of HIV-1 from plasma of seropositive individuals and from factor VIII concentrates. *AIDS* **5,** 675–681.
3. Kaneko, S., Murakami, S., Unoura, M., and Kobayashi, K. (1992) Quantitation of hepatitis C virus by competitive polymerase chain reaction. *J. Med. Virol.* **37,** 278–282.
4. Whitby, K. and Garson, J. A. (1995) Optomisation and evaluation of a quantitative chemiluminescent polymerase chain reaction assay for hepatitis C virus RNA. *J. Virol. Methods.* **51,** 75–88.
5. Hawkins, A., Davidson, F., and Simmonds, P. (1997) Comparison of plasma virus loads among individuals infected with hepatitis C virus (HCV) genotypes 1, 2, and 3 by Quantiplex HCV RNA Assay versions 1 and 2, Roche Monitor Assay and an in-house limiting dilution method. *J. Clin. Microbiol.* **35,** 187–192.
6. Lauwers, S., Bissay, V., and Rombaut, B. (1997) Development of an enterovirus specific PCR method for the quantification of enterovirus genomes in blood of diabetes patients. *Clin. Diagn. Virol.* **9,** 135–139.
7. Whitby, K. and Garson, J. A. (1997) A single tube two compartment reverse transcription polymerase chain reaction system for ultrasensitive quantitative detection of hepatitis C virus RNA. *J. Virol. Methods* **66,** 15–18.

8. Kashanchi, F., Melpolder, J. C., Epstein, J. S., and Sadaie, M. R. (1997) Rapid and sensitive detection of cell-associated HIV-1 in latently infected cell lines and in patient cells using sodium-n-butyrate induction and RT-PCR. *J. Med. Virol.* **52,** 179–189.
9. Liuzzi, G., Chirianni, A., Clementi, M., Bagnarelli, P., Valenza, A., Cataldo, P. T., et al. (1996) Analysis of HIV-1 load in blood, semen and saliva: Evidence for different viral compartments in a cross-sectional and longitudinal study. *AIDS* **10,** F51–F56.
10. Mayerat, C., Burgisser, P., Lavanchy, D., Mantegani, A., and Frei, P. C. (1996) Comparison of a competitive combined reverse transcription PCR assay with a branched-DNA assay for hepatitis C virus. *J. Clin. Microbiol.* **34,** 2702–2706.
11. Gilliland, G., Perrin, S., Blanchard, K., and Bunn, H. K. (1990) Analysis of cytokine mRNA and DNA: Detection and quantitation by competitive polymerase chain reaction. *Proc. Natl. Acad. Sci. USA* **87,** 2725–2729.
12. Higuchi, R., Krummel, B., and Saiki, R. K. (1988) A general method of in-vitro preparation and specific mutagenesis of DNA fragments: study of protein and DNA interactions. *Nucleic Acids Res.* **16,** 7351–7367.
13. Vallette, F., Mege, E., Reiss, A., and Adesnik, M. (1989) Construction of mutant and chimeric genes using the polymerase chain reaction. *Nucleic Acids Res.* **17,** 723–733.
14. Simmonds, P., Zhang, L. Q., Watson, H. G., Rebus, S., Ferguson, E. D., Balfe, P., et al. (1990) Hepatitis C quantification and sequencing in blood products, haemophiliacs, and drug users. *Lancet* **336,** 1469–1472.
15. van Kerckhoven, I., Fransen, K., Peeters, M., de Beenhouwer, H., Piot, P., and van der Groen, G. (1994) Quantification of human immunodeficiency virus in plasma by RNA PCR, viral culture, and p24 antigen. *J. Clin. Microbiol.* **32,** 1669–1673.

33

Quantitative PCR for cAMP RI Alpha mRNA

Use of Site-Directed Mutation and PCR Mimics

John M. S. Bartlett

1. Introduction

Precise and accurate determination of mRNA expression levels in tissues and model systems is a central methodology in a wide range of research applications. Expression of many genes is currently assessed by northern blotting, RNAse protection assays, Serial Analysis of Gene Expression (SAGE), and many other techniques; however, for single transcripts, especially where tissue is limited or abundance is low, quantitative polymerase chain reaction (PCR) is the method of choice. However, where quantitative PCR is to be used, the reproducibility, accuracy, and detection limits of the technique must be clearly defined.

We address this issue in the context of breast cancer by establishing a quantitative PCR technique for the measurement of cAMP RI alpha-binding proteins, the regulatory subunitis of cAMP dependent protein kinase, using PCR mimics. The technique was evaluated for interassay and intraassay variation and precision. This approach is applicable to any marker of interest where quantitation of RNA or DNA levels by PCR is attempted. For details of the construction of the mimics. see Bartlett et al. ***(1)***; however, a more robust and reproducible method of mimic construction is described within this volume (*see* **Chapter 7**).

2. Materials

1. RI alpha PCR primers (100 µ*M*, *see* **Note 1**).
   RI alpha 430 sense: 5′-GCATAACATTCAAAGCACTGC-3′
   RI alpha antisense: 5′-CTTGCTGAATCACAGTCTCTCC-3′
2. RI alpha control (430 base pairs) in pCRII vector: Dilute the control plasmid to 100, 10, 1, and 0.1 pg of insert per µL (*see* **Note 2**).
3. RI alpha MIMIC (430 base pairs) in pCRII vector (*see* **Note 3**): Dilute the mimic plasmid to 100, 10, 1, and 0.1 pg of insert per µL.
4. Extracted RNA (1 µg in 5 µL per sample; *see* **Subheadings 2.1.3.–2.1.5.**, *ibid*).
5. Random hexamer N6 (100 µ*M*, Life Technologies, Paisley, UK, *see* **Note 4**).
6. dNTP mix: 2 m*M* each of dATP, dTTP, dCTP and dGTP in distilled water, aliquoted, and stored at –20°C (Life Technologies, Paisley, UK).
7. MMLV Reverse transcriptase and buffer (200 units/µL; Life Technologies, Paisley, UK).

From: *Methods in Molecular Biology, Vol. 226: PCR Protocols, Second Edition*
Edited by: J. M. S. Bartlett and D. Stirling © Humana Press Inc., Totowa, NJ

Table 1
Example Assay Layout

Control	Sample					
	10 pg of RIα	100 pg of RIα	Sample 1	Sample 2	Sample 3	Etc.
0.1 pg of RIα mutant						
1.0 pg of RIα mutant						
10 pg of RIα mutant						
100 pg of RIα mutant						

8. Human placental ribonuclease inhibitor (40 U/μL, Pharmacia, UK).
9. Radioactive dNTP mix: 1.25 m*M* each of dATP, dTTP, dGTP, and 0.5 m*M* dCTP in distilled water, aliquoted, and stored at –20°C (Life Technologies, Paisley, UK). Before use, add 0.1 μCi^{32}P dCTP (Amersham, UK)/5 μL dNTP mix to be used in PCR (*see* **Note 5**).
10. *Taq* polymerase (5 units/μL) and buffer (Applied Biosystems, UK) 10× buffer: 500 m*M* potassium chloride, 100 m*M* Tris-HCl, 1% Triton-X and 25 m*M* magnesium chloride (*see* **Note 6**).
11. Paraffin oil (*see* **Note 7**).
12. Sodium chloride (100 m*M*) in sterile distilled water.
13. *Eco*RV restriction enzyme 10 units/μL (Pharmacia UK).
14. PAGE electrophoresis system (e.g., Protean II, Bio-Rad UK).
15. Gel fixative: 5% glacial acetic acid, 40% methanol, 10% glycerol in water.
16. Gel dryer.

3. Methods

3.1. Reverse Transcription

1. Add 1 μL of random hexamer (100 ng) to 1 μg of RNA and make up to a total volume of 9.5 μL with distilled water in a 0.2-mL thin-walled PCR tube.
2. Heat to 65°C for 10 min and cool on ice.
3. Prepare reverse transcription mix as follows: 4 μL of 5× reverse transcriptase buffer, 5 μL of 2 m*M* dNTP mix, 1 μL of reverse transcriptase (200 units), 0.5 μL of human placental ribonuclease inhibitor per reverse transcription reaction.
4. Incubate at 42°C for 1 h, inactivate reverse transcriptase at 80°C for 5 min, and store cDNA at –20°C until required for quantitative PCR (*see* **Note 8**).

3.2. Quantitative PCR

1. Construction of standard reactions: Duplicate reactions were set up as follows: to separate aliquots of 10 pg of control RI alpha (in 10 μL), add 0.1, 1, 10, and 100 pg of mutant RI alpha (in 10 μL). Also, to separate aliquots of 100 pg of control RI alpha (in 10 μL), add 0.1, 1, 10, and 100 pg of mutant RI alpha (in 10 μL). Prepare separate tubes only containing 10 pg of mutant plasmid or 10 pg of control plasmid (restriction digest controls).
2. For each sample (unknown concentration), take 1 μL of cDNA from the reverse transcription reaction plus 9 μL distilled water and to separate aliquots add 0.1, 1, 10, and 100 pg of mutant RI alpha (in 10 μL; *see* **Table 1** for assay layout; also, *see* **Note 9**).

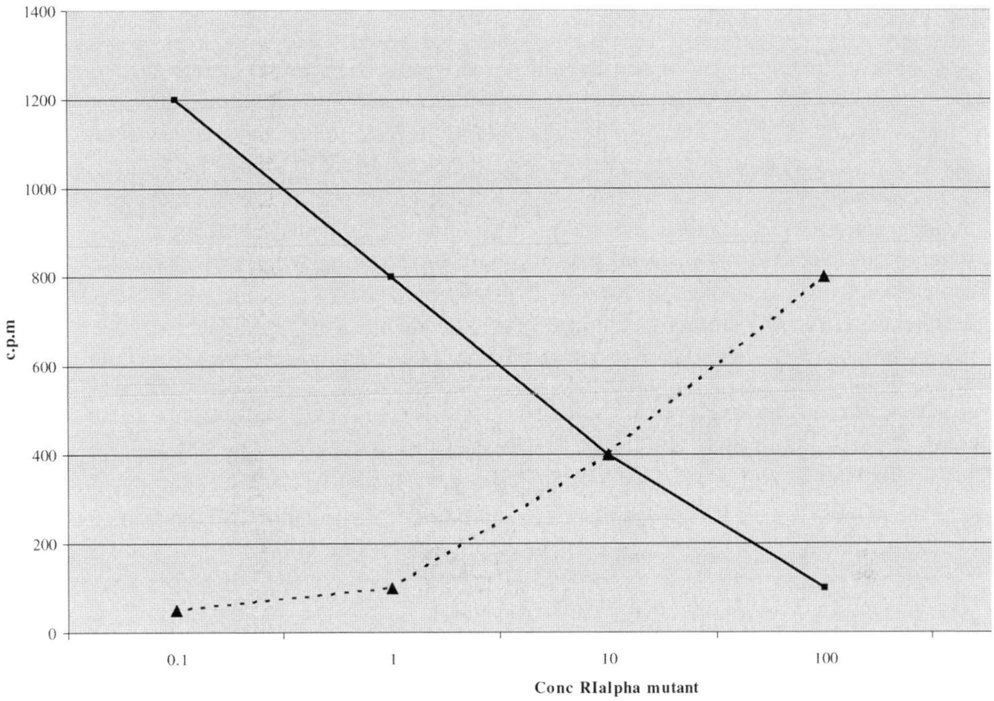

Fig. 1. Example of a crossover plot for a sample containing 10 pg of RI alpha cDNA. Vertical axis, counts per minute. Horizontal axis concentration of RI alpha mutant cDNA added. Solid curve = mutant RI alpha plasmid, dashed curve = sample. Reproduced with permission from Bartlett et al. *(1)*.

3. Prepare PCR master mix as follows: per tube, add To each tube 5 µL of 10× reaction buffer, 5 µL of 25 m*M* $MgCl_2$, 0.1 µL of *Taq* polymerase, 5 µL of radioactive dNTP mix, and 14.9 µL of distilled water.
4. Add 30 µL of PCR master mix to each tube set up in steps 1 and 2. If not using a heated lid PCR block, overlay with 100 µL of paraffin oil.
5. Perform PCR amplification for 26 cycles (94°C for 40 s, 55°C for 60 s, 72°C for 70 s) followed by extension at 72°C for 5 min.

3.3. Restriction Digestion

1. Add 5.0 µL of 100 m*M* sodium chloride to each reaction followed by 1.0 µL of *Eco*RV restriction enzyme (10 units; *see* **Note 10**).
2. Incubate at 37°C for 2 h to ensure complete digestion of mutant RIα product.
3. Labeled PCR products were separated on a 6% polyacrylamide gel at 30 mA for 2 to 3 h using a Protean II vertical electrophoresis system (Bio-Rad UK).
4. Gels were fixed for 30 to 60 min in fixative and dried using a flat bed gel dryer and heating to 80°C for 1 to 2 h.
5. Gels were exposed to preflashed X-OMAT (Kodak UK) film for 1 to 8 h using radioactive ink to orient the gels.
6. Bands corresponding to normal (430 base pair) and mutant component (215 base pair) component of each reaction were excised and ^{32}P incorporation determined by Cerenkov counting (*see* **Note 11**).

3.4. Calculation of Results

1. The results of a typical assay are illustrated in **Fig. 1**. After restriction digestion of the co-amplified mutant and normal RI alpha, two bands are clearly visible, representing products of 430 and 215 base pairs. If no normal cDNA is added, all the product is digested to give only a 215-bp band. In the illustrated assay, mutant RI alpha cDNA is co-amplified in a range of concentrations with known (100 or 10 pg) or unknown (patient sample) concentrations of unmutated RI alpha cDNA. At high concentrations of mutant plasmid, the lower band is the most intense. As the concentration of mutant cDNA is decreased, the relative intensity of the lower 215 bp band decreases and that of the larger 430 bp band increases. Where concentrations are equivalent, each product is produced at the same intensity. Thereafter, the upper 430-bp band becomes more intense. The point of equivalence of concentration is therefore represented by a crossover between the lower and upper band intensities.
2. For each sample, the counts per minute determined by Cerenkov ***(1)*** counting for the mutant and normal bands are plotted against the concentration of mutant plasmid added. The point at which the two curves cross represents the point at which the concentration of mutant and normal RI alpha are equivalent, thus allowing the concentration in unknown samples to be determined from this crossover point (*see* **Fig. 2**). In this example, the counts for the mutant and normal RI alpha bands from the PCR assay are plotted against the concentration of mutant RI alpha added. The curves cross over at 10 pg, indicating a concentration in the test sample of 10 pg RI alpha cDNA equivalent to 3.4 fmol RI alpha mRNA in the sample.
3. The sensitivity of the assay was assessed using a range of cDNA concentrations from 34 fmol to 0.0034 attomol (10^{-14} to 10^{-21} mol, 30 cycles of PCR were used for this lower limit). The sensitivity under these conditions was 0.002 attomol (2×10^{-21} mol or approx 1000 copies of mRNA; *see* **Note 12**).
4. Intra and interassay variation for the quantitative PCR assay were determined as 8.0 and 14.3%, respectively (*see* **Note 13**).
5. Calculation of confidence intervals for results: As the co-efficients of variation are known for each stage of the reverse transcription (RT)-PCR assay, we were able to calculate the confidence intervals for sample concentrations determined by this assay technique. Where samples are measured within the same assay this is calculated as follows (*see* **Note 13**):

$$C_{\mathrm{imax}} = C_{\mathrm{o}}(1 + E_{\mathrm{i}})(1 + V_{\mathrm{i}}) \text{ and } C_{\mathrm{imin}} = C_{\mathrm{o}}(1 - E_{\mathrm{i}})(1 - V_{\mathrm{i}})$$

Where C_{imax} is the maximum estimated concentration and C_{imin} is the minimum. C_{o} is the observed concentration, E_{i} is the intra-assay variation for reverse transcriptase, and V_{i} is the intra-assay variation for PCR.

If necessary interassay confidence limits can be defined from the above values using the following formula:

$$C_{\mathrm{bmax}} = C_{\mathrm{imax}}(1 + E_{\mathrm{b}})(1 + V_{\mathrm{b}}) \text{ and } C_{\mathrm{bmin}} = C_{\mathrm{imin}}(1 - E_{\mathrm{b}})(1 - V_{\mathrm{b}})$$

Where C_{bmax} is the maximum estimated concentration and C_{bmin} is the minimum and E_{b} is the inter-assay variation for reverse transcriptase and V_{b} is the intra-assay variation for PCR.

3.5. Discussion

By assessing the variation at each step during the RT-PCR procedure, this method defines the variation as a result of the reverse transcription and PCR steps and show that for samples assayed within a single assay the variation can be kept within acceptable limits. Furthermore, although it is possible, using the pCRII vector, to generate mRNA as an additional control, the low intra assay variation defined in this system allows this simpler procedure to be followed.

Interassay and intra-assay variations were similar to those obtained by conventional radioimmunoassays ***(2,3)***, suggesting this technique could be robustly applied to clinical diagnostic problems, such as the determination of viral load for either RNA or DNA viruses (omitting the RT step for the latter).

The sensitivity of this technique is such that low copy number genes could be assayed in 100 s or at most a few thousand cells and can be applied to patient tissue samples and small cell cultures. In addition, by allowing absolute concentrations to be determined, this assay will facilitate comparisons between laboratories previously hampered by semiquantitative approaches to PCR ***(4,5)***, and this precision is maintained even over most applications of fluorescence real-time PCR ***(6)***.

4. Notes

1. The cDNA sequence for human cAMP-dependent protein kinase subunit was retrieved from the Gembl database (accession no. M33336; *7*). Using this sequence, primers were designed that amplified bases 159 to 589. This 430-bp fragment codes entirely for mRNA, which is subsequently translated into protein.
2. The mimic is inserted in the 3900-bp pCRII vector by calculating the length of the vector containing the insert and dividing this by the length of the insert the molecular weight fraction made up by the insert is calculated and the Mwt corrected to reflect that of the PCR insert only. The molar equivalent for RI alpha mRNA added to the assay system was calculated as follows: 1 pg RI alpha plasmid is equivalent to 0.34 fmols RI alpha mRNA.
3. The PCR mimic was constructed as described elsewhere. Using site-directed mutagenesis, a single bp change was introduced within the RI alpha PCR product to introduce a *Eco*RV restriction site. For researchers wishing to evaluate this quantitative PCR approach, this mimic is available on request from the author.
4. Reverse transcription with a random hexamer will target all RNA species. Specific primers for the product of interest, out with the PCR product sequence, or polyA primers to target mRNA may be substituted as required.
5. Because of the high activity of ^{32}P, alternatives such as ^{33}P or biotin-dCTP may be considered.
6. Magnesium chloride concentrations may need to be altered for different PCR products. We therefore recommend using buffer with separate magnesium chloride.
7. Using heated lids avoids the need for paraffin oil overlay, which reduces evaporation.
8. To evaluate the reproducibility of the reverse transcription reaction we recommend including 0.5 μCi [^{35}S]dATP. This small amount of [^{35}S]dATP does not affect subsequent quantitation of the PCR. In our previous experiments, we demonstrated an intra-assay variation of 17% between samples in RT reactions and a 9% mean inter assay variation.
9. The sensitivity of the assay can be varied by varying the standard range; decreasing the range of mutant and control plasmid from 0.1 to 100 pg to 0.01 pg to 1.0 pg increased the sensitivity of detection. For these experiments, PCR was performed over 30 cycles.
10. The restriction enzyme *Eco*RV has an optimal digestion in buffered 50 m*M* KCl, 10 m*M* NaCl, and addition of 10% 100 m*M* NaCl to the PCR mix best approximates these conditions and avoids the need for purification of the PCR product at this step. The use of amplified control and mutant plasmids provides a high degree of control over the efficiency of the restriction digestion reaction.

11. Cerenkov counting is a simple principle where excised gel fragments are counted in the absence of scintillation fluid. Image analysis of the exposed autoradiograph can also be used at this stage.
12. The assay has not been evaluated below this limit and theoretically could function at sensitivities of tens or hundreds of mRNA copies; at these low concentrations, stochastic errors may be introduced.
13. Where all samples from a given individual or experiment are included in the same assay, interassay variation can be ignored.

References

1. Bartlett, J. M. S., Hulme, M. J., and Miller, W. R. (1996) Analysis of cAMP RI-alpha messenger-RNA expression in breast cancer—Evaluation of quantitative polymerase chain-reaction for routine use. *Br. J. Cancer* **73,** 1538–1544.
2. Bartlett, J. M., Wu, F. C. W., and Sharpe, R. M. (1987) Enhancement of Leydig cell testosterone secretion by isolated seminiferous tubules during co-perifusion in vitro: comparison with static co-culture systems. *Intl. J. Androl.* **10,** 603–617.
3. Bartlett, J. M. S., Weinbauer, G. F., and Nieschlag, E. (1989) Quantitative analysis of germ cell numbers and relation to intratesticular testosterone following vitamin a-induced synchronization of spermatogenesis in the rat. *J. Endocrinol.* **123,** 403–412.
4. Touraine, P., Martini, J. F., Zafrani, B., Durand, J. C., Labaille, F., Malet, C., et al. (1998) Increased expression of prolactin receptor gene assessed by quantitative polymerase chain reaction in human breast tumors versus normal breast tissues. *J. Clin. Endocrinol. Metab.* **83,** 667–674.
5. Sugimoto, T., Tsukamato, F., Fujita, M., and Takai, S. (1994) Ki-ras and c-myc oncogene expression measured by coamplification polymerase chain reaction. *Biochem. Biophys. Res. Commun.* **201,** 574–580.
6. Bieche, I., Olivi, M., Champeme, M. H., Vidaud, D., Lidereau, R., and Vidaud, M. (1998) Novel approach to quantitative polymerase chain reaction using real-time detection: application to the detection of gene amplification in breast cancer. *Intl. J. Cancer* **78,** 661–666.
7. Sandberg, M., Skalhegg, B., and Jahnsen, T. (1990) The two mRNA forms for the type I alpha regulatory subunit of cAMP-dependent protein kinase from human testis are due to the use of different polyadenylation site signals. *Biochem. Biophys. Res. Commun.* **167,** 323–330.

34

Quantitation of Multiple RNA Species

Ron Kerr

1. Introduction

Real-time quantitative polymerase chain reaction (PCR) methods *(1)* can be further optimized for various purposes by performing quantitative PCR for multiple RNA species in one sample *(2)*. Advantages of this method not only include the elimination of differences in reaction mix volumes and conditions that may occur in separate samples, but importantly it also allows reference of RNA quantitation to an internal control (also *see* **Notes 1** and **2**). Internal controls are RNA species for which the quantity of RNA does not change across different cell types or conditions. They allow correction of RNA quantitation for different starting quantities of total RNA. Widely used examples of internal controls are 18S ribosomal RNA and glyceraldehyde-3-phosphate dehydrogenase (GAPDH) *(3)*, and commercial standard kits for these are available. Different internal controls may be appropriate depending on the cell types or conditions being studied, and it is advisable to try several different controls when setting up a new assay. Quantitation of multiple RNA species is possible through the use of fluorescent probes, such as Taqman™ probes, as previously described.

When multiple RNA species are to be quantified in a single sample, probes with fluorophores of different wavelengths are used. As with any assay that relies on the measurement of multiple fluorescent signals, it is important to ensure that the wavelengths of the emitted signals are as distant as possible to reduce overlap in signal detection (**Fig. 1**). The materials and methods for quantitation of mRNA for the fibrinolytic protein tissue plasminogen activator (t-PA) using 18S ribosomal RNA as an internal control are shown below as a worked example of the quantitation of two RNA species in a single sample.

2. Materials

All reagents may be obtained from Applied Biosystems, Warrington, Cheshire, UK, unless otherwise stated.

1. 100 ng/µL RNA sample (*see* **Chapters 9–11** for RNA isolation method).
2. Reverse Transcriptase mastermix (9 µL per reaction): 2.85 µL of DEPC distilled H_2O, 1 µL of 10× Taqman buffer, 2.2 µL of 25 m*M* $MgCl_2$, 2 µL of dNTPs, 0.2 µL of RNase Inhibitor, 0.25 µL of multiscribe RT, and 0.5 µL of random hexamers.

From: *Methods in Molecular Biology, Vol. 226: PCR Protocols, Second Edition*
Edited by: J. M. S. Bartlett and D. Stirling © Humana Press Inc., Totowa, NJ

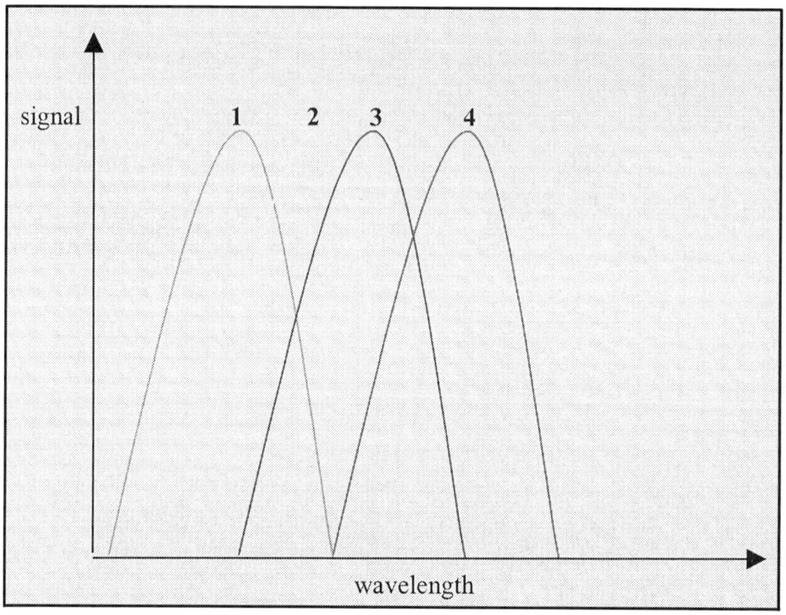

Fig. 1. Fluorescent probes 1 to 4 have different wavelengths as shown. For quantitation of multiple RNA species in a single sample, probes with the greatest difference in wavelengths should be selected to minimize overlap in signal detection. In this example probes 1 and 4 would be selected.

3. Primers/probes: Strict conditions must be adhered to for primer/probe design when using the ABI 7700™ quantitative PCR system. These are shown in **Table 1** and may be designed using Primer Express™ software. Primer and probe concentrations used are 10 pmol/µL and 5 pmol/µL, respectively.

   t-PA forward primer: GCAGGCTGACGTGGGAGTAC

   t-PA reverse primer: CCTCCTTTGATGCGAAACTGA

   t-PA probe: TGATGTGCCCTCCTGCTCCACCT

   These primers generate an amplicon of 91 base pairs, which spans intron 9 (1249 bp) of the t-PA gene. Because of the large intron size, only t-PA mRNA and not genomic DNA is amplified, removing the need for a DNase step. The t-PA Taqman probe has a FAM fluorescent label.
4. 18S rRNA primer-probe mix (supplied premixed). The 18S rRNA probe has a VIC fluorescent label.
5. PCR mastermix (Stratagene, La Jolla, California): 100 µL (500U) hot start *Taq* polymerase, 1.7 mL of 10× PCR buffer, 1.44 mL of 50 m*M* $MgCl_2$, 400 µL of 5 m*M* dNTPs, 6 µL of reference dye, and 6.354 mL of DEPC distilled H_2O. This can be stored at –20°C until needed.
6. Standard wall 0.6-mL capped conical tubes.
7. 96-well reaction plate.
8. ABI Prism™ 7700 Sequence detection system.

3. Methods

1. Reverse Transcriptase mastermix (9 µL) is added to 0.6-mL conical tubes.
2. To this, 1 µL (approx 100 ng/µL) of total RNA is added.

Table 1
Conditions for Primer/Probe Design for Quantitative PCR Using the ABI 7700™ Quantitative PCR System

Primer
T_m (melting temperature) 58–60°C
20–80% of nucleotides GC
Length 9–40 bases
<2°C difference in T_m beween the two primers
Maximum of 2/5 G or C at 3′ end
Probe
T_m 10°C higher than primer T_m
20–80% of nucleotides GC
length 9–40 bases
No G on the 5′ end
<4 contiguous G's
Must not have more G's than C's
Amplicon
50–150 bp in length
3′ end of primer as close to the probe as possible without overlapping

3. Reverse transcription is performed on a thermal cycler at the following temperatures: –25°C for 1 h, 48°C for 45 min, then 95°C for 5 min.
4. The cDNA produced is diluted 1 in 2.5 by adding 15 µL of DEPC distilled H_2O per tube.
5. The following mix is added to labeled tubes (67.5 µL/tube): 37.5 µL PCR mastermix, 1.125 µL 18S rRNA (or alternative internal control) primer-probe mix, 7.875 µL of DEPC distilled H_2O, 9 µL of t-PA forward primer, 9 µL of t-PA reverse primer, and 3 µL of t-PA Taqman probe.
6. cDNA (7.5 µL) from **Step 4** is added to each tube and mixed thoroughly by pipetting up and down.
7. Aliquots (3 × 25 µL) from each tube are then added to the 96-well reaction plate, sealed, and PCR performed on the ABI 7700.
8. Also included are samples with DEPC distilled H_2O substituted for RNA–non template controls.
9. Standards with RNA concentrations from 10× (approx 1 µg/µL RNA) diluted with doubling dilutions down to 0.156× are also included. The exact concentration of target RNA in the standards is not important if we are quantifying relative rather than absolute mRNA (*see* **Note 3**).

4. Notes

1. To ensure quantitation of the desired target RNA, the amplicons should also be run on an 8% acrylamide gel, sequenced, and the sequence checked to exclude homology with other proteins.
2. In addition to the advantages of performing PCR for multiple RNA species in the same tube as mentioned in the introduction, there are a number of additional advantages that the use of quantitative PCR as described above has over previously used methods, such as competitive template PCR. The rigorous criteria for primer and probe selection allow for the use of a standard PCR mix so that the concentration of magnesium and buffers used does not have to be optimized. It also allows the use of a closed system so that tubes are not opened

after the PCR. This reduces the occurrence of contamination of subsequent assays by previous PCR products, which is an important consideration for any PCR system.

A disadvantage of this system is that it may not be suitable for all target proteins. We have found that for some target RNAs trial of a large number of primer sets may be necessary before suitable primers are found. Also, the short amplicon length leads to an increased chance of homologous products amplified compared with PCR using primers for longer amplicons. We have found that the strict conditions may result in it not being possible to find suitable primers/probe for some RNA species. In our experience, these have been coagulation proteins, which are known to have a high degree of homology with other proteins.

3. The above protocol will generate results as the cycle at which a set fluorescence threshold (Ct) is reached for each standard/unknown for both t-PA mRNA (FAM) and 18S rRNA (VIC). The standards are used to ensure that the difference between the Ct of the test RNA species and the Ct of the 18S rRNA internal control (ΔCt) remains constant across all mRNA dilutions in the range chosen for the standards (10 to 0.156× above). We can ensure that this is the case by plotting the ΔCt for each of the standards against the log total RNA for each of the standards. If the gradient of the slope of this line is less than ± 0.1, this indicates that the ΔCt is remaining constant and that the "ΔΔCt∀ method can be used to quantitate the mRNA for the unknowns as follows. The relative mRNA quantity of various unknowns compared with a nominal baseline unknown are calculated by using the formula $2^{-\Delta\Delta ct}$, where ΔΔCt is the difference in the ΔCt of the test sample and the ΔCt of the nominal baseline sample.

   The use of the ΔΔCt method is advantageous when a large number of repeated samples have to be analyzed because it removes the need for running a set of standards each time, which is both time-consuming and increases costs. However, for many purposes, it seems sensible to generate a standard curve each time to ensure the validity of the results.

References

1. Heid, C. A., Stevens, J., Livak, K. J., and Williams, P. M. (1996) Real time quantitative PCR. *Genome Res.* **6,** 986–994.
2. Desjardin, L. E., Chen, Y., Perkins, M. D., Teixeira, L., Cave, M. D., and Eisenach, K. D. (1998) Comparison of the ABI 7700 system (TaqMan) and competitive PCR for quantification of IS6110 DNA in sputum during treatment of tuberculosis. *J. Clin. Microbiol.* **36,** 1964–1968.
3. Goidin, D., Mamessier, A., Staquet, M. J., Schmitt, D., and Berthier-Vergnes, O. (2001) Ribosomal 18S RNA prevails over glyceraldehyde-3-phosphate dehydrogenase and beta-actin genes as internal standard for quantitative comparison of mRNA levels in invasive and noninvasive human melanoma cell subpopulations. *Anal. Biochem.* **295,** 17–21.

V

Transcriptome Analysis

35

Differential Display

A Technical Overview

John M. S. Bartlett

1. Introduction

Since the completion of the human genome-sequencing project, scientists are now able to read the code of all human genes stored on the 46 chromosomes of the human genetic library. However, we are far from reaching an understanding of the functional relationships existing between more than a tiny fraction of these genes. The value of the human genome-sequencing project, beyond the simple collation of data, has been to teach us that it is possible to take a highly intensive analytical approach to the study of human systems in health and disease ***(1–4)***.

The aim of our research has now shifted from the study of individual genes, isolated from the cellular environment in which they play their roles, to the investigation of the hugely complex interactions between gene and gene families. We are shifting from the observation of individual trees to an evaluation of the wood itself. To accomplish this, novel techniques have emerged that simultaneously allow the analysis of entire pathways or indeed entire cellular transcription patterns. The challenge this provides is both a molecular and mathematical one. Experiments now yield vast amounts of data that must be sifted through to formulate novel hypotheses. Conversely, we are now able to see how whole families of genes are regulated and interact rather than having to slowly piece together information often from quite different experimental approaches. The continuing development of biomathematical modeling systems ***(5–9)*** is essential to this approach.

Over the next decade, scientists face the challenge of transforming the knowledge gained of the human genome sequence into a practical and functional understanding of complex biological systems in health and disease. It is clear that analysis of gene expression represents a highly significant pointer to the altered function of transcripts identified by the human genome project whose function is largely unknown. The ability to select candidate genes from expression libraries from different tissues and disease states (and stages) for rapid investigation represents the single most important driver behind the current explosion in expression library analysis. It is therefore critical that we understand both the potentials and limitations of technologies available for expression analysis of entire transcriptomes.

From: *Methods in Molecular Biology, Vol. 226: PCR Protocols, Second Edition*
Edited by: J. M. S. Bartlett and D. Stirling © Humana Press Inc., Totowa, NJ

As with any experimental approach, the researcher must have a clear goal and hypothesis in mind at the outset of the experimental procedure such that the techniques selected to achieve that goal are appropriate. The attraction, and pitfall, of transcriptome analysis is that these are extremely powerful tools for the identification of transcripts implicated in altered tissue functions. There are therefore a few basic principles that bear stating at this stage before the examination of the techniques available in this area.

First and most importantly, a point commonly overlooked by those researching RNA expression is that proteins are effectors, and RNA is a blueprint. When examining RNA expression profiles, insight is gained into the regulation of expression and stability of the RNA species under examination, and these data must be extrapolated, usually by the assumption that expression of RNA relates to expression of protein. This extrapolation must be made with caution. However, the existence of the blueprint does not prove the material for which it codes has been produced and, conversely, failure to detect the template does not mean the protein is not present. It is therefore essential to link RNA analyses with analyses of protein expression and function; this requirement is frequently overlooked in transcriptome analyses.

Second, in common with all analytical techniques, it is imperative that the composition of the tissue under examination be understood before correct interpretation of results can be performed. Although many transcriptome analyses are performed in monoclonal cell systems, frequently the analysis uses tissue, malignant or otherwise, as a starting point. Where mRNA is extracted from entire tissues, there exists a strong possibility of incorrectly assigning expression to the wrong cellular component. The presence of blood vessels, inflammatory cells, stromal components, and a mixture of diseased and normal tissue cells can complicate analyses. It is imperative, at some stage of the investigation, to determine the tissue source of the mRNA detected to avoid ascribing an incorrect origin to a protein or mRNA species.

Each approach described within this section presents specific problems that must be sufficiently addressed to allow accurate and reliable data to be gathered if the effort expended is to be worthwhile. However, almost all techniques aimed at global analysis of RNA transcripts rely upon reverse transcription of mRNA, and most incorporate polymerase chain reaction (PCR). Reverse transcription (RT) comprises the transcription of the mRNA of choice into DNA (more accurately termed copy DNA or cDNA). This stage of the reaction is performed using RNA virus enzymes whose role in vivo is to transcribe viral RNA genome into a template for host transcription systems. The earliest RT enzymes used were derived from the Molony murine leukemia virus (Mo-MLV reverse transcriptase) and the avian myeloblastosis virus (AMV reverse transcriptase). More recently, genetically modified forms of these enzymes with enhanced activity have become the agents of choice. The enzymes are relatively heat labile (particularly in comparison with *Taq* polymerase) and reactions are performed at 42°C with mRNA and nucleotides. The reaction must be primed because the enzymes are dependent on double-stranded nucleic acid template. The primer is usually allowed to anneal to the RNA after a short incubation (in the absence of enzyme) at 85°C to denature RNA tertiary structure. Primer selection is very important. Either a primer specific for the target sequence may be used or, more commonly, primers targeted at copying the entire mRNA population can be used. The use of poly (dT) primers will target mRNA poly A tails and transcribe from this point. In this case between 2 to 4 kb of cDNA can be routinely copied, which may place a constraint on PCR primer design. RNA can be transcribed with random hexanucleotide primers (more efficient than poly dT and produce cDNAs from the entire RNA pool); however, in this case, separation of mRNA from ribosomal and transfer RNAs is recommended as these will also be copied diluting the target sequences ***(10,11)***. The cDNA produced is then used for multiple PCRs using specific sequence primers. A further novel development has been the use of rTth DNA polymerase (from the bacterium *Thermus thermophilus*) allowing reverse transcription and PCR to be performed in a single reaction tube using this thermostable DNA polymerase with RT activity. Use of higher temperatures for the RT step with this system allows more efficient transcription, particularly from mRNAs with high guanine-cytosine content ***(12,13)***. The quality of mRNA extracted is an important confounder of many experimental approaches; even with conventional and proprietary mRNA extraction techniques, some carryover of polymerase or transcriptase inhibitors can be observed. In extreme cases, this may reduce the efficiency of the RT to an extent that causes errors in quantification, which may bias comparisons between experimental samples. Therefore, every effort should be made to ensure that mRNA preparations for quantification are as pure as possible.

2. Systems for Transcriptome Analysis

2.1. Differential Display

The use of differential display technologies has rapidly expanded over the last decade as this technique has become established as a potent tool for the simultaneous analysis of multiple mRNA species ***(14)***. Differential display techniques have become as varied as the questions they are used to answer; however. in general they combine the following three separate techniques to address this single question.

1. Production of cDNA from mRNA by RT.
2. Design of arbitrary primers to allow parts of the cDNA (tags) to be amplified by PCR.
3. Use of sequencing quality resolution by acrylamide electrophoresis.

This approach builds up a fingerprint of the RNA species expressed in different tissue or cell samples. Comparison of these fingerprints identifies those genes that are upregulated or downregulated in different tissues. The technique is quite elegant in both its simplicity and power. However, as with many such techniques, the secret lies in the careful design and interpretation of the results.

The model is best suited to the study of gene regulation in tissue culture where conditions can be varied under careful control to avoid artifacts. Even then extreme care must be taken to ensure that results are reproducible between experiments, the control of tissue culture conditions is in fact the most critical phase of the experiment in the production of accurate differential display results. Use of tissue samples makes the approach all the more complex. Although there is obvious value in investigations seeking to identify metastasis related gene expression, care must be taken to ensure that, for example, differences between premetastatic and postmetastatic tissues are related to metastasis and not a consequence of altered tumor–stroma interactions, or differences between primary tumors. In addition, extreme care must be taken to ensure samples to be compared are treated exactly the same to avoid artifacts being introduced during tissue storage, mRNA extraction, or subsequent amplification.

Even once novel regulatory changes are identified using differential display techniques, it remains common to double-check these changes using a second system, such as representational differential analyses (RDA), PCR. or Northerns, to confirm the result. This is an important confirmatory step before investing significant effort into the study of novel genes or gene transcripts identified by differential display ***(14)***. Even with these caveats, differential display has already proven itself as a highly valuable system for the study of gene expression changes during neoplastic progression.

There are many similarities between differential display technologies and serial analysis of gene expression (SAGE). Indeed, SAGE might be seen as a logical progression from differential display. Both use degenerate primer sequences to produce a fingerprint of mRNA species for analysis. In differential display. these sequences may range from the highly selective (members of a particular gene family) to the more inclusive (polyT based) ***(14–21)***. Both methods use tagged primers, differential display for the purpose of detection, and SAGE for the purpose of capture and ligation. The fundamental difference between these approaches is that differential display continues to use a semiquantitative measure of expression and although more sensitive than microarrays in the detection of low copy number changes in expression, signal strength remains a determining factor of sensitivity. Again, identification of transcripts with altered expression relies on further analysis, either by sequencing of a representative clone or by blotting with a selective probe. A strength of this approach is illustrated by its ability to be targeted at specific genes or gene families. However, the number of transcripts able to be analyzed is limited by the electrophoretic separation required for the analysis of the results. Although this still has the capacity to recognize many hundreds of distinct transcripts, it is unlikely that the capacity of differential display can match that of either gene microarrays or SAGE. This section includes two different differential display protocols, one using targeted primers (AU Motifs) and the other being a more global protocol. Further techniques are available in an associated volume in this series ***(13)***.

2.2. Microarrays

DNA Chips or microarrays are becoming much more widely applied to the investigation of both model systems and whole tissues ***(14,22–25)***. However, these approaches are not, strictly speaking PCR related and therefore we have included an overview of arrays purely for completeness. DNA microarrays have the advantage of being data rich, that is to say it is possible to analyze many thousands of genes simultaneously. Using this approach it is possible to identify transcripts that are markedly upregulated or downregulated after experimentation or, indeed, as recently reported in the simple classification of cancers. This approach, in common with many conventional methods of gene expression analysis (northerns, RDA, etc.) relies on the measurement of signal intensity resulting from nucleic acid hybridization. Therefore, the efficiency of detection is a compound of the efficiency of labeling and hybridization of the individual clone. The resulting data give a semiquantitative estimate of changes in expression either up or down.

The major weakness of the microarray approach is that it is firstly a semiquantitative approach and that it is therefore not optimal for the detection of low copy number gene transcripts. The major strength is the large number of genes (10^3–10^4) that can be examined simultaneously in a single experiment and the rapidity with which information can be gathered from multiple samples.

2.3. SAGE

In SAGE, short expressed sequence tags are produced from each mRNA expressed. Thus far, the technique is very similar to that of expressed sequence tag (EST) library screening. The advantage of SAGE is that the tags used are 9- to 10-bp long, and through a number of steps (see chapter 40 by Oien, these tags are concatenated into clones containing multiple (up to 40+) sequence tags before sequencing. Therefore, the expression profile is generated by sequencing many such clones and simple counting of the number of times each tag is represented *(26)*.

Therefore, the chief difference between DNA microarrays, differential display, and SAGE is that SAGE produces a digital count of the number of clones representing each sequence expressed. This is, however, only one of the differences between these approaches. Microarrays can identify large numbers (10,000 or more) of expressed genes, and determine, semiquantitatively, alterations in their expression. However, genes whose expression is modulated only marginally are unlikely to be identified by this approach. In SAGE, each expressed gene may be sequenced and thus single copy changes in expression may be detected if sufficient clones are sequenced. In SAGE, the sensitivity is determined not by the detection system but by the amount of effort given to identification of sequence tags. A further advantage of SAGE analysis over microarrays is that it is not necessary to have to hand a clone representing the gene(s) of interest; because the detection of genes is performed by sequencing, no hybridization matrix is required. SAGE analysis is therefore limited only by the amount of sequencing that can be performed in a cost-effective manner. Conventional sequencing of ESTs relies on the analysis of several hundred bps to identify each gene. In SAGE the sequence tag used to identify the expressed sequence is 10 bp. Theoretically, the discriminatory power of a 10-bp region of cDNA is 1 in 1,048,576 (4^{10}). By linking together the sequence tags from many different genes into a concatemeric clone, a single sequencing run of 800 bp can be used to identify 50 or more different sequences. Given the high throughput available in today's 96-lane sequencing platforms, approx 5000 gene transcripts may be analyzed from a SAGE library in a single sequencing run. In addition, SAGE has the advantage that digital data is more suited to statistical analysis. Finally, and perhaps of greatest significance as increasingly experiments become more time consuming and costly to perform the quantitative nature of SAGE library analysis potentially allows direct comparisons between laboratories and between libraries, subject only to the constraints of the initial experimental variables. Initial data comparisons support the premise that comparisons between laboratories performing SAGE can provide valuable and valid information. The caveat that must be applied is that although such comparisons are at present validated by internal controls and performed in expert laboratories with rigorous controls, there are also specific pitfalls in SAGE analysis relating to identification of clones and PCR biases in construction of tags *(26)*. Although some groups are already extrapolating from this data to compare data from different experimental approaches *(27)*, in the light of some concerns raised and the possible impact on gene expression of even small changes in pH, nutrient, etc., this approach is at present premature, not just for SAGE but for all expression analyses.

Notwithstanding these caveats, SAGE analysis is a highly powerful experimental tool on which is increasingly applied to transcriptome analysis. The power of SAGE analysis is the product of an extremely complex experimental protocol, one that relies on the construction of a representative cDNA library for each RNA sample to be analyzed. In addition, the number of clones that must be sequenced for even a simple comparison between two libraries, despite the use of short sequence tags, is high (between 800–4000 has been suggested). Furthermore, although there is a high probability that a short sequence tag of 10 bp will identify a unique sequence, it is undeniably more complex to identify gene transcripts from such tags. Proponents of SAGE point out, with some justification, that the effort may well repay the cost. Nonetheless, the prospect of performing even a limited analysis of clinical material (50–100 tumors) using this method is a daunting one and many will be tempted to take a simpler, albeit less quantitative approach, such as microarray or differential display analysis, for their first steps in transcriptome analysis. More importantly, SAGE is an undirected technique that produces a global transcriptome map. Both microarrays and differential display techniques can, however, be readily tailored to the particular experimental question and hypothesis under investigation. As discussed at the outset of this chapter, careful consideration will be required to select the method most appropriate to the research question, expertise, and resources available to the investigator.

References

1. Zhang, L., Zhou, W., Velculescu, V. E., Kern, S. E., Hruban, R. H., Hamilton, S. R., et al. (1997) Gene expression profiles in normal and cancer cells. *Science* **276,** 1268–1272.
2. Lennon, G. G. (2000) High-throughput gene expression analysis for drug discovery. *Drug Discovery Today* **5,** 59–66.
3. Loging, W., Lal, A., Siu, I.-M., Loney, T., Wikstrand, C., Marra, M., et al. (2000) Identifying potential tumour markers and antigens by database mining and rapid expression screening. *Genome Res.* **10,** 1393–1402.
4. Szallasi, Z. (1998) Gene expression patterns and cancer. *Nat. Biotechnol.* **16,** 1292–1293.
5. Larsson, M., Stahl, S., Uhlen, M., and Wennborg, A. (2000) Expression profile viewer (ExProView): A software tool for transcriptome analysis. *Genomics* **63,** 341–353.
6. Lash, A. E., Tolstoshev, C. M., Wagner, L., Schuler, G. D., Strausberg, R. L., Riggins, G. J., et al. (2000) SAGEmap: A public gene expression resource. *Genome Res.* **10,** 1051–1060.
7. http://www.ncbi.nlm.nih.gov/geo/.
8. www.ncbi.nlm.nih.gov/sage.
9. http://www.ebi.ac.uk/arrayexpress/.
10. Ji, H. J., Zhang, Q. Q., and Leung, B. S. (1990) Survey of oncogene and growth factor/receptor gene expression in cancer cells by intron-differential RNA/PCR. *Biochem. Biophys. Res. Commun.* **170,** 569–575
11. Jindal, S. K., Ishii, E., Letarte, M., Vera, S., Teerds, K. J., and Dorrington, J. H. (1995) Regulation of transforming growth factor alpha gene expression in an ovarian surface epithelial cell line derived from a human carcinoma. *Biol. Reprod.* **52,** 1027–1037.
12. White, B. A., ed. (1984) *PCR Protocols: Current Methods & Applications.Volume 15, Methods in Molecular Biology.* Humana Press, Totowa, NJ.
13. Liang, P. and Pardee, A. B., eds. (1998) *Differential Display Methods & Protocols.Volume 85, Methods in Molecular Biology.* Humana Press, Totowa, NJ.
14. Jurecic, R. and Belmont, J. W. (2000) Long-distance DD-PCR and cDNA microarrays. *Curr. Opin. Microbiol.* **3,** 316–321.

15. Marguiles, E. and Innis, J. (2000)eSAGE: Managing and analysing data generated with serial analysis of gene expression (SAGE). *Bioinformatics* **16,** 650–651.
16. Liang, P. and Pardee, A. B. (1998) Differential display: A general protocol. *Mol. Biotechnol.* **10,** 261–267.
17. Jorgensen, M., Bevort, M., Kledal, T. S., Hansen, B. V., Dalgaard, M., and Leffers, H. (1999) Differential display competitive polymerase chain reaction: An optimal tool for assaying gene expression. *Electrophoresis* **20,** 230–240.
18. Matz, M. V. and Lukyanov, S. A. (1998) Different strategies of differential display: Areas of application. *Nucleic Acids Res.* **26,** 5537–5543.
19. Lapointe, J. and Labrie, C. (1999) Identification and cloning of a novel androgen-responsive gene, uridine diphosphoglucose dehydrogenase, in human breast cancer cells. *Endocrinology* **140,** 4486–4493.
20. Cirelli, C. and Tononi, G. (1999) Differences in brain gene expression between sleep and waking as revealed by mRNA differential display and cDNA microarray technology. *J. Sleep Res.* **8,** 44–52. Supplement.
21. Aittokallio, T., Ojala, P., Nevalainen, T. J., and Nevalainen, O (2000) Analysis of similarity of electrophoretic patterns in mRNA differential display. *Electrophoresis* **21,** 2947–2956.
22. Lockhart, D., Dong, H., Byrne, M., Follettie, M., Gallo, M., Chee, M., et al. (1996) Expression monitoring by hybridization to high-density oligonucleotide arrays. *Nat. Biotechnol.* **14,** 1675–1680.
23. Schena, M., Shalon, D., Davis, R., and Brown, P. (1995) Quantitative monitoring of gene expression patterns with a complementary cDNA microarray. *Science* **270,** 467–470.
24. Lomax, M. I., Huang, L., Cho, Y., Gong, T.-W., and Altschuler, R. A. (2000) Differential display and gene arrays to examine auditory plasticity. *Hearing Res.* **147,** 293–302.
25. Sun, Y. (2000) Identification and characterization of genes responsive to apoptosis: Application of DNA chip technology and mRNA differential display. *Histol. Histopathol.* **15,** 1271–1284.
26. Bartlett, J. M. S. (2001) Technology evaluation: SAGE, Genzyme molecular oncology. *Curr. Opin. Mol. Ther.* **3,** 85–96.
27. Kal, A. J., Van Zonneveld, A. J., Benes, V., Van den Berg, M., Koerkamp, M. G., Albermann, K., et al. (1999) Dynamics of gene expression revealed by comparison of serial analysis of gene expression transcript profiles from yeast grown on two different carbon sources. *Mol. Biol. Cell* **10,** 1859–1872.

36

AU-Differential Display, Reproducibility of a Differential mRNA Display Targeted to AU Motifs

Orlando Dominguez, Lidia Sabater, Yaqoub Ashhab, Eva Belloso, and Ricardo Pujol-Borrell

1. Introduction

AU-rich elements (AREs) are found in 3′ untranslated regions (3′ UTR) of many highly unstable mRNAs for mammalian early-response genes. The minimal AU sequence core within the ARE is the heptamer WAUUUAW, although from a functional point of view, several pentanucleotides clustered in close proximity are the key sequence motif that mediates mRNA degradation *(1)*.

Genes containing AREs are of potential biological and pharmacological interest because they often code for inflammatory mediators, cytokines, proto-oncoproteins, and transcription factors *(2–5)*.

A targeted differential display named AU-motif directed display (AU-DD) is described here. It allows the isolation of cDNA fragments from ARE-containing mRNAs with minimal sequence information. AU-DD combines a high specificity gained at the polymerase chain reaction (PCR) level with the advantages of direct comparison between samples at different stages of activation to detect differentially expressed genes *(6,7)*.

Previous examples of targeted differential display have used longer and well-conserved motifs from multigene families. These include zinc finger motifs *(8,9)*, plant MADS boxes *(10)*, and motifs specific for heat shock proteins *(11)*. They employ longer conserved sequences (from 6 to 8 codons) to design primers that can predictably work well on PCR.

The PCR step of AU-DD uses primers which contain two distinctive domains, a double AU motif at the 3′ end (AU2 in short) and a guanosine rich 5′ (**Table 1**). This primer core sequence is previously incorporated as a 5′ tag in the retrotranscription primer (*see* below). The 5′ domain configures a sticky anchor that increases Tm and stabilizes primer annealing whereas the 3′ confers motif specificity. Both 5′ and 3′ domains contribute in a cooperative way to primer annealing. When the natural primer-binding site was identified on cloned products, it was always found to be a true AU-rich site. The use of the AU2 domain-based primers does not restrict; however, the spectrum of amplified mRNAs to clustered AU-pentamer containing genes. In fact, cDNAs

From: *Methods in Molecular Biology, Vol. 226: PCR Protocols, Second Edition*
Edited by: J. M. S. Bartlett and D. Stirling © Humana Press Inc., Totowa, NJ

Table 1
Primer Sequences for Both AU-DD and Control Genes

Protocol	Primer Designation	Sequence
(A) Retrotranscription (RT) and PCR Primers Used in AU-DD		
1	G7AU2dT (RT)	GGGGGGGTATTTATTTA(ACGT)TTTTTTTTT TTTTTT(ACG)
2	G7AU2	GGGGGGGTATTTATTTA
	G7AU2A	GGGGGGGTATTTATTTAA
	G7AU2C	GGGGGGGTATTTATTTAC
	G7AU2G	GGGGGGGTATTTATTTAG
	G7AU2T	GGGGGGGTATTTATTTAT
3	GTGAU2dT (RT)	GGTGGGTGGTATTTATTTA(ACGT)TTTTTTT TTTTTTTT(ACG)
4	GTGAU2	GGTGGGTGGTATTTATTTA
	GTGAU2A	GGTGGGTGGTATTTATTTAA
	GTGAU2C	GGTGGGTGGTATTTATTTAC
	GTGAU2G	GGTGGGTGGTATTTATTTAG
	GTGAU2T	GGTGGGTGGTATTTATTTAT
(B) Primers Used in Control PCR Experiments		
IFNα (J00207)	IFNα513s	GGCCTTGACCTTTGCTTTA
	IFNα915as	CTTCATCAGGGGAGTCTCTGT
GAPDH (M33197)	GAPDH19s	TCTTCTTTTGCGTCGCCAG
	GAPDH390as	AGCCCCAGCCTTCTCCA

(A) 1 and 3 are alternative retrotranscription (RT) primers; 2 and 4 list two series of PCR primers compatible with RT primers 1 and 3, respectively. (B) Specific primers for IFNA and GAPD, respectively.

carrying single AU motifs were often amplified, and only 15% of cloned cDNAs contained a complete double AU element ***(12)***.

On its part, the reverse transcriptase primer incorporates a significant feature by including the PCR primer as a 5′ tag. This allows the use of a single primer in the PCR, providing the possibility of a fine titration of the annealing temperature to avoid primer artifacts generated when two oligonucleotides are used in low stringency conditions.

The cDNA is synthesized from total RNA trying to ensure initiation from true poly-A tails of mRNA, which is not always easy. With our protocol, only 8% of cloned products lacked the poly-A tail ***(12)***.

For two mRNA species to be effectively sampled by DD, the following two conditions are critical: (1) comparable amounts of template must be used and (2) cDNA concentration must be above a threshold to obtain reproducible results. During all the initial processes, RNA concentration and integrity and the absence of contaminating DNA has to be monitored carefully. Even trace amounts of genomic DNA can significantly alter the final profile, yielding fake products and turning the technique unreliable. RNA should be checked for quantity and integrity both after purification and after DNase treatment. The use of good-quality DNase does not guarantee the removal of all DNA, and this process requires also monitoring. The assessment of

DNase treatment efficiency and titration and normalization of cDNA samples were performed by PCR as described *(12)*.

In general, it is always easy to get amplification products. However, this does not guarantee that those products are specific and reproducible. When the cDNA concentration is below a minimal threshold, the final products belong to nonspecific templates that were selected and amplified. This is why both minimal amounts of RNA to be retrotranscribed and cDNA normalization are fundamental. The minimal amount ensures that specific targets can be easily found and selected during the first cycles of PCR, and the normalization ensures that DD profiles between samples activated and resting can be compared.

The major procedures described in the protocol refer to RNA preparation and DNase treatment, cDNA synthesis, AU-DD, and confirmation of differential expression.

2. Materials

In general, any reagent of analytical grade was considered of sufficient purity for general procedures, such as electrophoresis or initial stages of RNA purification. Otherwise, reagents added to enzymatic reactions or intended to dissolve RNA were of molecular biology or ultrapure grades.

2.1. RNA Preparation

1. Lysis solution: GCS (Guanidinium thiocyanate (GuTh), Citric acid, Sarkosyl) is a modification of Sacchi and Chomczynski solution D *(14)*, that is, 4 *M* GuTh, 50 m*M* sodium citrate (citric acid adjusted to pH 3.8 with NaOH), and 0.5% Sarkosyl. The solution can be stored for up to no more than 1 mo at 4°C.
2. Double-distilled phenol (Sigma). Phenol was saturated and equilibrated in ultrapure water and treated with 8-hydroxyquinoline (Sigma) *(13)* yielding unbuffered acidic phenol (pH 4 to 5). Aliquots containing a water overlay were dispensed in sterile polypropylene tubes and stored frozen at –20°C. After thawing, they were kept at 4°C and discarded after 2 wk.
3. Complete Lysis solution (GCSMP): Mix GCS containing beta-2 mercaptoethanol up to 2% with 1 volume of acidic phenol to give GCSMP immediately before use.
4. Phase lock heavy gel (PLG, Eppendorf) stored at room temperature.
5. Glycogen as coprecipitant (Roche Molecular Biochemicals) stored at –20°C.
6. Chloroform:isoamyl alcohol: Chloroform should be mixed with isoamyl alcohol in a proportion of 98:2 v/v (CI), respectively, in a Pyrex bottle with minimal air volume and kept at 4°C.
7. DNAse I RQ1 from Promega at 0.2 U/µL in a solution containing 10 m*M* Bis-Tris-HCl, pH 6.5, 1 m*M* EDTA, 5 m*M* $MgCl_2$, 5 m*M* DTT, and 1 U/µL RNAsin.
8. RNasin Ribonuclease Inhibitor (40 U/µL) from Promega stored at –20°C.
9. Ethanol 96% and 75%.
10. 3 *M* Potassium Acetate pH 5.0 (KAc).
11. Isopropanol.
12. Guanidinium thiocyanate (4 *M*) in water.

2.2. cDNA Synthesis and Further Treatment

1. Reverse transcriptase: SuperScript II RNase H- (Gibco-BRL). Store at –20°C.
2. dNTPs (Amersham Biosciences) at 10 m*M* each.
3. RNasin Ribonuclease Inhibitor 40 U/µL from Promega (Madison, WI) stored at –20°C.

4. Oligonucleotides used as reverse transcription primers (HPLC purified; Genset, Paris, France) are listed in **Table 1**. There are two variants used in independent but complementary experiments, G7AU2dT and GTGAU2dT.
5. RNase H (Gibco BRL) stored at –20°C.
6. Qiaquick PCR purification kit columns (Qiagen).
7. EE (10 m*M* EPPS), 0.1 m*M* EDTA, pH 8.2 adjusted with 0.1 *N* NaOH was used throughout as a general elution/dilution solution.
8. Reagents for agarose gel electrophoresis.
9. GAPDH primers (*see* **Table 1**).

2.3. AU-DD

1. α-^{32}P-dATP at 3000 Ci/mmol/DuPont NEN.
2. Platinum *Taq* DNA polymerase (Gibco-BRL).
3. PCR buffer 1×: 10 m*M* Tris-HCl, pH 8.8, 1.5 m*M* $MgCl_2$, 50 m*M* KCl, 0.1% Triton X-100.
4. dNTPs at 2 m*M*.
5. AU-DD primers are listed in **Table 1**.
6. For confirmation of differential expression by standard PCR, specific primers were derived from sequences of cloned AU-DD products with the program Oligo v5.0 (National Biosciences Inc.) to have a Tm of 63 to 65°.

2.4. Gel Electrophoresis

1. S2 GibcoBRL electrophoresis apparatus for 0.8-mm thick gels.
2. 40% (38:2) Polyacrylamide-bispolyacrylamide (Serva, Germany).
3. TEMED (*N*,*N*,*N*′,N′-tetramethylethylethlenediamine).
4. 10% Ammonium persulfate.
5. Whatman 3 MM paper (gel blotting paper).
6. TBE buffer (89 m*M* Tris Base, 89 m*M* boric acid, 1 m*M* EDTA).
7. 20% methanol-10% acetic acid.

3. Method

A schematic representation of AU-DD method is shown in **Fig. 1**.

3.1. RNA preparation (see Note 1)

1a. Solid tissue is pulverized in liquid nitrogen with mortar and pestle (*see* **Note 2**).
1b. Cultured cells are pelleted at 800*g* for 5 min at 4°C, and put on ice as dry pellets.
1c. For culture flasks with adherent cells, decant, rinse with ice-cold phosphate-buffered saline and decant by aspiration.
2. Add GCSMP in a proportion of 2 mL per 50 mg of tissue or 10^7 cells.
3. Homogenize lysates at 25,000 rpm for 30 s with a mechanical homogenizer (Ultraturrax T25, Ika) to ensure both lysis and a complete DNA shearing. Stand for 2 min at room temperature (*see* **Note 3**).
4. Add 0.4 volumes of CI (chloroform:isoamyl alcohol) with respect to the complete lysis solution—including phenol volume—and shake energetically for 10 s.
5. Pour on a prepacked PLG tube. Empty prepacked PLG tubes are 14-mL PPN round-bottom centrifuge tubes containing 1.5 mL of PLG-heavy and centrifuged 2 min at 1500*g*; it should then be left for 10 min on ice.
6. Spin 10 min at 2000*g* on bench-top centrifuge with swinging bucket rotor. Save upper aqueous phase to high speed tubes and discard PLG interphase and bottom phase.

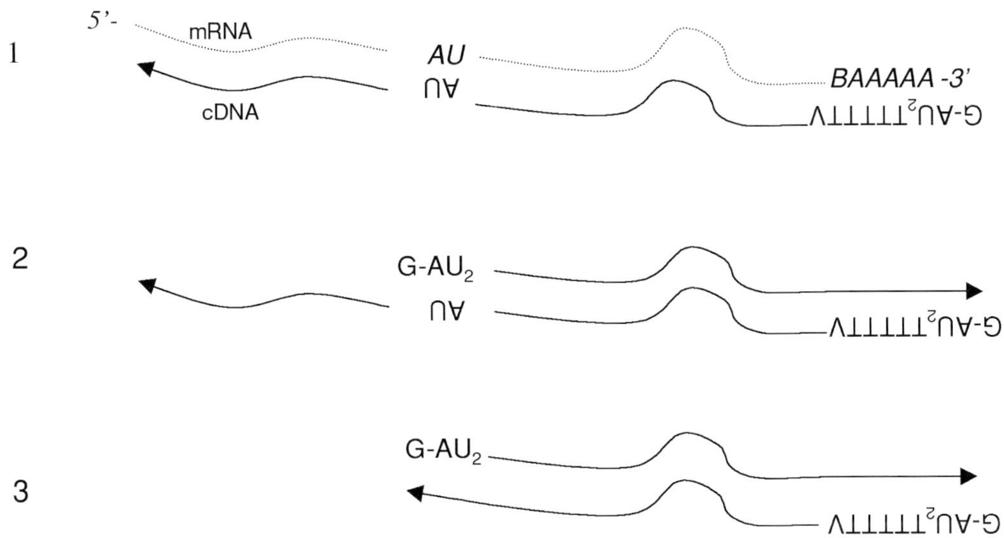

Fig. 1. Schematic representation of AU-DD. **(1)** Retrotranscription. Dotted line represents mRNA template. The RNA molecule at the top contains an AU-rich sequence (AU) and a poly-A tail. G-AU2TTTTTV represents a generic RT primer, with a 5′ anchor of rich on Guanosines, an AU2 sequence and an oligo(dT). **(2)** First round of PCR with a single primer G*N*AU2. **(3)** Final products of AU-DD amplification.

7. Precipitate by mixing with 1 volume of isopropanol. Incubate 1 h at –20°C and centrifuge 30 min at 10,000*g*. Wash with 75% ethanol.
8. Dissolve pellets in 4 *M* guanidinium thiocyanate. Use at least 0.2 volumes of the lysis solution used initially. Heat if necessary (pulses of 5 min at 60°C) until there are not pellet particles left in solution. Be careful because the pellet particles become transparent and it is difficult to see them. Coprecipitant may be incorporated here adding 1 µL of glycogen (20 mg/mL) and mix.
9. Repeat the precipitation step with isopropanol *(9)*.
10. Wash with 75% ethanol. Dissolve in ultrapure water. The volume of water depends on the RNA concentration desired. Typically, 50 µL is appropriate to run out all the following experiments.
11. Total RNA concentration is measured by spectrophotometry and gel electrophoresis. Not less than 3 µg (*see* **Note 4**) or up to 50% of RNA prep is DNase I treated (next step) in a separate tube, storing the other half as backup or for other purposes at –80°C after mixing with 3 volumes of 95% ethanol (*see* **Note 5**).
12. RNA integrity and concentration is assessed by titration using *Escherichia coli* rRNAs (Sigma) as standard in 1% TBE agarose gel electrophoresis and ethidium bromide staining *(13)*. Different concentrations of commercial *E. coli* rRNAs (800, 400, 200 ng) are compared in intensity with different dilutions of the RNA sample.
13. Typically, 5 to 10 µg of total RNA (*see* **Note 6**) is incubated at 0.25 µg/µL with DNAse I for 30 min at 37°C. One microliter is then taken to check the absence of DNA (*see* **Note 7**).
14. RNA is precipitated by adding 0.1 volumes of 3 *M* KAc, pH 5.0; 1 µL of glycogen; and 3 volumes of 95% ethanol (*see* **Note 8**). After standing for 5 min at room temperature, the pellet is recovered by 10 min of centrifugation in a microcentrifuge at 10,000*g* and washed in 75% ethanol (*see* **Note 9**).

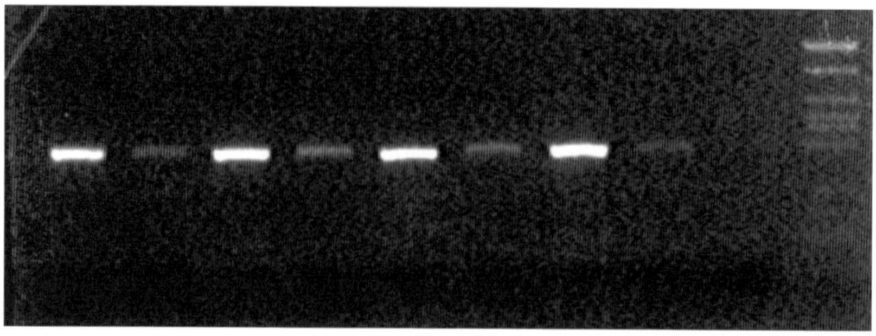

1 2 3 4 5 6 7 8 9 M

Fig. 2. cDNA normalization for their concentration in GAPDH gene. Lanes 1, 3, 5, and 7 cDNA diluted 1:5 are different samples and. Lanes 2, 4, 6, and 8 are the respective cDNA diluted 1:50. Lane 9 is a PCR-negative control.

3.2. cDNA Synthesis and Further Treatment

1. A single reverse transcription with any of the two RT primers shown in **Table 1** (G7AU2dT and GTGAU2dT) produced cDNA for a complete set of AU-DD reactions with matching PCR primers. These two different RT primers are anchored oligo$(dT)_{15}$ primers with a 5′ tag that accommodates the sequence of any of the AU-DD primers that will be used at the PCR step. The tag defines the particular RT primer and the set of AU-DD primers to be used. The 15 T stretch was found to anneal more specifically on poly-A tails at the conditions used than others of 25 or longer. The tag was designed with the intention of generating the weaker secondary structure as possible. Standard oligo$(dT)_{15}$ was used to retrotranscribe control RNA samples (Jurkat and U937). First-stranded cDNA was prepared with Superscript II (Gibco-BRL) following manufacturer instructions with minor modifications indicated below.
2. DNase treated total RNA (2–3 μg) in water and 10 pmoles of the chosen RT primer (**Table 1**) for a reaction volume of 20 μL were denatured at 72°C for 3 min on a PCR machine and chilled on ice for 1 min (*see* **Note 10**).
3. The other reagents for a reaction volume of 20 μL except the enzyme are supplemented at room temperature (*see* **Note 11**). Annealing was allowed to proceed for 10 min at room temperature.
4. 200 U SuperScript II (Gibco-BRL) were then added and the mixture was incubated for 1 h at 42°C.
5. Reverse transcription was stopped by heating at 90°C for 2 min and RNAse H was used as recommended by its supplier (Gibco-BRL). 1.8 U RNase H, 20 min at 37°C.
6. To validate cDNA synthesis, normalization was performed by amplifying 1:10 and 1:500 cDNA dilutions for GAPDH in a 25-cycle PCR adjusted to an annealing temperature of 60°C (primers in **Table 1B**). Product concentration was estimated by visual inspection of ethidium bromide stained gels, and cDNAs were normalized by dilution according to their GAPDH equivalents if required. Typically no adjustment was needed (**Fig. 2**, *see* **Note 12**).
7. Free nucleotides and primer were washed out in Qiaquick columns (Qiagen). cDNA was eluted in 50 μL of EE. The eluates were used directly in AU-DD.

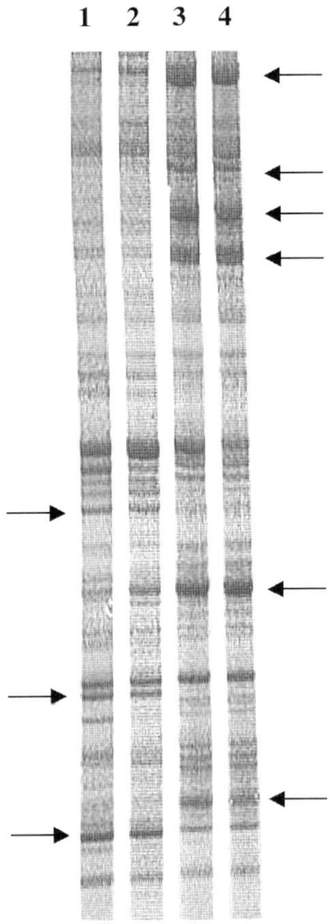

Fig. 3. Typical AU-DD gel. Results from adherent monocytes by using a single PCR (not nested) are shown. Lanes 1 and 2 show fingerprints from activated cells. Lanes 3 and 4 show resting cells. Net cDNA was used for lanes 1 and 3 and diluted (1 : 5) for 2 and 4. Arrows signal fragments from differentially expressed genes.

3.3. AU-DD

At this stage, there are two desalted cDNA samples ready to be compared, the experimental and the control. Every reaction on a template is made in duplicate to detect variability caused by the sample concentration only. In this way, two cDNA concentrations, net and 1 : 5, are used in independent PCR tubes with otherwise identical primer and conditions (*see* **Note 13**). Reactions are prepared for 10-µL volumes and contain 2 µL of cDNA, primer at 3 µ*M* (1.5 µ*M* in nested reactions as proposed in **Note 13**), 10 m*M* Tris-HCl, pH 8.8; 1.5 m*M* $MgCl_2$, 50 m*M* KCl, 0.1% Triton X-100, dNTPs each at 0.2 m*M* (*see* **Note 14**), 0.25 µL of α-^{32}P-dATP and 50 mU/µL Platinum DNA polymerase (*see* **Note 15**). Reactions are run in a PTC-100 thermocycler (MJ Research) for 40 cycles with the following profile: 94°C for 40 s; 42°C for 1 min, 20 s; and 70°C for 40 s.

3.4. Gel Electrophoresis

1. AU-DD products (2 μL) are mixed with nondenaturing DNA loading buffer and separated in native 0.8-mm thick, 6% TBE polyacrylamide gels at 12 V/cm. Voltage was chosen such that it did not generate higher temperature than 45°C during the run to avoid heat denaturation. Every set of four reactions coming from the same cell line, both resting and activated, and at each of the two cDNA concentrations are run in adjacent lanes.
2. Gels are fixed in 20% methanol: 10% acetic acid for 30 min before drying under vacuum at 80°C about 1 h with a Whatman 3 MM as support and autoradiographed as for standard sequencing gels (***13***, *see* **Note 16**).
3. Individual bands **(Fig. 3)** are selected when unique or more intense in activated cells (*see* **Note 17**).
4. The autorad is clamped against the dried gel by superimposing background signals with corresponding well line and gel edges. Edges of selected bands are punctured through the autorad with a hypodermic needle, leaving a mark in the dried gel. Gel slices containing those bands are cut out of the dried gel with a sterile scalpel knife.
5. Dried gel pieces are rehydrated in 0.2 mL of EE for 3 h at 50°C. A further 1:20 dilution in EE was used as template for reamplification by 40 cycles of PCR and subsequent cloning.

3.5. Confirmation of Differential Expression

To confirm the differential expression, a semiquantitative RT-PCR was performed (**Fig. 4**, *see* **Subheading 2.3., step 6**). Specific PCR primers to the selected fragments were derived from the fragment sequences and were used in 40 cycles of PCR with paired resting and stimulated cDNAs. These cDNAs are normalized as described above, under **Subheading 2.2., step 5**.

4. Designing a Primer Directed to AU Motifs

A variety of AU-DD primers directed to single AU motifs were tested in preliminary experiments (not reported here). Their poor specificity and performance led us to design oligos with double AU pentamer motifs, which are often found in rich AU containing 3′UTR of tightly controlled genes.

AU-DD features reside in those specially designed primers that were used from reverse transcription to PCR. Results from experiments using two primers which differ in their 5′ anchors are presented: GTGAU2 and G7AU2 (**Table 1**). The 5′ anchors influence the annealing to natural AU containing sites and, therefore, the array of cDNAs that are selected and amplified during the PCR. In this way AU2 primers with different 5′ anchors sample distinct subsets of genes.

The correctness of the rationale on the primer design was demonstrated in experiments which used an interleukin (IL)-2 gene fragment containing the AU2 sequence as template. In otherwise similar conditions the GTGAU2 primer allowed the detection of 10^3 times less molecules of IL-2 cDNA than G7AU2 (not shown). A better anchoring of GTGAU2 on IL-2 does not necessarily imply that it also anneals better on other cDNAs. Therefore, both primers were used in parallel experiments, as other anchors can be tested, to increase the number of genes sampled.

To keep structural simplicity, anchor domains containing only G or G and T were used, but variations can be expected to work as well as long as primer Tm is maintained.

In an attempt to increase the affinity of the anchor, we tested inosine containing primers in different sequence configurations. When used alone, these primers reduced

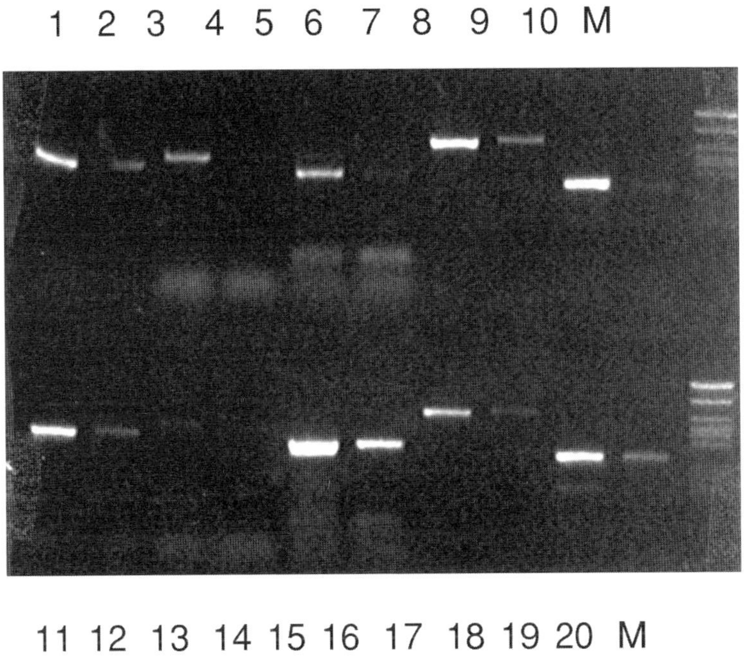

Fig. 4. Confirmation of the differential expression by specific RT-PCR (primers derived from sequences of cloned AU-DD products) in agarose gels of a few AU-DD fragments taken as examples. Odd lanes are from activated cells and even lanes are the counterpart resting cells.

the system sensitivity and reproducibility on the IL-2 cDNA. However, when used to supplement standard primers, a small but consistent gain in the number of products obtained from total cDNA was noticed, even when sensitivity to the IL-2 standard was not improved (not shown). Profile reproducibility among replicas was excellent for the primers and conditions described.

5. Notes

1. Different RNA preparation methods were examined in function of copurifying RNase activities, contaminating DNA, and yield. Ultracentrifugation in Cs salts (CsCl, CsTFA) gave an acceptable low level of DNA, but it was discarded because of its relatively poor yield. Phase lock heavy gel was found to improve both the yield in RNA and the partitioning of the genomic DNA to the lower phenolic phase. The single step acid-phenol method ***(14)*** was used with minor modifications intended to ensure RNA integrity and an efficient removal of both RNase activities and of contaminating genomic DNA. The procedure is described since this latter factor is considered relevant to the main method.
2. Fresh tissue is snap frozen in liquid nitrogen and then stored at –80°C or lower until needed. Ceramic mortar is precooled by pouring on it liquid nitrogen where pestle is submerged; it is advised to wear gloves during this procedure. Initial bubbling will cease when the tool is cool enough. Then, the tissue is submerged and ground till reduced to a powdered state. The remaining liquid nitrogen is left to evaporate until the tissue is just wet, like a paste. Then, it is poured to a polypropylene 50-mL conical centrifuge tube. As soon as the tissue starts looking dry, add the lysis solution, mix by vortexing and homogenize.

3. If the container is not a polypropylene (PPN) tube, transfer lysates to a PPN centrifuge tube. Choose the tube so that the lysate volume is not larger than one third of its capacity.
4. Because lower RNA amounts negatively affect the reliability of DD techniques, not less than 3 μg are processed.
5. The RNA concentration in the storage ethanol solution is one fourth of the original concentration before adding the alcohol. In absence of salts, this solution forms a relatively uniform suspension of RNA, which is easy to pipet. To recover an aliquot, transfer the volume that contains the desired amount to a fresh tube. Supplement and mix with both coprecipitant (glycogen) and 0.1 vol of KAc, pH 5.0. Store an additional hour at –80°C and gather the sediment by centrifugation at 12,000*g*. After a single wash in 75% ethanol, the pellet is ready for any downstream application.
6. To assure a complete removal of contaminating DNA, trial tests were performed. RNA from human thymus, which copurifies with relatively high content of DNA, was used for these assays. The best conditions were found when RNA (substrate) and DNase (enzyme) were kept as described.
7. The absence of residual contaminating DNA was demonstrated by the failure to amplify the genomic locus of interferon-α, a multicopy gene, by PCR. A program of 40 cycles with an annealing temperature of 60°C was performed, with primers at 1 μ*M* and reaction in 10 μL (primers in **Table 1**). The inclusion of negative controls to test for PCR contamination is advisable. If DNA is detected, the DNAse-treated RNA sample is discarded.
8. After DNase treatment, ethanol precipitation is apparently enough to inactivate any significant DNase activity without the need of any further treatment such as phenol.
9. The RNA pellet can be stored indefinitely in this wash solution of 75% ethanol. It is left there before reverse transcription until it is confirmed by PCR that no residual DNA contamination has survived (*see* **Note 7**).
10. Total RNA was always used because poly A selection from low amounts (tens of micrograms) of RNA was found inappropriate. The overall losses were significant, and an unquantifiable amount made the DD unreproducible and difficult to normalize.
11. Given the 3′ location of the AU motifs there was a special interest in retrotranscribing true 3′ ends. The separate step of annealing at room temperature was found to reduce primer extensions from false poly A tails.
12. In this setting, the 1:500 dilution always gives comparable results among samples; 4 μL checked by electrophoresis usually contains 5 to 10 ng of GAPDH amplimer.
13. Because it is of interest to sample as many genes as possible different conditions are used side by side to increase throughput. Limited sample availability can also benefit from these variations. Primers of the series G7AU2 (**Table 1**) are used on cDNAs initiated from G7AU2dT. Similarly, GTGAU2 primers are *more efficient* on cDNA primed from GTGAU2dT. Both types of primers give different profiles, and different designs of the nonspecific 5′ domain also give different patterns. Proposed variations of conditions for a given cDNA to yield different profiles are as follows: (1) use the five primers of any series in 40 cycles of a single round of PCR; (2) run first a nonradioactive PCR of 30 cycles with either G7AU2 or GTGAU2, then dilute products 1:50 and run a series of nested PCR of 30 cycles with the remaining four inner primers (G7AU2N or GTGAU2N) in four separate radioactive reactions; (3) reduce the concentration of cold dNTPs; (4) change the reaction buffer (different profiles can be achieved just by changing the Mg concentration in 1 m*M* differences); (5) combine any of these conditions.
14. A cold dNTP concentration of 0.2 m*M* on radioactive reactions gives stronger signals and lower background than lower concentrations.

15. To reduce artifacts, it is important to use some hot start techniques or a preblocked polymerase. Some brands offer this kind of product. Those products that require extra PCR cycles for activation have been avoided. Note that different polymerases can also give different band patterns and sensitivity.
16. Although even when fixed and washed radioactive gels still tend to give background signals on the autorad that assists the orientation, it may be preferred to use luminescent labels (such as the Glogos labels from Stratagene) for this purpose.
17. According to the scientist's interests, genes that are downregulated in experimental samples might also be of major interest. They would be isolated from the control sample since its concentration there would be higher.

Acknowledgments

Pepi Caro is warmly thanked for her expert technical support and patience during the development of this work.

This work has been supported in part by the CDTI as Concerted Projects 95/0184 and 97/0313.

References

1. Xu, N., Chen, C. Y., and Shyu, A. B. (1997) Modulation of the fate of cytoplasmic mRNA by AU-rich elements: key sequence features controlling mRNA deadenylation and decay. *Mol. Cell Biol.* **17,** 4611–4621.
2. Shaw, G. and Kamen, R. (1986) A conserved AU sequence from the 3′ untranslated region of GM-CSF mRNA mediates selective mRNA degradation. *Cell* **46,** 659–667.
3. Caput, D., Beutler, B., Hartog, K., Thayer, R., Brown-Shimer, S., and Cerami, A. (1986) Identification of a common nucleotide sequence in the 3'-untranslated region of mRNA molecules specifying inflammatory mediators. *Proc. Natl. Acad. Sci. USA* **83,** 1670–1674.
4. Balmer, L. A., Beveridge, D. J., Jazayeri, J. A., Thomson, A. M., Walker, C. E., and Leedman, P. J. (2001) Identification of a novel AU-Rich element in the 3′ untranslated region of epidermal growth factor receptor mRNA that is the target for regulated RNA-binding proteins. *Mol. Cell Biol.* **21,** 2070–2084.
5. King, L. M. and Francomano, C. A. (2001) Characterization of a human gene encoding nucleosomal binding protein nsbp1. *Genomics* **71,** 163–173.
6. Liang, P. and Pardee, A. B. (1992) Differential display of eukaryotic messenger RNA by means of the polymerase chain reaction. *Science* **257,** 967–971.
7. Welsh, J., Chada, K., Dalal, S. S., Cheng, R., Ralph, D., and McClelland, M. (1992) Arbitrarily primed PCR fingerprinting of RNA. *Nucleic Acids Res.* **20,** 4965–4970.
8. Stone, B. and Wharton, W. (1994) Targeted RNA fingerprinting: the cloning of differentially-expressed cDNA fragments enriched for members of the zinc finger gene family. *Nucleic Acids Res.* **22,** 2612–2618.
9. Donohue, P. J., Alberts, G. F., Guo, Y., and Winkles, J. A. (1995) Identification by targeted differential display of an immediate early gene encoding a putative Serine/Threonine kinase. *J. Biol. Chem.* **270,** 10,351–10,357.
10. Fischer, A., Saedler, H., and Theissen, G. (1995) Restriction fragment length polymorphism-coupled domain directed differential display: a highly efficient technique for expression analysis of multigene families. *Proc. Natl. Acad. Sci. USA* **92,** 5331–5335.

11. Joshi, C. P., Kumar, S., and Nguyen, H. T. (1996) Application of modified differential display technique for cloning and sequencing of the 3′ region from three putative members of wheat HSP70 gene family. *Plant Mol. Biol.* **30,** 641–646.
12. Dominguez, O., Ashhab, Y., Sabater, L., Belloso, E., Caro, P., and Pujol-Borrell, R. (1998) Cloning of ARE-containing genes by AU-motif-directed display. *Genomics* **54,** 278–286.
13. Sambrook, J., Fritsch, E. F., and Maniatis, T. (1989) *Molecular Cloning: A Laboratory Manual.* Cold Spring Harbor Laboratory Press, Plainview, NY.
14. Chomczynski, P. and Sacchi, N. (1987) Single-step method of RNA isolation by acid guanidinium thiocyanate–phenol–chloroform extraction. *Anal. Biochem.* **162,** 156–159.

37

PCR Fluorescence Differential Display

Kostya Khalturin, Sergej Kuznetsov, and Thomas C. G. Bosch

1. Introduction

Differential display of mRNA via polymerase chain reaction (DD-PCR) has become a powerful procedure for the quantitative detection of differentially expressed genes in distinct cell populations *(1–4)*.

The standard procedure includes selective reverse transcription of polyadenylated RNA using specific anchored oligo(dT) primers, PCR amplification of cDNA using the oligo(dT) primer and an arbitrary upstream primer, resolution of PCR products on denaturing sequencing gels, and radioactive detection methods. To avoid hazardous radioisotopes, several nonradioactive methods for identification of differential display cDNAs have been reported, including ethidium bromide visualization in agarose gels *(5)*, silver staining *(6,7)* and chemiluminescent detection *(8,9)* of cDNA bands.

Several protocols have been published reporting the use of automated DNA sequencers for differential display of cDNAs labeled with infrared dyes *(10)* or fluorescent tags *(11–13)*. A major drawback for using automated sequencers for this purpose is the inability to recover the amplified cDNAs from the gel. As an alternative, the programmable GenomyxLR™ DNA sequencer (Genomyx, Foster City, CA), which allows high resolution of cDNAs and easy localization and excision of radiolabeled bands from the gel, is becoming widely used in differential display studies *(3)*. For nonradioactive detection of differential cDNA bands on the GenomyxLR DNA sequencer, cDNAs can be fluorescently labeled by using tetramethylrhodamine (TAMRA)-anchored primers. Alternatively, cDNAs can be fluorescently labeled by using TAMRA-dUTP (Perkin–Elmer), which is incorporated into extending cDNA during the PCR *(14)*. Fluorescently labeled PCR products are separated on a GenomyxLR DNA sequencer (Genomyx), detected by the GenomyxSC™ Fluorescent Imaging Scanner, and directly excised from the gel for further characterization using an actual and virtual grid system. The flowchart shown in **Fig. 1** outlines the experimental procedure.

2. Materials

1. peqGold RNAPure™ kit (PEQLAB Biotechnologie GmbH, Germany) or other reagents for total RNA/mRNA extraction.
2. DEPC water.

From: *Methods in Molecular Biology, Vol. 226: PCR Protocols, Second Edition*
Edited by: J. M. S. Bartlett and D. Stirling © Humana Press Inc., Totowa, NJ

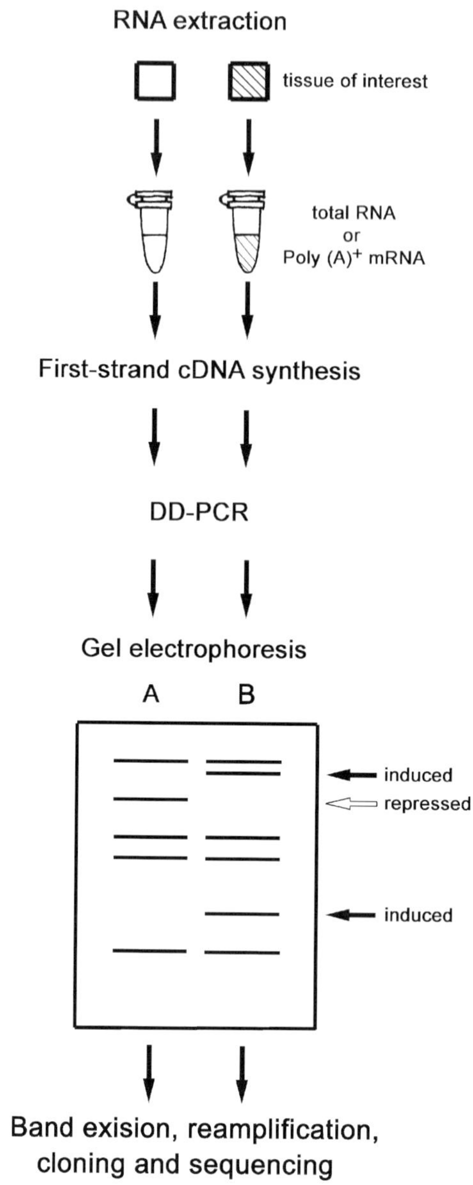

Fig. 1. Flowchart of the fluorescent differential display procedure.

3. First-Strand cDNA synthesis Kit (Amersham Pharmacia Biotech Inc.).
4. Tailing primers T_{12}2NN (e.g., T_{12}CA) at concentration 25 μ*M* (*see* **Note 2**).
5. Arbitrary 10-mer primers (e.g., OPA Kit by Operon Technologies Ltd., Amelada, CA) at concentrations of 5 μ*M*.
6. dNTPs stock solution (1 m*M* of each).
7. Taq DNA Polymerase and 10, PCR buffer (Amersham Pharmacia Biotech Inc.).
8. TAMRA-dUTP (400 μ*M* stock solution, Perkin–Elmer).
9. $MgCl_2$ (25 m*M* stock solution).
10. GeneQuantII photometer (Pharmacia) or equivalent device for measuring RNA concentration.

11. Thermal cycler with a ramping speed not more then 2.5°C/s. (or with programmable ramping speed).
12. GenomyxLR™ DNA sequencer and GenomyxSC™ Fluorescent Imaging Scanner (Genomyx, Foster City, CA).

3. Methods

3.1. RNA Isolation (see Note 1)

1. Isolate total RNA using peqGold RNAPureKit from approx 1×10^6 cells according to the manufacturer's protocol. The final volume of total RNA suspension should be 20 µL.
2. Measure OD_{260} of 2 µL of RNA suspension in 80 µL of DEPC water (1:40 dilution). The total yield should be approx 300 µg of total RNA.

3.2. Reverse Transcription (see Note 2)

1. Reverse transcribe 3 µg of total RNA using First-Strand cDNA synthesis Kit.
2. Place 3 µg of total RNA in a microcentrifuge tube and add RNAse-free (DEPC) water, if necessary, to bring the suspension volume to 7.75 µL.
3. Heat the mixture for 10 min at 70°C and then place on ice immediately.

Components (concentration of stock solution)	Amount	Final concentration
Bulk First-Strand cDNA Reaction Mix (3×)	5 µL	1×
DTT solution (200 m*M*)	1 µL	13.3 m*M*
Tailing primer, e.g. 5′-T_{12}CA-3′ (25 µ*M*)	1.25 µL	2 µ*M*
RNA suspension (3 µg)	7.75 µL	0.2 µg/µL
Final volume	**15 µL**	

4. Mix the following components on ice according to **Table 1** for one reaction.
5. Incubate for 1 h at 37°C.
6. Incubate the tube at 95°C for 2 min to inactivate reverse transcriptase.
7. Dilute the reaction mix 1:25. This quantity of template is sufficient for approx 180 DD-PCRs in the volume of 10 µL.

3.3. PCR (see Note 3)

1. Set up the PCR mix according to Table 2.

Components of stock solution)	Amount	Final concentration
cDNA (1:25 dilution of RT-reaction mix)	2 µL	–
10× buffer	1 µL	1×
$MgCl_2$ (25 m*M*)	0.625 µL	3 m*M*
random 10-mer primer (5 µ*M*)	0.5 µL	0.5 µ*M*
tailing primer (25 µ*M*)	1 µL	2.5 µ*M*
dNTP (1 m*M* of each)	2 µL	200 µ*M*
TAMRA-dUTP (400 µ*M*)	0.1 µL	4 µ*M*
Taq DNA Polymerase (5 U/µL)	0.1 µL	0.5 U
H_2O	2.675 µL	–
Final volume	**10 µL**	

2. Prepare all the samples in duplicates to check the reproducibility of the DD-PCR banding pattern.
3. Perform the PCR with the following settings: initial denaturing 2 min at 95°C; 40 cycles 30 s at 94°C (denaturation), 30 s at 40°C (annealing), and 30 s at 72°C (extension).
4. Use the thermal cycler with a ramping speed not more then 2.5°C/s (*see* **Note 4**).

3.4. Electrophoresis

1. Before electrophoresis, add 7 µL of sample buffer (95% formamide, 0.25% dextran blue, 10 m*M* EDTA) to each 10-µL sample.
2. Denature amplified labeled fragments at 95°C for 3 min and load onto a high-resolution denaturing polyacrylamide gel (HR-1000 4.5% matrix, Genomyx, Foster City, CA).
3. Perform the electrophoresis for 2.5 h in parallel using the GenomyxLR DNA sequencer (according to the manufacturer's instructions).

3.5. Detection of PCR Products, Elution, and Cloning

1. cDNA bands are detected using the GenomyxSC Fluorescent Imaging Scanner (Genomyx). A typical gel with several differentially expressed transcripts of the freshwater polyp *Hydra vulgaris* is shown in **Fig. 2** (filled and open arrows). Fluorescently labeled cDNA fragments usually range from about 100 to 2000 bp.
2. For further characterization, DNA from differentially expressed bands is localized using the grid coordinate system provided with the GenomyxSC Fluorescent Imaging Scanner. After excision from the gel, cDNA in gel slices can easily be recovered and reamplified using described procedures *(7,**14**)*.

4. Notes

In our minds, the most crucial steps for performing a successful differential display screening are as follows: (1) quality of the initial RNA taken for the cDNA preparation; (2) the quantity of cDNA used for the DD-PCR; and (3) the PCR cycling profile, which depends greatly on the type of thermal cycler used.

1. In our experience DD-PCR works equally well in using either total RNA or mRNA. Kits for the isolation of poly(A)$^+$ mRNA, however, seem to be more reliable. The reason is that when total RNA is extracted, DNAse I treatment is usually needed to eliminate the DNA. However, this procedure sometimes can lead to considerable loss of mRNA.

   If extraction of total RNA is the method of choice, it is crucial to check the quality of the RNA obtained not only by the OD_{260} and OD_{280} measurement but also by electrophoresis of the RNA in an agarose gel. In the ethidium bromide-stained gel, the total RNA should appear as a smear of approx 500 to 2000 bp with two prominent bands of 28S rRNA and 18S rRNA. If any RNA degradation or traces of high molecular weight DNA are observed, the RNA samples should not be used in DD-PCR.
2. We prefer to synthesize cDNA using $T_{(12)}$NN tailing primers (where NN stands for one of the 12 possible combinations of 4 nucleotides) instead of standard *Not*I primer. Using the same tailing primer both in reverse transcription and DD-PCR gives more distinct bands in comparison with the case when cDNA is made with *Not*I primer and then one of $T_{(12)}$NN primers is used in a DD-PCR. A slight increase in the reaction temperature (40°C instead of the usual 37°C) during reverse transcription improves the specificity of cDNA synthesis. DD-PCR banding patterns obtained with the same random primer, but two different tailing primers (e.g., –5′-T_{12}CA-3′ and 5′-T_{12}AC-3′) should be considerably different.
3. High final concentrations of dNTP (from 25 to 250 µ*M* of each) and Mg^{2+} (from 1.5 to 4 m*M*) may help increasing the quantity and quality of bands. For optimal results, it is always necessary to try different dilutions of the cDNA sample (e.g., 1:25 → 1:50). A low final concentration of cDNA in DD-PCR can lead to nonreproducible banding patterns and even to differences within pares of tubes containing identical cDNA templates. One typical problem encountered when doing DD-PCR is that the tailing primer does not anneal during PCR, resulting in fragments flanked by random primer only. In that case,

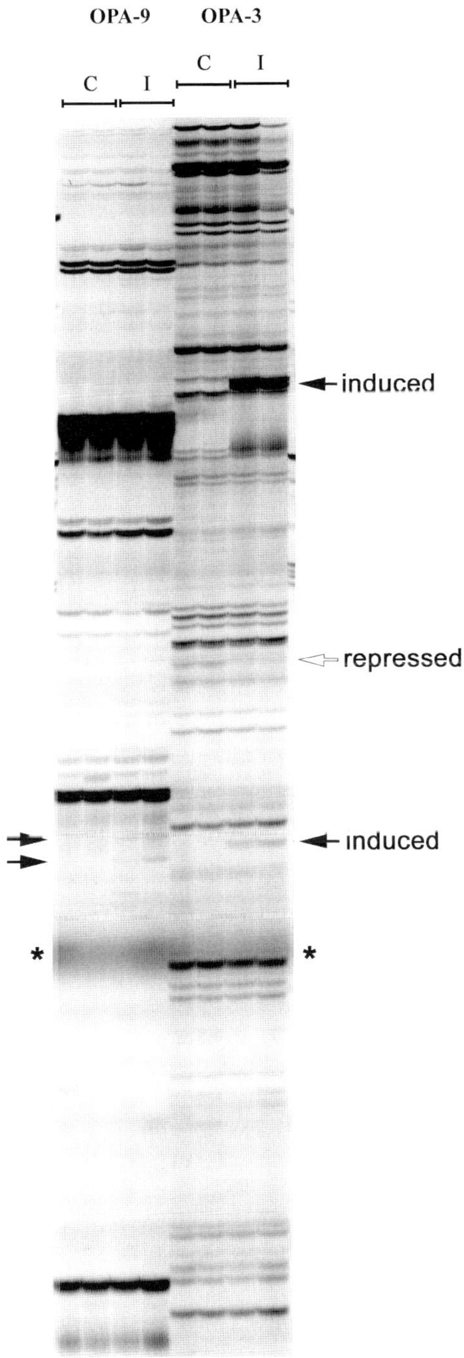

Fig. 2. Typical picture of DD-PCR of *H. vulgaris* cDNA using two different random primers (OPA-9 and OPA-3) and 5′-$T_{(12)}$CA-3′ tailing primer. Fluorescently labeled PCR products were visualized by GenomyxSC Fluorescent Imaging Scanner. Filled arrows, induced transcripts; open arrows, repressed transcripts; C, control animals; I, immuno challenged animals; *, unincorporated TAMRA-dUTP.

the annealing temperature should be decreased from 40°C down to the 35°C. Besides that, the ratio between random 10-mer primer and tailing primer should be adjusted. Using fluorescent TAMRA-dUTP in the reaction mix adds additional complexity to the DD-PCR approach because it can be destroyed by frequent freeze-thaw cycles. PCR products labeled with new TAMRA-dUTP are nearly 10 times brighter than those labeled using TAMRA-dUTP that had undergone 3 to 4 freeze-thaw cycles. Because of that, TAMRA-dUTP should be always kept in 2- to 3-μL aliquots. Usually, it is difficult to estimate the optimal concentration of TAMRA-dUTP for effective labeling. Therefore, a different dTTP:TAMRA-dUTP ratio should be tried (for example 30:1, 100:1, 200:1). Unincorporated TAMRA-dUTP results in the fluorescent cloud in the gel indicated by the asterisk in **Fig. 2**.

4. One of the major variables during the DD-PCR experiment appears to be the fact that thermal cyclers vary greatly in their ramping time and other technical details. In our laboratory, we use two types of PCR machines: Omn-E (Hybaid) and RoboCycler (Stratagene). The former has a Peltier-element with the ramping time of approx 2.5°C/s (while cooling from 94° to 70°C), the latter nearly no ramping since a robotic arm transfers the samples from one thermoblock to another. When examining the influence of the ramping time on the DD-PCR and using the same temperature and time profiles in both machines (as well as identical components of the PCR mix) we observed considerably more fragments produced using the Omn-E thermal cycler in comparison with the Robocycler. The difference is not caused by the destruction of TAMRA-dye because the same result is observed in silver stained gels *(7)* as well as on those viewed by the fluorescent scanner. It seems the rapid change of the temperature provided by the RoboCycler prevents the efficient synthesis of PCR products when using standard PCR protocol. An increase of the annealing time from 30 to 90 s may overcome that drawback.

Acknowledgments

The work in the laboratory is supported by grants from the DFG (to T.C.G.B.) and by the Daimler-Benz Foundation (S.K.)

References

1. Liang, P., Averboukh, L., Keyomarsi, K., Sager, R., and Pardee, A. B. (1992) Differential display and cloning of messenger RNAs from human breast cancer versus mammary epithelial cells. *Cancer Res.* **52,** 6966–6968.
2. Liang, P. and Pardee, A. B. (1992) Differential display of eukaryotic messenger RNA by means of the polymerase chain reaction. *Science* **257,** 967–971.
3. Martin, K. J., Kwan, C. P., O'Hare, M. J., Pardee, A. B., and Sager, R. (1998) Identification and verification of differential display cDNAs using gene-specific primers and hybridization arrays. *BioTechniques* **24,** 1018–1026.
4. McClelland, M., Mathieu-Daude, F., and Welsh, J. (1995) RNA fingerprinting and differential display using arbitrarily primed PCR (review). *Trends Genet.* **11,** 242–246.
5. Rompf, R. and Kahl, G. (1997) mRNA differential display in agarose gels. *BioTechniques* **23,** 28–32.
6. Bosch, T. C. G. and Lohmann, J. (1996) Non-radioactive differential display of messenger RNA, in *Fingerprinting Methods Based on Arbitrarily Primed PCR*, Springer Lab Manual (Micheli, M.R. and Bova, R., eds.), Springer Verlag, Heidelberg.
7. Lohmann, J., Schickle, H. P., and Bosch, T. C. G. (1995) REN, a rapid and efficient method for non-radioactive differential display and isolation of mRNA. *BioTechniques* **18,** 200–202.

8. An, G., Luo, G., Veltri, R. W., and O'Hara, S. M. (1996) Sensitive, nonradioactive differential display method using chemiluminescent detection. *BioTechniques* **20,** 342–346.
9. Ross, R., Ross, X.-I., Rueger, B., Laengin, T., and Reske-Kunz, A. B. (1999) Nonradioactive detection of differentially expressed genes using complex RNA or DNA hybridization probes. *BioTechniques* **26,** 150–155.
10. Motlik, J., Carnwath, J. W., Hermann, D., Terletski, V., Anger, M., and Niemann, H. (1998) Automated recording of RNA differential display patterns from pig granulosa cells. *BioTechniques* **24,** 148–153.
11. Bauer, D., Müller, H., Reich, J., Riedel, H., Ahrenkiel, V., Warthoe, P., et al. (1993) Identification of differentially expressed mRNA species by an improved display technique (DDRT-PCR). *Nucleic Acid Res.* **21,** 4272–4280.
12. Jones, S. W. (1997) Generation of multiple mRNA fingerprints using fluorescence-based differential display and an automated DNA sequencer. *BioTechniques* **22,** 536–543.
13. Luehrsen, K. R. (1997) Analysis of differential display RT-PCR products using fluorescent primers and Genescan™ software. *BioTechniques* **22,** 168–175.
14. Reinhardt, B., Frank, U., Gellner, K., Lohmann, J., and Bosch, T. C. G. (1999) High-resolution, fluorescence-based differential display on a DNA sequencer followed by band excision. *BioTechniques* **27,** 268–271.

38

Microarray Analysis Using RNA Arbitrarily Primed PCR

Steven Ringquist, Gaelle Rondeau, Rosa-Ana Risques, Takuya Higashiyama, Yi-Peng Wang, Steffen Porwollik, David Boyle, Michael McClelland, and John Welsh

1. Introduction

RNA arbitrarily primed polymerase chain reaction (RAP-PCR) has been used extensively to identify differentially regulated genes *(1–6)*. The RAP-PCR method begins with conversion of RNA into cDNA, followed by arbitrarily primed PCR. The technique uses arbitrarily primed PCR *(7–11)* to amplify cDNA stretches lying between sequences that, by chance, match arbitrarily chosen oligonucleotide primers well enough to initiate primer extension. In earlier applications of the method, the complex mixture of products was resolved by polyacrylamide gel electrophoresis (PAGE), yielding highly reproducible fingerprints characteristic of the RNA source. Differences between fingerprints resulting from differentially expressed genes were verified by Northern blot analysis or reverse transcription (RT)-PCR.

Here, a combination of RAP-PCR and cDNA array technology is described (*see* **Note 1**, **refs.** *12–14*), which provide dramatic improvement in detection sensitivity and ease of analysis when compared with PAGE. Typically, PAGE analysis of a RAP-PCR examines ~50 to 100 genes, whereas microarray analysis of RAP-PCRs examines ~10 to 20% of the genes represented on the microarray, simultaneously. With PAGE, characterization of differentially regulated genes requires laborious purification of the band, cloning, and sequencing *(15)*. With microarrays, the identity of the gene is known immediately because the arrayed sequences are known. As is generally understood with microarrays, the microarray-based RAP-PCR approach can quickly narrow all mRNAs down to a subset that exhibits regulation under the conditions examined. However, the RAP-PCR method selectively examines a different population of mRNAs than does probe from mRNA or total RNA: RAP-PCR selectively samples the complex, rare class of mRNA, presumably as the result of the greater likelihood of the arbitrary primers encountering a sufficiently good match in the complex class to enable primer extension. This enables the method to measure changes in RNA abundances for rare transcripts with greater ease than with simple oligo $(dT)_n$ priming of mRNA for probe synthesis. Iteration of the method using different arbitrary primers allows greater coverage of the mRNA population to be achieved.

From: *Methods in Molecular Biology, Vol. 226: PCR Protocols, Second Edition*
Edited by: J. M. S. Bartlett and D. Stirling © Humana Press Inc., Totowa, NJ

In this chapter, protocols are presented for the synthesis of RAP-PCR probes for application to microarrays. Arrays spotted on nylon membranes can also be used. The manufacture of microarrays, themselves, is discussed only briefly, and the reader is reminded that there are a number of different approaches. For example, diverse chemistries for adhering cDNA sequences to glass slides, PCR-amplified vs. synthesized spotted sequences, different fluorescent dye-labeled nucleotides, different robotic spotting strategies, and other variables have been explored. The reader is encouraged to seek current technical information in this rapidly moving field.

2. Materials

2.1. Special Reagents and Supplies

1. RNeasy Mini Kit (Qiagen #74106, Valencia, CA).
2. QIAshredder columns (Qiagen #79656, Valencia, CA).
3. QIAquick PCR Purification Kit (Qiagen, #74106, Valencia, CA).
4. Microcon YM-30 (Millipore #42410, Bedford MA).
5. Human cDNA clones (I.M.A.G.E; Research Genetics, Huntsville, AL).
6. Vector-specific primers (Genosys Biotechnologies, The Woodlands, TX): Forward: 5′-ctgcaaggcgattaagttgggtaac-3′, Reverse:5′-gtgagcggataacaatttcacacaggaaacagc-3′
7. First strand cDNA primer: oligo$(dT)_{20}$.
8. Arbitrary primers: primer A: (5′-ACGAAGAAGAAGAG), primer B (5′-GTGACAGACA; Genosys Biotechnologies, The Woodlands, TX).
9. [α-^{32}P] dCTP 10 Ci/mL (ICN, Irvine, CA).
10. Random hexamers (NNNNNN; Genosys Biotechnologies, The Woodlands, TX).
11. Cy3-dCTP and Cy5-dCTP (Amersham, Piscataway, NJ).

2.2. Cell Culture

1. Fibroblast (ATCC #CRL 2091, Manassas, VA).
2. Cell-culture dishes (150 cm; Nunc, Rochester, NY).
3. Cell culture media (DMEM, Irvine Scientific #9031 Irvine, CA), plus 10% fetal bovine serum (Omega scientific #FB-01, Tarzana, CA).
4. Penicillin (1000 U/mL).
5. Streptomycin (1 mg/mL).

2.3. Enzymes

1. M-MLV reverse transcriptase (200 U/µL; Promega, Madison, WI).
2. AmpliTaq DNA polymerase Stoffel fragment (10 U/µL; Perkin–Elmer Cetus, Norwalk, CT).
3. RNase-free DNase (10 U/µL) (Roche Molecular Biochemicals, Indianapolis, IN).
4. RNase inhibitor (40 U/µL) (Roche Molecular Biochemicals, Indianapolis, IN).

2.4. Common Reagents, Supplies, Buffers, and Equipment

1. DMSO.
2. Distilled, deionized H_2O.
3. 1× TE (pH 8.0).
4. Formamide dye solution.
5. 4% polyacrylamide, 8 *M* urea sequencing-style gels, prepared with 1× TBE buffer.
6. 1% agarose minigels (1× TBE).
7. 5× DNase I digestion buffer: 100 m*M* Tris-HCl, 50 m*M* $MgCl_2$, pH 8.0.

8. 5× reverse transcription mixture: 250 m*M* Tris-HCl, pH 8.3, 375 m*M* KCl, 15 m*M* $MgCl_2$, 100 m*M* DTT, 1.0 m*M* each dNTP, 2.5 μ*M* oligo$(dT)_{20}$, and 100 U M-MLV reverse transcriptase.
9. 2× RAP-PCR mixture: 20 m*M* Tris-HCl, pH 8.3, 20 m*M* KCl, 6 m*M* $MgCl_2$, 0.35 m*M* each dNTP, 2 μ*M* each arbitrary primer (see text); 2 μCi [α-^{32}P] dCTP, and 0.5 U/μL AmpliTaq DNA polymerase Stoffel fragment.
10. 2× Klenow reaction mixture: 20 m*M* Tris-HCl, pH 7.5, 10 m*M* $MgCl_2$, 15 m*M* DTT, 0.050 m*M* dATP, 0.050 m*M* TTP, 0.050 m*M* dGTP, 0.018 m*M* dCTP, 40 U Klenow.
11. 2× terminal transferase buffer: 400 m*M* potassium cacodylate; 50 m*M* Tris-HCl, pH 7.2; 500 mg/mL bovine serum albumin, and 3 m*M* $CoCl_2$.
12. Blocking solution: 10 μg/μL oligo $(dA)_{20}$, 10 μg/μL yeast tRNA, 10 μg/μL human Cot-1 DNA.
13. 20× SSC.
14. Human Cot-1 DNA (Invitrogen #15279, Carlsbad, CA).
15. Klenow (New England BioLabs #MO210S, Beverly, MA).
16. 96-well microtiter plates.
17. 96-well PCR amplification plates.
18. Ultraviolet spectrophotometer.
19. Oven for baking slides.
20. Vacuum desiccator for slide storage.
21. Kodak BioMax X-Ray (Kodak #8715187, Rochester, NY).
22. Succinic anhydride (Sigma #S-7626, St. Louis, MO).
23. 1-methyl-2-pyrrolidinone (Sigma #6762, St. Louis, MO).
24. poly-L-lysine (Sigma #P8920, St. Louis, MO).

2.5. Special Equipment

1. PCR amplification machine (GeneAmp® PCR System 9700, Perkin-Elmer Cetus, Norwalk, CT).
2. High-speed robotic arrayer (OmniGrid, GeneMachine, San Carlos, CA).
3. Ultraviolet irradiation device (Stratagene, La Jolla, CA).
4. ScanArray 5000 (GSI Lumonics, Billerica, MA).
5. QuantArray analysis software (GSI Lumonics, Billerica, MA).
6. GoldSeal glass slides (Fisher Scientific, Pittsburgh, PA).

Molecular biology grade, RNase free reagents are used. Sterile disposable polypropylene was used rather than glassware.

3. Methods

3.1. Preparation of RNA

1. Fibroblasts were grown in culture to about 7×10^6 cells per plate and harvested by scraping in the presence of RTL buffer lysis buffer (RNeasy kit) and homogenized through Qiashredder columns, both according to the manufacturer's instructions. Total RNA was isolated using the RNeasy total RNA purification kit and eluted into 50 μL of distilled water.
2. To 49 μL total RNA was added 6 μL of DNase I digestion buffer, 5 μL of DNase I, and 0.5 μL of RNase inhibitor for 30 min at 37°C.
3. DNase-treated total RNA was purified again using the RNeasy total RNA kit according to manufacturer's instructions and eluted into 50 μL of distilled water.
4. RNA concentration was determined by spectrophotometry in TE buffer (10 m*M* Tris-HCl, pH 8.4; 0.1 m*M* EDTA) assuming that one OD unit was equivalent to 40 μg/mL of RNA.

5. Total RNA samples were adjusted to 400 ng/μL with distilled water, checked for the integrity of the rRNAs by 1% agarose gel electrophoresis, and stored at –20°C.

3.2. cDNA Synthesis

1. Reverse transcription was performed on total RNA using at least three twofold dilutions of total RNA per sample between 1000 and 125 ng per 10 μL of reaction (*see* **Note 1**). cDNA synthesis uses oligo$(dT)_{20}$ for first strand priming, or alternatively, an arbitrary sequence primer can be used under exactly the same conditions. The reaction mixture contained 20 μL of purified, DNase I-treated RNA plus 5 μL of 5× reverse transcription mixture. cDNA reactions were performed at room temperature (usually 23°C) for 15 min followed by incubation at 37°C for 1 h.
2. The reactions were stopped by heating for 5 min in a boiling water bath followed by cooling on ice. Reactions should be diluted fourfold with 75 μL of distilled water and stored at –20°C, or one can proceed to the next step. (*see* **Note 2**).

3.3. RAP-PCR

1. 2× RAP-PCR mixture was prepared using different pairs of arbitrary primers (*see* **Note 3**). In **Fig. 1**, primers A and B were used.
2. Diluted cDNAs (10 μL) were mixed with the same volume of 2X RAP-PCR mixture. Thermocycling was performed using an initial 3-min incubation at 94°C followed by 35 cycles of 94°C for 1 min, 35°C for 1 min, and 72°C for 2 min.

3.4. Polyacrylamide Gel Electrophoresis

1. An aliquot of the amplification products (2 μL) was mixed with 9 μL of formamide dye solution, denatured at 85°C for 4 min, and chilled on ice. A sample of 2 μL was loaded onto a 4% polyacrylamide, 8 *M* urea gel, prepared with 1× TBE buffer. The PCR products resulting from different concentrations of the same RNA template were loaded side by side on the gel.
2. Electrophoresis was performed at 1700 V or at a constant power of 50 to 70 W until the xylene cyanol tracking dye reached the bottom of the gel (about 4 h). The gel was dried under vacuum and either placed on Kodak BioMax X-Ray film for 16 to 48 h or used to expose a phosphor screen (Molecular Dynamics, Sunnyvale, CA) and analyzed on a Storm 820 (Molecular Dynamics, Sunnyvale, CA). The results are shown in **Fig. 1.**

3.5. Fluorescent Labeling

1. Up to 10 μg of PCR product from the RAP-PCR can be purified using a QIAquick PCR Purification Kit (Qiagen, Valencia, CA), which removes unincorporated bases, primers, and primer dimer less than 40 bp. DNA was recovered in 50 μL of 10 m*M* Tris-HCl (pH 8.3), usually yielding 2 to 3 μg of DNA per RAP-PCR amplification. Recovered PCR products were quantified by spectroscopy assuming 33 μg/mL/OD unit.
2. Klenow labeling of RAP-PCR DNA was performed by random primed synthesis using up to 1 μg of PCR product per reaction. PCR product was mixed with 12 μg total of 6-mer or 9-mer random sequence primer and boiled for 4 min, cooled to room temperature, and spun briefly in a microfuge. The volume was adjusted to 20 μL, 25 μL of 2× Klenow reaction mixture, and 5 μL of either 1 m*M* Cy3-dCTP or 1 m*M* Cy5-dCTP (Amersham, Piscataway, NJ) was added and the mixture was incubated at 37°C for 4 h. The reaction was stopped by incubation at 70°C for 10 min to inactivate the Klenow fragment. Alternatively, addition of EDTA pH 8 to a final concentration of 10 m*M* can also be used. Probes from oligo $(dT)_n$ can be prepared by standard means to examine more abundant transcripts (*see* **Note 4**).

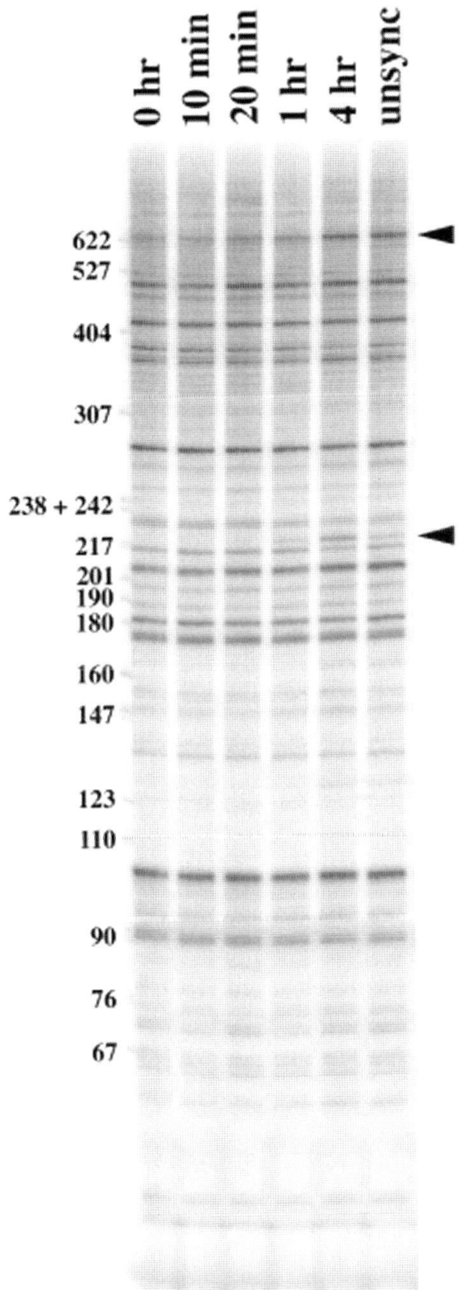

Fig. 1. RAP-PCR fingerprints. Products from RAP-PCR were resolved by electrophoresis as described. DNA molecular weights were determined using ^{32}P-labeled DNA fragments from *Msp*I-digested pBR322 plasmid DNA, lane 1. Lanes 2 through 25 contain RAP-PCR products prepared from human fibroblast total RNA. Fingerprints were prepared from fibroblasts synchronized by serum starvation for 48 h and then treated by addition of serum containing media for the indicated times prior to isolation of total RNA, including isolation of total RNA from nonserum-starved cells. Fingerprinting of each sample was performed using three different concentrations of total RNA in the cDNA reaction, that is, 500, 250, and 125 ng per 10 µL of reaction. The fourth (empty) lane in each sample is the reverse transcriptase minus control.

3. Purification of fluorescently labeled material from unincorporated dye was performed by centrifuging the sample in a microfuge using Microcon columns twice, using 500 μL TE (pH 8.0) washes each time, according to the manufacturer's instructions.
4. Labeling of PCR products can also be performed by Terminal Transferase-catalyzed addition of Cy3-dCTP or Cy5-dCTP, as well as other fluorescent-labeled or modified deoxynucleotide triphosphates. 1 μg of PCR product in 20 μL was combined with 25 μL 2× terminal transferase buffer and 5 μL of 1 m*M* Cy5-dCTP or 1 m*M* Cy3-dCTP. The reaction was performed at 37°C for 1 h and purified using the Microcon columns as described previously.
5. After purification over microcon columns, fluorescent probe was dried under vacuum and dissolved in 8 μL of 1× TE buffer. Preparation of probe in hybridization solution was performed by addition of 8 μL of probe solution to 2 μL of blocking solution, 2.1 μL of 20× SSC, and 0.4 μL of 10% SDS. The solution was then incubated 1 min in a boiling water bath, and allowed to cool on the bench top for 30 min.

3.6. Microarray Hybridization

1. Preparation of microarrays followed the protocol of Eisen and Brown ***(16)*** and used polylysine-treated slides and PCR products generated from sequenced human cDNA libraries; please see Eisen and Brown ***(16)*** for details on preparing polylysine-coated slides. Inserts from the IMAGE library, supplied by Research Genetics, were PCR amplified by standard means using the vector-specific primers. Printing of PCR products was done using an Omnigrid robot to spot PCR products onto GoldSeal glass slides using a 16 pin print head. After printing, slides were rehydrated briefly over an 80°C waterbath and snap heated for about 6 s by placing the slides on an 80°C heat block. Ultraviolet treatment was performed using a Stratalinker 2400 (Stratagene, La Jolla, CA) at 60 mJ for 1 min. The slides were then baked at 80°C for 2 h and stored.
2. Immediately before hybridization, slides were treated with succinic anhydride as described by Eisen and Brown ***(16)***. The succinic anhydride blocking solution must be prepared fresh by placing 6 g of succinic anhydride into a dry beaker followed by addition of 335 mL of 1-methyl-2-pyrrolidinone. After the succinic anhydride was dissolved 15 mL of 1 *M* sodium borate, pH 8.0 was added. The solution was stirred for 5 s to mix the solutions. The mixture was then poured into a clean glass chamber and the slides placed quickly inside and incubated for 15 min.
3. Immediately after succinic anhydride blocking, slides were placed in a chamber with boiling water (the volume should be at least twice that of the blocking solution). The slides were plunged up and down and shaken for 2 min, and then the slides were immediately transferred to a chamber containing 95% ethanol, agitated and then dried for 37°C for 15 min.
4. Hybridization was performed by placing one glass chamber with distilled water in a 65°C oven for at least 30 min before doing the hybridization. One hybridization chamber was cleaned with pressurized air, and 2 drops (10 μL) of 3XSSC (450 m*M* sodium chloride, 45 m*M* sodium citrate, pH 7.0) were placed in the wells. The slides were then cleaned with pressurized air and placed in the hybridization chamber. The probe was then pipetted onto the slide, usually in 40 μL, taking care to avoid bubbles. The coverslip was then cleaned with pressurized air and placed over the probe. This volume of probe usually spreads easily along the surface; if not, the overslip can be pressed to help the spreading. The chamber was then sealed and submerged in a 65°C water bath for 4 to 16 h.
5. The hybridization was stopped by washing in 2× SSC, 0.1% SDS, followed by washing in 1× SSC. Samples were incubated for 2 min with 0.2× SSC. The washing steps were performed at room temperature. Drying should be avoided in all the washing steps. Prepare

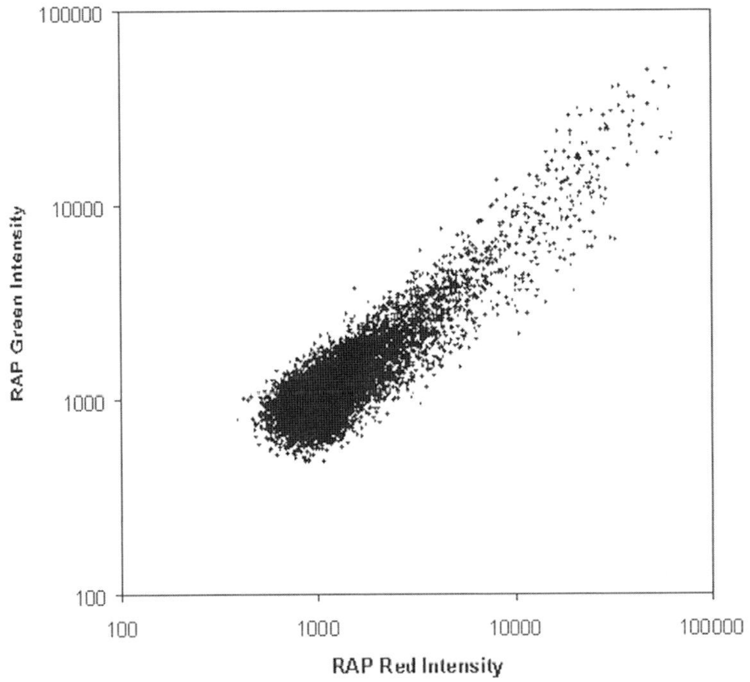

Fig. 2. Scatter plot of fluorescent intensity of RAP-PCR products analyzed by microarray hybridization. Samples were RAP-PCR products (1 μg) from duplicate fingerprint reactions labeled using Cy3-dCTP (green fluorescence) or Cy5-dCTP (red fluorescence), random primer and Klenow as described. The microarray contained 6400 double spotted cDNAs of known identity.

400 mL of each solution and place them in three different glass chambers. To minimize the transfer of solutions, a different container should be used each time and only the slides should be transferred. The solution should then be spun down 5 min to dry the slides.

3.7. Analysis of Microarray Data

Analysis of the microarray hybridization was performed using a ScanArray 5000 or equivalent instrument to monitor fluorescent signal. Data was gathered first from the Cy5 dye followed by scanning of the Cy3 fluorescent signal. Typical instrument settings were laser power at 50 and photomultiplier tube at 100. Microarrays were scanned using 10-μm resolution. Quantification of microarray data was performed using QuantArray software or an equivalent software package. The results are shown in a scatter plot in **Fig. 2.** More than 95% of the signals that are significant relative to negative controls from salmonella are within twofold.

4. Notes

1. Knowing the total RNA concentration is critical for RAP-PCR. cDNA is synthesized at several concentrations ranging between 1000 ng and 125 ng of total RNA per 10-μL reaction. At lower concentrations of input total RNA there sometimes arise spurious PCR products, which overwhelm the amplification step. Also, DNA contaminants are monitored by the inclusion of a reverse transcriptase-free control in initial RAP-PCR experiments. This reaction is set up exactly as a RAP-PCR, but reverse transcriptase is omitted and replaced with water. The result should be a gel lane with no bands.

2. The present method is based closely on the protocol described by Trenkle et al. ***(12,14)***, in which RAP-PCR was used to generate probes for differential screening of cDNA arrays on nylon membranes. Each array contained 18,432 cDNA clones from the IMAGE consortium and hybridization detected approx 1000 cDNA clones using each RAP-PCR probe. Different RAP-PCR fingerprints gave hybridization patterns having very little overlap (less than 3%) with each other or with hybridization patterns from total cDNA probes. Thus, repeated application of RAP-PCR probes allows a greater fraction of the message population to be screened on this type of array than can be achieved with a radiolabeled total cDNA probe.
3. In general, there are no constraints on the primers except that they contain at least a few C or G residues, that the 3′-ends are not complementary with themselves or the other primer in the reaction, to avoid primer dimer, and that primer sets are chosen that are different in sequence so that the same parts of mRNA are not amplified in different fingerprints. PCR with a combination of arbitrary and oligo(dT) primers, as has been used in differential display ***(17–20)***, can also be used to generate an effective probe for cDNA arrays. The use of different oligo(dT) anchor primers with the same arbitrary primer results in considerable overlap among the genes sampled by each probe. This can be avoided by using different arbitrary primers with each oligo(dT) anchor primer ***(13)***.
4. Total RNA as well as poly(A) purified mRNA can also be labeled during reverse transcription for microrray analysis. Methods for preparing fluorescent cDNA from RNA have been presented by Eisen and Brown ***(16)***.

References

1. Welsh, J., Chada, K., Dalal, S. S., Cheng, R., Ralph, D., and McClelland, M. (1992) Arbitrarily primed PCR fingerprinting of RNA. *Nucleic Acids Res.* **20,** 4965–4970.
2. Vogt, T. M., Welsh, J., Stolz, W., Kullmann, F., Jung, B., Landthaler, M., et al. (1997) RNA fingerprinting displays UVB-specific disruption of transcriptional control in human melanocytes. *Cancer Res.* **57,** 3554–3561.
3. Vogt, T., Stolz, W., Welsh, J., Jung, B., Kerbel, R. S., Kobayashi, H., et al. Overexpression of Lerk-5/Eplg5 messenger RNA: A novel marker for increased tumorigenicity and metastatic potential in human malignant melanomas. *Clin. Cancer Res.* **4,** 791–797.
4. Ralph, D., McClelland, M., and Welsh, J. (1993) RNA fingerprinting using arbitrarily primed PCR identifies differentially regulated RNAs in mink lung (Mv1Lu) cells growth arrested by transforming growth factor beta 1. *Proc. Natl. Acad. Sci. USA* **90,** 10,710–10,714.
5. McClelland, M. and Welsh, J. (1994) RNA fingerprinting by arbitrarily primed PCR. *PCR Methods Appl.* **4,** S66–S81.
6. McClelland, M., Mathieu-Daude, F., and Welsh, J. (1995) RNA fingerprinting and differential display using arbitrarily primed PCR. *Trends Genet.* **11,** 242–246.
7. Welsh, J. and McClelland, M. (1991) Genomic fingerprinting using arbitrarily primed PCR and a matrix of pairwise combinations of primers. *Nucleic Acids Res.* **19,** 5275–5279.
8. Welsh, J., Petersen, C., and McClelland, M. (1991) Polymorphisms generated by arbitrarily primed PCR in the mouse: application to strain identification and genetic mapping. *Nucleic Acids Res.* **19,** 303–306.
9. Welsh, J. and McClelland, M. (1990) Fingerprinting genomes using PCR with arbitrary primers. *Nucleic Acids Res.* **18,** 7213–7218.
10. McClelland, M. and Welsh, J. (1994) DNA fingerprinting by arbitrarily primed PCR. *PCR Methods Appl.* **4,** S59–S65.
11. Williams, J. G., Kubelik, A. R., Livak, K. J., Rafalski, J. A., and Tingey, S. V. (1990) DNA polymorphisms amplified by arbitrary primers are useful as genetic markers. *Nucleic Acids Res.* **18,** 6531–6535.

12. Trenkle, T., Welsh, J., Jung, B., Mathieu-Daude, F., and McClelland, M. (1998) Non-stoichiometric reduced complexity probes for cDNA arrays. *Nucleic Acids Res.* **26,** 3883–3891.
13. Trenkle, T., Welsh, J., and McClelland, M. (1999) Differential display probes for cDNA arrays. *BioTechniques.* **27,** 554–560,562,564.
14. Trenkle, T., Mathieu-Daude, F., Welsh, J., and McClelland, M. (1999) Reduced complexity probes for DNA arrays. *Methods Enzymol.* **303,** 380–392.
15. Mathieu-Daude, F., Cheng, R., Welsh, J., and McClelland, M. (1996) Screening of differentially amplified cDNA products from RNA arbitrarily primed PCR fingerprints using single strand conformation polymorphism (SSCP) gels. *Nucleic Acids Res.* **24,** 1504–1507.
16. Eisen, M. B. and Brown, P. O. (1999) DNA arrays for analysis of gene expression. *Methods Enzymol.* **303,** 179–205.
17. Liang, P. and Pardee, A. B. (1992) Differential display of eukaryotic messenger RNA by means of the polymerase chain reaction. *Science* **257,** 967–971.
18. Liang, P., Zhu, W., Zhang, X., Guo, Z., O'Connell, R. P., Averboukh, L., et al. (1994) Differential display using one-base anchored oligo-dT primers. *Nucleic Acids Res.* **22,** 5763–5764.
19. Liang, P. and Pardee, A. B. (1998) Differential display. A general protocol. *Mol. Biotechnol.* **10,** 261–267.
20. Martin, K. J. and Pardee, A. B. (1999) Principles of differential display. *Methods Enzymol.* **303,** 234–258.

39

Oligonucleotide Arrays for Genotyping

Enzymatic Methods for Typing Single Nucleotide Polymorphisms and Short Tandem Repeats

Stephen Case-Green, Clare Pritchard, and Edwin Southern

1. Introduction

Much of modern genetics is based on analysis of DNA sequence. Therefore, there is great pressure to scale up sequence analysis while decreasing its cost. The most promising platforms are based on the use of oligonucleotide arrays (DNA chips), which perform many analyses in parallel *(1)*. Arrays comprise libraries of oligonucleotides or polynucleotides attached to solid supports at defined locations. Assays can be performed on the whole library simultaneously, increasing both the speed and accuracy of analysis of the target nucleic acid.

In this chapter, we describe the fabrication and some uses of oligonucleotide arrays that are under development in our laboratory and give an idea of the flexibility of the array platform. Practical array-based assays will need to be individually optimized to the desired target nucleic acid. Although arrays can be made from large fragments of DNA *(2)*, this chapter concentrates on arrays of synthetic oligonucleotides. Analytic methods for single nucleotide polymorphisms (SNPs) and short tandem repeat (STR) measurement are described. Other uses have been made of DNA arrays, such as expression monitoring *(3)* and antisense oligonucleotide optimization *(4)*.

Most methods for DNA sequence analysis are based on gel electrophoretic separation of DNA fragments. Automation and miniaturization of these techniques has been difficult to achieve. A major advantage of arrays is the potential for automation at all stages of manufacture and application. **Figure 1** shows a scheme for an assay using an oligonucleotide array.

1.1. Fabrication of Arrays

Two methods of array fabrication are in general use, listed as follows:

1. Presynthesized oligonucleotides can be applied to a support.
2. Oligonucleotide synthesis can be performed *in situ* at specific sites on the support.

From: *Methods in Molecular Biology, Vol. 226: PCR Protocols, Second Edition*
Edited by: J. M. S. Bartlett and D. Stirling © Humana Press Inc., Totowa, NJ

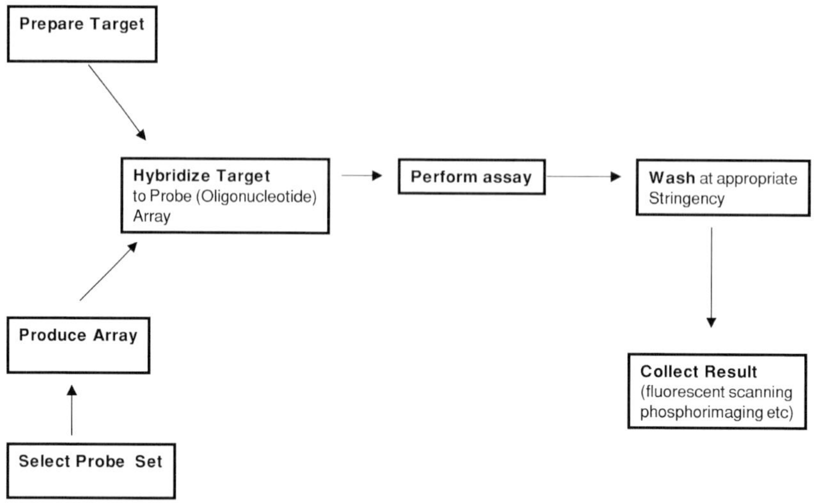

Fig. 1. General scheme for assays using oligonucleotide arrays.

Both methods have advantages and disadvantages, as summarized here.

- *In situ* fabrication allows a truly combinatorial synthesis of oligonucleotide probes to be performed, whereas presynthesis involves synthesis of every probe and its individual application to the array.
- The combinatorial fabrication route may be best suited to producing low numbers or single copies of many different arrays, whereas presynthesis allows enough pure oligonucleotide to be produced to make many copies of the same array with speed and reliability.
- Presynthesized oligonucleotides can be purified before addition to the array, whereas *in situ* fabrication depends on the reliability and high yields of conventional solid phase synthesis to ensure that most of the array probe products are correct.

In our laboratory, we have developed barrier-based methods for combinatorial *in situ* array synthesis using standard oligonucleotide synthesis reagents *(5)* confined within reaction cells (**Fig. 2**).

These techniques allow combinatorial synthesis of multiple oligonucleotides using four basic steps *(6)*. Different shapes of cells are used to create arrays with different organizations and uses. For example, a square-shaped cell can be used for the synthesis of an array containing overlapping oligonucleotide sequences that have all the possible complements of a target sequence up to a chosen length of oligonucleotide. Alternatively, using a block with channels cut into it allows serial synthesis of oligonucleotides of any sequence in a series of stripes *(7,8)*. If the oligonucleotides on an array are arranged in parallel lines, it may be used in more than one reaction if the target samples are hybridized in lines perpendicular to the oligonucleotides *(8)* or the array cut into strips.

Assays described in this chapter are performed on amino functionalized polypropylene support (Beckman) *(7,9)*. This surface is suitable as a support for solid-phase oligonucleotide synthesis. All syntheses in this chapter use 3′-dimethoxytrityl-5′-*N,N′-diisopropylcyanoethylphosphoramidites (reverse phosphoramidites)*. Deprotection

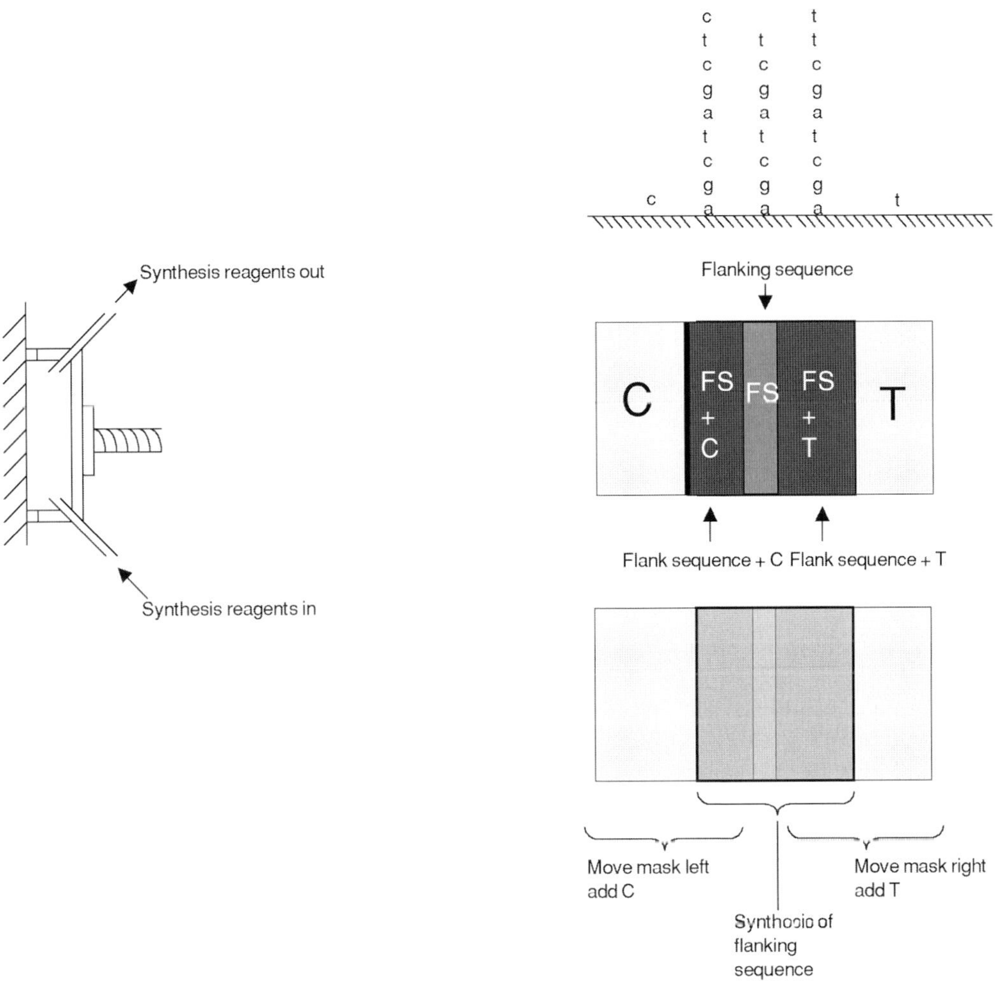

Fig. 2. The layout of reaction cells used for the synthesis of arrays. The synthesis support is pressed to the reaction cell by a clamp. The inlet and outlet ports are attached to a DNA synthesizer. DNA synthesis on the support is achieved using the synthesizer to add and remove reagents from the reaction cell; the reagents flood the surface of the support. After the required sequence has been added the clamp is released and the cell moved and reclamped to allow further synthesis. Apparatus described in **ref. *6*** can be used to aid this process.

leaves the oligonucleotides attached to the polypropylene via the initial, ammonia stable, phosphoramidite linkage to the 5′ end of the oligonucleotide and the amine of the polypropylene support.

1.1.1. Arrays for Biallelic (SNP) Typing

The simplest fabrication method involves the synthesis of each allele specific oligonucleotide (ASO) one at a time at a separate site on the surface. A patch of the first oligonucleotide is synthesized on the support using a cell to direct the synthesis reagents. The cell is then moved to a position adjacent to the first oligonucleotide and the second oligonucleotide synthesized. A long narrow cell (**Fig. 2**, left) minimizes reagent use and allows the array to be cut into strips in a direction perpendicular to the array synthesis.

Alternatively, a sequence that flanks the variable base is synthesized; the cell is displaced by half a cell's width and a base corresponding to one of the alleles is added, the cell is then moved to cover the other half and the other varying base added. The cell can then be returned to its original position and any further bases common to both allele specific probe oligonucleotides added.

1.1.2. STR Array Fabrication

The basic requirements for an array suitable for use in STR typing are shown in **Fig. 3**. All the oligonucleotides of the array include a sequence complementary to the region immediately adjacent to the repeats. This registration sequence forms duplex between the target and the array oligonucleotides and aligns the start of the repeat of the oligonucleotides with that in the target. The array oligonucleotides vary in the number of repeat units that they contain on top of the flanking registration sequence. Synthesis of these arrays can be carried out in a combinatorial manner. **Figure 4** shows a scheme for the synthesis of an array for typing the FES locus ***(10)***.

1.2. Preparation of Target DNA

To achieve good hybridization yields and allow efficient extension of the array oligonucleotides, a single-stranded target is required. Depending on the type of assay the target may be either labeled or unlabeled. Because the DNA for analysis is usually genomically derived material, an amplification step is normally performed.

For hybridization assays, the most common target species is body labeled RNA, prepared by carrying out a polymerase chain reaction (PCR) of the required region using a primer that includes an RNA polymerase promoter sequence and transcribing the product with the appropriate enzyme. The reaction mixture also contains labeled nucleotide triphosphate.

For assays involving enzymes, a single-stranded unlabeled DNA target is required. A two-step procedure can be used where the sample is first amplified using the PCR and the unwanted strand digested away using an exonuclease ***(11)***. The primer for the strand to be retained is synthesized with approximately the last five internucleotide linkages at the 5′ end as phosphorothioate ***(12)***, which protect the strand against nuclease degradation. After PCR, T7 gene 6 exonuclease is added to the product to digest away the unwanted, (unphosphorothioated) strand. The resulting single-stranded DNA can often be used without further purification but, if required, standard purification techniques could be used, for example, Sephadex or Qiaquick columns.

1.3. Hybridization Assays

Target alleles are distinguished by their ability to hybridize to complementary allele specific oligonucleotides (ASOs) on the array (**Fig. 5A**) *(8)*. Hybridization yield and discrimination depend on temperature, salt concentration, time, and target concentration. The sequence of the oligonucleotide also affects both yield and discrimination. Many alleles can be simultaneously examined on the same array, but this requires careful choice of both oligonucleotide sequence and hybridization and washing conditions to achieve the maximum discrimination under one set of conditions. Computer-based

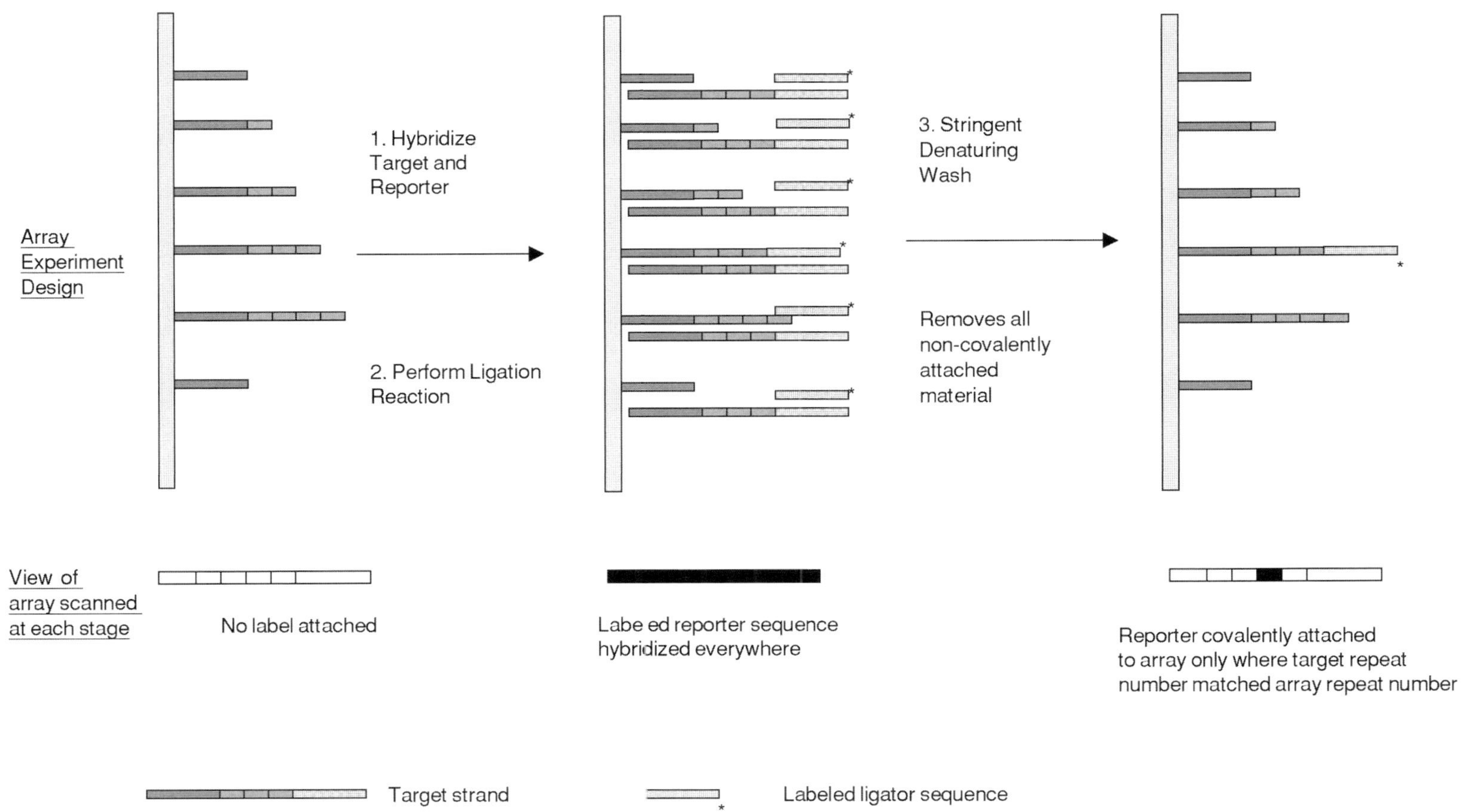

Fig. 3. General scheme for the typing of STRs using oligonucleotide reporter ligation to immobilized oligonucleotides. The addition of nucleotide triphosphates catalyzed by polymerase can be used in a similar manner.

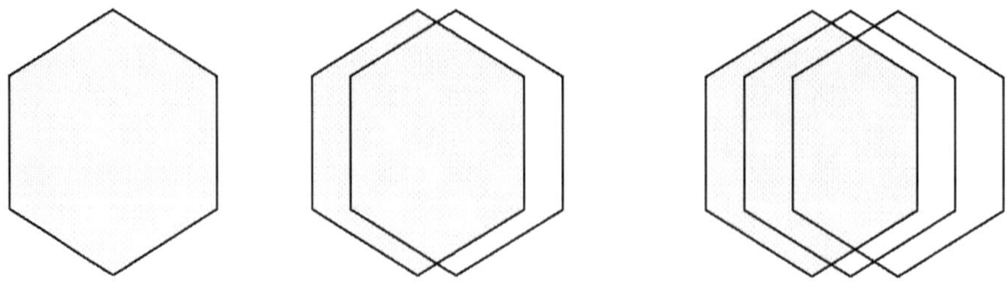

Fig. 4. Synthesis of an array for typing F13 locus.

methods have proved inadequate for choosing the oligonucleotide sequences because many factors, such as the position of mismatches and secondary structure in both the oligonucleotides and target, affect the duplex yield. An empirical solution to this problem is to test various conditions using combinatorially synthesized arrays containing many of the possible oligonucleotides complementary to the allele being investigated *(13,14)*.

1.4. Enzyme-Aided Assays

1.4.1. SNPs

Several variants of the hybridization/extension method are suitable for the analysis of (biallelic) SNPs (**Fig. 5B**,**C**) *(15,16)*. In the simplest form, two oligonucleotides are synthesized that differ in their terminal base; the two bases complementary to the bases to be analyzed. Hybridization of the target sequence will occur in similar yields to both oligonucleotides because of the poor discrimination at terminal bases *(17)*. Ligases and polymerases can distinguish the two duplexes and are able to extend only oligonucleotides with the matching base. Labeled nucleotide triphosphate, complementary to the first base after the allelic position in the target, is incorporated only on oligonucleotides, which form perfect duplex. The position of the label on the support indicates the sequence that is complementary to the target (**Fig. 5C**). Alternatively, a common primer can be used and labeled nucleotide added complementary to either of the alleles of interest in the target. Only when the nucleotide added is complementary to the biallelic position will the primer be extended. (**Fig. 5B**) Ligation can also be used in these assays, replacing the nucleotide triphosphates with labeled reporter oligonucleotides having terminal bases complementary to the alleles to be analyzed.

1.4.2. STR Typing

A method for measurement of STR length is shown in **Fig. 3**. As with SNP typing, assays using both DNA ligases and DNA polymerases are possible. In the basic ligation assay, the target sequence plus a reporter oligonucleotide complementary to the distal

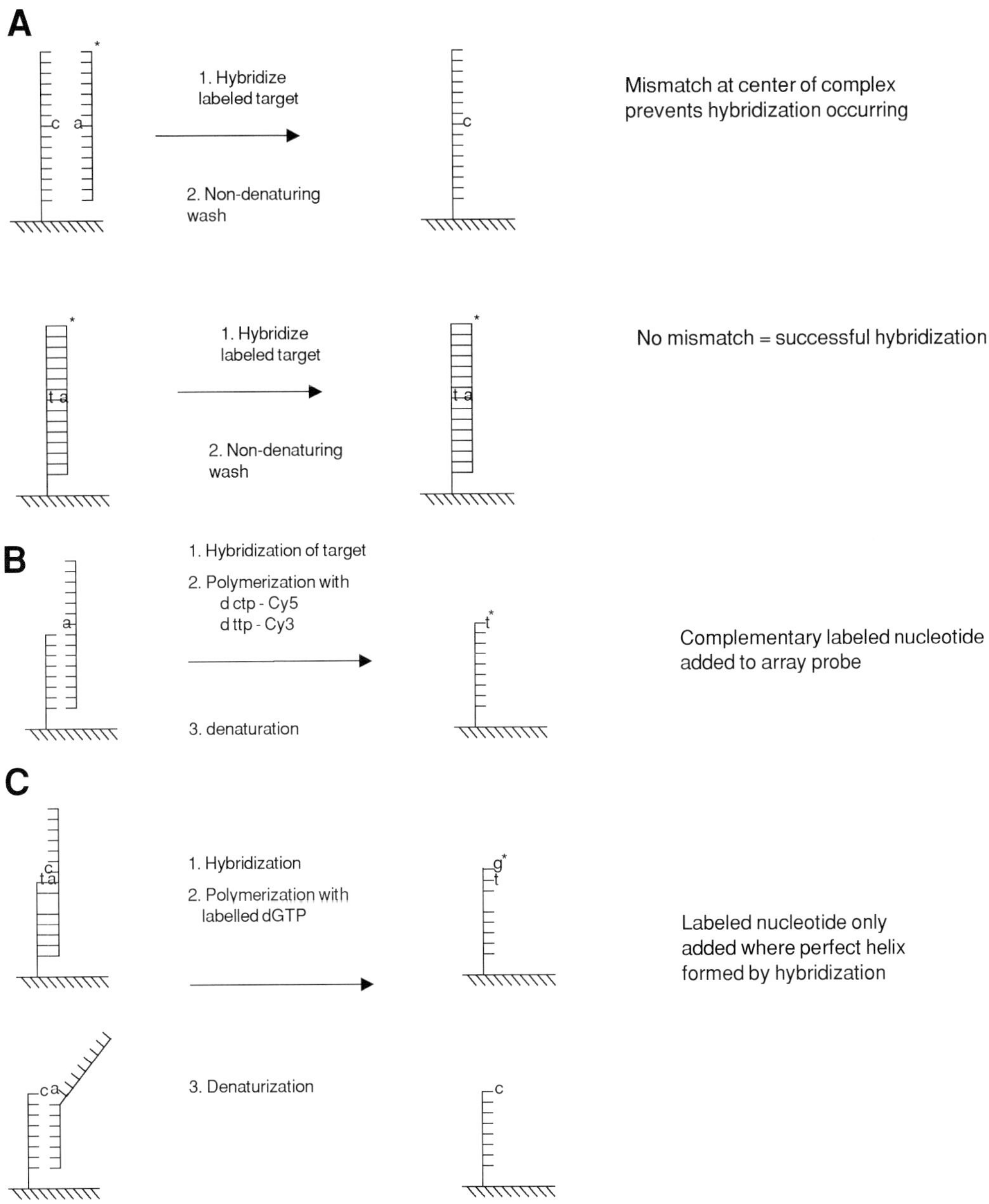

Fig. 5. Three schemes for typing of single nucleotide polymorphisms. The target varies at the middle of the three bases and can be a G or A. **(A)** Hybridization to ASOs. **(B)** Polymerase-catalyzed extension to a common, locus-specific oligonucleotide primer using nucleotide triphosphates complementary to one or other of the alleles. In this case, the bases may be labeled with different fluorophores. **(C)** Polymerase catalyzed extension to ASO primers using a common nucleotide triphosphate. The ligation of labeled oligonucleotides catalyzed by ligase can also be used in a similar manner.

flanking region to the repeat is hybridized to an array of oligonucleotides that contain a guide registration sequence plus a varying number of repeats. The ligase can only join the labeled oligonucleotide to the oligonucleotides of the array where a perfect duplex junction is formed. This only occurs for oligonucleotides with repeat length equal to those of the target ***(18)***.

The polymerase method is similar. The STRs must have a base of the repeat unit that differs from the first base in the flanking sequence following the repeats. Labeled nucleotide triphosphate complementary to the first base after the repeat and DNA polymerase is added. The labeled nucleotide is incorporated only where the repeat number on the array matches that of the target. As confirmation, the reaction can be performed using a labeled nucleotide triphosphate complementary to the first base of the repeat. Only array oligonucleotides shorter than the target repeat size are extended in this case.

1.5. Discussion

Three broad classes of assays useful with oligonucleotide arrays are described above: allele-specific hybridization; primer extension by polymerase (minisequencing); and ligase assay. All have been used in solution and adapted to arrays. Each assay incorporates an initial hybridization of target nucleic acid (usually PCR product) to the oligonucleotide array.

Allele-specific hybridization and the related technique of sequencing by hybridization for sequencing and resequencing nucleic acids has been under development and in use for some time ***(8,14)***. Hybridization has several well-documented complications, the major one being the variability of hybridization yield between different oligonucleotide probes against the nucleic acid target. This variability has two causes, detailed as follows:

- Sequence-dependent hybridization efficiency. Where oligonucleotides of the array are of the same length, then the stability of the duplex formed with the nucleic acid target will depend on the sequence of the oligonucleotides and the type of target ***(19)***. This effect of base composition in DNA/DNA duplexes is lessened by the use of salts, such as tetramethylammonium chloride as hybridization buffer ***(20)***. The stabilities of each set of duplexes could be measured by thermal denaturation in solution but this is extremely time consuming and therefore impractical.
- Differences in accessibility of the target to oligonucleotide probes. Secondary and tertiary structure in both the target and oligonucleotides can prevent hybridization. The problem can be alleviated by degrading the target to short fragments.

Selecting a set of oligonucleotide probes that give similar hybridization yields between loci and similar differences in hybridization yield between the allele specific oligonucleotides at each locus, under the same set of hybridization conditions, is difficult. In practice, the problem can be minimized by building redundancy into oligonucleotide arrays ***(14)***. Each locus and each allele is represented by oligonucleotides with different lengths so that at least one pair of allele-specific oligonucleotides will exhibit the correct hybridization characteristics.

In enzymatic assays, the hybridization step is used to capture the target. This step can be done at low stringency, allowing efficient hybridization at all ASOs. Alleles are then discriminated by either polymerase or ligase, which add a labeled reporter group to the array at any position where probe/target duplex contains no mismatches. The use of enzymes that produce a covalently bonded modification to the array also allows the assays to be thermally cycled. This increases the signal to noise ratio.

Primer extension/minisequencing/genetic bit analysis has the great advantage that the reporter (labeled dideoxynucleotide triphosphates) is simple and readily commercially available. At most four different reporters are needed for each assay. This contrasts with ligase assays where a unique oligonucleotide reporter is required for each locus studied. Primer extension can also be adapted for resequencing, using a tiling path of complementary oligonucleotides.

2. Materials

2.1. Array Fabrication

2.1.1. SNP Array Synthesis

1. DNA synthesizer (Perkin–Elmer/ABI).
2. Synthesis cell (**Fig. 2**, left).
3. Glass or perspex sheet.
4. Deoxynucleotide 3′-DMT, 5′-phosphoramidites (Glen Research).
5. Reagents for oligonucleotide synthesis (Perkin–Elmer/ABI, Pharmacia, Cruachem).
6. Ammonium hydroxide solution (30%) (BDH).
7. Duran bottle.
8. Aminated polypropylene support (Beckman).
9. Scalpel.

2.1.2. STR Array Synthesis

1. *See* **Subheading 2.1.1.**
2. Synthesis cell (**Fig. 2**, right).

2.2. Target Preparation

2.2.1. In Vitro Transcription of RNA

1. PCR reagents.
2. PCR primer for the target containing a T7 or SP6 promoter sequence.
3. T7 or SP6 RNA polymerase (Promega).
4. Transcription buffer 5μ: 200 m*M* Tris-HCl, pH 7.9, 30 m*M* $MgCl_2$, 10 m*M* spermidine, 50 m*M* NaCl; Promega.
5. DTT (100 m*M*; Promega).
6. RNase Inhibitor (Recombinant RNasin; Promega).
7. [α-^{32}P]UTP (3000 Ci/mmol; Amersham).
8. NTPs: ATP, CTP, and GTP as 10 m*M* stock solutions, and UTP at 250 μmol (all in nuclease free distilled water; Pharmacia).

2.2.2. Single-Stranded DNA Preparation

1. PCR reagents.
2. PCR primer for the target strand containing 6 phosphorothioate linkages at the 5′-most 6 phosphate linkages.
3. Thermal cycler (MJ Research).
4. T7 gene 6 exonuclease (Amersham).
5. PCR purification kit (optional, Qiagen).

2.3. Hybridization

2.3.1. Hybridization Assay

1. Buffer: 1 *M* NaCl, 10 m*M* Tris-HCl, pH 8.0, 1 m*M* EDTA, 0.01% SDS.
2. Petri dish.
3. Tissues.
4. Forceps.
5. Strip cut from oligonucleotide array (30 × 2 mm).
6. Target single-stranded RNA (*see* **Subheading 3.2.1.**).

2.3.2. Hybridization of Target for Polymerase Assay

1. Buffer: 1 *M* NaCl.
2. Petri dish.
3. Tissues.
4. Forceps.
5. Strip cut from oligonucleotide array (30 × 2 mm).
6. Target single-stranded DNA (*see* **Subheading 3.2.2.**).

2.3.3. Hybridization of Target and Reporter for Ligation Assays

1. Buffer: 1 *M* NaCl.
2. Petri dish.
3. Tissues.
4. Forceps.
5. Labeled reporter oligonucleotide containing 5′ phosphate group.
6. Target single-stranded DNA (*see* **Subheading 3.2.2.**).
7. Strip of oligonucleotide array (30 × 2 mm).

2.4. Enzyme-Aided Assays

2.4.1. Ligation

1. 5μ ligation buffer: 100 m*M* Tris-HCl, pH 8.3, 0.5% Triton X-100, 50 m*M* $MgCl_2$, 250 m*M* KCl, 5 m*M* NAD^+, 50 m*M* DTT, 5 m*M* EDTA (Advanced Biotechnologies Ltd.).
2. *Thermus thermophilus* DNA ligase (Tth DNA ligase; Advanced Biotechnologies Ltd).
3. Petri dish set up as humid chamber.
4. Forceps.

2.4.2. Primer Extension with Polymerase

1. 5μ Polymerase buffer: 200 m*M* Tris-HCl, pH 7.5, 100 m*M* $MgCl_2$, 250 m*M* NaCl (Amersham).
2. Polymerase dilution buffer: 50 m*M* Tris-HCl, pH 7.5; 10 m*M* DTT (Amersham).
3. DTT (100 m*M*, Promega).
4. T7 Sequenase V2. 13 U/μL (Amersham).
5. Dideoxynucleotidetriphosphate [α-^{33}P]ddNTP (450 μCi/mL; Amersham).
6. Humid chamber.
7. Forceps.
8. Buffer: 10 m*M* Tris-HCl, pH 8.0, 1 m*M* EDTA, pH 8.

3. Methods

3.1. Array Synthesis (Notes 1–3)

3.1.1. SNP Array Synthesis

1. Clamp a reaction cell (**Fig. 2**, left, 4 × 40 mm) to a piece of aminated polypropylene using a G-clamp and a glass or perspex backing plate.
2. Attach ports to the synthesizer and synthesize a patch of oligonucleotide corresponding to the first allele.
3. Displace the cell by 4 mm so that the cell is alongside the previously synthesized patch.
4. Synthesize the oligonucleotide corresponding to the other allele.
5. Repeat as necessary for other alleles.
6. Place the array in a Duran bottle, add 30% ammonia solution, seal the bottle securely, and heat at 55°C for 6 h.
7. After allowing the bottle and contents to cool, remove the array, wash with ethanol and dry. Store at –20°C.
8. The array is cut into strips (1- to 2-mm wide) in a direction perpendicular to the stripes of oligonucleotides using a scalpel and straight edge.

3.1.2. STR Array Synthesis

1. Clamp a cell (**Fig. 2**, right, 30 × 40 mm) to a piece of aminated polypropylene using a G-clamp and a glass backing-plate.
2. Attach ports to the synthesizer and synthesize a patch of oligonucleotide corresponding to the complement of the flanking sequence before the repeats of the target single stranded DNA.
3. On top of this add a sequence complementary to the smallest number of repeats that may be present in the target DNA.
4. Displace the cell by 2 mm and synthesize a patch corresponding to one repeat.
5. Repeat **step 4** until the maximum required number of repeats has been reached.
6. Place the array in a Duran bottle, add 30% ammonia solution, and heat at 55°C for 6 h.
7. After allowing the bottle and contents to cool, remove the array from bottle, wash with ethanol, and dry. Store at –20°C.
8. The array is cut into strips (1- to 2-mm wide) in a direction perpendicular to the stripes of oligonucleotides.

3.2. Single-Stranded Target Production

3.2.1. In Vitro Transcription of RNA

1. Perform PCR under conditions that give the required product.
2. Dilute this product to give a concentration of 0.5 to 1.0 μg/μL.
3. In a mircofuge tube at room temperature add the following reagents in order: 5μ transcription buffer (4 μL); 100 m*M* DTT (2 μL); RNasin (20 units); 10 m*M* ATP, CTP, and GTP (1 μL each); 250 m*M* UTP (1 μL); Template DNA (2 μL); [α-^{32}P] UTP (2 μL) T7 or SP6 RNA polymerase (20 units); and water to give final volume of 20 μL.
4. Mix and Incubate at 37°C for 1 h.
5. Remove unincorporated label by Sephadex G-25 or G-50 chromatography.

3.2.2. Single-Stranded DNA Preparation

1. Perform PCR under conditions that give the required product.
2. Add T7 Gene 6 exonuclease to give a concentration of 2 U per μL.

3. Incubate at 25°C for 1 h.
4. Incubate at 80°C for 5 min.
5. If desired, purification can be performed using a commercially available purification kit.

3.3. Hybridization (see Note 4)

3.3.1. Hybridization Assay (see **Notes 5** and **6**)

1. Make up a solution of the target (50–100 fmol, *see* **Subheading 3.2.1.**) in 100 μL of buffer.
2. Soak a tissue in 1X buffer and place around the inner perimeter of a Petri dish to create a 100% humidity chamber.
3. Place the solution as a puddle in the middle of the Petri dish and place a strip of array (30 × 2 mm) on top.
4. Incubate for 1hr at the required temperature.
5. Wash the array in fresh buffer solution and blot dry.

3.3.2. Hybridization Reaction for Polymerase Assay (see **Note 7**)

This method is the same as used in **Subheading 3.3.1.** except that a solution of 1 to 5 pmol of single-stranded DNA target (*see* **Subheading 3.3.2.**) in a 1 *M* NaCl solution is made in **step 1**.

3.3.3. Hybridization for Ligation Assay (see **Note 8**)

This method is the same as used in **Subheading 3.3.1.** except that a solution of 1 to 5 pmol of single-stranded DNA target (*see* **Subheading 3.3.2.**) and 5 to 10 pmol of reporter oligonucleotide in a 1 *M* NaCl solution is made in **step 1**.

3.4. Enzyme-Catalyzed Assays (see Notes 9–11)

3.4.1. Ligase Assay

1. Make up a solution of DNA ligase (1 U/μL) in 1μ ligase buffer (100 μL) and place as a puddle in a converted Petri dish.
2. Place a strip of array (30 × 2 mm) that has undergone hybridization according to **Subheading 3.3.3.** face down in the puddle, ensuring that the surface is completely covered by the solution.
3. Incubate at 37°C (or 65°C for STR arrays) for 1 to 8 h.
4. Heat in 50/50 water/formamide v/v solution at 100°C for 5 min.
5. Blot dry and image.

3.4.2. Polymerase Assay (see **Note 12**)

1. Make up 40 μL of a solution containing sequenase (0.5 U/μL), DTT (2.5 m*M*), polymerase buffer (1μ), and the appropriate ddNTP (25 nCi/μL).
2. Soak a tissue in water and place round the inside of a Petri dish.
3. Add the solution as a puddle to the middle of the Petri dish.
4. Float a strip of prehybridized array (*see* **Subheading 3.3.2.**, 1 × 30m) face down on the solution.
5. Incubate at 37°C for 1 h.
6. Wash away unbound material with Tris-EDTA (TE) buffer at 65°C for 5 min.
7. Blot dry and image.

4. Notes

1. The synthesis cells are produced by milling or moulding blocks of PTFE. The cavity is generally 0.5- to 0.75-mm deep and the wall thickness 0.3 to 0.5 mm.
2. Polymerase-catalyzed extension requires the oligonucleotides to be attached through their 5′ ends. For ligation, the oligonucleotides on the array can be attached in either orientation, using either the method described in **Subheadings 2.1.** or **2.2.** Thus, deoxynucleotide 5′-dimethoxytrityl-3′- phosphoramidites replace the deoxynucleotide 3′-dimethoxytrityl-5′-phosphoramidites. A requirement for ligation with oligonucleotides is that the 5′ end must be phosphorylated. This can be achieved by the reaction of a chemical phosphorylating reagent (Phosphate-On phosphoramidite available from most DNA synthesis reagent suppliers). This is added as the last step in the chemical synthesis of the oligonucleotides. Phosphate can also be added post synthetically using polynucleotide kinase and ATP (this reaction works on both oligonucleotides in solution and array supported oligonucleotides).
3. The use of a linker between the surface and the oligonucleotides improves both hybridization yield and access to the oligonucleotides by enzymes. Addition of phosphoramidites based on polyethylene glycol ***(21)*** or addition of deoxythymidine phosphoramidites ***(22)*** have both been used to add linkers before oligonucleotide synthesis.
4. Performing hybridizations and reactions of the array strips in puddles allows the volume of liquid to be minimized. However, the chamber has to be made as humid as possible to prevent evaporation, especially at elevated temperatures. The methods described use Petri dishes with wet tissues inside, but the tissues can also be draped over the top of the dish. The Petri dishes can be placed in sealed plastic containers. Sealed tubes may also be used and although the volumes are often greater than in puddles of solution, they can be more easily incubated in hot blocks or in water baths.

   Both the hybridization and washing buffer and temperature can be varied. Conditions should be found that give high yields of duplex and good discrimination between matched and mismatched duplexes.
5. For hybridizations below 37°C, care must be taken not to warm the plates by touching with fingers because this can cause the melting of short duplexes. For hybridizations below room temperature, all the apparatus that comes in contact with the array must be cooled to the hybridization temperature or below.
6. When using large RNA targets, secondary and tertiary structures can become a problem preventing hybridization. One method to alleviate this is to randomly cleave the RNA in base.
7. For assays, including reactions with enzymes, both the buffer and temperature of the hybridization reaction can be varied. Using a two-step hybridization followed by reaction procedure the hybridization buffer need not be the buffer required for the enzymatic reaction. Also, because many of the enzymes used do require specific temperatures, using a two-step hybridization/reaction the hybridization temperature can be completely different from the enzyme-catalyzed reaction.
8. The reporter oligonucleotide is typically 5′ end labeled with a radio label using standard methods. If fluorescence labeling is used, the label is incorporated during oligonucleotide synthesis.
9. In our methods, radiolabels were used. These are convenient for many applications because they are readily commercially available and are easily detected using a phosphorimager. Other label types can be used and in most cases there are similar ready-labeled reagents available. We have used Cy5 labels on oligonucleotides in ligation assays and nucleotide triphosphates in polymerase extension assays. This label can be detected using a STORM phosphorimager. Because of the large fluorescent background of the polypropylene support at short wavelengths, we have found that fluorescein is not useful with polypropylene.

10. Instead of performing the assays in two steps, a hybridization followed by a ligation or polymerization, an all-in-one reaction can be used. There are several advantages to conducting all-in-one reactions, such as the saving time and effort and the potential to cycle the temperature allowing the target DNA to be used as a template for more than one ligation or polymerization. Conducting all-in-one reactions puts additional constraints on the conditions because the majority of enzymes work in only a small range of temperatures and salt concentrations. The target single-stranded DNA generally needs to be purified.
11. The methods described use Thermus Thermophilus DNA ligase for ligation reactions and Sequenase in extension reactions. Other enzyme types can also be used and have advantages if particular temperatures are required, particularly in all in one reactions. Therefore, for example, *Taq* and Thermosequenase have all been used successfully.
12. The extension method described adds a dideoxynucleotidetriphosphate (ddNTP). For most applications, appropriate deoxynucleotidetriphosphate (dNTP) can be used.

References

1. Case-Green, S. C., Mir, K. U., Pritchard, C. E., and Southern, E. M. (1998) Analysing genetic information with DNA arrays. *Curr. Opin. Chem. Biol.* **2,** 404–410.
2. Schena, M., Shalon, D., Davis, R. W., and Brown, P. O. (1995) Quantitative monitoring of gene expression patterns with a complementary DNA microarray. *Science* **270,** 467–470.
3. Wodicka, L., Dong, H., Mittmann, M., Ho, M. H., and Lockhart, D. J. (1997) Genome-wide expression monitoring in *Saccharomyces cerevisiae*. *Nat. Biotechnol.* **15,** 1359–1367.
4. Milner, N. M., Mir, K. U., and Southern, E. M. (1997) Selecting effective antisense reagents on combinatorial oligonucleotide arrays. *Nat. Biotechnol.* **15,** 537–541.
5. Brown, T. and Brown, D. J. S. (1991) Modern machine-aided methods of oligodeoxyribonucleotide synthesis. In *Oligonucleotides and Analogues: A Practical Approach* (Eckstein, R., ed.), IRL Press, Oxford, UK, pp. 1.
6. Southern, E. M., Case-Green, S. C., Elder, J. K., Johnson, M., Mir, K. U., Wang, L., et al. (1994) Arrays of complementary oligonucleotides for analysing the hybridisation behaviour of nucleic acids. *Nucleic Acids Res.* **22,** 1368–1373.
7. Matson, R. S., Rampal, J., Pentoney, S. L., Anderson, P. D., and Coassin, P. (1995) Biopolymer synthesis on polypropylene supports: Oligonucleotide arrays. *Anal. Biochem.* **224,** 110–116.
8. Maskos, U. and Southern, E. M. (1993) A novel method for the parallel analysis of multiple mutations in multiple samples. *Nucleic Acids Res.* **21,** 2269–2270.
9. Matson, R. S., Rampal, B., and Coassin, P. J. (1994) Biopolymer synthesis on polypropylene supports. *Anal. Biochem.* **217,** 306–310.
10. Polymeropoulos, M. H., Rath, D. S., Xiao, H., and Merril, C. R. (1991) Tetranucleotide repeat polymorphism at the human c-fes/fps proto-oncogene (FES). *Nucleic Acids Res.* **19,** 4018.
11. Nikiforov, T. T., Rendle, R. B., Goelet, P., Rogers, Y.-H., Kotewics, M. L., Anderson, S., et al. (1994) Genetic bit analysis: a solid phase method for typing single nucleotide polymorphisms. *Nucleic Acids Res.* **22,** 4167–4175.
12. Zon, G. and Stec, W. J. (1991) In Phosphorothioate oligonucleotides. In *Oligonucleotides and Analogues: A Practical Approach* (Eckstein, F., ed.), IRL Press, Oxford, UK, pp. 87.
13. Maskos, U. and Southern, E. M. (1993) A novel method for the analysis of multiple sequence variants by hybridisation to oligonucleotides. *Nucleic Acids Res.* **21,** 2267–2268.
14. Kozal, M. J., Shah, N., Shen, N., Yang, R., Fucini, R., Merigan, T., et al. (1996) Extensive polymorphisms observed in HIV-1 clade B protease gene using high-density oligonucleotide arrays. *Nat. Med.* **2,** 753–759.

15. Schumaker, J. M., Metspalu, A., and Caskey, C. T. (1996) Mutation detection by solid phase primer extension. *Human Mutat.* **7,** 346–354.
16. Pastigen, T., Kurg, A., Metspalu, A., Peltonen, L., and Syvanen, A. C. (1997) Minisequencing: A specific tool for DNA analysis and diagnostics on oligonucleotide arrays. *Genome Res.* **7,** 606–614.
17. Case-Green, S. C. and Southern, E. M. (1994) Studies on the base pairing properties of deoxyinosine by solid phase hybridisation to oligonucleotides. *Nucleic Acids Res.* **22,** 131–136.
18. Pritchard, C. E. and Southern, E. M. (1996) Spatially addressable ligation assays: application of oligonucleotide arrays to DNA fingerprinting. In *Innovation and Perspectives in Solid Phase Synthesis and Combinatorial Libraries* (Epton, R., ed.), Mayflower Scientific, Birmingham, UK, pp. 499.
19. Ratmeyer, L., Vinayak, R., Zhong, Y. Y., Zon, G., and Wilson, W. D. (1994) Sequence specific thermodynamic and structural properties for DNA-RNA duplexes. *Biochemistry* **33,** 5298–5304.
20. Maskos, U. and Southern, E. M. (1993) A study of oligonucleotide reassociation using large arrays of oligonucleotides synthesised on a glass support. *Nucleic Acids Res.* **21,** 4663–4669.
21. Shchepinov, M. S., Case-Green, S. C., and Southern, E. M. (1997) Steric factors influencing hybridisation of nucleic acids to oligonucleotide arrays. *Nucleic Acids Res.* **25,** 1151–1161.
22. Guo, Z., Guilfoyle, R. A., Thiel, A. J., Wang, R., and Smith, L. M. (1994) Direct fluorescence analysis of genetic polymorphisms by hybridization with oligonucleotide arrays on glass supports. *Nucleic Acids Res.* **22,** 5456–5465.

40

Serial Analysis of Gene Expression

Karin A. Oien

1. Introduction

Serial analysis of gene expression (SAGE™) is a patented, large-scale mRNA-profiling technology that produces comprehensive, quantitative, and reproducible gene expression profiles (originally described in **refs.** *1* and *2*). Unlike the alternative technologies of differential display and subtractive hybridization, SAGE produces a full catalog of all transcripts, not only differentially expressed genes, and unlike smaller arrays, SAGE needs no assumptions about the genes that are likely to be expressed, thus allowing the identification of novel genes (are among many excellent reviews, *see* **refs.** *3–5*). SAGE is based on generating clones of concatenated (linked) short sequence tags derived from mRNA from the target cells or tissue (**Fig. 1**).

Each tag is 9- or 10-bp long and represents one mRNA and each clone insert contains up to 40 tags joined serially. Therefore, sequencing of multiple concatenates describes the pattern and abundance of mRNA, with an improvement in efficiency of up to 40-fold compared with conventional analysis of expressed sequence tags. The mRNA transcript corresponding to the short SAGE tag is identified from genetic databases using appropriate software.

The SAGE method was developed in 1995 by Velculescu et al. *(1)* of the Kinzler and Vogelstein laboratory at Johns Hopkins University, from where the SAGE protocol, software, and updates are available to academic investigators for noncommercial use via http://www.sagenet.org/sage_protocol.htm. For commercial purposes, the user should contact Genzyme Corporation, who own and license the SAGE™ technology. This and other useful web sites are listed in **Table 1.** Since the first report of SAGE, many technical modifications have been described. Some enable the use of smaller amounts of starting material *(3,6–10)*, whereas others have improved the efficiency of intermediary SAGE reactions *(11–14)*. Invitrogen has also released a kit called I-SAGE™, which provides all of the numerous reagents required in a high-quality form.

The Johns Hopkins protocol is 27 pages long, and Invitrogen's online I-SAGE™ instructions contain 73 pages. This is because SAGE involves many stages, each using well-established and relatively straightforward molecular biological techniques, but any of which can go wrong! This article thus provides concise instructions with an emphasis

From: *Methods in Molecular Biology, Vol. 226: PCR Protocols, Second Edition*
Edited by: J. M. S. Bartlett and D. Stirling © Humana Press Inc., Totowa, NJ

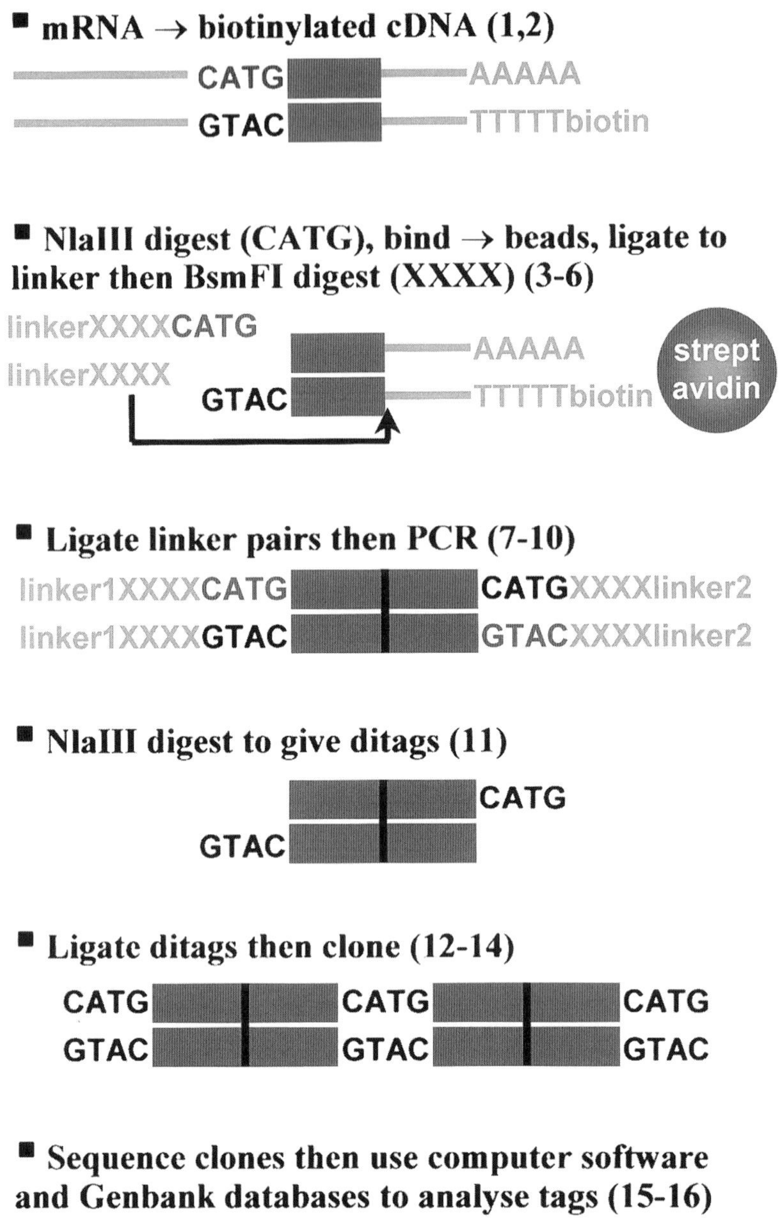

Fig. 1. A schematic diagram of the SAGE method: the numbers refer to the corresponding stages in **Subheading 3.**

on common problems (well discussed in **ref.** *15*), but it is strongly recommended that one also consult the full protocols and references supplied in this chapter.

The software used to analyze the tags has been regularly updated, and alternative programs have been developed, including eSAGE *(16)*, USAGE *(17)*, and ExProView *(18)*. The web-based bioinformatics facilities of the National Center for Biotechnology Information (NCBI) are extremely useful *(19)*. There, through SAGEmap *(20)*, numerous publicly available SAGE libraries can be accessed online both for global

Table 1
Useful Web Sites for SAGE

URL	Contents
http://www.sagenet.org/sage_protocol.htm	Johns Hopkins SAGE protocol and software for non-commercial use, plus conferences, publications and other information *(1,2)*
http://www.genzyme.com/sage/	Genzyme Molecular Oncology: information on SAGE and its commercial use
http://www.invitrogen.com	Invitrogen including I-SAGE™ protocol
http://www.dsv.cea.fr/thema/get/sade.html	SADE: a SAGE Adaptation for Downsized Extracts *(6)*
http://www.ambion.com	Ambion website with advice on RNA work
http://www.ncbi.nlm.nih.gov/SAGE/	NCBI's SAGEmap *(20)*
http://www.ncbi.nlm.nih.gov/UniGene/index.html	NCBI's UniGene ("unique gene": clusters all transcripts of one gene under one name) *(19)*
http://www.geneontology.org	Gene Ontology databases: information on gene function and cellular location *(25)*

comparisons and for investigation of the expression of individual transcripts, which can be further studied via Unigene (see **Table 1**, **ref.** ***19***). The statistical basis for designing and analyzing SAGE experiments has also been discussed in detail ***(21–24)***. Here, however, the focus is on the wet laboratory work.

2. Materials

The reagents specified have been used successfully, but many equivalents are available and could be used if preferred (*see* **Note 1**). This description is based on the Johns Hopkins protocol (version 1.0d); the most helpful modifications are also described.

2.1. RNA Work (see Note 2)

1. Aerosol-resistant pipet tips (Greiner Labortechnik Ltd, Gloucestershire, UK).
2. DEPC-treated water (Diethylpyrocarbonate: Sigma, Dorset, UK).
3. (Optional) RnaseZap® (Ambion (Europe) Ltd, Cambridgeshire, UK).

2.2. Kits for Purification of RNA and cDNA Synthesis

1. Preparation of total RNA: TRIzol® Reagent (Invitrogen Life Technologies, Paisley, UK).
2. Purification of polyA+ mRNA: Poly(A)Purist™ Kit (Ambion).
3. (Alternative protocols: streptavidin-coated tubes obtained with mRNA Capture Kit or separately (Roche Diagnostics Ltd, East Sussex, UK), instead of **steps 1** and **2** (*see* **Note 3** and **Subheadings 2.4.** and **2.6.**).
4. cDNA synthesis: SuperScript™ Choice System for cDNA Synthesis (Invitrogen).

2.3. Purification of DNA

1. Phenol:chloroform (P/C): phenol:chloroform:isoamyl alcohol (25:24:1; Ambion).
2. Glycogen, molecular biology grade (20 mg/mL) (Roche Diagnostics Ltd).
3. Ammonium acetate (10 *M*).

4. 100% and 70% ethanol.
5. LoTE: 3 m*M* Tris-HCl (pH 7.5), 0.2 m*M* EDTA (pH 7.5).
6. TE: 10 m*M* Tris-HCl (pH 7.5), 1 m*M* EDTA (pH 7.5).
7. (Optional) QIAquick® DNA Cleanup System: QIAquick Nucleotide Removal and PCR Purification Kits; and, for higher through-put, QIAquick 8 PCR Purification Kit with QIAvac 6S (Qiagen Ltd, West Sussex, UK).

2.4. Oligonucleotides

1. Biotinylated oligo dT: 5′ [biotin] T_{18} (Alternative protocols, *see* **Note 3**).
2. SAGE Linker 1A: 5′ TTT GGA TTT GCT GGT GCA GTA CAA CTA GGC TTA ATA GGG ACA TG 3′
3. SAGE Linker 1B: 5′ TCC CTA TTA AGC CTA GTT GTA CTG CAC CAG CAA ATC C [amino mod. C7] 3′
4. SAGE Linker 2A: 5′ TTT CTG CTC GAA TTC AAG CTT CTA ACG ATG TAC GGG GAC ATG 3′
5. SAGE Linker 2B: 5′ TCC CCG TAC ATC GTT AGA AGC TTG AAT TCG AGC AG [amino mod. C7] 3′
6. SAGE Primer 1: 5′ [biotin] GGA TTT GCT GGT GCA GTA CA 3′ ***(11)***
7. SAGE Primer 2: 5′ [biotin] CTG CTC GAA TTC AAG CTT CT 3′ ***(11)***
8. M13 Forward: 5′ GTA AAA CGA CGG CCA GT 3′
9. M13 Reverse: 5′ GGA AAC AGC TAT GAC CAT G 3′

The working concentration of all primers, except biotinylated oligo dT, is 350 ng/µL. Obtain linkers and biotinylated oligos gel-purified: recommended suppliers include Integrated DNA Technologies, Inc. (IA, USA) and Oswel Research Products Ltd. (Southampton, UK). Before use, linkers must be kinased, either biochemically during synthesis or later enzymatically (*see* **Note 4**), and annealed to the complementary linker.

2.5. Restriction and Modifying Enzymes

1. *Nla*III (10 U/µL), supplied with NEBuffer 4 and 100× bovine serum albumin (New England Biolabs [NEB] Inc, MA,) stored at –70°C).
2. *Bsm*FI (2 U/µL) (NEB).
3. *Sph*I (5 U/µL) (NEB).
4. T4 Polynucleotide Kinase 10U/µL (NEB) (or obtain linker oligonucleotides already kinased, *see* **Note 4**).
5. T4 DNA Ligase (5 U/µL, or 20 U/µL if preferred; NEB). Aliquot ligase enzyme and buffer to prevent cross-contamination from PCR products.
6. DNA polymerase I large fragment (Klenow); 5 U/µL (NEB).

2.6. Magnetic Beads and Related Materials

1. Dynabeads® M-280 Streptavidin (-coated) (Dynal AS, Oslo, Norway).
2. Magnetic particle concentrator (magnetic stand to immobilize beads) and sample mixer (Dynal AS).
3. 2× Binding and washing (B+W) buffer: 2 *M* NaCl, 10 m*M* Tris-HCl (pH 7.5), 1 m*M* EDTA. Also prepare 1× B+W buffer solution.
4. (Optional) Nonstick RNase-free tubes (Ambion) have been recommended for use with magnetic beads to minimize adhesion of beads to the tube walls.
5. (Alternative protocols: Dynabeads® Oligo dT_{25} (Dynal AS; *see* **Note 3** and **Subheadings 2.2.** and **2.4.**).

2.7. Gel Materials

1. 40% polyacrylamide (37.5 : 1 acrylamide : bis) (Bio–Rad Ltd, Hertfordshire, UK).
2. 40% polyacrylamide (19 : 1 acrylamide : bis) (Bio–Rad).
3. DNA molecular weight markers, for example, 10- and 100-bp DNA ladders (Invitrogen).
4. SYBR® Green I nucleic acid gel stain (Molecular Probes Inc., OR) or a 0.5 µg/mL solution of Ethidium Bromide (Sigma).
5. Gel-loading buffer containing Bromophenol Blue and Xylene Cyanol (e.g., Ambion).
6. Apparatus for vertical polyacrylamide gel electrophoresis, for example, Atto Maxi Slab with 16- × 16-cm glass gel plates, 1.5-mm spacers, and 12- and 20-well combs (Genetic Research Instrumentation, Essex, UK).
7. 50× Tris-Acetate buffer: 242 g of Tris base, 57.1 mL of glacial acetic acid, and 100 mL of 0.5 *M* EDTA (pH 8.0) made up to 1 L with dH_2O.
8. 10% Ammonium persulfate.
9. TEMED (*N,N,N′,N′*-Tetramethylethylenediamine; Sigma).
10. Sterile scalpel.
11. 21-gauge needles.
12. Spin-X filter microcentrifuge tubes (Corning Ltd, Buckinghamshire, UK).
13. Materials, buffers, and apparatus for agarose gel electrophoresis

2.8. Polymerase Chain Reaction (PCR)

1. Option 1: Johns Hopkins protocol. PLATINUM *Taq* DNA Polymerase (5 U/µL; Invitrogen); 10× PCR buffer: 166 m*M* $(NH_4)_2SO_4$, 670 m*M* Tris-HCl (pH 8.8), 67 m*M* $MgCl_2$, 100 m*M* β-mercaptoethanol; Dimethyl sulphoxide (Sigma). Option 2: own modification. HotStarTaq DNA Polymerase Kit and *Taq* PCR core kit (Qiagen).
2. dNTPs (10 m*M*; Invitrogen; included in some kits).
3. Thermal cycler of choice (e.g., Thermo Hybaid, Middlesex, UK).
4. Mineral oil (Sigma). Ensure that it is free from PCR contamination.

2.9. Cloning

1. For cloning of concatemers: Zero Background™ Cloning Kit (Invitrogen).
2. ELECTROMAX™ DH10B™ Cells (Invitrogen).
3. S.O.C. Medium (Invitrogen).
4. Low-salt LB medium and agar plates with Zeocin™, prepared according to instructions in Zero Background™ Cloning Kit.
5. Electroporator and electroporation cuvettes (Thermo Hybaid).

2.10. Sequencing

1. Sequencing as performed locally, for example, use BigDye Primer Kit and ABI automated sequencer (both Perkin–Elmer from Applied Biosystems, Warrington, UK).

2.11. SAGE Bioinformatics

Johns Hopkins SAGE program (downloaded from web site after registration), in combination with NCBI's GenBank® databases and Microsoft® Access and Excel programs, or other software (see **Subheading 1.**, last paragraph, and **Table 1**).

3. Methods (*see* Notes 1 and 2)

3.1. mRNA Preparation

1. Prepare total RNA from cells or tissue by standard methods, for example, TRIZOL Reagent.

2. Purify polyA$^+$ mRNA from total RNA using, for example, Poly(A)Purist kit. This original protocol requires 2.5 to 5 μg of mRNA for the next step (but *see* **Note 3**).

3.2. cDNA Synthesis

1. Synthesize double-stranded cDNA using SuperScript® Choice System. For the first strand synthesis, use 2.5 μg of the biotinylated oligo dT purchased separately, instead of that supplied with the kit. Synthesize the second strand, proceed to the next step in this SAGE protocol. (Do not perform *Eco*RI adapter addition).
2. Phenol:chloroform (P/C) extract and ethanol precipitate (*see* **Notes 5–7**). Resuspend in 20 μL of LoTE.

3.3. Cleavage of Biotinylated cDNA with Anchoring Enzyme Nla*III* to Create CATG Sticky-End

1. To 10 μL (half) of the biotinylated cDNA, add 74 μL of LoTE, 10 μL of 10× NEBuffer 4, 1 μL of 100× BSA, and 5 μL of *Nla*III. Incubate at 37°C for 1 h.
2. P/C extract, ethanol precipitate, and resuspend in 20 μl LoTE.

3.4. Binding of Biotinylated cDNA to Magnetic Beads

1. Add 100 μL of streptavidin Dynabead slurry to each of two 1.5-mL microcentrifuge tubes. Immobilize beads with Dynal magnet and remove supernatant.
2. Wash beads with 200 μL of 1× B+W buffer then remove supernatant.
3. Add 100 μL of 2× B+W buffer, 90 μL of dH_2O, and 10 μL of cleaved biotinylated cDNA to each of the two tubes. Incubate with gentle mixing for 15 min at room temp.
4. Immobilize beads with magnet and remove supernatant. Wash three times with 200 μL of 1× B+W buffer and once with 200 μL of LoTE. Proceed immediately to next step.

3.5. Ligating CATG Sticky-Ended Linkers to Bound cDNA

1. Linkers must be kinased and annealed in advance (*see* **Note 4**).
2. Use one of the two tubes containing washed magnetic beads for linker 1 and the other for linker 2. Immobilize beads with magnet and remove supernatant.
3. Add 29 μL of LoTE, 5 μL of annealed linker 1 or 2, and 4 μL of 10× ligase buffer. Resuspend the beads with gentle mixing (flick tube with finger).
4. Heat at 50°C for 2 min then allow to cool to room temp over 15 min.
5. Add 2 μL of T4 DNA Ligase. Incubate at 16°C for 2 h with intermittent gentle mixing.
6. After ligation, wash beads three times with 200 μL of 1× B+W buffer. Transfer to new tubes. Wash once with 200 μL of 1× B+W buffer and twice with 200 μL of 1× NEBuffer 4.

3.6. Creation of cDNA Tags and Their Release from Magnetic Beads Using Tagging Enzyme Bsm*FI*

1. Remove buffer then add 87 μL of LoTE, 10 μL of 10× NEBuffer 4, and 1 μL of 100× BSA to each tube. Pre-incubate at 65°C for 2 min.
2. Add 2 μL of *Bsm*FI. Incubate at 65°C for 1 h with intermittent gentle mixing.
3. Immobilize beads with magnet. This time, collect supernatant and transfer to two new tubes. Wash beads with 100 μL of 1× NEBuffer 4. Collect buffer and add to previous supernatant, discarding beads.
4. P/C extract, ethanol precipitate, and resuspend in 10 μL of LoTE.

3.7. Blunt-Ending-Released cDNA Tags

1. To each of the two tubes (from linkers 1 and 2) containing released cDNA tags, add 31 μL dH_2O, 5 μL of 10× EcoPol buffer, 0.5 μL of 100× BSA, 2.5 μL of 10 m*M* dNTPs, and 1 μL of DNA polymerase I large fragment (Klenow).
2. Incubate at 37°C for 30 min, and then pool both tubes of blunt-ended tags.
3. P/C extract, ethanol precipitate, and resuspend in 12 μL of LoTE.

3.8. Ligating Blunt-Ended Tags to Form 102-bp Ditags

1. Set up two new tubes, ideally of small size (0.2 mL). One tube is for the ditag ligation reaction. The other is a negative ligation control, to exclude cross-contamination at the next, PCR, step: set this tube up first. To each of the two tubes, add 4 μL of blunt-ended tags, 0.8 μL dH_2O and 0.6 μL of 10× ligase buffer.
2. To the negative control tube, add 0.6 μL of dH_2O.
3. To the ditag reaction, add 0.6 μL of T4 DNA ligase.
4. Cover with a drop of mineral oil to avoid evaporation and incubate at 16°C overnight.

3.9. PCR Amplification of 102-bp Ditags

The PCR aims to produce sufficient 102-bp ditag DNA for subsequent isolation and concatemerization of 26 bp ditags but may itself be problematic (*see* **Note 8**).

1. After ligation, add 14 μL of LoTE to increase volume to 20 μL and mix. Take 1 μL and dilute 100-fold with LoTE. Use 1 μL of the dilution in a 50- or 100-μL PCR with biotinylated SAGE Primers 1 and 2 ***(11)***. To avoid cross-contamination, set up the two negative control reactions (no template and no ligase) first.
2. Step 2, option 1: Johns Hopkins protocol. Per 50-μL reaction, use 30.5 μL dH_2O, 5 μL of 10× SAGE PCR buffer, 3 μL of DMSO, 7.5 μL of 10m*M* dNTPs, 1 μL of each of SAGE Primers 1 and 2, and 1 μL PLATINUM *Taq* DNA Polymerase. The cycling parameters, optimized for a Hybaid thermal cycler, are as follows: 94°C for 1 min; 26 to 30 cycles of 94°C for 30 s, 55°C for 1 min and 70°C for 1 min; then 70°C for 5 min. Optimize with different template dilutions (1/50, 1/100, or 1/200 per reaction). Step 2, option 2: own modification. The HotStarTaq™ DNA Polymerase Kit routinely works well with 1 μL, or often less, of the 1/100 dilution with no need for further adjustment of template dilution. Per 100-μL reaction, use: 60.5 μL of dH_2O, 10 μL of 10× Qiagen PCR buffer, 5 μL of 25 m*M* $MgCl_2$, 20 μL of 5× Q-Solution, 2 μL of 10 m*M* dNTPs, 0.5 μL of each of SAGE Primers 1 and 2, and 0.5 μLHotStarTaq™ DNA Polymerase. The cycling parameters are: 94.5°C for 15 min; 26 to 30 cycles of 94.5°C for 30 s; 56°C for 1 min and 72°C for 1 min; then 72°C for 5 min.
3. Optimize the cycle numbers between 26 and 30. More than 30 cycles usually results in high molecular weight smearing with less of the desired product.
4. After PCR, load 10 μL of each reaction on a 12% polyacrylamide gel with a 20-bp ladder. Run at 160 V for 2.5 h until the bromophenol blue dye front has run out of the gel and the xylene cyanol is 1 to 2 cm from the bottom, then stain (*see* **Note 9**). The amplified ditags should produce a 102-bp band. Background bands are common: the brightest runs at 80 bp and contains amplified ligated linkers without tags. The negative controls should contain no product.
5. After optimization, perform large-scale PCR by preparing then distributing a master-mix into three 96-well PCR plates with 50 or 100 μL in each well.
6. After PCR, pool the reactions. P/C extract and ethanol precipitate, scaling up as needed. The large volumes can be dealt with using either multiple 1.5-mL microcentrifuge tubes or

50-mL conical tubes. Resuspend in a total of 250 µL of LoTE. (*see* **Note 10** for optional DNA quantitation at this stage.)

3.10. Isolation of 102-bp Ditags by Gel Purification

1. Load pooled PCR products on three 12% polyacrylamide gels. Run and stain as before.
2. Cut out the 102-bp band of amplified ditags.
3. Fragment gel by placing cut-out bands in 0.5 mL of microcentrifuge tubes that have previously been pierced through the base with a needle, and insert into a 2-mL tube. (Depending on bandwidth, 3–4 tubes are used per gel.) Centrifuge at full speed for 2 min.
4. Elute DNA from gel fragments by adding 250 µL of LoTE and 50 µL of 10 *M* ammonium acetate to each 2-ml tube. Ensure that gel fragments are covered by buffer (add more if necessary). Vortex tubes then incubate at 65°C for 2 h.
5. For each 0.5-mL tube, prepare two Spin-X filter microcentrifuge tubes by placing 5 µL of LoTE on the filter. Transfer contents of each 0.5-mL tube to two Spin-X tubes. Centrifuge at full speed for 5 min. (*see* **Note 7** and **ref.** ***12***).
6. Pool samples and ethanol precipitate. Resuspend in a total of 100 µL of LoTE.

3.11. Isolation of 26-bp Ditags by NlaIII Digestion and Removal of Linkers Using Magnetic Beads and Gel Purification

1. To the pooled PCR products, add 58 µL of LoTE, 20 µL of 10× NEBuffer 4, 2 µL of 100× BSA and 20 µL of *Nla*III. Incubate at 37°C for 1 h.
2. During *Nla*III digestion, add 100 µL of streptavidin Dynabeads to each of two 1.5-mL tubes. Immobilize beads with magnet and remove supernatant. Wash beads with 200 µL of 1× B+W buffer. Remove buffer when *Nla*III digestion is complete.
3. To each of the two tubes containing streptavidin Dynabeads, add 100 µL of 2× B+W buffer and 100 µL (half) of the *Nla*III digestion ***(11)***.
4. Incubate with gentle mixing for 15 min at room temp.
5. Immobilize beads with a magnet. Collect supernatant.
6. Wash beads once with 200 µL of 1× B+W buffer and once with 200 µL of LoTE. Collect supernatant in each case.
7. Pool supernatants and keep on ice to prevent ditag denaturation ***(13)***. Discard beads.
8. P/C extract at 4°C. Ethanol precipitate with centrifugation at 4°C ***(13)***. Resuspend in 15 µL of TE (not LoTE).
9. Load on two lanes of a 12% polyacrylamide gel with a 20-bp ladder. Run at 160 V for 2 h until the bromophenol blue dye front is 3 cm from the bottom. Stain. (Purified ditags run at 22–26 bp. Released linkers run at 40 bp. Bands between 60–100 bp result from incomplete *Nla*III digestion.)
10. Cut out ditag band running at 22–26 bp.
11. Elute DNA from gel fragments as before, but incubate at 37°C, not 65°C.
12. Ethanol precipitate with centrifugation at 4°C ***(13)***. Resuspend in 6.4 µL of LoTE.

3.12. Ligation of Sticky-Ended 26-bp Ditags to Form Concatemers Then Gel Purification of Concatemers

1. To the purified ditags, add 0.8 µL of 10× ligase buffer and 0.8 µL of T4 DNA Ligase. Incubate at 16°C for 2 h (or longer, e.g., overnight, if desired).
2. Add loading buffer directly to concatemer ligation. Heat at 65°C for 15 min then chill on ice for 5 min ***(14)***. Load on 8% polyacrylamide gel in one lane with a 100-bp ladder. Run at 130 V for 3 h, until the bromophenol blue is 3 cm from the bottom. Stain.
3. Cut out DNA smear over 500 bp in size.
4. Elute concatemer DNA from gel fragments as before but incubate at 65°C.
5. Ethanol precipitate and resuspend in 6 µL of LoTE.

3.13. Cloning Concatemers

Clone concatemers using the Zero Background™ Cloning Kit. The pZErO®-1 vector contains a lethal gene which is disrupted by insertion of DNA, thus only positive recombinants should grow (this is the theory: in practice, some colonies do lack inserts).

1. To linearize the vector, mix 1 µL of pZErO®-1 (1 µg/µL), 7 µL of dH_2O, 1 µL of NEBuffer 2 and 1 µL of *Sph*I. Incubate at 37°C for 15 to 30 min (not over 30 min).
2. P/C extract and ethanol precipitate. Resuspend in 30 µL of LoTE.
3. To 1 µL of *Sph*I-linearized pZErO®, add the 6 µL of purified concatemers, 1 µL of 10× ligase buffer and 1 µL of T4 DNA Ligase. Include no insert (omit concatemers) and no ligase (omit concatemers and ligase) controls. Incubate at 16°C for 2 h.
4. P/C extract and ethanol precipitate. Resuspend in 6 µL of LoTE.
5. Transfect 2 µL of DNA into ElectroMAX™ DH10B™ *Escherichia coli* cells by electroporation, according to manufacturer's instructions.
6. Plate one-tenth of transfected bacteria onto each 13-cm Zeocin™-containing low-salt LB agar plate. Keep all plates at 4°C until inserts are checked. If inserts are of appropriate size, plates may be used for large-scale sequencing.

3.14. Screening of Transformants by PCR to Identify Long Concatemer Inserts

Perform PCR with vector-specific primers to determine insert size in each bacterial colony. Tubes (0.5 mL) may be used initially, but 96-well PCR plates are more efficient for large-scale screening. The 25-µL reaction volume can be reduced, for example, to 16 µL.

1. Step 1, option 1: Johns Hopkins protocol. Per 25 µL of reaction, use: 2.5 µL of 10× SAGE PCR buffer; 1.25 µL of DMSO; 1.25 µL of 10 m*M* dNTPs; 0.5 µL of each of M13 forward and reverse Primers; 19 µL of dH_2O; and 0.2 µL of Platinum *Taq* DNA Polymerase. The cycling parameters are 95°C for 2 min; 25 cycles of 95°C for 30 s, 56°C for 1 min and 70°C for 30 s; then 70°C for 5 min. Step 1, option 2: own modification. The colony PCRs are robust and work well with the *Taq* PCR core kit. Per 100 µL of reaction volume, use: 61.5 µL of dH_2O, 10 µL of 10× Qiagen PCR buffer, 5 µL of 25 m*M* $MgCl_2$, 20 µL of 5× Q-Solution, 2 µL of 10 m*M* dNTPs, 0.5 µL of each of M13 forward and reverse Primers, and 0.5 µL of *Taq* DNA Polymerase. The cycling parameters are 94.5°C for 1.5 min; 30 cycles of 94.5°C for 30 s, 52°C for 1 min and 72°C for 1 min; then 72°C for 5 min.
2. Touch a single colony with a new pipet tip then dip into reaction mix and shake. Repeat as necessary. Perform PCR.
3. Run 5 µL of each reaction on a 2% agarose gel with a 100-bp ladder.

3.15. Sequencing of SAGE Concatemer Inserts

Select and sequence the PCR products of 500 bp in size or over because these should contain at least 15 tags (226 bp of flanking pZErO®-1 vector plus 12 to 13 bp per tag).

1. Before sequencing, purify the PCR product (partly to remove primers because M13 forward is used again). Individual phenol chloroform extraction and ethanol precipitation is one option. However, Qiagen's QIAquick 8 PCR Purification Kit with the QIAvac 6S are more efficient for large-scale sequencing (*see* **Note 7**).
2. Sequence according to local preference. For example, use the BigDye Primer Kit with one-half to one-tenth of the purified PCR product per sequencing reaction and the M13

Forward Primer, and then run the reaction on an ABI automated sequencer. In some laboratories, DNA purification and sequencing can be fully automated.

3.16. Analysis of SAGE Sequence Files

Within the concatemer sequences, the linked ditags of approx 26 bp are separated by CATG, which is the recognition site of *Nla*III. The SAGE software uses the CATG sequence to identify and extract the ditags, which are then halved into individual tags. The software then quantifies the number of times the tag occurs within a given population of clone inserts and creates a report of the abundance of each tag. The report can be linked to genetic databases for identification of the gene(s) corresponding to the tags and used to compare different SAGE libraries.

Use the downloaded Johns Hopkins SAGE software according to the instructions provided, in combination with NCBI's Genbank databases, and Microsoft Access and Excel. Other SAGE programs are also available (*see* **Subheading 1** and **Table 1**).

3.17. Discussion

The I-SAGE instructions suggest that the whole procedure takes at least 9 d and longer to screen and sequence the selected clones. A few weeks, or more likely, months, is a more realistic estimate, especially if setting up SAGE on your own without a kit. Here I have emphasized commonly encountered problems, but the more detailed Johns Hopkins and I-SAGE protocols contain excellent trouble-shooting sections, and I-SAGE in particular describes verification steps to check the success of each stage.

Once the SAGE libraries have been produced and analyzed, individual SAGE tags may be selected for further study, through NCBI's SAGEmap ***(20)*** and Unigene ***(19)***, and the developing Gene Ontology databases ***(25)***, among other resources. This process is straightforward where the tag clearly corresponds to one gene but may be more difficult where either no matching gene or multiple matches exist. This problem can be addressed by reverse-transcription PCR with the short SAGE tag as a primer ***(26–28)***. This generates longer, more specific, 3′ cDNA fragments that facilitate investigation of the gene and can also be used to check whether the tag is truly differentially expressed between samples of interest ***(26)***.

To conclude, SAGE is an excellent method of mRNA expression profiling. Although it is time-consuming and laborious, and requires expertise in molecular biology, the resulting libraries are extremely valuable, providing data that are truly comprehensive and quantitative and that enable the identification of novel genes.

4. Notes

1. General instructions. These concise instructions assume knowledge of and experience in standard molecular biological techniques: purification and manipulation of RNA and DNA, including phenol:chloroform extraction and ethanol precipitation; cDNA synthesis; restriction enzyme digestion; PCR; agarose and polyacrylamide gel electrophoresis; cloning; and sequencing. For further information, consult the references and web sites listed and, of course, Sambrook et al. ***(29)***. Where kits are used, follow the manufacturer's protocol. Most other reagents, including enzymes and magnetic beads, are also supplied with detailed instructions. Alternatively, use Invitrogen's I-SAGE kit and protocol throughout. The availability of standard laboratory equipment is also assumed: 0.2-, 0.5-, 1.5-, and 2- mL microcentrifuge tubes; 96-well PCR plates; a microcentrifuge, preferably

refrigerated; a vortex mixer; 50-mL tubes and an appropriate centrifuge; wet and dry ice; water baths; and incubators, including one for bacterial culture.

2. Working with RNA. The quality and quantity of input RNA is critical to the success or failure of SAGE. Working with RNA may be difficult, and advice is given in Maniatis and on Ambion's web site. Materials should be RNase-free: solutions may be DEPC treated and RnaseZap® or other RNase inhibitors may be used on equipment. Check RNA quality by agarose gel electrophoresis: 0.5 µg of total RNA should yield two clear bands of ribosomal RNA (4.5 kb and 1.9 kb).
3. Protocols for smaller amounts of starting material (*see* **Subheadings 2.2.**, **2.4.**, **2.6.**, and **3.2.**). The Johns Hopkins protocol as described here (and used personally) requires a relatively large amount of input material: ideally, at least 2.5 µg of mRNA, broadly equivalent to 250 µg of total RNA, 250 mg tissue, or 2.5×10^7 cultured cells. This protocol therefore cannot be used to generate expression profiles where RNA is limited, for example, tissue biopsies. Various technical modifications now enable SAGE to be applied to smaller quantities of RNA ***(3)***: at least 100-fold (and up to 5000-fold) less may be needed. SADE (a SAGE Adaptation for Downsized Extracts) ***(6)*** uses Dynal's oligo dT-coated magnetic beads to capture polyA+ mRNA directly from the total RNA or cell lysate. (This substitutes for mRNA purification then cDNA synthesis with biotinylated oligo dT followed by capture onto streptavidin-coated Dynabeads.) All steps from mRNA isolation to tag release are then performed directly on the beads. This procedure significantly reduces sample loss and has been adopted in Invitrogen's I-SAGEkit. Oligo dT is used to similar effect in microSAGE ***(7)*** and miniSAGE ***(8)***, in the form of a coating inside microcentrifuge tubes (Roche's Streptavidin-Coated Tubes). The references contain the experimental protocols. Further modifications include additional PCR steps. In SADE and microSAGE, the ditags generated by the first round of large-scale PCR amplification are then re-amplified using extra PCR cycles ***(6,7)***. In contrast, SAGE-Lite ***(9)*** and PCR-SAGE ***(10)*** have adapted Clontech's SMART™ system to generate PCR-amplified cDNA, to increase the amount of input material before proceeding to SAGE proper.
4. Enzymatic kinasing of linker oligonucleotides. Dilute linkers to 350 ng/µL. Set up two tubes, one for linker pair 1 and the other for pair 2. Mix 9 µL of Linker B (either 1B or 2B) with 8 µL of LoTE, 2 µL of 10× ligase buffer (works with kinase and contains ATP) and 1 µL of T4 polynucleotide kinase. Incubate at 37°C for 30 min then heat inactivate at 65°C for 10 min. Add 9 µL of Linker 1A to the 20 µL of kinased Linker 1B, and do likewise for Linkers 2. To anneal linkers, heat to 95°C for 2 min, then place at 65°C for 10 min, 37°C for 10 min, and room temp for 20 min. Store at –20°C. The final linker concentration is 200 ng/µL. Test the kinase reaction by self-ligating 200 ng of each linker pair. Run on a 12% polyacrylamide gel. Kinased linkers should result in linker-linker dimers of 80 to 100 bp, whereas unkinased linkers should not self-ligate. Only linker pairs resulting in over 70% self-ligation should be used in further steps.
5. Phenol:chloroform extraction. If the sample volume is below 200 µL, increase it to 200 µL with LoTE in a 1.5-mL microcentrifuge tube. Add an equal volume of P/C. Vortex and centrifuge at full speed for 5 min at room temp. Transfer the aqueous (top) phase to another tube.
6. Ethanol precipitation. For a 200-µL sample volume in a 1.5-mL tube, add 3 µL glycogen (as a carrier), 100 µL 10 *M* ammonium acetate, and 700 µL 100% ethanol. (Scale volumes up or down as required.) Mix well. Precipitate on dry ice (or at –20°C) for at least 15 min. Centrifuge at full speed for 15 min, preferably at 4°C. Wash pellet with 70% ethanol and respin. Resuspend in LoTE.
7. QIAquick columns (*see* **Subheading 2.3.**). In this protocol, DNA is prepared by P/C extraction and ethanol precipitation. At some stages, QIAquick® columns can be substituted, saving time and possibly providing purer samples for SAGE ***(12)***.

8. PCR. With SAGE, achieving the initial 102-bp ditag PCR product takes time. Thereafter, however, equally strenuous efforts are necessary to avoid PCR cross-contamination, hence the inclusion of negative controls. The Johns Hopkins protocol now recommends three separate PCR areas, for pre-PCR assembly of reagents and negative controls; the addition of ligated ditag template; and the manipulation of post-PCR products. Ultraviolet PCR preparation hoods may also be useful.
9. Polyacrylamide gel electrophoresis (PAGE). The SAGE PCR products and ditags are isolated by 12% PAGE and the concatemers are separated by 8% PAGE. 12% PAGE uses 14 mL of 40% polyacrylamide (19:1 acrylamide:bis) and 31.3 mL of dH_20. 8% PAGE requires 9.3 mL of 40% Polyacrylamide (37.5:1 acrylamide:bis) and 36 mL of dH_20. To either, add 930 µL of 50× Tris acetate buffer, 470 µL of 10% ammonium persulfate, and 30 µL of TEMED. Mix and pour in a vertical gel apparatus. Allow to polymerize for 30 min. Run gel as described in the text (the time is approximate). (Note that bromophenol blue is dark blue and runs ahead of xylene cyanol, which is turquoise.) After electrophoresis, stain gel with SYBR® Green I, according to manufacturer's instructions, or with ethidium bromide. Visualize bands under ultraviolet light. The Johns Hopkins and I-SAGE protocols and some of the methodological papers contain useful photographs of sample gels at each stage.
10. DNA quantitation. The Johns Hopkins and I-SAGE protocols recommend assessment of DNA yield by dot quantitation at various stages during PCR purification and ditag isolation. In general, however, this is not routinely required.

Acknowledgments

Thanks to the Cancer Research Campaign and the University of Glasgow for research funding and to Dr. Kenneth W. Kinzler for SAGE protocols, software and advice.

References

1. Velculescu, V. E., Zhang, L., Vogelstein, B., and Kinzler, K. W. (1995) Serial analysis of gene expression. *Science* **270,** 484–487.
2. Zhang, L., Zhou, W., Velculescu, V., Kern, S., Hruban, R., Hamilton, S., et al. (1997) Gene expression profiles in normal and cancer cells. *Science* **276,** 1268–1272.
3. Velculescu, V. E., Vogelstein, B., and Kinzler, K. W. (2000) Analysing uncharted transcriptomes with SAGE. *Trends Genet.* **16,** 423–425.
4. Madden, S. L., Wang, C. J., and Landes, G. (2000) Serial analysis of gene expression: from gene discovery to target identification. *Drug Discov. Today* **5,** 415–425.
5. Polyak, K. and Riggins, G. J. (2001) Gene discovery using the serial analysis of gene expression technique: implications for cancer research. *J. Clin. Oncol.* **19,** 2948–58.
6. Virlon, B., Cheval, L., Buhler, J. M., Billon, E., Doucet, A., and Elalouf, J. M. (1999) Serial microanalysis of renal transcriptomes. *Proc. Natl. Acad. Sci. USA* **96,** 15,286–15,291.
7. Datson, N. A., van der Perk-de Jong, J., van den Berg, M. P., de Kloet, E. R., and Vreugdenhil, E. (1999) MicroSAGE: a modified procedure for serial analysis of gene expression in limited amounts of tissue. *Nucleic Acids Res.* **27,** 1300–1307.
8. Ye, S. Q., Zhang, L. Q., Zheng, F., Virgil, D., and Kwiterovich, P. O. (2000) MiniSAGE: Gene expression profiling using serial analysis of gene expression from 1 µg total RNA. *Anal. Biochem.* **287,** 144–152.
9. Peters, D. G., Kassam, A. B., Yonas, H., O'Hare, E. H., Ferrell, R. E., and Brufsky, A. M. (1999) Comprehensive transcript analysis in small quantities of mRNA by SAGE-lite. *Nucleic Acids Res.* **27,** e39.
10. Neilson, L., Andalibi, A., Kang, D., Coutifaris, C., Strauss, J. F., Stanton, J. A. L., et al. (2000) Molecular phenotype of the human oocyte by PCR-SAGE. *Genomics* **63,** 13–24.

11. Powell, J. (1998) Enhanced concatemer cloning—a modification to the SAGE (Serial Analysis of Gene Expression) technique. *Nucleic Acids Res.* **26,** 3445–3446.
12. Angelastro, J. M., Klimaschewski, L. P., and Vitolo, O. V. (2000) Improved NlaIII digestion of PAGE-purified 102 bp ditags by addition of a single purification step in both the SAGE and microSAGE protocols. *Nucleic Acids Res.* **28,** E62.
13. Margulies, E. H., Kardia, S. L., and Innis, J. W. (2001) Identification and prevention of a GC content bias in SAGE libraries. *Nucleic Acids Res.* **29,** E60.
14. Kenzelmann, M. and Muhlemann, K. (1999) Substantially enhanced cloning efficiency of SAGE (Serial Analysis of Gene Expression) by adding a heating step to the original protocol. *Nucleic Acids Res.* **27,** 917–918.
15. Yamamoto, M., Wakatsuki, T., Hada, A., and Ryo, A. (2001) Use of serial analysis of gene expression (SAGE) technology. *J. Immunol. Methods* **250,** 45–66.
16. Margulies, E. H. and Innis, J. W. (2000) eSAGE: managing and analysing data generated with Serial Analysis of Gene Expression (SAGE). *Bioinformatics* **16,** 650–651.
17. van Kampen, A. H. C., van Schaik, B. D. C., Pauws, E., Michiels, E. M. C., Ruijter, J. M., Caron, H. N., et al. (2000) USAGE: a web-based approach towards the analysis of SAGE data. *Bioinformatics* **16,** 899–905.
18. Larsson, M., Stahl, S., Uhlen, M., and Wennborg, A. (2000) Expression profile viewer (ExProView): A software tool for transcriptome analysis. *Genomics* **63,** 341–353.
19. Wheeler, D. L., Church, D. M., Lash, A. E., Leipe, D. D., Madden, T. L., Pontius, J. U., et al. (2001) Database resources of the National Center for Biotechnology Information. *Nucleic Acids Res.* **29,** 11–16.
20. Lal, A., Lash, A. E., Altschul, S. F., Velculescu, V., Zhang, L., McLendon, R. E., et al. (1999) A public database for gene expression in human cancers. *Cancer Res.* **59,** 5403–5407.
21. Audic, S. and Claverie, J.-M. (1997) The significance of digital gene expression profiles. *Genome Res.* **7,** 986–995.
22. Stollberg, J., Urschitz, J., Urban, Z., and Boyd, C. D. (2000) A quantitative evaluation of SAGE. *Genome Res.* **10,** 1241–1248.
23. Man, M. Z., Wang, X. N., and Wang, Y. X. (2000) POWER_SAGE. Comparing statistical tests for SAGE experiments. *Bioinformatics* **16,** 953–959.
24. Kal, A. J., van Zonneveld, A. J., Benes, V., van den Berg, M., Koerkamp, M. G., Albermann, K., et al. (1999) Dynamics of gene expression revealed by comparison of serial analysis of gene expression transcript profiles from yeast grown on two different carbon sources. *Mol. Biol. Cell.* **10,** 1859–1872.
25. Ashburner, M., Ball, C. A., Blake, J. A., Botstein, D., Butler, H., Cherry, J. M., et al. (2000) Gene ontology: tool for the unification of biology. The Gene Ontology Consortium. *Nat. Genet.* **25,** 25–29.
26. van den Berg, A., van der Leij, J., and Poppema, S. (1999) Serial analysis of gene expression: rapid RT-PCR analysis of unknown SAGE tags. *Nucleic Acids Res.* **27,** e17.
27. Chen, J. J., Rowley, J. D., and Wang, S. M. (2000) Generation of longer cDNA fragments from serial analysis of gene expression tags for gene identification. *Proc. Natl. Acad. Sci. USA* **97,** 349–353.
28. Matsumura, H., Nirasawa, S., and Terauchi, R. (1999) Transcript profiling in rice (Oryza sativa L.) seedlings using serial analysis of gene expression (SAGE). *Plant J.* **20,** 719–726.
29. Sambrook, J., Fritsch, E. F., and Maniatis, T. (1987) *Molecular Cloning.* Cold Spring Harbor Laboratory Press, Cold Spring Harbor, NY.

VI

Mutations and Polymorphisms

41

Mutation and Polymorphism Detection

A Technical Overview

Joanne Edwards and John M. S. Bartlett

1. Introduction

Analysis of DNA variation (polymorphism and mutations) is one of the most common challenges faced by molecular biologists. Studies of polymorphisms and mutations as molecular markers of or underlying causes of disease have confirmed the importance of mutation and polymorphism detection. With mutation detection currently being so important for the study of genetic diseases, gene discovery, and solving problems of basic biology, there is a large demand for quick and relatively cheap methods for mutation detection. Therefore, many different methods have been developed for detecting new mutations and screening populations for known mutations or polymorphisms. Traditional mutation detection systems, such as restriction fragment length polymorphism and denaturing gradient gel electrophoresis, have their limitations. With restriction fragment length polymorphism, the mutational event needs to either create or destroy a restriction site *(1)*. With denaturing gradient gel electrophoresis, although a change at a restriction digest site is not required, this method may only detect about 50% of possible mutations and polymorphisms *(1)*. The use of polymerase chain reaction (PCR)-based mutation and polymorphism detection systems increase the sensitivity and accuracy of the screening methods. This chapter will introduce some of the available detection methods and discuss the problems associated with these techniques.

2. Detection Methods

2.1. Detection of Single Nucleotide Polymorphisms (SNPs)

2.1.1. Single-Strand Conformation Polymorphism (SSCP)

SSCP can detect up to 90% of single base changes and relies on the different mobilities of DNA strands containing single bp differences when run on a denaturing polyacrylamide gel *(2)*. SSCP analysis is widely used to screen large numbers of samples for mutations. If a mutation is detected, direct sequence analysis is often then used to determine the exact location and base change of the mutation. SSCP can be performed using either fluorescent or radioactive technology *(2,3)*. The advantages of fluorescent

From: *Methods in Molecular Biology, Vol. 226: PCR Protocols, Second Edition*
Edited by: J. M. S. Bartlett and D. Stirling © Humana Press Inc., Totowa, NJ

technology over radioactive are (1) an internal molecular weight standard is used in each lane to align data, thereby eliminating lane-to-lane variability; (2) strands of DNA in the PCR amplicon can be labeled with different-colored labels; And (3) there is the ability to multiplex products from different PCRs in one lane to increase sample throughput. SSCP is further discussed in Chapter 48. However, although SSCP is a useful screening technique, it yields no information on the location of the mutation. A band shift will identify the presence of the mutation within the PCR product, but no sequence information on the exact base that has been mutated can be determined without sequencing.

2.1.2. Mutational Detection Using Cleavage Systems

2.1.2.1. Gel-Based Method

An alternative screening method for mutations or polymorphisms that does provide some information on the location of the mutation is enzymatic cleavage of DNA. Enzymatic cleavage systems rely on enzymes such as resolvases (e.g., the enzyme T4 endonuclease VII; **ref. *4***). These enzymes cleave double-stranded DNA at sites where a "bubble" is formed because of miss-pairing of bases. Mutation analysis is performed by mixing PCR products from the test samples and samples with known sequences (either normal alleles or wild type DNA). The PCR products are melted and the DNA strands allowed to re-anneal. In the presence of either a mutation or a different allele from the wild type, a miss-match in sequence provides a target for the resolvase enzyme. The enzyme scans along the double-stranded DNA and binds to the single-stranded "bubble" at the site of the miss-match, and the DNA is then cleaved in this region. The resultant fragments are then separated by electrophoresis, and the presence of cleavage products indicates the presence of a mutation ***(4,5)***. The size of the cleaved fragments indicates the approximate location of the mutation, which is an advantage over SSCP.

2.1.2.2. Nongel-Based Method

The method just discussed uses a gel-based detection system. An example of a cleavage method that is not gel based is the use of a structure-specific 5′ nuclease to cleave sequence-specific structures in each of two cascading reactions ***(6)***. The cleavage structure forms when two synthetic oligonucleotides hybridize in tandem to a target. One of the oligonucleotides cycles on and off the target and is cut by the nuclease only when the appropriate structure forms. The cleaved probes then participate in a second reaction involving a dye-labeled fluorescence resonance energy-transfer probe. Cleavage of this probe generates a signal, which can be analyzed by fluorescence microtiter plate readers ***(6)***.

2.1.3. Oligonucleotide Microarrays

Current screening methods are rapid and when combined with gel-based fluorescent DNA sequencing technology can accurately locate and identify mutations and polymorphisms. However, SNPs are now being uncovered and assembled into large SNP databases. Although this will be invaluable for linking SNPs with disease, it will require analysis of thousands of polymorphisms ***(7)***. Currently, there is no technique available that is fast, specific, sensitive, reliable, and cost-effective enough for genome-wide polymorphism analysis. Oligonucleotide microarrays are currently under intensive development and if this technique is successful should enable the investigation of thousands of SNPs in parallel *(8)*. However, currently there are still limitations associated with it. A recent mapping study analyzed 500 SNPs using array technology and only 70% of sites analyzed were genotyped correctly *(9)*. This limitation has been broached by using enzymatic reactions with the oligonucleotide arrays to enhance specificity of hybridization, for example, array primer extensions are being developed that use the principle of allele specific single-base primer extension. It has been demonstrated that this method is more specific at identifying unknown mutations in a known sequence (deletions, transversions, and up to two-base insertions) that can be readily found, using arrayed primer extension reactions *(10)*. However, promising mutational analysis using oligonucleotide microarrays still remains in the developmental stage.

2.2. Detection of Large Inserts or Deletions

Large inserts or deletions may be detected by the methods discussed above, such as SSCP, enzyme cleavage systems, or amplification refractory mutation system discussed in detail in Chapter 47. However, currently the most commonly used detection system is either radioactive or fluorescent-based automated sequencing. Although many of the above mentioned available screening methods detect in the region of 90 to 98% of mutations (whether SNPs, deletions, or insertions) there are still mutations that are missed. Therefore, currently the only way to assure that every mutation is found is to sequence the region of interest, although this method is both costly and time consuming and requires high-quality pure DNA.

2.3. Detection of Loss of Heterozygosity (LOH) and Replication Error Phenotype

Microsatellite loci have a high degree of polymorphism that is caused by problems associated with copying repetitive sequences of DNA. Microsatellites are popular genetic markers because of their abundance and high level of allelic variation, and expansion of microsatellite trinucleotide repeats have been demonstrated to cause several human genetic disease *(11)*. Microsatellite analysis is also useful for detection of LOH and replication errors in tumors *(12,13)*. LOH is detected as a reduced intensity or total loss of one or more bands in the tumor DNA compared with normal DNA form the same individual *(12)*. Replication errors are detected as changes in the length of the microsatellite sequences in the tumor DNA compared with normal DNA *(13)*. Mutation polymorphism detection of microsatellites basically involves amplifying the microsatellite loci and estimating the size of the PCR product. The most common techniques used to estimate the size of the microsatellite product involves separation of products by gel electrophoresis. The most common method used to visualize PCR products on an agarose gel is ethidium bromide staining. Staining with ethidium bromide is not a routine method used for microsatellite analysis because it has low sensitivity and does not provide a permanent record. Microsatellite analysis is more commonly performed using polyacrylamide gels because they give better separation than agarose gels. Visualization of DNA on a polyacrylamide gel may be achieved by silver staining, radioactive-labeled DNA visualized by autoradiography, or laser-activated fluorescent-labeled DNA detected by an automated sequencer. Silver staining can be difficult to control, and use of radiation within the laboratory setting is becoming less popular in current years because of the associated health risks. Therefore, fluorescent-based methods are currently the method of choice, that is, labeling of DNA using fluorescent-labeled primers or nucleotides. Fluorescent technology eliminates the handling of radioactivity, makes it easier to interpret stutter bands and laser-activated detection of fluorescent products during electrophoresis, allows immediate detection of signal, permits rapid data analysis, and provides permanent records of results. The use of fluorescent technology has allowed development of techniques that greatly increase the speed by which microsatellite analysis can be performed by increasing throughput. For example, a rapid way to screen for microsatellites is described in Chapter 42. This chapter discusses multiplex touchdown PCR, which allows amplification of multiple microsatellite loci simultaneously. Use of different fluorescent-labeled primers then allows identification of each loci. Similarly, Chapter 43 describes the use of fluorescent technology to increase throughput of LOH analysis by labeling PCR products with differently labeled fluorescent primers.

3. Technical Constraints

All of the above methods require good-quality DNA; therefore, in most cases before mutational analysis can begin RNA or DNA must be extracted from the sample and a PCR performed to provide DNA of sufficient purity and quantity for mutational analysis.

3.1. Problems Associated with Quality and Quantity of Extracted DNA or RNA

If the amount of available DNA or RNA is limited and therefore insufficient for use in mutational analysis, PCR or RT-PCR may be used to increase the quantity of DNA. If multiple regions of interest are to be investigated, then it may be necessary to globally amplify DNA by degenerate oligonucleotide-primed PCR (DOP-PCR) ***(14)***. The use of this technique in mutational analysis is described in Chapter 45.

The purity of extracted DNA or RNA for subsequent mutational analysis is extremely important. Contamination with other cellular components, such as protein, or chemicals, such as ethanol, can easily be removed after the extraction process using various cleanup protocols. However, contamination with foreign DNA is not dealt with as easily. Contamination at this stage of the process with either foreign DNA from external sources or from within the sample will be amplified as the process proceeds and can result in mutational analysis being conducted on the wrong DNA. Precautions should be taken to limit cross-contamination and external contamination as with any PCR. Methods for increasing the purity of RNA and decreasing the introduction of contamination during RT-PCR are discussed in Chapter 46. Contamination may also be a problem from within the tissue, for example, if studying cancer genetics, normal cells may contaminate a tumor cell population. Microdissection is a technique that allows separation of normal and tumor cells, and Chapter 43 discusses use of this method in microsatellite analysis.

For mutational analysis, DNA is required to be of high quality because poor-quality DNA, for example, degraded or nicked, could result in false results. If performing RT-PCR from RNA, RNA should be extracted from fresh tissue or from tissue snap frozen on removal to decrease RNA degradation. Although kits now are available for extraction of RNA from archival material, the quality of RNA retrieved is often poor.

DNA may be extracted from fresh, frozen, or archival material. High-quality DNA should be easily retrieved from fresh or frozen material, for example, tissue or blood. Extraction of good-quality DNA from archival material is more difficult; however, the most abundant source of clinically available tissue is archival formalin-fixed paraffin-embedded tissue. This process not only drastically decreases the yield of DNA in comparison with fresh or frozen material but also damages the DNA, often making it unsuitable as a DNA template. Chapter 43 discusses how archival DNA can be used for microsatellite analysis by keeping the PCR product size small.

In summary, for successful mutational analysis a significant quantity of high-quality, pure DNA is required. Low quantities of DNA make analysis difficult, and poor-quality or contaminated DNA may cause false results.

3.2. PCR Artifacts

Although PCR is the recognized method for increasing the quantity of DNA required for polymorphism or mutational analysis, PCR can in itself introduce problems. Contamination by foreign DNA and cross-contamination is often introduced at this stage of the process. With appropriate care, that is, gloves, aseptic technique, and clean pipette tips between every sample, this should be limited. However, this is a particular problem with DOP-PCR. DOP-PCR is very sensitive to contamination because degenerate primers in the reaction will amplify DNA from any source present in the tube. However, the presence of DNA in the blank control after DOP-PCR does not always indicate contamination, and the low temperature in the DOP-PCR can result in DOP primers attaching to each other and replicating ***(15)***. If this occurs, then a second PCR amplifying only the region of interest should not be possible. It is therefore recommended that blanks should be tested by sequential PCR after DOP-PCR before samples are discarded.

Other problems created by PCR are the addition of an extra base (normally adenine) at the end of the DNA sequence and the creation of shadow bands (discussed in Chapter 44). In summary, although PCR solves the problem of insufficient DNA for analysis, with its use comes an assortment of additional problems.

4.1. Applications

Mutational analysis has many applications, for example, markers of disease, cancer research, and genotyping. Mutations are responsible for diseases, such as sickle cell anemia ***(1)***, cystic fibrosis ***(1)***, and adrenal leukodystrophy ***(1)***. An expansion of the CAG repeat in exon 1 of the androgen receptor alone can result in Kennedy's disease, spinocerebellar ataxia, or fragile X syndrome ***(15)***. Therefore, polymorphism and mutational analysis are of use in diagnosing these diseases.

Detection of polymorphism and mutational analysis is also widely used in the field of cancer research. Mutation detection has demonstrated that *p53* function is lost in approximately 50% of all cancers and that the loss of function is caused by point mutations ***(16)***. It also demonstrates that a mutation in *BRCA 1* or *BRCA 2* increases susceptibility to breast and ovarian cancer ***(17)***. Microsatellite analysis is used in cancer research to study LOH and microsatellite instability. LOH studies have demonstrated most genetic alterations that occur in bladder cancer are on chromosome 9 ***(18)*** and that microsatellite instability is commonly found in colorectal cancer ***(13)***.

Stepwise mutation models and microsatellite polymorphisms have been used to determine relationships among primates *(19)*. Such studies have investigated the relationship between humans, chimpanzees, and gorillas *(19)*. There is also an ongoing debate about proper interpretation of DNA sequence polymorphisms and their ability to reconstruct human population history *(20)*.

Mutational analysis is also used in genotyping. It was use of mutational analysis that demonstrated that birds are not monogamous because genotyping performed on eggs in the same nest demonstrated different parentage *(21)*. Genotyping is also used when breeding animals in captivity to determine their most appropriate mate *(21)*. In summary, polymorphism and mutational analysis is a major tool in science today, without which much of our current understanding of genetics would not have been possible.

References

1. Alberts, B., Bray, D., Lewis, J., Raff, M., Roberts, K., and Watson, J. D. (1989) Recombinant DNA Technology, in *Molecular Biology of The Cell*, 2nd ed., chapter 4, (Robertson, M., ed.), Garland Publications, New York, pp. 180–196.
2. Iwahana, H., Yoshimoto, K., and Itakura, J. (1994) Multiple fluorescence-based PCR-SSCP analysis. *Biotechniques* **16,** 296–304.
3. Nataraj, A. J., Olivos-Glander, I., Kusukawa, N., and Highsmith, W. E. (1999) Single-strand conformation polymorphism and heteroduplex analysis for gel-based mutation detection. *Electrophoresis* **20,** 1177–1185.
4. Youil, R., Kemper, B. W., and Cotton, R. G. (1995) Screening for mutations by enzyme mismatch cleavage with T4 endonuclease VII. *Proc. Natl. Acad. Sci. USA* **92,** 87–92.
5. Del Tito, B. J., Poff, H. E., Noviotny, M. A., Cartlidge, D. M., Walker, R. I., Earl, C. D., et al. (1998) Automated fluorescent analysis procedure for enzymatic mutation detection. *Clin. Chem.* **44,** 731–739.
6. Cotton, R. G. (1999) Mutation detection by chemical cleavage. *Genet. Anal.* **14,** 165–168.
7. Gray, I. C., Campbell, D. A., and Spurr, N. K. (2000) Single nucleotide polymorphisms as tools in human genetics. *Hum. Mol. Genet.* **9,** 2403–2408.
8. Kikoris, M., Dix, K., Moynihan, K., Mathis, J., Erwin, B., Grass, P., et al. (2000) High throughput SNP genotyping with the mass code system. *Mol. Diagn.* **5,** 329–340.
9. Faham, M., Baharloo, S., Tomitaka, S., DeYoung, J., and Freimer, N. B. (2001) Mismatch repair detection: high-throughput scanning for DNA variations. *Hum. Mol. Genet.* **10,** 1657–1664.
10. Fan, J. B., Chen, X., Halushka, M. K. Berno, A., Huang, X., Ryder, T., et al. (2000) Parallel genotyping of human SNPs using generic high-density oligonucleotide tag arrays. *Genome Res.* **10,** 853–860.
11. Rohrbach, H., Hass, C .J., Baretton, G. B., Hirschmann, A., Diebold, J., Behrendt, R. P., et al. (1999) Microsatellite instability and loss of heterozygosity in prostatic carcinomas: comparison of primary tumours, and of corresponding recurrences after androgen-deprivation therapy and lymph node metastases. *Prostate* **40,** 20–27.
12. Niederacher, D., Picard, F., van Roeyen, C., An, A. X., Bender, H. G., and Beckmann, M. W. (1997) Patterns of Allelic loss on chromosome 17 in sporadic breast carcinomas detected by fluorescent labelled microsatellite analysis. *Genes Chromosomes Cancer* **18,** 181–192.
13. Dietmaier, W., Riedlinger, W., Kohler, A., Wegele, P., Beyser K., Sagner, G., et al. (1999) Detection of microsatellite instability and loss of heterozygosity in colorectal tumours by fluorescence based multiplex microsatellite PCR. *Biochemica* **2,** 43–46.

14. Kuukasjarvi, T., Tanner, M., Pennanen, S., Karhu, R., Visakorpi, T., and Isola, J. (1997) Optimising DOP-PCR for universal amplification of small DNA samples in comparative genomic hybridisation. *Genes, Chromosomes Cancer* **18,** 94–101.
15. Chamberlain, N. L., Driver, E. D., and Mainsfeld, R. L. (1994) The length and location of CAG trinucleotide repeats in the androgen receptor N-terminal domain affect trans activation function. *Nucleic Acids Res.* **22,** 3181–31886.
16. Lane, D. P. (1992) TP53, the guardian of the genome. *Nature* **358,** 15–16.
17. Rohlfs, E. M., Learning, W. G. Friedman, K. J., Cough, F. J., Weber, B..L., and Silverman, L. M. (1997) Direct detection of mutations in the breast and ovarian cancer susceptibility gene BRCA1 by PCR mediated site directed mutagenesis. *Clin. Chem.* **43,** 24–29.
18. Knowles, M. A. (1998) Molecular genetics of bladder cancer: Pathways of development and progression. *Cancer Surveys* **31,** 49–53.
19. Goldstein, D. B., Ruis Linares, A., Cavalli-Sforsa, L. L., and Feldman, M. W. (1995) Genetic absolute dating based on microsatellite and the origin of modern humans. *Proc. Natl. Acad. Sci. USA* **92,** 6723–6727.
20. Cann, R. L. (2001) Genetic clues to dispersal in human populations: retracing the past from the present. *Science,* **291,** 1742–1748.
21. Alberts, B., Bray, D., Lewis, J., Raff, M., Roberts, K., and Watson, J. D. (1989) The organisation and evolution of the nuclear genome, in *Molecular Biology of The Cell,* 2nd ed, chapter 10 (Robertson, M., ed.), Garland Publications, New York, pp. 599–609.

42

Combining Multiplex and Touchdown PCR for Microsatellite Analysis

Kanokporn Rithidech and John J. Dunn

1. Introduction

An improved nonradioactive polymerase chain reaction (PCR)-based method for simultaneous amplification of multiple loci of microsatellites has been developed as a rapid way to screen for microsatellites *(1)*. The approach, termed multiplex-touchdown PCR (MT-PCR), is performed in a single PCR tube by combining touchdown *(2–5)* and multiplex *(6)* PCR protocols. The touchdown format is used to improve the specificity and the quality of amplification, that is, only DNA bands of an expected size are present as the major PCR bands observed on the nondenaturing polyacrylamide gels, thereby overcoming the presence of differently sized background bands (a ladder-like problem). The multiplex strategy is used so that simultaneous amplification of multiple microsatellite loci is achieved. In this chapter, we describe the MT-PCR strategy that has been successfully used for simultaneous amplification of up to three mouse microsatellites by choosing primer pairs with the corresponding touchdown-PCR parameters. The MT-PCR is very useful for genotyping hybrid mice, provided the allelic size difference between two parental genotypes is amenable to separation by gel electrophoresis. In principle, this MT-PCR should be applicable to similar studies in other species, including humans.

2. Materials

1. Sterile deionized water, for example, MilliQ water.
2. 10× PCR buffer (10 m*M* Tris-HCL, pH 8.3, 50 m*M* KCL).
3. 10 m*M* of each of dNTP (A,T,C, and G).
4. Microsatellite primers.
5. *Taq* polymerase.
6. $MgCl_2$.
7. Template DNA (5 µg/mL, in sterile water).
8. Polyacrylamide (premixed 30:0.8 acrylamide:bisacrylamide, Owl Separation System Portsmouth, NH).
9. Slab gel, for example, Mini-protean cell from Bio-Rad, large format single sided vertical system (20 cm × 20 cm) from Owl separation System.
10. 1× TBE (Tris borate EDTA) gel running buffer.

From: *Methods in Molecular Biology, Vol. 226: PCR Protocols, Second Edition*
Edited by: J. M. S. Bartlett and D. Stirling © Humana Press Inc., Totowa, NJ

11. Ethidium bromide (0.5 µg/mL).
12. DNA size standards.

3. Methods

Basically, two basic principles are involved in this protocol, that is, 1) the repeat sequences are amplified by PCR by using primers specific for the flanking genes and 2) the amplicons are then sized on nondenaturing polyacrylamide gels. The protocol outlined below has been used successfully for genotyping 50 mouse microsatellite markers, included in our study on leukemogenesis and hepatoma, of BALB/cJ X CBA/CaJ hybrid mice. The sizes of these markers are 84 to 270 bp with allelic size differences of 8 to 30 bp. The following steps are required.

3.1. Step 1

Set up PCR Master Mix. An example given in below is for a single PCR amplification of three microsatellite markers. To make a total volume of 15 µL Master Mix, combine the appropriate set of components in order listed in below in a 0.5-mL thin-walled PCR tube.

	Volume (µL)	Final Concentration
Sterile Milli Q Water	1.19	
10× PCR buffer	1.5	1X
dATP at 10 mM	0.3	200 µM
dCTP at 10 mM	0.3	200 µM
dGTP at 10 mM	0.3	200 µM
dTTP at 10 mM	0.3	200 µM
Microsatellite marker 1:		
Forward Primer (6.6 µM)	1.14	0.5 µM
Reverse Primer (6.6 µM)	1.14	0.5 UM
Microsatellite marker 2:		
Forward Primer (6.6 µM)	1.14	0.5 µM
Reverse Primer (6.6 µM)	1.14	0.5 µM
Microsatellite marker 3:		
Forward Primer (6.6 µM)	1.14	0.5 µM
Reverse Primer (6.6 µM)	1.14	0.5 µM
$MgCl_2$ (25 mM)	1.2	2.0 mM
▲template DNA (5 µg/mL)	3.0	1.0 µg/mL
¶Taq polymerase (5Units/µL)	0.072	2.5 Units/100 µL

NOTES:

1. All chemicals, except primers, are purchased from Perkin-Elmer, Norwalk, CT.
2. All primers are purchased from Research Genetics, Inc. Huntsville, AL.

▲ Add template DNA after a drop of mineral oil (Sigma, St. Louis, MO) has been placed into the tube.

¶ Add after "Hot Start" in Step 2.

3.2. Step 2

Hot-start at 96°C for 3 min. Then, add 0.072 µL of *Taq* polymerase (5 U/µL).

3.3. Step 3: Touchdown

A typical PCR protocol for the initial Touchdown cycle is denaturation at 96°C for 30 s, annealing at 65°C for 30 s, and extension at 72°C for 30 s.

3.4. Step 4: Standard PCR Cycles

After 5 to 10 touchdown cycles (depending on the targeted annealing temperature), 20 standard PCR cycles are performed under the following conditions: denaturation at 96°C for 30 s, annealing at 55°C (or 60°C) for 30 s, extension at 72°C for 30 s, and final extension at 72°C for an additional 5 min (*see* **Note 3**).

3.5. Step 5: Gel Electrophoresis

Typically, 6 and 10% nondenaturing polyacrylamide gels are used for separating microsatellites with allelic size differences of >12 bp and <12 bp, respectively. The gels are 1.5-mm thick. For the 6% gel, electrophoresis is at 75 Volts (constant, room temperature) for 1 h. However, electrophoresis is at 150 to 250 Volts (constant, room temperature) for 2 to 3 h for the 10% gel. Subsequently, DNA bands are visualized by soaking the gels in 0.5 μg/mL ethidium bromide for 15 min, followed by a brief rinse in deionized water, and observing fluorescence with long wavelength ultraviolet light.

4. Notes

1. The initial annealing temperature of 65°C gives good results for all microsatellites included in our study. This temperature is approx 5 to 10°C higher than the calculated annealing temperature for each primer pair, that is, 2°C [A + T] plus 4°C [G + C] *(7)*. During the touchdown phase, the annealing temperature is decreased at the rate of 1°C for every cycle of the amplification reaction until the targeted annealing temperature is reached. Denaturation and extension conditions for each touchdown cycle are 96°C for 30 s and 72°C for 30 s, respectively.
2. An increase in the number of regular PCR cycles (>20 cycles) results in deterioration of the PCR specificity leading to an increase in the background of spurious bands. Under the proper conditions, several amplicons from one round of MT-PCR can be analyzed simultaneously and each amplicon can be identified unambiguously on the same lane of nondenaturing PAGE. By choosing primer pairs with the corresponding T-PCR parameters, the MT-PCR is highly efficient for concurrent identification of three microsatellite loci (**Fig. 1**, lane 1).
3. A soak file at 4°C can be set after the final extension step.
4. A Mini-protean cell from Bio-Rad is used for the 6% nondenaturing polyacrylamide gel. A large format single sided vertical system (20 cm × 20 cm) from Owl separation System is used for the 10% nondenaturing polyacrylamide gel.
5. Beside speed and ease, an additional advantage of MT-PCR is that it is notably useful when only small quantities of DNA template are available. Amounts of DNA template as low as 10 ng can be used in one simultaneous reaction with three different pairs of primers. DNA samples from different tissues of the mouse, that is, bone marrow cells, spleen, liver, and tail, have been used successfully as templates in MT-PCR. Normally, the simple salting out procedure for isolating genomic DNA described by Miller et al. *(8)* routinely gives, in our laboratory, high-quality DNA templates from these tissues. The only drawback to the use of MT-PCR for the genotyping of microsatellites is that each protocol depends largely on primer compatibility, that is, having similar reaction kinetics (amenable to the same T-PCR protocol).

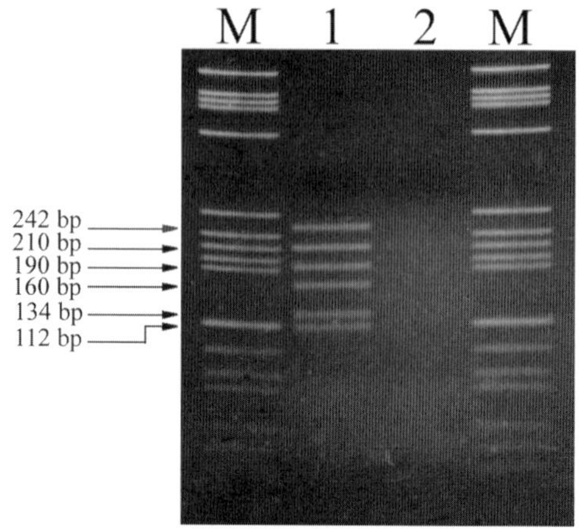

Fig. 1. PCR amplification of the mouse microsatellites: D2Mit43 (242-210 bp), D2Mit107 (134-112 bp, and D2Mit126 (160–190 bp) using genomic DNA isolated from the bone marrow cells of C3HeB/HeJ X C57BL/6J F1 Hybrid. Products were separated in a 6% nondenaturing PAGE (30:0.8 acrylamide:bisacrylamide) run at 150 V (constant) for 3 h. The gel was stained with ethidium bromide (0.5 mg/mL) and photographed under ultraviolet light. The arrows indicate the position of the expected amplicons. The molecular markers (*Hae*III digested pBR332 DNA) are shown in lanes M. The PCR products were obtained using five Touchdown cycles, and 20 cycles of constant annealing temperature (60°C). Lane 2 represents a control PCR without added mouse DNA. The sequences of primers are as follows:

D2Mit43:

Forward:	5'-GGG AGG GGT CAG AAT TCA AT-3'
Reverse:	5'-GTG CAG GAT ACT TGA TGT CTT CC-3'

D2Mit107:

Forward:	5'-GGG AGT GAA GCC AGC ATA AG-3'
Reverse:	5'-AAC TGA CTG AGT TTC AAA GTG CC-3'

D2Mit126:

Forward:	5'-GCT GAA CTG AGC AAA TTC TGG-3'
Reverse:	5'-TGA CAA TGG GAA GTT ATG TGT ATG-3'

Acknowledgments

This work was support by the U.S.DOE Contract ER62487/0001684.

References

1. Rithidech, K., Dunn, J. D., and Gordon, C. R. (1997) Combining multiples and touchdown PCR to screen murine microsatellite polymorphisms. *Biotechniques* **23,** 36–44.
2. Chou, Q., Russell, M., Birch, D. E., Raymond, J., and Bloch, W. (1992) Prevention of pre-PCR mis-priming and primer dimerization improves low-copy-number amplifications. *Nucleic Acids Res.* **20,** 1717–1723.
3. Don, R. H., Cox, P. T., Wainright, B. J., Baker, K., and Mattick, J. S. (1991) Touchdown PCR to circumvent spurious priming during gene amplification. *Nucleic Acids Res.* **19,** 4008.
4. Scrimshaw, B. J. (1992) A simple nonradioactive procedure for visualization of (dC-dA)n dinucleotide repeat length polymorphisms. *BioTechniques* **13,** 189.

5. Mellersh, C. and Sampson, J. (1993) Simplifying detection of microsatellite length polymorphisms. *BioTechniques* **15(4),** 582–583.
6. Chamberlain, J. S., Gibbs, R. A., Ranier, J. E., Nguyen, P. N., and Caskey, C. T. (1988) Detection screening of the Duchenne muscular dystrophy via multiplex DNA amplification. *Nucleic Acids Res.* **16,** 11,141–11,156.
7. Thein, S. L. and Wallace, R. B. (1986) The use of synthetic oligonucleotides as specific hybridization probes in the diagnosis of genetic disorders, in *Human genetic diseases: a practical approach* (Davis, K. E., ed.), IRL Press, Herndon, Virginia, pp. 33–50.
8. Miller, S. A., Dykes, D. D., and Polesky, H. F. (1989) A simple salting out procedure for extracting DNA from human nucleated cell. *Nucleic Acids Res.* **16,** 1215.

43

Detection of Microsatellite Instability and Loss of Heterozygosity Using DNA Extracted from Formalin-Fixed Paraffin-Embedded Tumor Material by Fluorescence-Based Multiplex Microsatellite PCR

Joanne Edwards and John M. S. Bartlett

1. Introduction

Microsatellites are widely distributed highly repetitive DNA sequences composed of di-, tri-, or tetranucleotide repeats *(1)*. They are spread over the whole human genome and are located between and within genes. Physiologically, they exhibit high levels of polymorphism, relative to different chromosomal loci, and within different individuals, different microsatellite lengths can even be noted between two alleles of the same gene *(1)*. Although the role of microsatellites in the genome and their evolutionary mechanisms are still incompletely understood, they are widely used tools in genetic mapping studies and as markers for prediction of disease *(2,3)*.

The length of microsatellite repeats may be used to predict the presence of disease, for example, the length of the CAG repeat in the *AR* receptor gene (40–65 repeats) is a marker of Kennedy's disease *(3)*. Microsatellites also exhibit a form of genetic instability characterized by the gain or loss of repeat units at multiple independent loci. Such alterations have been observed to accumulate in cells with defective DNA repair mechanisms and are commonly known as microsatellite instability (MIS) *(1,2)*. The DNA mismatch repair system plays an important role in controlling the accumulation of somatic mutations and therefore there is an association between DNA repair defects and carcinogenesis, and the presence of MIS may be used as a marker of certain cancers *(4)*. MIS has been associated with familial cancer syndrome hereditary nonpolyposis colorectal cancer, sporadic ovarian, prostate. and pancreatic cancer *(1,4,5)*. Cancer research also uses microsatellites when investigating inactivation of tumor suppressor genes. Loss of heterozygosity (LOH) of microsatellites located in or close to a tumor suppressor gene is an indirect way of testing for inactivation of tumor suppressor genes in accordance to the classical two-hit model, that is, a recessive mutation is uncovered by loss of the second copy of the gene, resulting in inactivation of the tumor suppressor gene *(6)*.

From: *Methods in Molecular Biology, Vol. 226: PCR Protocols, Second Edition*
Edited by: J. M. S. Bartlett and D. Stirling © Humana Press Inc., Totowa, NJ

Studies that involve microsatellites usually involve amplification of DNA using polymerase chain reaction (PCR). The three methods for detection of microsatellites currently in use are: radioactive PCR visualized by autoradiography; PCR visualized by silver-stained gels; and fluorescent PCR visualized on an automated DNA fragment analyzer *(7)*. The current method of choice is fluorescent PCR coupled with analysis of fluorescent-labeled DNA fragments on polyacrylamide gel electrophoresis in an automated DNA fragment analyzer. This method considerably increases the throughput of samples because it uses detection of laser-activated fluorescent products during electrophoresis and data are then immediately available for analysis. A large number of microsatellites may be studied at once because as many as 20 PCR products may be run down one lane of the gel at any one time. This method is also of use if there is a limited amount of DNA template, for example, small biopsy samples, because it enables the detection of very low concentrations of PCR products. The fluorescence detection system has been proven to be at least 10 times more sensitive than autoradiography and silver staining because of the use of an internal standard the fluorescent system can more accurately size PCR products and it minimizes stutter bands. This method also enables detection of LOH even if complete allele loss is not apparent because of contamination from other cell types. There are also less associated hazards when using this technique because no handling of radioactivity is required.

LOH and MIS studies commonly are used in cancer research; therefore, tumor and normal tissue must be separated before DNA can be extracted. Microdissection of histologically characterized cells from fresh-frozen or paraffin-embedded tissue sections has become an important technique for the analysis of genetic alterations occurring in tumors, allowing the separation of normal and tumor cells. Microdissection is a technique where by tissue sections approx 5-μm thick are stained in toluidine blue and viewed using a light microscope, and areas of normal/tumor cells are dissected using a micromanipulator. DNA from the different cells is extracted from the dissected tissue.

The most abundant source of tumor tissue with clinical information is formalin-fixed paraffin-embedded tissue. Hospitals routinely archive tumors in this fashion, and although this is acceptable for most pathological processes, it does cause nicks to form in DNA, this damage to the DNA affects the ability to use it as a PCR template. The yield of DNA extracted from formalin fixed paraffin embedded tissue is also significantly diminished, (by about 70% compared to the yield from frozen tissue) *(8)*. To overcome the problems associated with using archival DNA as a PCR template, the PCR product should be kept to a minimum size when designing primers, preferably under 300 bp. To overcome the concentration problem, fluorescent PCR with 40 or more amplification cycles should be used. Hot start *Taq* polymerase should also be used to decrease nonspecific binding of primers to the DNA template and the formation of primer dimers. With hot start *Taq*, the amplification reaction only starts after the initial denaturation step. Therefore, nonspecific PCR products, primer dimer formation, and background are minimized, maximizing the yield of specific PCR product from limited template.

The following method is an example of how MIS and LOH can be determined for a panel of nine microsatellites spanning chromosome 9 using only 9 μL of DNA (approx 15 μg/μL) microdissected from formalin-fixed paraffin-embedded archival transitional cell carcinomas of the urinary bladder tissue.

Table 1
Sequence of the Forward and Reverse Primers for Each Microsatellite and the Annealing Temperature Used in the PCR

Primer name	Primer sequence	Annealing temp
D9S126		
Forward	ATT GAA ACT CTG CTG AAT TTT CTG	50°C
Reverse	CAA CTC CTC TTG GGA ACT GC	
D9S259		
Forward	GGC ATC ATT GCA CCA T	50°C
Reverse	GGA TGG ATC TTA TGG GTG GAA	
D9S275		
Forward	CAG GAA CTT GTC CAT TCT C	47°C
Reverse	TCT ATT ATT GCC TTA CTC ACA G	
D9S195		
Forward	AGC TCA GCA CGG AGG G	53°C
Reverse	AGG GCA GGT TCC TAC AAA	
D9S258		
Forward	GCT AGA GAT GCC CTTCGAG TG	53°C
Reverse	AGG ATT TAT AGA AAG TCC AAA ACC C	
D9S102		
Forward	ATA GAC TTC CAG ACA GAT AG	50°C
Reverse	CCT CTC TCA TTC CTG GTA CT	
GSN		
Forward	CAG CCA GCT TTG GAG ACA AC	50°C
Reverse	TCG CAA GCA TAT GAC TGT AA	
D9S1199		
Forward	AAA AAT CAT GTG CAT CAA TTC C	53°C
Reverse	CCA GAG AAG CAG AAC CAA CG	
D9S1198		
Forward	TGG GAG AGG GAA AAT GCT ATC	47°C
Reverse	GTA CTC CAG CCT GGG TGG	

2. Materials

1. Xylene (BDH, UK).
2. 100%, 95% and 70% alcohol (BDH, UK).
3. 0.05% Toluidine blue dissolved in distilled water (Sigma Aldrich, UK).
4. Leitz microscope and Leica micromanipulator (Leica, UK).
5. Microdissecting needle and needle holder (Leica, UK).
6. Protein digestion buffer (0.5% Tween in 50 m*M* Tris-EDTA, pH 8.5).
7. 0.5 mg/mL proteinase K, add 450 µL of protein digestion buffer to 50 µL of 10 mg/m proteinase K dissolved in water (Sigma Aldrich, UK).
8. 37°C water bath.
9. MJ Research PTC-225 Peltier Thermal Cycler (Genetic Research Instrumentation, UK).
10. 2 µ*M* forward and reverse fluorescent-labeled primers per reaction (MWG-Biotech, UK Ltd). Primers are present in the reaction at 0.2 µ*M*. For individual primer sequences, refer to **Table 1**.

11. Hot start *Taq* polymerase, 5 units/μL, reaction concentration is 0.5 units/reaction (Qiagen Ltd, UK).
12. 10× reaction buffer and 25 m*M* $MgCl_2$ (Qiagen Ltd, UK); reaction concentration is 1× reaction buffer and 3 m*M* $MgCl_2$.
13. dNTPs (10 m*M*; Advanced Biotech); reaction concentration is 200 μm of each dNTP.
14. PCR-grade water (Sigma Aldrich, UK).
15. Formamide (BDH, UK).
16. Loading buffer (50 mg/mL blue dextran, 25 m*M* EDTA).
17. Genescan 400 HD size standard (rox), (ABI Perkin–Elmer, UK).
18. Acrylamide (Sigma Aldrich, UK).
19. Bisacrylamide (Sigma Aldrich, UK).
20. Automated laser-activated fluorescent DNA sequencer, ABI 377 sequencer (Perkin Elmer, UK).
21. Genescan™ and Genotyping™ software (Perkin Elmer, UK).

3. Methods

3.1. Microdissection of Archival DNA

1. Archival formalin-fixed, paraffin-embedded TCCs are cut into 5-μm thick sections and put on to glass slides.
2. Tissue sections are dewaxed by incubating in xylene (2 × 10 min).
3. Tissue is rehydrated by incubating in a series of alcohol solutions, 100% alcohol (2 × 2 min), 95% alcohol (2 min), and 70% alcohol (2 min).
4. Sections are stained in 0.05% Toluidine blue for 30 s.
5. Areas of tumor cell are microdissected from 5-μm sections using a dissecting microscope and a micromanipulator.
6. Dissected tissue is placed in an RNA/DNA free tube containing 12 μL of protein digestion buffer.
7. DNA is extracted by addition of a 13 μL of protein digestion buffer containing proteinase K and incubating for 4 to 7 d at 37°C in a water ***(9)***.
8. On removal from the water bath, proteinase K is inactivated by heating for 10 min at 95°C using a thermal cycler (*see* **Note 1**).
9. Normal tissue is also dissected from the same section as the tumor, thus providing normal DNA to be used as a control in MIS and LOH analysis.
10. LOH and MIS analysis is conducted using 1-μL aliquots of DNA without further purification.

3.2. Primers and Loci Analyzed

Primer sequences used for amplification of 9 microsatellites are found in **Table 1**. Two microsatellites (D9S126 and D9S259) are located on the short arm of chromosome 9 spanning the 9p21 region, five microsatellites (D9S275, D9S195, D9S258, D9S103, and GSN) are located on the long arm of chromosome 9 spanning the 9q 32-33 region and two microsatellites (D9S1199 and D9S1198) are located on the long arm spanning the 9q 34 region. Primers were purchased from MWG-Biotech UK Ltd, one primer from each pair was fluorescently labeled at the 5′ end.

3.3. PCR

The target sequences were amplified by PCR in 10-μL reactions, with each reaction containing the following.

1. Archival DNA template (1 µL).
2. 1× reaction buffer (1 µL of 10× reaction buffer); because the reaction buffer contained 15 m*M* $MgCl_2$, this gave a reaction concentration of 1.5 m*M* $MgCl_2$.
3. 2 µ*M* reverse and forward primer (1 µL) to give a reaction concentration of 0.2 µ*M* for each primer.
4. 0.6 1 µL of 25 m*M* $MgCl_2$ to give a final reaction concentration of 3 m*M* $MgCl_2$.
5. 0.2 1 µL of 10 m*M* dNTP to give an reaction concentration of 200 µL each of dATP, dCTP, dGTP, and dTTP.
6. 0.1 1 µL of 5 units/1 µL hot star *Taq* polymerase to give 0.5 units per reaction.
7. Volume is made up to 10 1 µL using PCR-grade water.

The reaction was started after a 15-min denaturation of DNA at 95°C. DNA amplification was performed in a thermal cycler as follows: 45 cycles of denaturation at 94°C for 30 s; annealing at 47, 50, or 53°C (**Table 1**) for 30 s and extension at 72°C for 1 min; followed by a final extension for 15 min at 72°C.

The PCR is repeated using the same DNA template for each of the primer sets. On removal from the thermal cycler 2 µL of each of the 9 PCR products from each primer set are combined for each template and mixed thoroughly by vortexing.

3.4. PCR Analysis

1. PCR products are then prepared for gel electrophoresis.
2. The following are combined in a 200-µL DNA/RNA free microfuge tube and vortexed: .5 µL of combined PCR products; 1 µL of formamide; 0.5 µL of loading buffer (50 mg/mL blue dextran, 25 m*M* EDTA); and 0.5 µL of commercial standard (Genescan 400 HD (rox), ABI Perkin–Elmer, Norwalk, CT).
3. The mixture is then incubated at 96°C for 4 min and cooled on ice.
4. The mixture (1.5 µL) is applied to a 4% arcrylamide/bisacrylamide gel and electrophoresed for 2 h on an automated laser-activated fluorescent DNA sequencer (Perkin–Elmer ABI 377 sequencer).
5. Fluorescent gel data were collected automatically during electrophoresis and analyzed using Genescan software. An example of the gel image is shown in **Fig. 1**. The data gained from Genescan is then exported into Genotyper and further analyzed.

3.5. Assessment of Allele Loss and Microsatellite Instability

Allele loss should be assessed as described by Dietmaier et al. ***(10)***. In heterozygous individuals, two alleles, that is, two PCR products of different size can be detected in normal DNA. Because PCR fragments of different sizes are amplified with different efficiencies, the ratio of allele peak heights is calculated in matched normal and tumor DNA. Peak heights of the longer length allele peaks are divided by the peak heights of the shorter length allele peaks (*see* **Note 2**). The ratio obtained in tumor DNA divided by the allele peak ratio of paired normal DNA gives a result range of 0.00 to 1.00 (*see* **Note 3**), that is, (tumor allele 1 peak height/tumor allele 2 peak height)/(normal allele 1 peak height/normal allele 2 peak height).

Theoretically a complete allele loss results is a value of 0 and both alleles retained results in a ratio of 1. In cases where the shorter length allele is lost the ratio obtained is greater than 1, this is therefore inverted (1/×) to obtain values within the 0.00 to 1.00 range. A ratio below 0.65 represents an allele signal reduction of 35%, this is considered to be indicative of allele loss (*see* **Note 4**). This limit was chosen because

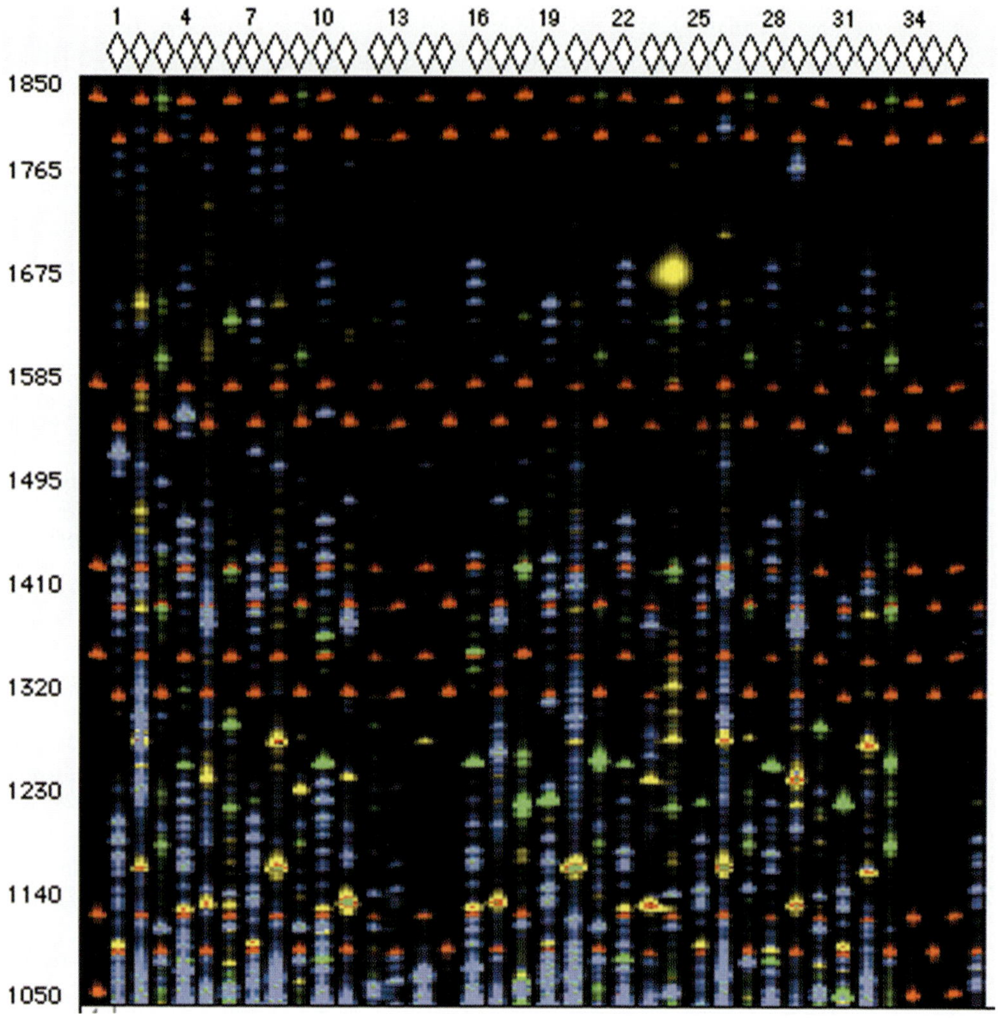

Fig. 1. Example of a gel image.

the tumor cell content after microdissection is assessed to be greater than 70% and inter assay variations of the detection system is below 5%. An example of LOH detection using the Geneotype® Software can be seen in **Fig. 2**.

A locus is described as unstable if a novel peak is present or if a peak has undergone a size shift after PCR amplification of tumor DNA compared with PCR amplification of paired normal DNA. A tumor is described as exhibiting MIS if at least 40% of the loci investigated are found to be unstable.

4. Notes

1. Inactivation of proteinase K is a very important step because failure to inactivate proteinase K will cause PCR to fail, with the proteinase K destroying the *Taq* polymerase.
2. When analyzing Genotyper data, more than one peak may be present for each allele because of polymerase artifact stutter bands. It is therefore difficult to decide the size and the height of the peak to include in the calculation. In general, however, the sizes

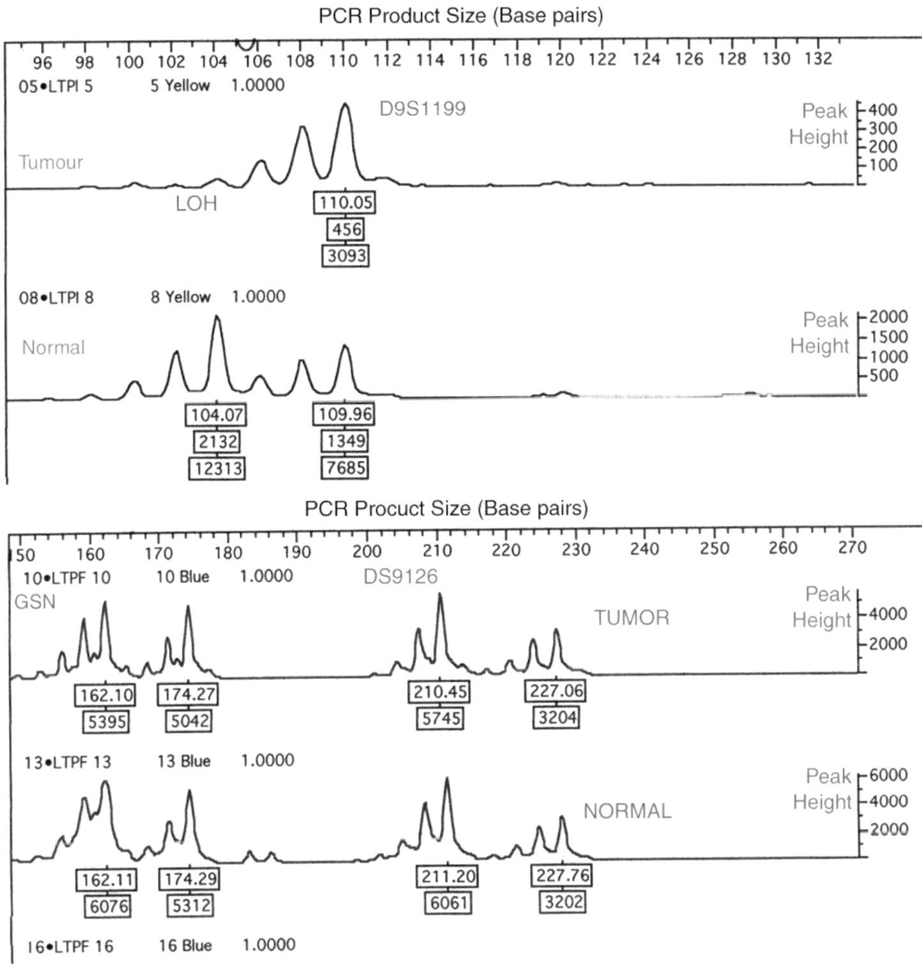

Fig. 2. The top graph shows LOH, as the peak at 104 base pairs is missing. However, the bottom graph demonstrates retention of both alleles.

of the two alleles are assigned to the peaks of greatest height and the smaller peaks are interpreted as stutter bands.

3. Tumor cells are not completely separated from adjacent normal cells by microdissection. Therefore, a major advantage of the fluorescence-based method is that the loss of alleles can be determined precisely by calculation of the ratio of the peak heights of normal and tumor alleles.
4. Assays of samples with allele ratios in the border line range of 0.6 to 0.7 were repeated at least three times to ensure accuracy of results. Most repeated assays gave consistent results, and all informative markers gave similar allele ratios in the same sample.

References

1. King, B. L., Carcangiu, M.-L., Carter, D., Kiechle, M., Pfisterer, J., and Kaoinski, B. M. (1995) Microsatellite instability in ovarian neoplasms. *Br. J. Cancer* **72,** 376–382.
2. Sardi, I., Bartoletti, R., Occhini, I., Piazzini, M., Travaglini, F., Guazzelli, R., et al. (1999) Microsatellite alterations in superficial and locally advanced transitional cell carcinoma of the bladder. *Oncol. Reports* **6,** 901–905.

3. Sartor, O. and Sheng, A. (1997) Determinatin of CAG repeat length in the androgen receptor gene using frozen serum. *Urology* **49,** 301–304.
4. Bronner, C. E., Baker, A. M., Morrison, P. T., Warren, G., Smith, L. G., Lesloe, M. K., et al. (1994) Mutation in the DNA mismatch repair gene homologue hMLH1 is associated with hereditary non-polyposis colon cancer. *Nature* **368,** 258–261.
5. Rohrbach, H., Haas, J. C., Baretton, G. B., Hirschmann, A., Diebold, J., Behrendt, R. P., et al. (1999) Microsatellite instability and loss of heterozygosity in prostatic carcinomas: Comparison of primary tumor and of corresponding recurrences after androgen deprivation therapy and lymph node metastases. *Prostate* **40,** 20–27.
6. Niederacher, D., Picard, F., van Roeyen, C. An, H. X., Bender, H. G., and Beckmann, W. (1997) Patterns of allelic loss on chromosome 17 in sporadic breast carcinomas dectected by fluorescent labelled microsatellite analysis. *Genes Chromosomes Cancer* **18,** 181–192.
7. Christensen, M., Sunde, L., Bolund, L., and Orntorft, T. F. (1999) Comparison of three methods of microsatellite detection. *Scand. J. Clin. Lab. Invest.* **59,** 167–178.
8. Serth, J., Kuczyk, M. A., Paeslack, U., Lichtinghagen, R., and Jonas, U. (2000) Quantitation of DNA extracted after micropreparation of cells from frozen and formalin-fixed tissue sections. *Am. J. Pathol.* **156,** 1189–1196.
9. Going, J. J. and Lamb, R. F. (1996) Practical histological microdissection for PCR analysis. *J. Pathol.* **79,** 121–124.
10. Dietmaier, W., Riedlinger, W., Kohler, A., Wegele, P., Beyser, K., Sagner, G., et al. (1999) Detection of microsatellite instability (MSI) and loss of heterozygosity (LOH) in colorectal tumours by fluorescence based multiplex microsatellite PCR. *Biochemica* **2,** 43–46.

44

Reduction of Shadow Band Synthesis During PCR Amplification of Repetitive Sequences from Modern and Ancient DNA

Wera M. Schmerer

1. Introduction

Repetitive sequences like short tandem repeat (STR) loci are generally referred to as slippery DNA *(1)*. They owe this nickname to a characteristic leading to slippage within the primer-template complex during PCR elongation of the new strand *(2,3)*, resulting in the synthesis of byproducts shortened by one repeat unit compared with the original sequence. The generation of these so-called shadow bands *(4)* is a well-known problem connected with the amplification of repetitive DNA, complicating the genotype analysis of modern (e.g., **ref.** *5*), forensic *(6)*, and ancient *(7,8)* specimens. In some applications, the occurrence of this artifact makes it necessary to develop guidelines for allele designation *(6,9)*.

The intensity of these byproducts increases with the degradation of the target DNA *(8,10)*, and therefore represents a particular problem concerning the analysis of highly degraded DNA as in genotyping of forensic (e.g., **refs.** *11,12*) and ancient DNA *(7,13,14)*. In amplification products of highly degraded or ancient DNA, the intensity of a shadow band can exceed the peak height or band intensity of the original allele. As artifact alleles *(7,15)* they can lead to mistyping of amplification products (cf. also **refs.** *8,16*). Because an amplification product is not necessarily affected in each case, this phenomenon can even result in seemingly different genotypes for independent amplification products of the same sample *(7,15,16)*.

To improve the reproducibility of amplification results and consequently decrease the probability of mistyping by reducing the generation of this artifact, the optimization of the PCR amplification process itself represents the most important strategy besides optimization of the extraction of the DNA used as template *(17–19)*.

The findings presented in the following resulted from a study within the scope of which different strategies to optimize PCR amplification of repetitive DNA were investigated with reference to their effect on the generation of shadow bands (cf. **ref.** *19* Schmerer, manuscript in preparation). The model locus investigated was HUMVWA31/A *(20)*, the amplifications of which show a high tendency to accumulate shadow bands compared with other STR loci *(11,12,21)*. Amplifications were performed on DNA Thermal Cycler (TC1, Perkin–Elmer Cetus). For a detailed presentation of

From: *Methods in Molecular Biology, Vol. 226: PCR Protocols, Second Edition*
Edited by: J. M. S. Bartlett and D. Stirling © Humana Press Inc., Totowa, NJ

the investigation concerned, please refer to Schmerer ***(19)*** and Schmerer (manuscript in preparation).

2. Materials

1. InViTAQ® DNA polymerase (Invitek).
2. NH_4 reaction buffer (Invitek). 10× buffer: 160 m*M* $(NH_4)_2SO_4$, 500 m*M* Tris-HCl (pH 8.3 at 25°C), Tween® 20.
3. Betaine (3 *M* solution with sterile water also used for set up of the reaction mix).
4. Betaine (2 *M*) + 10% dimethyl sulfoxide (DMSO; solution with sterile water).
5. Bovine serum albumin (BSA; 125 µg/mL solution with sterile water).
6. dNTP-mix composed according to the sequence amplified (e.g., for HUMVWA31/A with a A/T:G/C ratio of 1.9:1, a stock of 121 µ*M* each of dGTP, dCTP, and 220 µ*M* each of dATP and dTTP were used).

3. Methods

Each of the following variations of a standard PCR amplification protocol resulted in reduced accumulation of shadow bands. They might be applied either soldiery or in any combination.

3.1. Denaturation Time

A reduction of denaturation time to 15 to 30 s results in a 28% decrease in shadow bands compared with the standard denaturation time of 1 min by decreasing the occurrence of additional degradation of the template DNA (cf. **ref. *22***) because of the shorter incubation period at the high temperature of 94°C.

3.2. Elongation Step

Applying an elongation time of 1 min, which is the standard used for synthesis of a PCR product up to 1 kb, the lowest intensity of shadow bands was found with an elongation temperature of 68°C, resulting in a 20.4% reduction compared with the generally applied temperature of 72°C. This reduction could be even increased to 24.7% by doubling the time for synthesis. The highest reduction of the generation of this artefact (–30.9%) could be achieved with an elongation at 70°C for 2 min.

3.3. Polymerase and Composition of Reaction Buffer

Comparing different polymerases and polymerase mixes respective, the lowest intensity of shadow bands was observed with InViTAQ® (Invitek) and the CombiPoil® polymerase mix (Invitek) in combination with OptiPerform™ Buffer (Invitek) with a reduction of the artefact by up to 24 and 21.4% compared with AmpliTaq™ Gold (Perkin–Elmer) in combination with GeneAmp® buffer (Perkin–Elmer). Applying InViTaq, a replacement of the generally used KCl-reaction buffer, by an optimized NH_4 buffer resulted in a 22% reduction in artefact accumulation applying the same polymerase. Further differences of the two buffer systems consist in a slightly elevated pH value of 8.8 (+0.5) and an addition of 0.01% Tween-20 in case of the NH_4 buffer.

The use of a polymerase displaying a 3′→5′ exonuclease proof-reading ability showed no positive effects concerning the reduction of shadow band accumulation, neither applied alone, nor in combination with a *Taq* polymerase.

3.4. *A/T*: *G/C Ratio of the dNTP-Mix*

Changing the composition of the dNTP-mix from the usual equimolar concentration of nucleotides to a A/T:G/C ratio of 1.9:1 corresponding to the composition of the general sequence of HUMVWA31/A, the locus amplified, resulted in a decrease in shadow band generation by 7.3%. Changes in amplification efficiency were not observed, neither in processing modern nor ancient DNA. Equability of the amplification of both alleles belonging to a heterozygous genotype was slightly improved.

3.5. *Betaine* (N,N,N-*trimethyl glycine)*

The presence of betaine in a concentration of 0.5 to 2 *M* reduced the accumulation of shadow bands, with a maximum reduction of 15.5% at 0.5 *M*. Concentrations lower than 0.5 showed an increase in artifact production. Concerning the amplification of ancient DNA, low concentrations of betaine (0.25–1 *M*) resulted in an increase in product yield because of the neutralizing effect of this reagent against inhibitory substances (cf. **ref.** ***23***) frequently present within ancient DNA extracts. Beyond 1.5 *M*, an addition of the inhibitory effect of betaine at elevated concentrations ***(24)*** and the inhibition caused by co-extracted impurities occurred, resulting in partial inhibition of amplification.

3.6. *Betaine Combined with DMSO*

Like betaine, low concentrations of both betaine and DMSO resulted in a slight increase in intensity of shadow bands. A reduction of the artifact was observed at concentrations of 0.4 to 1 *M* betaine combined with 2 to 5% DMSO, with a maximum decrease of shadow band intensity by 12.6% at 0.8 *M* betaine and 4% DMSO. In ancient DNA amplifications, the presence of these reagents increased product yield and reproducibility between multiple amplifications of the same sample (*see* **Note 1**).

3.7. *BSA*

The addition of BSA in a concentration of 10 to 25 μg/mL resulted in reduced intensity of artifact bands with a maximum at 25 μg/mL by 21.2% accompanied by an increased equability of amplification of a heterozygote genotype. In addition to this, the presence of BSA showed to increase the efficiency of ancient DNA amplifications by neutralizing inhibitory substances ***(25)*** that consequently resulted in higher yield of specific product. A concentration of 25 to 50μg/mL was determined as optimal for the amplification of ancient DNA.

4. Notes

1. Also investigated concerning their impact in the generation of shadow bands were further reagents commonly used as PCR-enhancing additives, such as DMSO, glycerol, and formamide. In amplifications of the locus concerned the addition of DMSO, as well as formamide in different concentrations resulted in an increase in shadow band accumulation. Glycerol, however, did not show any effect, neither on modern, nor ancient DNA amplification.

References

1. Kunkel, T. A. (1993) Nucleotide repeats. Slippery DNA and diseases. *Nature* **365,** 207–208.
2. Levison, G. and Gutman, G. A. (1987) Slipped-strand mispairing: a mayor mechanism for DNA sequence evolution. *Mol. Biol. Evol.* **4,** 203–221.
3. Schlötterer, C. and Tautz, D. (1992) Slippage synthesis of simple sequence DNA. *Nucleic Acids Res.* **20,** 211–215.
4. Weber, J. L. and May, P. E. (1989) Abundant class of human polymorphisms which can be typed using the polymerase chain reaction. *Am. J. Hum. Genet.* **44,** 388–396.
5. Bluteau, O., Legoix, P., Laurent-Puig, P., and Zucman-Rossi, J. (1999) PCR-based genotyping can generate artifacts in LOH analyses. *BioTechniques* **27,** 1100–1102.
6. Gill, P., Sparkes, R., and Kimpton, C. (1997) Development of guidelines to designate alleles using a STR multiplex system. *Forensic Sci. Int.* **89,** 185–197.
7. Schmerer, W. M., Hummel, S., and Herrmann, B. (1997) Reproduzierbarkeit von aDNA-typing. *Anthrop. Anz.* **55,** 199–206.
8. Ivanov, P. L. and Isaenko, M. V. (1999) Identification of human decomposed remains using the STR systems: effect on typing results, in *Proceedings of the second European symposium on human identification 1998*. Promega Corporation, Innsbruck, Austria.
9. Fourney, R. M., Fregau, C. J., Bowen, J. H., Bowen, K. L., Shutler, G. G., Elliot, J. C., et al. (1995) Interpretation guidelines for fluorescent automated detection of STRs: Defining the allele and the limits of detection, in *Proceedings from the Sixth International Symposium on Human Identification 1995*. Promega Corporation, Innsbruck, Austria.
10. Murray, V., Monchawin, C., and England, P. R. (1993) The determination of the sequences present in the shadow bands of a dinucleotide repeat PCR. *Nucleic Acids Res.* **21,** 2395–2398.
11. Kimpton, C., Fisher, D., Watson, S., Adams, M., Urquhard, A., Lygo, J., et al. (1994) Evaluation of an automated DNA profiling system employing multiplex amplification of four tetrameric STR loci. *Int. J. Leg. Med.* **106,** 302–311.
12. Lygo, J. E., Johnson, P. E., Holaway, D. J., Woodroffe, S., Whitaker, J. P., Clayton, T. M., et al. (1994) The validation of short tandem repeat (STR) loci for use in forensic casework. *Int. J. Leg. Med.* **107,** 77–89.
13. Ramos, M. D., Lalueza, C., Girbau, E., Perez-Perez, A., Quevedo, S., Turbon, D., et al. (1995) Amplifying dinucleotide microsatellite loci from bone and tooth samples of up to 5000 years of age: More inconsistency than usefulness. *Hum. Genet.* **96,** 205–212.
14. Schultes, T., Hummel, S., and Herrmann, B. (1997) Recognizing and overcoming inconsistencies in microsatellite typing of ancient DNA samples. *Ancient Biomol.* **1,** 227–233.
15. Schmerer, W. M. (1996) Reproduzierbarkeit von Mikrosatelliten-DNA-Amplifikation und Alleldetermination aus bodengelagertem Skelettmaterial. Unpublished diploma-thesis, Göttingen.
16. Taberlet, P., Griffin, S., Goossens, B., Questiau, S., Manceau, V., Escaravage, N., et al. (1996) Reliable genotyping of samples with very low DNA quantities using PCR. *Nucleic Acids Res.* **24,** 3189–3194.
17. Schmerer, W. M., Hummel, S., and Herrmann, B. (1999) Optimized DNA extraction to improve reproducibility of short tandem repeat genotyping with highly degraded DNA as target. *Electrophoresis* **20,** 1712–1716.
18. Schmerer, W. M., Hummel, S., and Herrmann, B. (2000) STR-genotyping of archaeological human bone: Experimental design to improve reproducibility by optimisation of DNA extraction. *Anthrop. Anz.* **58,** 29–35.
19. Schmerer, W. M. (2000) *Optimierung der STR-Genotypenanalyse an Extrakten alter DNA aus bodengelagertem menschlichen Skelettmaterial.* Cuvillier Verlag, Göttingen.

20. Kimpton, C. P., Walton, A., and Gill, P. (1992) A further tetranucleotide repeat polymorphism in the vWF gene. *Hum. Mol. Genet.* **1,** 287.
21. Bacher, J. and Schumm, J. W. (1998) Development of highly polymorphic pentanucleotide tandem repeat loci with low stutter. *Profiles DNA* **2,** 3–6.
22. Cheng, S., Fockler, C., Barnes, W. M., and Higuchi, R. (1994) Effective amplification of long targets from cloned inserts and human genomic DNA. *Proc. Natl. Acad. Sci. USA* **91,** 5695–5699.
23. Baskaran, N., Kandpal, R. P., Bhargava, A. K., Glynn, M. W., Bale, A., and Weissman, S. M. (1996) Uniform amplification of a mixture of deoxyribonucleic acids with varying GC content. *Genome Res.* **6,** 633–638.
24. Weissensteiner, T. and Lanchbury, J. S. (1996) Strategy for controlling preferentiual amplification and avoiding false negatives in PCR Typing. *Biotechniques* **21,** 1102–1108.
25. Pääbo, S., Gifford, J. A., and Wilson, A. C. (1988) Mitochondrial DNA sequences from a 7000-year old brain. *Nucleic Acids Res.* **20,** 9775–9787.

45

Degenerate Oligonucleotide-Primed PCR

Michaela Aubele and Jan Smida

1. Introduction

The amount of genomic DNA available for genetic studies can often be limiting. Degenerated oligonucleotide-primed polymerase chain reaction (DOP-PCR) is an appropriate method for overcoming these limitations by efficiently performing whole genome amplification. The DOP-PCR technique is increasingly being applied for simultaneous amplification of multiple loci in target DNA using oligonucleotide primers of partially degenerate sequences *(1)*. Contrary to other PCR-based general amplification methods (Alu-PCR or IRS-PCR), DOP-PCR is a species-independent technique for the amplification of small amounts of DNA.

Briefly, the use of one single degenerated primer allows random initial priming all over the target DNA during the first five cycles of PCR at low annealing temperatures. In subsequent cycles, more stringent conditions are then used to amplify the first PCR products. DOP-PCR technique is recently being applied to produce sufficient amounts of DNA from clinical tumor samples, in forensic analyses, prenatal diagnosis, and many other investigations. Slightly different protocols have been used and optimized worldwide. The protocol described here is adapted for use with microdissected, formalin-fixed, paraffin-embedded human tissue in comparative genomic hybridization analysis. A comparison of several DOP-PCR variations is also given by Larsen et al. *(2)*.

2. Materials

2.1. Deparaffination and Sampling of Material

1. Xylene (100%).
2. Ethanol (100%, 70%, 50%).
3. Distilled water.
4. Mayer's Hemalaun solution (Merck 109 249, Darmstadt, Germany).
5. Laser buffer: 100 m*M* Tris-HCl, pH 7.5.
6. Proteinase K: 10 mg/mL proteinase K, pH 7.5 (Merck, Darmstadt, Germany).
7. Qiagen DNA Mini Kit (Qiagen, Hilden, Germany).

From: *Methods in Molecular Biology, Vol. 226: PCR Protocols, Second Edition*
Edited by: J. M. S. Bartlett and D. Stirling © Humana Press Inc., Totowa, NJ

2.2. Chemicals Required for DOP-PCR

1. DOP-primer: 5′- CCG ACT CGA GNN NNN NAT GTG G - 3′, with N = A, C, G, or T in approx equal proportions, (Biometra, Göttingen, Germany).
2. Topoisomerase I (Life Technologies, Eggenstein, Germany).
3. *Taq* DNA Polymerase (Perkin–Elmer Life Sciences, Maryland).
4. DNTPs (10 m*M* each of dATP, dCTP, dGTP, and TTP; Sigma-Aldrich, Steinheim, Germany).
5. 10× amplification buffer: 500 m*M* KCl, 200 m*M* Tris-HCl, pH 8.4.
6. 50 m*M* $MgCl_2$.

3. Method

3.1. Preparation of Tumor Samples

1. Dewax 5- to 10-µm thick paraffin sections for 30 min in a Coplin jar with Xylene.
2. Transfer slides to a coplin jar with fresh xylene for 5 min.
3. Transfer slides to a Coplin jar with 100% EtOH for 5 min.
4. Transfer slides to a coplin jar with 70% EtOH for 5 min.
5. Transfer slides to a coplin jar with 50% EtOH for 5 min.
6. Transfer slides to a coplin jar with distilled water for 5 min. For sampling of small lesions or even single cells by microdissection, *see* **chapter 2.1.3**

3.2. DOP-PCR Procedure (see *Note 1*)

An optional Topoisomerase step may be performed (*see* **Note 2**).

1. Prepare PCR mix (50-µL volume, *see* **Note 3**):

Template	Pretreated cell sample in 20 µL of laser buffer or up to 100 ng isolated DNA
Primer 6-MW (0.1 nmol/µL, *see* **Note 4**)	1.0 µL
10× amplification buffer	5.0 µL
$MgCl_2$	3.4 µL
dNTP (40 m*M*)	1.2 µL
Taq Polymerase (5 U/µL)	0.8 µL
H_2O	Fill up to final volume of 50 µL

2. Set up cycler conditions: Start PCR in a thermal cycler in a 50-µL reaction volume using the following conditions (*see* **Note 5**).

Denaturation	10 min at 94°C
Followed by five cycles:	1 min at 94°C
(Low stringency)	1.3 min at 30°C
	3 min transition 30–72°C
	3 min extension at 72°C
Followed by 35 cycles:	1 min at 94°C
(High stringency)	1 min at 62°C
	3 min at 72°C
Final extension	10 min at 72°C

3. Store PCR product at –20°C or continue to next step.
4. Use a 5-µL aliquot of the PCR product to check the DNA fragment size on a 3% agarose ethidium bromide gel (*see* also **Fig. 1**),
5. Determine DNA amounts by fluorometric or photometric measurements.

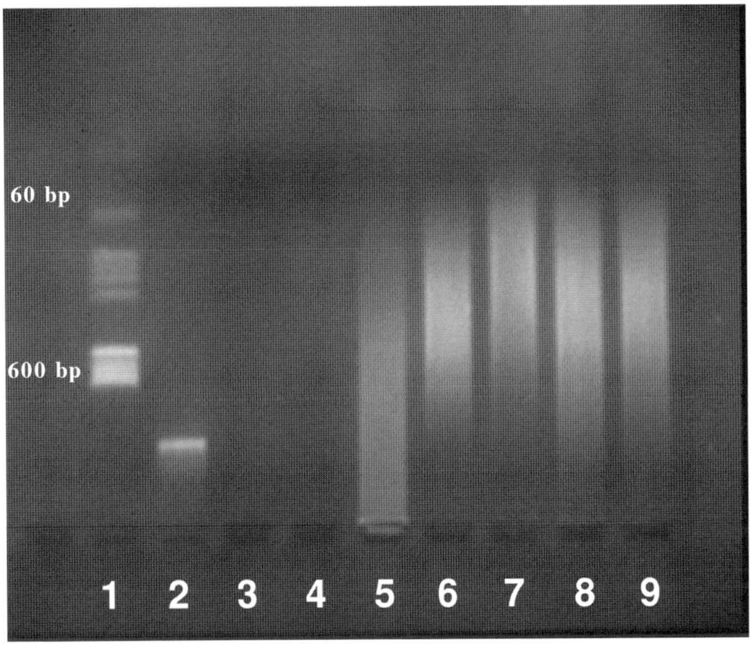

Fig. 1. DOP-PCR products reveal a typical smear ranging from about 100 to 2500 bp. Lane 1: marker pBR 322, *Hae*III; lane 2: positive control (PCR product with template DNA and a gene-specific primer (β-Actin); lane 3: negative control (PCR product without template and with degenerate primer); lane 4: empty; lanes 5 to 9: DNA samples from formalin-fixed, paraffin-embedded tissue that were amplified by DOP-PCR.

4. Notes

1. As in all preparations that require handling of small cell and/or small DNA amounts, the PCR mix should be set up in a laminar flow to avoid any contamination. Furthermore, because of the species-independent primer, a strict contamination-free working is prerequisite.
2. Tumor samples may be pretreated for 30 min with 2 units Topoisomerase I at 37°C to relax the template DNAs. Stop activation of topoisomerase at 90°C for 10 min and chill on ice, then add *Taq* DNA Polymerase and start PCR. This additional step is recommended for isolated fresh material but is not necessary for degraded DNA, as is the case for formalin-fixed, paraffin-embedded tissue.
3. All PCRs should be controlled for possible contamination by using at least one negative control (vial without template DNA), and one positive control (vial with template DNA and a gene specific primer, e.g., β-actin; **Fig. 1**).
4. The most commonly used DOP-PCR primer is 6-MW, originally described by Telenius *(1)*, a degenerate primer with 6 specific 3′ bases and a *Xho*I site at its 5′ end (the same primer has been described as UW4B, UN-1, or MTE1 by different authors).
5. More than 40 cycles in DOP-PCR should be avoided. Based on the authors' experiences, artificial results could arise. For this reason, labeling of DNA should be performed, for example, by Nick translation instead of a further DOP-PCR.

References

1. Telenius, H., Carter, N. C., Bebb, C. E., Nordenskjöld, M., Ponder, B. A. J, and Tunnacliffe, A. (1992) Degenerate oligonucleotide-primed PCR: General amplification of target DNA by a single degenerate primer. *Genomics* **13,** 718–725.
2. Larsen, J., Ottesen, A. M., Lundsteen, C., Leffers, H., and Larsen, J. K. (2001) Optimization of DOP-PCR amplification of DNA for high resolution comparative genomic hybridization analysis. *Cytometry* **44,** 317–325.

46

Mutation Detection Using RT-PCR-RFLP

Hitoshi Nakashima, Mitsuteru Akahoshi, and Yosuke Tanaka

1. Introduction

Genetic analysis by restriction fragment length polymorphism (RFLP) is one of the most common methods used to examine nucleic acids for the presence of known sequence variants. A segment that is to be searched for a mutation is amplified from genomic DNA or cDNA, digested by the appropriate restriction enzyme, and then separated by agarose gel electrophoresis. Although RFLP analysis is a highly sensitive method that is easy to apply for the screening of known sequence variants, many common polymorphisms are the result of single-base substitutions that fail to create or remove any restriction site and, therefore, these cannot be readily typed by simple polymerase chain reaction (PCR) and RFLP analysis. However, the use of a mismatch PCR primer to artificially create a restriction site in the amplified product make it possible to overcome this disadvantage ***(1,2)***. The mismatch primer contains a one- or two-base mismatch near its 3′ end such that the amplified product incorporates or removes a restriction site for the appropriate endonuclease in the presence of a base substitution ***(3–5)***. The protocol for this method is the same as standard PCR.

2. Materials

1. RNA extracted from cells or tissues of interest.
2. Primers: stock solutions are at 100 µ*M* in H_2O; working solutions are at 10 µ*M* in H_2O. The following primer pairs yield a 164-bp amplicon. The sites of mismatch are in capital letters ***(3)***. Forward mismatch primer: 5′-ctc cta ccc ctt gtc atg cag gAt-3′; reverse primer: 5′-gtt aaa aca ggg acc tgt ggc atg-3′.
3. 10× PCR buffer: 500 m*M* KCl, 100 m*M* Tris-HCl, pH 8.3.
4. $MgCl_2$ (25 m*M*) in distilled water.
5. dNTPs (10 m*M* each of dATP, dGTP, dCTP, and TTP).
6. RNase inhibitor (20 U/µL).
7. Random hexamers (50 µ*M*).
8. MuLV reverse transcriptase (50 U/µL).
9. AmpliTaq DNA polymerase (5 U/µL; Roche Molecular Systems, Inc. Branchburg, NJ).
10. Restriction enzyme *Fok*I (10 U/µL) and digestion buffer.
11. DEPC-treated sterile H_2O.
12. Sterile mineral oil.

From: *Methods in Molecular Biology, Vol. 226: PCR Protocols, Second Edition*
Edited by: J. M. S. Bartlett and D. Stirling © Humana Press Inc., Totowa, NJ

13. 80% ethanol.
14. PEG-NaCl solution (20% PEG6000, 2.5 *M* NaCl).
15. Microcentrifuge tube (0.5 mL).
16. Microcentrifuge tube (1.5 mL).
17. Equipment and reagents for 4% agarose gel electrophoresis.
18. TBE: 89 m*M* Tris, 89 m*M* boric acid, and 2 m*M* EDTA (pH 8.3).
19. Ethidium bromide (10 μg/mL).
20. Loading buffer: 0.4% bromophenol blue, 0.4% xylene cyanol FF, 50% glycerol in H_2O.

3. Method

To provide a DNA target for the PCR, cDNA is synthesized using reverse transcriptase.

1. Place 1 μg of RNA in a 0.5-mL microcentrifuge tube, and add 4 μL of 25 m*M* $MgCl_2$, 2 μL of 10× PCR buffer, 8 μL of 10 m*M* dNTP, 1 μL of RNase inhibitor, 1 μL of random hexamers, 1 μL of MuLV reverse transcriptase, and DEPC-treated sterile H_2O to form final total volume of 20 μL.
2. Mix and stand for 10 min at room temperature. Add two drops of mineral oil. Briefly spin down the reaction mixture.
3. Incubate the reaction mixture for 15 min at 42°C, then for 5 min at 99°C, and finally for 5 min at 4°C.
4. Add 4 μL of 25 m*M* $MgCl_2$, 8 μL of 10× PCR buffer, 65.5 μL of sterile H_2O, 0.5 μL of AmpliTaq DNA polymerase, and 1 μL of forward primer (10 μ*M*) and reverse primer (10 μ*M*) to the RT product and mix. Briefly spin down the reaction mix.
5. Perform the PCR by subjecting the reaction mixture to an initial denaturation of 3 min at 94°C, 35 cycles of 1 min at 94°C and 1 min at 60°C, followed by a final extension of 7 min at 72°C.
6. Discard the mineral oil and transfer the PCR product to a 1.5-mL microcentrifuge tube.
7. Add 60 μL of PEG-NaCl solution to the PCR product. Mix and stand for 10 min at 37°C.
8. Centrifuge at 12,000*g* for 10 min at 4°C.
9. The DNA precipitate should be visible as a pellet at the bottom of the tube. Remove the supernatant and wash the pellet with 1 mL of 80% ethanol.
10. Centrifuge at 7500*g* for 5 min at 4°C.
11. Carefully remove all the supernatant and air dry the pellet.
12. Dissolve the pellet in 100 μL of H_2O.

3.1. Restriction Enzyme Digestion

1. Mix 5 μL of PCR product, 1 μL of 10× reaction buffer, 1 μL of the appropriate restriction enzyme (1–10 U), and 3 μL of H_2O in a 1.5-mL microcentrifuge tube.
2. Incubate the reaction mixture for more than 2 h at the optimum temperature for enzyme activity.

3.2. Analysis of RFLP

1. Prepare an appropriate concentration of agarose gel containing 0.2 μg/mL of ethidium bromide for electrophoresis with TBE buffer.
2. Load the gel with 10 μL of digested PCR product mixed with 1 μL of 10× loading buffer, 5 μL of undigested PCR product mixed with 4 μL of H_2O, and 1 μL of 10× loading buffer and a DNA size standard.
3. Perform electrophoresis and inspect the gel under an ultraviolet transilluminator (**Fig. 1**).

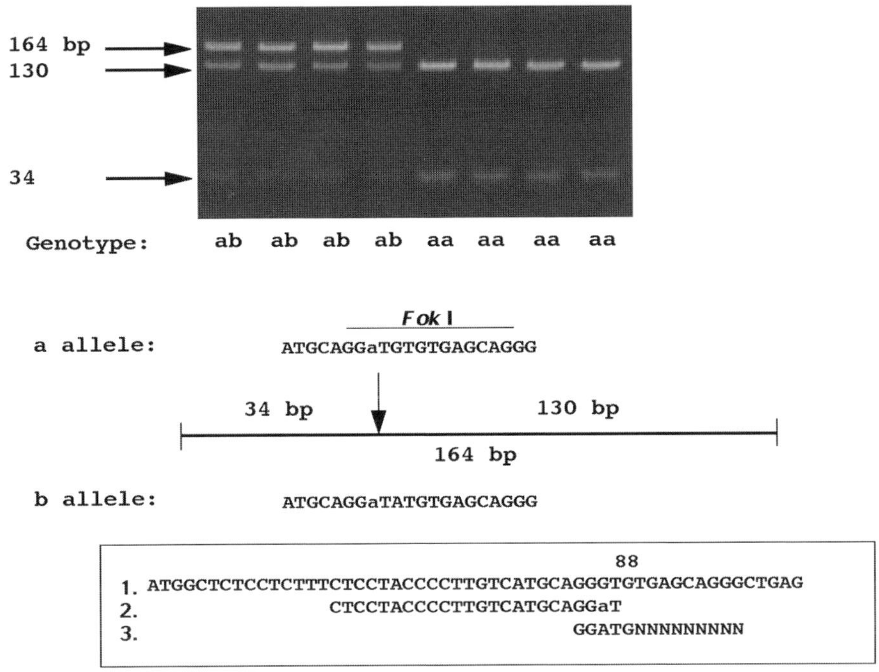

Fig. 1. G88A within *IFNGR1* is detected by *Fok*I digestion of the PCR product using a 1-bp mismatch primer. The upstream primer contains a 1-bp mismatch just proximal to its 3′ end such that the 164-bp amplified product incorporates a restriction site for *Fok*I in the presence of guanine at nucleotide 88 (a allele) but not in the presence of adenine (b allele). A PCR product of genotype ab is cleaved by endonuclease *Fok*I and shows three bands, 164, 130, and 34 bp, on the ethidium bromide-stained 4% agarose gel.

Lowercase nucleotides represent the mismatch base. Underlined nucleotides represent the polymorphism site. Line 1: Part of the registered *IFNGR1* cDNA sequence; line 2: the upstream primer for mismatch PCR; line 3: restriction enzyme *Fok* I recognition site.

4. Notes

We were searching for amino acid polymorphisms in cytokine receptors. At first we used the RT-PCR SSCP method for cDNA sequence. After confirmation of the base substitutions of SSCP-positive samples, we designed primers for population screening by RFLP analysis. Although RT-PCR RFLP is an easy and sensitive method for examining whether an already-known base substitution is present within the sample cDNA, there are some disadvantages involved. Concerning RT-PCR, the various kinds of amplicons resulting from alternative splicing could prove to be an obstacle for analysis. If the genomic DNA sequence is available, then analysis using PCR products templated with genomic DNA is recommended. Changing the primer setting position would be another solution. Concerning RFLP, this analysis is based on the fact that each restriction enzyme recognizes a specific DNA sequence, and incomplete digestion leads to a false-negative result. The causes of incomplete digestion are low enzyme activity (in cases with long-term usage of the restriction enzyme), an extremely small quantity of the enzyme compared with that of the DNA sample, reaction under the wrong conditions, and the use of the wrong reaction buffer. Performance of the same reactions using positive and negative control templates in parallel with sample examinations is recommended in order to check for incomplete digestion.

References

1. Russ, A. P., Maerz, W., Ruzicka, V., Stein, U., and Gross, W. (1993) Rapid detection of the hypertension-associated Met235-Thr allele of the human angiotensinogen gene. *Hum. Mol. Genet.* **2,** 609–610.
2. Hingorani, A. R. and Brown, J. M. (1995) A simple molecular assay for the C1166 variant of the angiotensin II type 1 receptor gene. *Biochem. Biophys. Res. Commun.* **213,** 725–729.
3. Tanaka, Y., Nakashima, H., Hisano, C., Kohsaka, T., Nemoto, Y., Niiro, H., et al. (1999) Association of the interferon-g receptor variant (Val14Met) with systemic lupus erythematosus. *Immunogenetics* **49,** 266–271.
4. Shibata, S., Asano, Y., Yokoyama, T., Shimoda, K., Nakashima, H., Okamura, S., et al. (1998) Analysis of the granulocyte colony-stimulating factor receptor gene structure using PCR-SSCP in myeloid leukemia and myelodysplastic syndrome. *Eur. J. Haematol.* **60,** 197–201.
5. Iwata, I., Nagafuchi, S., Nakashima, H., Kondo, S., Koga, T., Yokogawa, Y., et al. (1999) Association of polymorphism in the NeuroD/Beta2 gene with type 1 diabetes in the Japanese. *Diabetes* **48,** 416–419.

47

Multiplex Amplification Refractory Mutation System for the Detection of Prothrombotic Polymorphisms

David Stirling

1. Introduction

First described by Newton and colleagues in 1989 *(1)*, amplification refractory mutations system (ARMS) has become a standard technique that allows the discrimination of alleles that differ by as little as 1 bp. The system is simple, reliable, and nonisotopic. It clearly distinguishes heterozygotes at a locus from homozygotes for either allele. The system requires neither restriction enzyme digestion, nor allele-specific oligonucleotides as conventionally applied, nor the sequence analysis of polymerase chain reaction (PCR) products. The basis of the system is that oligonucleotides with a mismatched 3′-residue will not function as primers in the PCR under appropriate conditions.

A standard ARMS PCR consists of two complementary reactions (two tubes) and uses 3 primers. One primer is constant and complementary to the template in both reactions, and the other primers differ at their 3′ terminal residues and are specific to either the wild-type DNA sequence or the mutated sequence at a given base—only one of these primers is used per tube. If the sample is homozygous mutant or homozygous wild-type amplification will only occur in one of the tubes, if the sample is heterozygous amplification will be seen in both tubes.

Here, we report our protocol for the multiplex ARMS detection of the polymorphisms in clotting factor V and II associated with increased risk of thrombosis *(2,3)*. A portion of the factor IX gene is amplified as an internal positive control.

2. Materials

1. Thermal cycler.
2. Plate mixer.
3. Vertical PAGE system.
4. Electrophoresis Power Pack.
5. Thermowell 96-well plate and covers.
6. Round tips (200 µL).
7. Flat-cap PCR tubes (0.5 mL).
8. Thermostable DNA *Taq* polymerase.

From: *Methods in Molecular Biology, Vol. 226: PCR Protocols, Second Edition*
Edited by: J. M. S. Bartlett and D. Stirling © Humana Press Inc., Totowa, NJ

9. TAE (20×): 484 g of Tris, 114 mL of glacial acetic acid, 20 mL of 0.5 *M* EDTA. Dissolve in 3 L of distilled water then make up to 5 L with distilled water.
10. TAE (1×): 1:20 dilution of 20× TAE in distilled water.
11. 8% polyacrylamide gel.
12. dNTPs: supplied separately as four different types at concentrations of 100 m*M*. For use at 5 m*M*, take 100 µL of each stock dNTP and add to one tube 7600 µL of sterile distilled water. Aliquot and store at –20°C.

Oligo F2 Wild	5′-CAC TGG GAG CAT TGA GGA TC-3′
Oligo F2 Mutant	5′-CAC TGG GAG CAT TGA GGA TT-3′
Oligo F2 Consensus	5′-TCT AGA AAC AGT TGC CTG GC-3′
Oligo F5 Wild	5′-CAG ATC CCT GGA CAG ACG-3′
Oligo F5 Mutant	5′-CAG ATC CCT GGA CAG ACA-3′
Oligo F5 Consensus	5′-TGT TAT CAC ACT GGT GCT TAA-3′
Oligo F9 Forward	5′-CTC CTG CAG CAT TGA GGG AGA TGG ACA TT-3′
Oligo F9 Reverse	5′-CTC GAA TTC GGC AAG CAT ACT CAA TGT AT-3′

13. Oligonucleotides. The oligonucleotides must be diluted to 100 pmol/µL on arrival. Aliquot in 20-µL volumes and store at –20°C.

3. Methods

1. Using sterile pipet tips, take 1.0 µL of each sample into identified wells in plate, one labeled 'W' (wild type), the other 'M' (mutant). Ensure that the DNA sample is pipetted directly into the bottom of the respective well.
2. Prepare PCR mastermixes W and M with the following: DNTP (n × 1.5 µL); 10× polymerase reaction buffer (n × 2.5 µL); oligonucleotides (each (n × 0.5 µL); $MgCl_2$ (25 m*M*; (n × 2 µL); *Taq* polymerase (n × 0.2 µL); Sterile distilled water (n × 15.4 µL); where n = number of samples/controls + 2.
3. Mix mastermixes W and M and add 24 µL to each corresponding well. Cover and seal tightly with adhesive film and mix well but carefully by agitation on a plate shaker.
4. Place plate into thermal cycler and run the following program: 94°C for 5 min (94°C for 15 s, 55°C for 15 s, 72°C for 30 s × 35 cycles), and 72°C for 10 min.
5. When cycles are complete, add 4 µL of loading buffer to each sample well and agitate to mix.
6. Load samples (approx 20 µL) into the wells of an 8% polyacrylamide gel in the vertical electrophoresis unit and electrophorese at 250 volts for approx 1.25 to 1.5 h along with a strategically placed molecular weight marker.
7. Remove gel from tank and stain for 2 to 5 min in discarded buffer from upper tank containing 20 µL of ethidium bromide solution.
8. Interpretation.

The amplification generates three fragments: Prothrombin (F.II), 340 bp; Factor IX (control), 250 bp; and Factor V, 174 bp. Each set of two tubes, wild type and mutant, will show bands in the pattern below depending upon the allele detected.

	Wild type	Mutant
Homozygous negative	–	
Heterozygous	–	–
Homozygous positive		–

The Factor IX control fragments must be present in both the wild and mutant tubes to interpret the P20210A and V Leiden status. Specimens should be reanalyzed if be repeated if (1) the Factor IX control bands are not present; (2) other control specimens are not as expected; (3) the result is homozygous/heterozygous positive; (4) any dubiety exists with the overall interpretation.

4. Notes

1. Extreme care should be exercised to prevent the possibility of cross-contamination of samples with amplified DNA.
2. PCR setup should be performed in a physically separated area from product analysis.
3. Samples of known genotype (heterozygous and homozygous negative) should be analyzed along with unknowns.

References

1. Newton, C. R., Graham, A., Heptinstall, L. E., Powell, S. J., Summers, C., Kalsheker, N., et al. (1989) Analysis of any point mutation in DNA. The amplification refractory mutation system (ARMS). *Nucleic Acids Res.* **17,** 2503–2516.

48

PCR-SSCP Analysis of Polymorphism

A Simple and Sensitive Method for Detecting Differences Between Short Segments of DNA

Mei Han and Mary Ann Robinson

1. Introduction

Polymerase chain reaction single-strand conformation polymorphism (PCR-SSCP) *(1)* is a simple method that allows one to rapidly determine whether there are sequence differences between relatively short stretches of DNA. Coupled with sequence analysis, SSCP is an extremely useful method for both identifying and characterizing genetic polymorphisms and mutations. The theory of SSCP is that the primary sequence and the length of a single stranded DNA fragment determine its conformation when it is resolved in a nondenaturing polyacrylamide gel. Even single-base differences can cause different secondary conformations and thus result in different migration rates of the DNA strands. Radioactive nucleotides are incorporated into the DNA strands by PCR, making it possible to detect the DNA by autoradiography. SSCP has been widely used to identify mutations in host genes such as *p53* *(2–5)* and in viruses, such as simian immuno-deficiency virus (SIV), during the course of infection *(6)*. SSCP has been used to identify and characterize polymorphisms in a variety of genes (7–9) and was effective in characterizing alleles of linked genes present in individual sperm *(10)*.

Orita and colleagues *(11)* developed the SSCP method in 1989 and since then, it has been applied to screen for sequence differences in either genomic or complementary DNA (cDNA) samples. **Figure 1** schematically shows the steps of the procedure. First, the region of interest (the gene or cDNA) can be PCR amplified by using primers corresponding to the desired sequence. Because migration differences are better resolved using shorter DNA fragments, success is more likely with primers selected to amplify fragments in 100- to 300-bp range. The PCR mixture contains ^{32}P-dCTP to radioactively label the amplification product, which is important for visualizing the migration differences in later steps of the procedure. The second step is to heat diluted amplified samples, which will denature the double-stranded DNA into single-stranded DNA. The samples are mixed with loading buffer containing formamide to hinder reannealing of the DNA and dye to visually follow the migration of samples through the gel. The third step is to resolve the single-stranded DNA samples by nondenaturing polyacrylamide gel electrophoresis. The length of time necessary for running the

From: *Methods in Molecular Biology, Vol. 226: PCR Protocols, Second Edition*
Edited by: J. M. S. Bartlett and D. Stirling © Humana Press Inc., Totowa, NJ

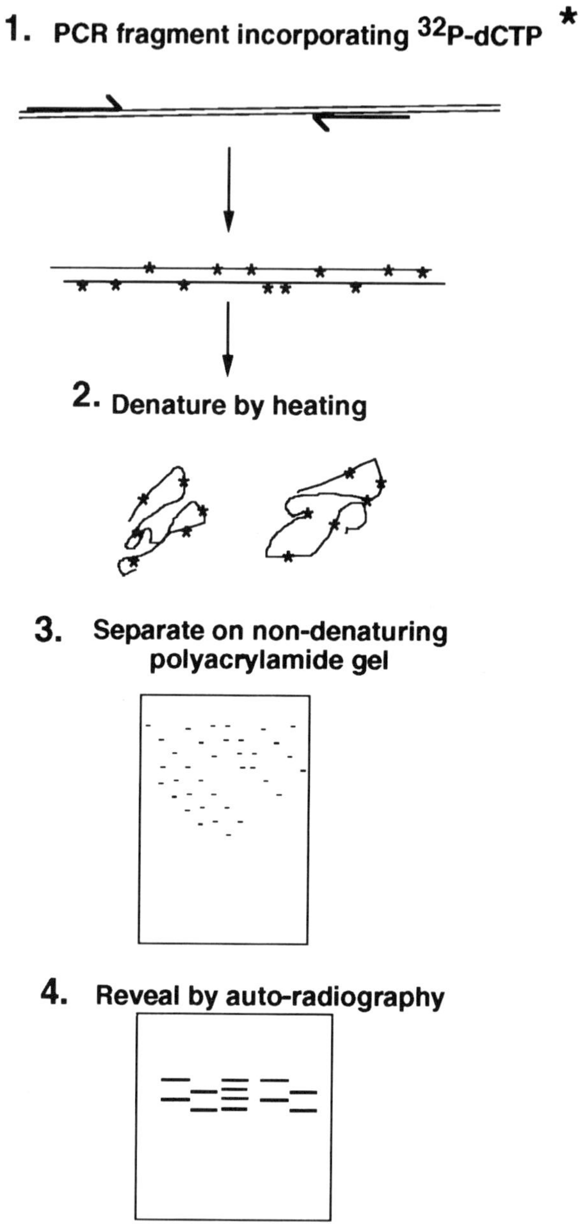

Fig. 1. Basic steps in SSCP Procedure. 1. The DNA or cDNA fragment of interest is amplified by PCR incorporating ^{32}P-dCTP. 2. The PCR amplified-labeled DNA fragment is denatured by heating. 3. Fragments are separated on an acrylamide gel. 4. Patterns are revealed by autoradiography.

gel is variable, dependent upon characteristics of the sequence amplified, and can be determined empirically. The final step is autoradiography to visualize the bands of radiolabeled DNA in the gel.

Analysis of SSCP autoradiograms is performed visually. Each sequence amplified in step one will often result in two bands on a corresponding autoradiogram. One band

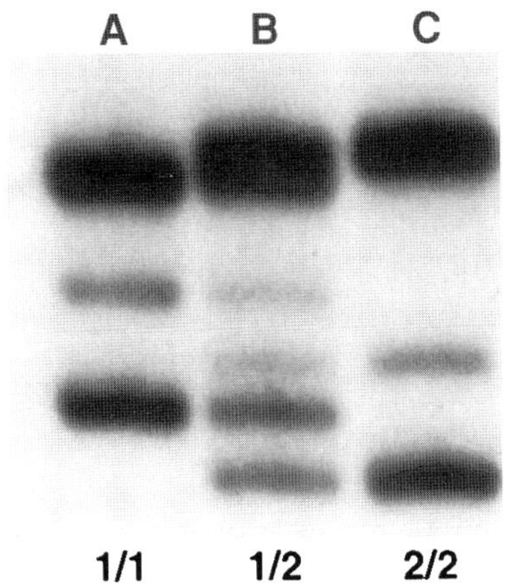

Fig. 2. Sample SSCP pattern. The human *CD1a* gene has two alleles. The DNA sample in band A derived from an individual homozygous for allele 1, in lane B from a heterozygous individual, and in lane C from an individual homozygous for allele 2. In this example, one of the DNA strands assumes two different conformations as it migrates through the gel and results in two bands. Notice the intensity of the top band compared with the lower bands showing that the top band corresponds to one strand in one conformation and the lower bands to two conformations for one strand. The pattern of the heterozygote is a composite of the homozygous patterns. The single-stranded DNA patterns are shown in this figure. Double-stranded DNA corresponding to nondenatured PCR product would be found migrating more rapidly on the gel.

corresponds to the upper strand of DNA and the other the lower strand of the amplified fragment. It is also possible for a segment of single stranded DNA to assume two different conformations as it migrates through the gel, and this will result in two bands for that strand of DNA on a SSCP autoradiogram. Examples of such SSCP patterns are shown in **Fig. 2**. There are two alleles of the human *CD1a* gene ***(9)***. DNA samples in lane A derives from an individual homozygous for allele 1, in lane C, the DNA is from an individual homozygous for allele 2, and in lane B the DNA is from an individual heterozygous for alleles 1 and 2. The bands corresponding to alleles 1 and 2 migrate at different rates, and each has one strand that assumes two different conformations. The pattern for a heterozygous individual is the composite of the two individual allele's patterns.

It was possible to make determinations about the assignment of CD1a alleles because the analysis was coupled with sequence analysis. To obtain material for sequencing, the same reaction mixture as used for the SSCP procedure is amplified using an amplification buffer that does not contain radioactivity for cloning. Many systems are commercially available for cloning and sequence analysis. In addition, SSCP bands can be excised from the gel and re-PCR amplified using the same set of primers, which provides a very convenient way to directly sequence the altered gene segment.

A successful approach to screen for genetic polymorphisms has been to examine DNA samples from 10 unrelated individuals of diverse ethnic backgrounds. Genetic polymorphism is likely to be found in such a sampling; however, this does depend upon allele frequencies and distributions. Although SSCP is quite efficient in identifying sequence differences between two segments of DNA, there are cases where differences are difficult to detect. Changing primers may make a difference. It is possible that the specific fragment will not show migration differences even if there are substitutions present. Analysis of overlapping fragments of the region of interest is one solution to circumvent this difficulty.

The techniques required for the SSCP method are not complicated, and the whole procedure is time saving (only one-round PCR is needed) compared with other methods for detecting mutations and DNA polymorphisms. Because of its simplicity and reliability, PCR-SSCP has been used extensively for identifying alleles and genotypes in both basic and clinical investigations.

2. Materials

1. Thermal cycling (PCR) machine.
2. Ampli*Taq*® DNA polymerase, 5 U/μL (stored at –20°C).
3. 10× PCR buffer stored at –20°C): 100 m*M* Tris-HCl, pH 8.3, 500 m*M* KCl, 15 m*M* $MgCl_2$.
4. dNTP (1.25 m*M*) stored at –20°C.
5. Primers diluted to a concentration of 20 μ*M* (*see* **Note 1**).
6. [α^{32}P]-dCTP (New England Nuclear, Boston, MA, BLU013H, 3000 Ci/mmol).
7. DNA templates (100 ng/μL genomic or 1 : 10 diluted cDNA).
8. 0.1% SDS and 10 m*M* EDTA, pH 8.0.
9. Loading buffer: 10 m*M* NaOH, 95% formamide, 0.05% bromphenol blue, 0.05% xylene cyanol.
10. Long Ranger™ 50% stock gel solution (FMC BioProducts, Rockland, ME; *see* **Note 2**).
11. 10× TBE buffer: 0.89 *M* Tris-borate; 0.02 *M* EDTA, pH 8.0.
12. TEMED.
13. 10% Ammonium Persulphate (APS, after dissolving in water, store at –20°C).
14. Glycerol (Ultra Pure, Life Technologies, Gaithersburg, MD).
15. Sigmacote (Sigma Chemical Co., St. Louis, MO).
16. Acrylamide gel electrophoresis supplies and equipment (*see* **Note 3**).
17. Whatman 3MM chromatography paper.
18. Plastic wrap (such as Saran Wrap).
19. Vacuum gel dryer.
20. Gel cassette with an enhancing screen.
21. X-ray films.
22. Lab safety area for working with radioactive materials (*see* **Note 4**).

3. Methods

3.1. PCR Amplification (see Note 5)

The PCR should contain the following ingredients (in μL): 10× PCR Buffer (2.0); 1.25 m*M* dNTP (3.2); 20 μ*M* 5′ primer (1.0); 20 μ*M* 3′ primer (1.0); 5 U/μL Ampli*Taq*® (0.1); Experimental template (2.0); [^{32}P]-dCTP (0.1); and H_2O (10.6) for a total of 20 μL.

PCR amplification is performed for 30 cycles (denaturation at 95°C for 30 s, annealing at 55°C (*see* **Note 6**) for 30 s, extension at 72°C for 90 s) with a final extension at 72°C for 7 min. PCR amplified DNA may be kept at 4°C or stored at –20°C.

3.2. SSCP Gel Preparation

1. Prepare a 6% nondenaturing polyacrylamide gel with 10% glycerol as following: Long Ranger gel solution (7.5 mL), glycerol (7.5 mL), 10× TBE (4.5 mL), H_2O (55.7 mL); 10% APS (0.4 mL), and TEMED (0.04 mL) for a total of 73.64 mL. Add TEMED and APS right before pouring the gel.
2. Place a few drops of Sigmacote on the inner side of the shorter plate (*see* **Note 3**) and spread evenly with a paper towel. Allow to air dry. Sigmocote forms a tight microscopically thin film of silicone on glass, which prevents the SSCP gel from sticking to this particular plate when you separate the two plates after the gel run is done, allowing the gel to remain smooth on the other plate. Set up the plates with 0.4-mm spacers on the sides and with a strip of Whatman chromatography paper at the bottom. Clamp three sides of the plates leaving the top open.
3. Immediately before pouring the gel, add TEMED and APS to the mixture, collect it into a 60-mL syringe, and push the solution into the plates slowly while holding the plates at an angle (*see* **Note 7**). Place either a 64-well or 36-well comb (*see* **Note 8**) into the top of the gel and clamp the top of the plates. The gel will be polymerized in 2 h.

3.3. Dilution and Denaturation of PCR Products (see Note 9)

1. Dilute 5 µL of amplified DNA in 45 µL of 0.1% SDS and 10 m*M* EDTA.
2. Mix 5 µL of above diluted PCR products with 5 µL of loading buffer. The sample is ready to load and may be stored at –20°C for 2 wk.

3.4. Loading Samples and Running Gels

1. Before running the gel, denature the diluted PCR samples by heating at 95°C for 2 min and cool rapidly in an ice bath.
2. Load 2 µL (for 64-well combs) or 5 µL (for 32-well combs) of the diluted PCR products onto gel. The leftover samples can be reused if they have been stored at –20°C for less than 2 wk.
3. Electrophoresis is performed by using 0.6× TBE buffer at a constant 30 watts (W) for 4 h at room temperature (*see* **Note 10**).

3.5. Drying Gels and Developing Films

1. After migration, the gel is transferred onto a double layer of Whatman Chromatography paper and covered with plastic wrap avoiding air bubbles.
2. The gel is vacuum dried in a gel dryer at 80°C for 1 h, and then the gel is placed facing up in a film cassette.
3. Expose gel to an X-ray film (the X-ray film should be put in between the enhancing screen and the gel) at –80°C for 16 to 20 h (*see* **Note 11**).
4. Warm up film or air dry film before developing. Cut a corner of the film before developing, which will help to figure out the sample orders to load. Develop film using an automatic film developer or using the dip method.

3.6. SSCP Gel Analysis

SSCP patterns are analyzed upon visual examination. Analysis is facilitated with knowledge of the gene sequence, which is often available because sequence information was necessary to derive primers. Inclusion of control samples of known sequence allow allele typing (*see* **Note 12** and **Subheading 1.**).

4. Notes

1. PCR-SSCP is a very sensitive method for detecting point mutations in a DNA fragment shorter than 300 bp. A single base change may effect the overall conformation more easily on small fragments than large fragments. Therefore, primers used in your experiment are better designed for amplifying a 100- to 300-bp DNA fragment. If long genomic or cDNA segments are to be screened, it is important to find comparable electrophoresis conditions in which mutations in long DNA fragments are efficiently detected (*see* **Note 10**). If the primers used do not allow detection of known substitutions, moving the primer sequence further 5′ or 3′ may make it possible to distinguish the fragments. The sequences of the oligo primers used to amplify the human CD1a gene, as shown in **Fig. 2**, are as follows: CD1a.ex2F1 AGACGGGCTCAAGGAGCCTC and CD1a.ex2R1 TCCAGTTCCTTC CACTCCTC.
2. Long Ranger stock gel solution contains acrylamide, a neurotoxin, which may cause cancer and/or heritable genetic damage. It is advisable to wear gloves while handling.
3. Acrylamide gel electrophoresis supplies and equipment format used in PCR-SSCP method may vary between laboratories. In our laboratory, we use a set of glass plates (44.5 × 37.5 cm and 42 × 37.5 cm) and corresponding electrophoresis supplies. Mini-gel format is also frequently used by some other investigators and which is often followed by silver-staining DNA bands rather than labeling with ^{32}P.
4. All procedures involved in dealing with [^{32}P]-dCTP should be performed in a lab safety area and handled by behind a protection screen. Radioactive waste should only be thrown into specialized trash cans.
5. The final concentration of 1 μ*M* primer in our PCR amplification provides clear migration bands, and we have never failed to detect known mutations using this primer concentration. However, it has been reported that lowering the upstream primer concentration improves DNA migration. Because the SSCP technique is not 100% effective in detecting any given substitution, it might be worthwhile to try lowering upstream primer concentrations if known base changes are not detected in your SSCP gel. The template can either be genomic or cDNA depending upon your interest. But do not forget to include DNA samples with known sequence in your PCR amplification, which will serve as a control SSCP pattern for reading allelic distributions of your test samples.
6. Annealing temperature may vary depending upon the primers used. Usually the annealing temperature is determined empirically with estimates made by the following formula: $[(G + C) \times 4]$ °C + $[(A + T) \times 2]$ °C = annealing temperature °C.
7. Some bubbles may appear between the plates if the plates are not very clean. You can get the bubbles out by further angling the plates with one hand and tapping on the plates with the other hand until the bubbles are out.
8. It is important to ensure that the comb is very clean when casting the gel. Leftover acrylamide present between the teeth of the comb may cause shorter wells or a gel surface that will not hold your samples.
9. Denaturation of amplified DNA samples can be performed by heating at 95°C or by both chemical agents (NaOH) and heating because the denaturation of the fragment to single strands is often incomplete.
10. The optimum balance for band separation versus gel run time depends upon the size of amplified DNA fragment. For a DNA fragment around 150 to 300 bp, 4 h would be ideal. The migration pattern of a certain single-strand conformer may be improved by modification of a number of electrophoresis conditions, including lowering the buffer pH, changing the temperature of the gel, or increasing the acrylamide concentration (up to 15%). Difficulties detecting certain mutations when using the routine protocol may be circumvented by adjusting one or more of these three gel conditions.

11. Usually the SSCP patterns can be read clearly after 16- to 20-h exposure. If the migration bands are still very weak after exposed for 20 h, exposure time should be prolonged to more than 48 h, sometimes even for as long as 2 wk.
12. SSCP gels are analyzed by observing the pattern of bands resulting on the autoradiograms. The number of bands present is dependent upon the sequence amplified. When one sequence is amplified, the most simple pattern that can be observed consists of two bands, one corresponding to the upper strand of DNA, and the other to the lower strand of the amplified fragment. If a segment of single-stranded DNA assumes more than one conformation as it migrates through the gel, the result will be multiple bands (equal to the number of conformers). Knowledge of the sequence being amplified provides a control and a point of reference. Amplification of more than one sequence results in a composite pattern of bands corresponding to those of the constituent sequences. A band that corresponds to double-stranded DNA may also be visible lower in the gel as double-stranded DNA migrates more rapidly. Inclusion of a sample of nondenatured DNA (not heated) serves as a marker for double-stranded DNA.

References

1. Orita, M., Suzuki, Y., Sekiya, T., and Hayashi, K. (1989) Rapid and sensitive detection of point mutations and DNA polymorphisms using the polymerase chain reaction. *Genomics* **5,** 874–879.
2. Bosari, S., Marchetti, A., Buttitta, F., Graziani, D., Borsani, G., Loda, M., et al. (1995) Detection of p53 mutations by single-strand conformation polymorphisms (SSCP) gel electrophoresis. A comparative study of radioactive and nonradioactive silver-stained SSCP analysis. *Diagn. Mol. Pathol.* **4,** 249–255.
3. Kurvinen, K., Hietanen, S., Syrjanen, K., and Syrjanen, S. (1995) Rapid and effective detection of mutations in the p53 gene using nonradioactive single strand conformation polymorphism (SSCP) technique applied on PhastSystem. *J. Virol. Methods* **51,** 43–53.
4. Neubauer, A., Brendel, C., Vogel, D., Schmidt, C. A., Heide, I., and Huhn, D. (1993) Detection of p53 mutations using nonradioactive SSCP analysis: p53 is not frequently mutated in myelodysplastic syndromes (MDS). *Ann. Hematol.* **67,** 223–226.
5. Ozcelik, H. and Andrulis, I. L. (1995) Multiplex PCR-SSCP for simultaneous screening for mutations in several exons of p53. *Biotechniques* **18,** 742–744.
6. Campbell, B. J. and Hirsch, V. M. (1994) Extensive envelope heterogeneity of simian immunodeficiency virus in tissues from infected macaques. *J. Virol.* **68,** 3129–3137.
7. Barron, K. S., Reveille, J. D., Carrington, M., Mann, D. L., and Robinson, M. A. (1995) Susceptibility to Reiter's syndrome is associated with alleles of TAP genes. *Arthritis Rheum.* **38,** 684–689.
8. Barron, K. S. and Robinson, M. A. (1994) The human T-cell receptor variable gene segment TCRBV6S1 has two null alleles. *Hum. Immunol.* **40,** 17–19.
9. Han, M., Hannick, L. I., DiBrino, M., and Robinson, M. A. (1999) Polymorphism of human CD1 genes. *Tissue Antigens* **54,** 122–127.
10. Day, C. E., Schmitt, K., and Robinson, M. A. (1994) Frequent recombination in the human T-cell receptor beta gene complex. *Immunogenetics* **39,** 335–342.
11. Orita, M., Iwahana, H., Kanashi, H., and, Sekiya, T. (1989) Detection of polymorphisms of human DNA by gel electrophoresis as single-strand conformation polymorphisms. *Proc. Natl. Acad. Sci. USA* **86,** 2766–2770.

VII

PCR-Based Sequencing

49

Sequencing

A Technical Overview

David Stirling

1. Introduction

The huge advances that have been made in human (and other species) genome projects have been lead by, and have themselves fueled, tremendous innovations in the field of sequencing. These innovations have lead to a specialization, with very expensive equipment, often in core sequencing centers or services. Until the cost of automated sequencing equipment falls considerably, the economics of scale mean that unless an investigator has a huge amount of sequencing to perform, they are most likely to be best served by using these services rather than by investing in the technology themselves. There is always a temptation in such situations to treat the sequencing as a black box, without worrying too much how the data are generated. This is a mistake. A clear understanding of the principals involved will not only aid the design of sequencing strategies but will also help in the interpretation of less than perfect data.

DNA/RNA sequence analysis has become fundamental to the understanding of biological processes. The foundations of this science were established in 1977 when Maxam and Gilbert *(1)* described a method for sequencing by base-specific chemical cleavage. Subsequently, Sanger and co-workers *(2)* developed a method for enzymatic sequencing using chain terminators. Both techniques produce populations of labeled DNA fragments that can be electrophoretically resolved to reveal the base sequence. Many different strategies have been developed to improve on these initial approaches and to make genome-sequencing projects feasible; all owe a great debt to those original protocols.

2. Chemical DNA Sequencing

In the original Maxam-Gilbert method of DNA sequencing, target DNA is radioactively labeled at one end (3′or 5′ end), which acts as the reference point for determining the positions of the remaining bases. This labeled DNA is processed with four base-specific reactions. First, a base-specific chemical modification, then a chain cleavage of modified bases.

These methods have fallen from popularity, both because of the toxicity of the reagents (dimethyl sulphate, hydrazine, potassium permanganate) and the development

From: *Methods in Molecular Biology, Vol. 226: PCR Protocols, Second Edition*
Edited by: J. M. S. Bartlett and D. Stirling © Humana Press Inc., Totowa, NJ

of simple and improved methods for enzymatic DNA. Although the chemical method is not widely used as the enzymatic method, it has some advantages and can be very useful in certain situations. Because it does not rely on the hybridization of a primer, very short sequences, such as oligonucleotides, can be analyzed. It is also useful for analyses of DNA modifications, such as methylation, and to study DNA–protein interactions (footprinting).

3. Enzymatic DNA Sequencing

Dideoxy sequencing reactions (Sanger method) are essentially primer extension reactions where ddNTPs are included in the mix. In the conventional dideoxy sequencing reaction, an oligo primer is annealed to a single-stranded DNA template and extended by DNA polymerase to synthesize a complementary copy of single-stranded DNA in the presence of four dNTPs, one of which is ^{35}S-labeled. Chain growth involves the formation of a phosphodiester bridge between the 3′-OH at the growing end of the primer and the 5′-phosphate group of the incorporated dNTP. Thus, overall chain growth is in the 5′→3′ direction. The reaction also contains one of four ddNTPs that terminate elongation when incorporated into the growing DNA chain. When a ddNTP is incorporated at the 3′ end of the growing primer chain, the elongation is terminated selectively at A, C, G, or T owing to the missing 3′ OH group of the primer chain. The enzymatic method is based on the ability of DNA polymerase to use both 2′ dNTPs and 2′,3′ ddNTPs as substrates. After completion of the sequencing reactions, the products are subjected to electrophoresis on a high-resolution denaturing polyacrylamide gel and then autoradiographed to visualize the DNA sequence.

For this form of sequencing, the purity and concentration of the template has to be very carefully controlled, and the template has to be rendered single stranded before the reaction will proceed. To reliably obtain good quality sequence from PCR products, these would first have to be cloned into a suitable vector. Any clone obtained is the product of a single molecule of PCR product, and so the likelihood of that sequence containing misincorporated bases is relatively high. Multiple clones have therefore to be sequenced to produce reliable data.

4. Cycle Sequencing

The introduction of thermostable DNA polymerase had the same revolutionary impact on sequencing protocols as in other areas *(3)*. The ability to function at higher temperature resulted in template remaining single stranded for longer periods, overcame many problems of secondary structure, and allowed more stringent primer annealing conditions resulting on less background noise in sequencing data. In addition, repeated cycles of primer annealing and extension amplifies (linearly rather than exponentially as for PCR) even small amounts of sample DNA to generate more template. The ability to sequence very low template concentrations means that even relatively impure material, such as PCR products, can be used, simply by diluting out the impurities.

5. Automation

One of the major advances in sequencing technology has been the development of automated DNA sequencers, which automate the gel electrophoresis step, detection of band pattern, and analysis of bands. These machines are based on the enzymatic, cycle sequencing approach and use fluorescent rather than radioactive labels. This has the advantages of greater safety, generation of machine-readable data, and greater reagent stability. The downside of this is the relative expense of the equipment required. Fluorescent dyes can be incorporated as labeled primers, dNTPs. or ddNTPs. The use of four different dyes, one for each of the ddNTPs (dye terminator sequencing), has allowed one of the biggest improvements in throughput of these systems. Now, rather than the base-specific reactions being kept separate and run down individual lanes of a gel, they can be performed in the same reaction tube and analyzed in one lane, with fragments being separated by size and distinguished by the wavelength of the fluorescent emission. Automated sequencers based on capillary electrophoresis have been developed, which dispense with the gel-making step. Many specialized centers now use robotic systems to extract DNA template, perform PCR and sequencing reactions, and load 96-capillary sequencers. Sequence data generated can then be directly imported into databases and processed with little or no hands-on intervention.

These developments still continue. Research is under way to develop the technology of mass spectrometry for DNA sequencing, and sequencing by hybridization is the subject of a great deal of development work.

6. Direct Sequencing of PCR Products

Unlike methods where the PCR product is cloned and a single clone is sequenced, direct sequencing of PCR products is usually unaffected by the relatively high error rate of Taq DNA polymerase because (unless there are only a few starting copies of template, and a misincorporation occurs in an early round of PCR) the vast majority of the amplified product will consist of the correct sequence.

Direct sequencing of PCR products has significant advantages over the cloning strategy. It is a simple procedure that can be easily standardized and only a single sequence needs to be determined for each sample. Indeed, the procedures are so well standardized that there has been an exponential growth in the number of laboratories offering core-sequencing services for PCR products. Although the majority of such core laboratories provide an excellent service, there are a number of factors that will affect their ability to produce good quality data.

7. Requirements for Good Quality Sequence from PCR Products

- Optimize the PCR reaction to yield only a single product. If the same oligos are used to sequence as primed the PCR, they will also sequence from any nonspecific products, interfering with data quality. Primer-dimers may not interfere with the identification of a PCR product on agarose gel electrophoresis, but they serve as efficient template for sequencing, resulting in characteristic noise in sequence data close to the primer.
- PCR should be performed with as little primer as possible. If labeled dideoxyterminator sequencing is used (probably the most common approach), residual PCR primer can serve as sequencing primer. Thus, even when only one PCR product is present, it is sequenced from each strand simultaneously, again interfering with data quality. There are a number of techniques available for the removal of unincorporated primers, from enzymatic digestion to column purification, but these are generally unnecessary if the primer concentration is limited in the PCR.
- If PCR products are isolated from a gel prior to sequencing, great care should be taken to minimize the amount of salt carried over from the isolation procedure. High-salt

concentrations interfere with the processivity of the polymerase reactions, resulting in very short read lengths.

- Primers should only be obtained from reliable sources and should be aliquoted to avoid repeated freeze thaw cycles. Any shorter primer species (e.g., n-1), in the PCR primer may not affect PCR efficiency. However, if the same primer is used for sequencing, a proportion of the fluorescent products will be primed from a shorter oligo. If the missing base is at the 5′ end of the oligo, the product will be shorter, and hence the sequencing results will be confounded. This often results in a characteristic 'shadow' sequence.

References

1. Maxam, A. M. and Gilbert, W. (1977) A new method for sequencing DNA. *Proc. Natl. Acad. Sci. USA* **74,** 560–564.
2. Sanger, F., Nicklen, S., and Coulson, A. R. (1977) DNA sequencing with chain-terminating inhibitors. *Proc. Natl. Acad. Sci. USA* **74,** 5463–5467.
3. Innis, M. A., Myambo, K. B., Gelfand, D. H., and Brow. M. A. (1988) DNA sequencing with Thermus aquaticus DNA polymerase and direct sequencing of polymerase chain reaction-amplified DNA. *Proc. Natl. Acad. Sci. USA* **85,** 9436–9440.

50

Preparation and Direct Automated Cycle Sequencing of PCR Products

Susan E. Daniels

1. Introduction

The polymerase chain reaction (PCR) is well known for being a rapid and versatile method for the amplification of defined target DNA sequences. This technique can be applied to a variety of research areas, such as the identification and typing of single nucleotide substitutions of DNA sequence polymorphisms, and genetic mapping *(1–4)*.

Since the introduction of PCR *(5)*, a variety of methods for sequencing PCR-generated fragments have been described. These are usually based on the Sanger chain-terminating dideoxynucleotide sequencing *(6)* rather than the Maxam and Gilbert chemical cleavage method *(7)*. Manual dideoxy sequencing methods are labor intensive, time-consuming, involve radioisotopes, and have limitations in sequence ordering. However, a technique combining the PCR and dideoxy terminator chemistry simplifies the process of sequencing and is known as cycle sequencing *(8)*. Automated or fluorescent DNA sequencing is a variation of the traditional Sanger sequencing using the cycle sequencing methodology, where fluorescent labels are covalently attached to the reaction products and data are collected during the polyacrylamide gel electrophoresis.

The introduction of fluorescently labeled dideoxynucleotides as chain terminators presented the opportunity for the development of reliable cycle sequencing for PCR products. The sequencing reaction with the dye terminators is performed in a thermal cycler, and each of the four dideoxynucleotide triphosphates (ddNTPs) is labeled with a different fluorescent dye. This allows the four chain extension reactions to be conducted within a single tube, sparing considerable labor *(9,10)*. The use of labeled chain terminators allows flexibility of sequencing strategy because the same primers can be used in the sequencing reaction. This eliminates the time and expense associated with a separate set of modified DNA sequencing primers and is well suited to high throughput sequencing.

Using this method, it is possible to amplify a target DNA sequence, purify the resulting fragment, and obtain sequencing data within 24 h. Also, it has been used

From: *Methods in Molecular Biology, Vol. 226: PCR Protocols, Second Edition*
Edited by: J. M. S. Bartlett and D. Stirling © Humana Press Inc., Totowa, NJ

to sequence a 500-bp PCR fragment on an automated DNA sequencer with 99.3% accuracy ***(10–12)***.

This chapter focuses on the techniques involved with the direct sequencing reactions rather than on the use of the machine because each automated DNA sequencer will be provided with an extensive manual for its operation.

2. Materials

All solutions should be made to the standard required for molecular biology. Use molecular-biology-grade reagents and sterile distilled water. The reagents for the cycle sequencing are available commercially.

2.1. Purification of PCR Products Before Cycle Sequencing

1. Ammonium acetate (4 *M*).
2. Isopropanol.
3. 70% (v/v) ethanol.
4. Tris-HCl (10 m*M*, pH 7.5), 1 m*M* EDTA.

2.2. Cycle Sequencing

Prism™ ready reaction DyeDeoxy terminator premix (1000 µL, Applied Biosystems [ABI]) consists of 1.58 m*M* A-dyedeoxy, 94.74 µ*M* T-dyedeoxy, 0.42 µ*M* G-dyedeoxy, 47.37 µ*M* C-dyedeoxy, 78.95 µ*M* dITP, 15.79 µ*M* dATP, 15.79 µ*M* dCTP, 15.79 µ*M* dTTP, 168.42 m*M* Tris-HCl (pH 9.0), 4.21 m*M* $(NH_4)_2SO_4$, 42.1 m*M* $MgCl_2$, 0.42 U/µL AmpliTaq DNA polymerase.

2.3. Purification of PCR Products After Cycle Sequencing

1. Chloroform.
2. Phenol:H2O:chloroform (16:18:14) at room temperature.
3. Sodium acetate (2 *M*, pH 4.5).
4. 100 and 70% (v/v) ethanol at room temperature.

2.4. 6% Polyacrylamide Sequencing Gels

1. 10× TBE: 890 m*M* Tris-borate, 890 m*M* boric acid, and 20 m*M* EDTA, pH 8.3.
2. Urea (40 g).
3. 12 mL of 40% (w/v) acrylamide stock solution (19:1 acrylamide/bis-acrylamide).
4. dH_2O (20 mL).
5. Mixed-bed ion-exchange resin (1 g).
6. TEMED.
7. 10% (w/v) ammonium persulfate, freshly made.

2.5. Loading Buffer

1. 50 m*M* EDTA, pH 8.0.
2. Deionized formamide.

3. Methods

3.1. Isopropanol Purification of PCR Products

It is essential to remove excess PCR primers before using DyeDeoxy terminators for cycle sequencing (*see* **Note 1**).

1. Aliquot an appropriate amount of the PCR into a 0.6-mL microfuge tube, and dilute to a total of 20 µL with distilled water.
2. Add 20 µL of 4 *M* ammonium acetate into the microfuge tube, mixing well.
3. Add 40 µL of isopropanol into the tube, mix well, leave at room temperature for 10 min, and centrifuge the microfuge tube for 10 min at 12,000*g*.
4. Carefully remove the supernatant and wash the pellet with 70% (v/v) ethanol. Then, briefly dry the pellet under vacuum.
5. Resuspend the pellet in 20 µL of TE buffer.

3.2. Cycle Sequencing of PCR Products

The amount of PCR product should be estimated on an agarose gel before sequencing. Approximately 1 µg of double-stranded DNA template or 0.5 µg of single-stranded DNA template is required for each sequencing reaction.

1. Mix the following reagents in a 0.6-mL microfuge tube: 5 mL of DNA template, 1 µL of primer (from a 3.2-pmol stock solution), and 4.5 µL of sterile dH_2O (*see* **Notes 4** and **5**).
2. Add 9.5 µL of ABI Prism ready reaction DyeDeoxy terminator premix.
3. Spin briefly to collect the reaction mix in the bottom of the tube and overlay with approx 50 µL of mineral oil.
4. Place the tubes in a thermal cycler (Perkin–Elmer Cetus [Warrington, UK] model 480 or 9600) that has been preheated to 96°C.
5. Immediately begin the cycle sequencing program, which is as follows: rapid thermal ramp to 96°C; 96°C for 30 s; Rapid thermal ramp to 50°C; 50°C for 15 s; Rapid thermal ramp to 60°C; and 60°C for 4 min for a total of 25 cycles.
6. Try to keep the samples in the dark at 4°C until further processing because they are sensitive to light.
7. Remove the excess DyeDeoxy terminators from the completed sequencing reactions.

3.3. Phenol: Chloroform Extraction of Cycle Sequencing Products

This step is essential to remove excess primers and unincorporated nucleotides.

1. To each sample, add 80 µL of sterile dH_2O.
2. Either add 100 µL of chloroform to dissolve the oil or remove the oil with a pipet.
3. Add 100 µL of phenol:H_2O:chloroform (68:18:14) to the sample and mix well by vortexing.
4. Centrifuge the sample at 12,000*g* for 1 min. Remove and discard the lower organic phase.
5. Re-extract the aqueous layer and transfer the aqueous upper layer to a clean tube.
6. Add 15 µL of 2 *M* sodium acetate and 300 µL of 100% ethanol to precipitate the extension products (*see* **Note 7**).
7. Centrifuge at 12,000*g* for 15 min at room temperature.
8. Carefully remove the supernatant and wash the pellet with 70% ethanol. Then, briefly dry the pellet under a vacuum (*see* **Note 2**).

3.4. Preparation of Samples for Loading

1. Add 4 µL of deionized formamide: 50 m*M* EDTA, pH 8.0 (5:1), to each sample tube, and mix well to dissolve the dry pellet.
2. Centrifuge briefly to collect the liquid at the bottom of the tube.
3. Before loading, heat the samples at 90°C for 2 to 3 min to denature. Then, transfer immediately onto ice.

4. Load all of the samples onto the automated DNA sequencer fitted with a 6% polyacrylamide gel using the manufacturer's software.

3.5. Preparation of 6% Polyacrylamide Sequencing Gel (see Note 3)

1. Place 40 g of urea, 12 mL of 40% acrylamide stock, 20 mL of dH_2O, and 1 g of mixed-bed ion-exchange resin into a beaker and stir gently while warming. Continue to stir the solution until all the urea crystals have dissolved (*see* **Notes 8** and **9**).
2. Filter the acrylamide through a 0.2-μ*M* filter, degas for 5 min, and transfer to a 100-mL cylinder.
3. Add 8 mL of filtered 10× TBE buffer and adjust the volume to 80 mL with dH_2O.
4. To polymerize the gel, add 400 μL of 10% APS (freshly made) and 45 μL of TEMED. Gently swirl to avoid adding air bubbles.
5. According to the instructions provided for the automated DNA sequencer, run the sequencing gel and analyze the readouts as indicated in **Fig. 1** (*see* **Note 10**).

4. Notes

1. Both purification steps can also be performed by spin columns, such as Centri-Sep or Quick Spin. Although they provide a quicker purification, they are an expensive alternative for those on a tight budget.
2. The PCR primers can be used as the DNA sequencing primers, and should be at least an 18 mer in length. Increasing the length will increase specificity and prevents priming at a secondary site. It will also decrease the chances for nonexact hybridization.
3. The GC content of the primer should be between 50 and 60%.
4. After ethanol or isopropanol precipitation, it is very important that the supernatant be carefully aspirated because the pellets are unstable and might be lost.
5. The dried sequencing pellet can be stored in the dark at 4°C for several days if required. However, once the loading buffer has been added, the samples should be loaded within a few hours.
6. The sequencing gel must polymerize for at least 1 h before use. A good time to prepare the gel is during the PCR of the cycle sequencing reactions.
7. Although there are various ready-made acrylamide solutions on the market, it is recommended that you make your own solutions because better resolution will be achieved. However, 40% acrylamide can be purchased ready-made and gives good results.
8. It has been shown that the use of formamide in the polyacrylamide gels can resolve compressions *(13)*.
9. During the phenol:chloroform extraction, if after the first spin two separate layers are not seen, then revortex the samples for 1 min and recentrifuge, after which an aqueous and organic phase should be obtainable.
10. If using the ABI 373A "Stretch" automated DNA sequencers, then it is better to use a 4.75% polyacrylamide gel, and from these machines, it is possible to obtain up to 1 kb of sequence with one run.
11. If the thermal cycler in your laboratory is not a Perkin–Elmer Cetus, then it will be necessary to optimize your thermal cycler for the sequencing reactions.
12. The addition of blue dextran to the loading buffer will make gel loading easier.

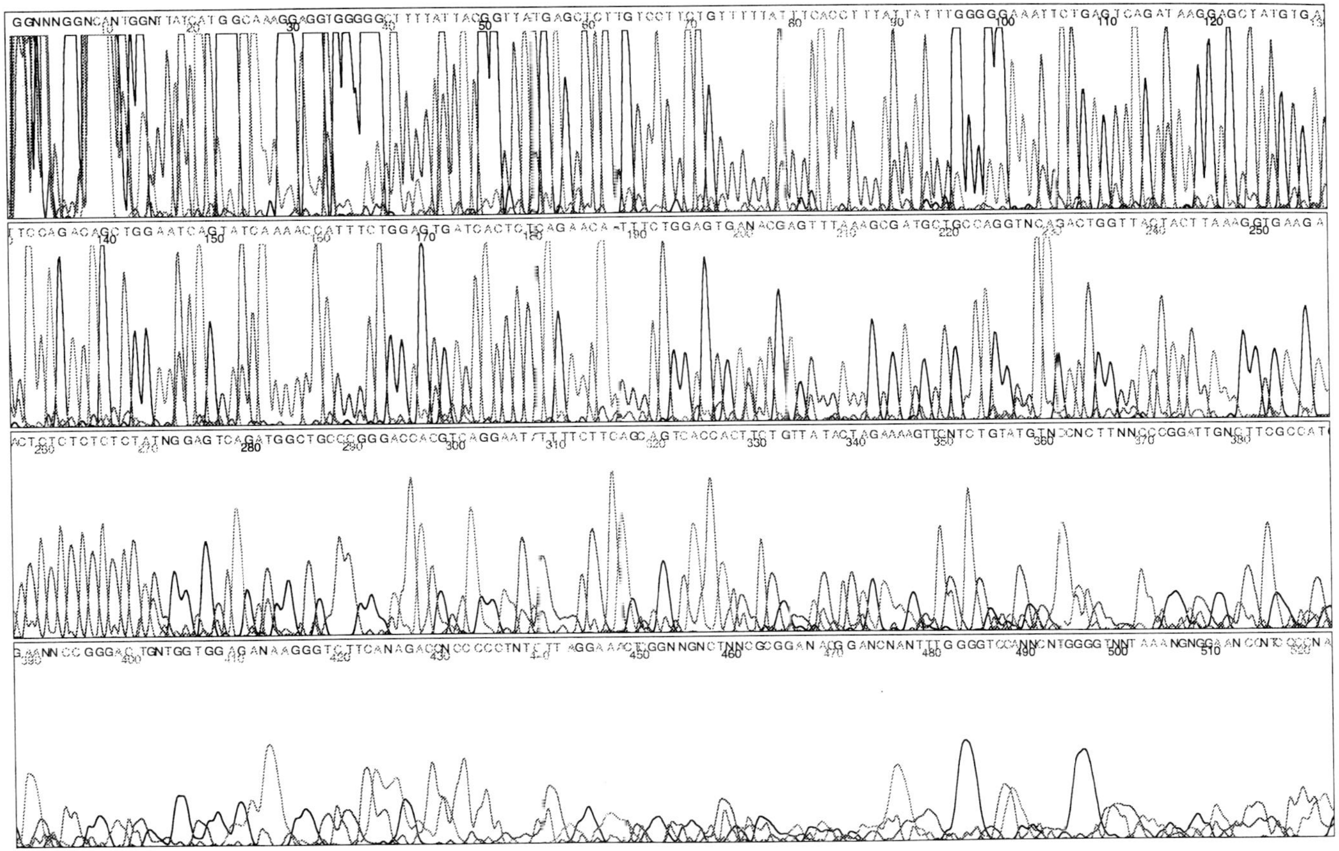

Fig. 1. Analyzed sequence data for a 350-bp PCR fragment amplified from genomic human DNA. The data were obtained using one of the primers used for PCR amplification.

References

1. Orita, M., Suzuki, Y., Sekeiya, T., and Hayashi, K. (1989) Rapid and sensitive detection of point mutations and DNA polymorphisms using the polymerase chain reaction. *Genomics* **5,** 874–879.
2. Kwok, P. Y., Carlson, C., Yager, T. D., Ankener, W., and Nickerson, D. A. (1994) Comparative analysis of human DNA variations by fluorescence based sequencing of PCR products. *Genomics* **23,** 138–144.
3. Martin-Gallardo, A., McCombie, W. R., Gocayne, J. D., Fitzgerald, M. G., Wallace, S., Lee, B. M. B., et al. (1992) Automated DNA sequencing and analysis of 106 kilobases from human chromosome 19q 13.3. *Nat. Genet.* **1,** 34–39.
4. NIH/CEPH Collaborative Mapping Group. (1992) A comprehensive genetic linkage map of the human genome. *Science* **258,** 67–86.
5. Mullis, K. B. and Faloona, F. A. (1987) Specific synthesis of DNA in vitro via a polymerase catalysed chain reaction. *Methods Enzymol.* **155,** 335–350.
6. Sanger, F., Nicklen, S., and Coulson, A. R. (1977) DNA sequencing with chain terminating inhibitors. *Proc. Natl. Acad. Sci. USA* **74,** 5463–5467.
7. Maxam, A. M. and Gilbert, W. (1977) A new method for sequencing DNA. *Proc. Natl. Acad. Sci. USA* **74,** 560–564.
8. Carothers, A. M., Urlaub, G., Mucha, J., Grunberger, D., and Chasin, L. A. (1989) Point mutation analysis in a mammalian gene: Rapid preparation of total RNA, PCR amplification of cDNA, and *Taq* sequencing by a novel method. *BioTechniques* **7,** 494–499.
9. McBride, L. J., Koepf, S. M., Gibbs, R. A., Nyugen, P., Salser, W., Mayrand, P. E., et al. (1989) Automated DNA sequencing methods involving polymerase chain reaction. *Clin. Chem.* **35,** 2196–2201.
10. Tracy, T. E. and Mulcahy, L. S. (1991) A simple method for direct automated sequencing of PCR fragments. *BioTechniques* **11,** 68–75.
11. Rosenthal, A. and Charnock Jones, D. S. (1992) New protocols for DNA sequencing with dye terminators. *DNA Sequence* **3,** 61–64.
12. Kelley, J. M. (1994) Automated dye terminator DNA sequencing, in *Automated DNA Sequencing and Analysis* (Adams, M. A., Fields, C., and Venter, J. C., eds.), Academic, London, pp. 175–181.
13. Hawkins, T. L. and Sulston, J. E. (1991) The resolution of compressions in automated fluorescent sequencing. *Nucleic Acids Res.* **19,** 2784.

51

Nonradioactive PCR Sequencing Using Digoxigenin

Siegfried Kösel, Christoph B. Lücking, Rupert Egensperger, and Manuel B. Graeber

1. Introduction

Techniques for direct sequencing of polymerase chain reaction (PCR) products are of central importance to contemporary research in molecular biology and genetics. The rapidly growing number of cloned human disease genes increasingly allows sequencing of PCR amplicons for diagnostic purposes. Nonradioactive sequencing protocols are of particular use because health, environmental, and administrative risks are minimized compared with conventional isotopic methods. The PCR-based nonradioactive cycle sequencing protocol described in this chapter has been successfully used to sequence mitochondrial and nuclear genes in Parkinson's and Alzheimer's disease brains using DNA extracted from formalin-fixed and paraffin-embedded neuropathological material *(1–3)*. This method, which allows sequence information of PCR products to be obtained within a single day, can be performed in a research or clinical laboratory using relatively inexpensive equipment. After initial PCR amplification, amplicons are purified using spin columns for affinity chromatography or ultrafiltration. Subsequently, cycle sequencing *(4)* is performed using 5′-digoxigenin end-labeled oligonucleotide primers. Because the nucleotide sequences of the PCR and sequencing primers can be identical, both reactions may be performed using the same thermalcycling protocol. This obviates the need for time-consuming optimization procedures. For visualization of sequencing results, sequencing reactions are separated on a standard sequencing gel, the gel is contact-blotted to a nylon membrane, and sequencing bands are visualized using alkaline phosphatase-conjugated antibodies (**Fig. 1**).

Potential pitfalls of our method are primarily related to the extreme sensitivity of PCR. The need for positive and negative sample controls cannot be overemphasized. We use different rooms and different pipets for setting up PCRs, pipetting sequencing templates, and thermal cycling *(5)*. In addition, aerosol-resistant pipet tips are always used.

2. Materials

2.1. Purification of Sequencing Templates

1. 10× TNE: 100 m*M* Tris-HCl, pH 7.4, 1.0 *M* NaCl, 10 m*M* EDTA (*see* **Note 1**).
2. Wizard PCR Preps DNA Purification System containing affinity chromatography spin columns, purification buffer, and resin (A7170, Promega, Madison, WI). Not contained

From: *Methods in Molecular Biology, Vol. 226: PCR Protocols, Second Edition*
Edited by: J. M. S. Bartlett and D. Stirling © Humana Press Inc., Totowa, NJ

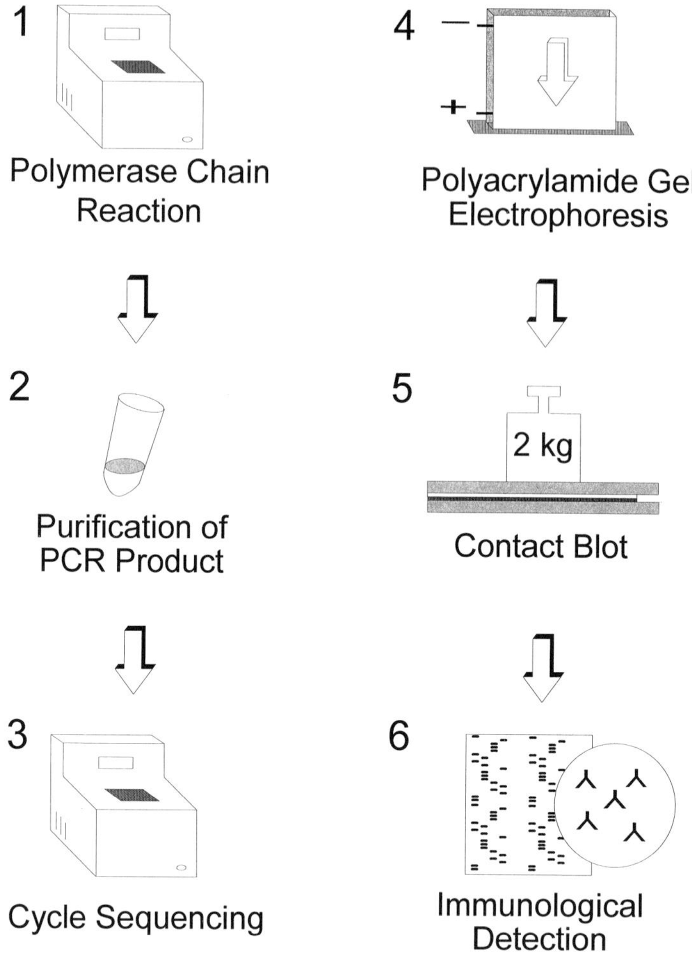

Fig. 1. Schematic drawing summarizing the essential steps of nonradioactive PCR sequencing using digoxigenin.

in the kit are 1× TE: 10 m*M* Tris-HCl, pH 8.0, 1 m*M* EDTA *(6)*; 80% isopropanol; and 2-mL disposable syringes (one per reaction) or, alternatively, Microcon-30 Concentrators (#42410, Amicon, Beverly, MA).

3. Eppendorf tubes (15 mL).
4. Eppendorf centrifuge (#5415C Eppendorf, Hamburg, Germany).
5. Fluorometer (TKO 100, Hoefer, San Francisco, CA) for quantification of template concentrations using DNA dye Hoechst No. 33258 and calf thymus standard DNA (supplied with the fluorometer).

2.2. Sequencing Reactions

1. Sequencing primer, 5′digoxigenin end-labeled (1 pmol/µL), desalted, and HPLC-purified (*see* **Note 2**).
2. Mineral oil (M-5904, Sigma, St. Louis, MO).
3. Dig *Taq* DNA Sequencing Kit (#1449443, Boehringer Mannheim, Mannheim, Germany), containing *Taq* DNA polymerase (3 U/µL); sequencing buffer (250 m*M* Tris-HCl,

pH 8.0, 50 mM $MgCl_2$); loading buffer containing formamide; and Termination mixtures containing dNTPs and the appropriate ddNTP:

ddATP (dATP, dCTP, dGTP, and dTTP, 25 μM each; 850 μM ddATP; 950 μM $MgCl_2$; pH 7.5).

ddCTP (dATP, dCTP, dGTP, dTTP, 25 μM each; 400 μM ddCTP; 500 μM $MgCl_2$; pH 7.5).

ddGTP (dATP, dCTP, dGTP, dTTP, 25 μM each; 75 μM ddGTP; 175 μM $MgCl_2$; pH 7.5).

ddTTP (dATP, dCTP, dGTP, dTTP, 25 μM each; 1275 μM ddTTP; 1370 μM $MgCl_2$; pH 7.5).

A second set of termination mixtures is shipped with the kit substituting 7-deaza-dGTP for dGTP.

4. 0.5-mL thin-walled reaction tubes (N801-0537, Applied Biosystems, Foster City, CA).
5. Eppendorf centrifuge.
6. Thermal cycler for cycle sequencing (*see* **Note 6** for reference).

2.3. Preparation of Sequencing Gel

1. 10× TBE: 0.9 M Tris, pH 8.3, 0.9 M boric acid, and 25 mM EDTA *(6)*.
2. 10% Ammonium persulfate (w/v) (should be prepared freshly before use).
3. Sigmacote (SL-2, Sigma).
4. GelMix 8 (#5545UA, Gibco BRL, Gaithersburg, MD), containing 7.6% acrylamide (w/v), 0.4% *N,N′*-methylene bis-acrylamide, 7.0 M urea, 100 mM Tris-borate, pH 8.3, 1 mM Na_2EDTA, 3 mM TEMED.
5. 10-mL disposable syringes with needles.
6. Scotch electrical tape 50 mm (FE-5000-0409-1, 3 M).
7. Plastic foil (e.g., Saran Wrap).

2.4. Gel Electrophoresis and Visualization of Sequencing Results

1. 1× Tris-buffered saline (TBS) 50 mM Tris-HCl, pH 7.5, 150 mM NaCl.
2. Alkaline phosphatase reaction buffer: 0.1 M Tris-HCl, pH 9.5, 50 mM $MgCl_2$, 0.1 M NaCl.
3. Stock solutions of 70% (v/v) and 100% *N,N*-dimethyl formamide.
4. Digoxigenin Detection Kit for Glycoconjugate and Protein Analysis (#1210220, Boehringer Mannheim), containing: antidigoxigenin antibodies conjugated to alkaline phosphatase; blocking reagent (purified casein fraction); nitroblue tetrazoliumchloride, 77 mg/mL in 70% *N,N*-dimethyl formamide; and X-Phosphate (5-bromo-4-chloro-3-indolylphosphate, 4-toluidinium salt), 50 mg/mL in 100% *N,N*-dimethyl formamide.
5. Nylon membrane, positively charged (#1417240, Boehringer Mannheim).
6. Whatman chromatography paper (#3030917, 3MM Whatman International Ltd., Maidstone, UK).
7. Plastic hybridization bags.
8. Scalpel blade.
9. Standard sequencing equipment: sequencing apparatus (Model S2, Gibco BRL) and high-voltage power supply (PS 9009, Gibco BRL).
10. Ultraviolet transilluminator (302 nm) for DNA crosslinking (Hoefer).
11. Eppendorf centrifuge.
12. Heating block or thermal cycler for denaturation of DNA.
13. Electric sealer for closing hybridization bags.

3. Methods

3.1. Purification of Sequencing Templates

1. PCR is performed according to established protocols *(7)*. After PCR amplification, PCR products are purified away from excess nucleotides and primers using either spin column chromatography (Wizard PCR Preps DNA Purification System, Promega, Madison, WI) or ultrafiltration (Microcon-30, Amicon). The purification systems are used according to manufacturer's recommendation (*see* **Note 3**).
2. Measure DNA concentration using the fluorometer and Hoechst dye No. 33258. The dye is dissolved in 1× TNE (prepare two stocks, 0.1 and 1 µg/mL, respectively). Calf thymus DNA (100 or 1000 ng/µL) is used as a standard for calibration (*see* **Note 4**).
3. Purification results may be checked by electrophoresis of samples through a 2% high-melting-point agarose gel.

3.2. Sequencing Reactions

1. Sequencing reactions are set up in a total volume of 20 µL containing the follow: 13 µL of DNA template solution (2–4 pmol); dilute with sterile, double-distilled water, if necessary; 3 µL of 5′-digoxigenin end-labeled primer (1 pmol/µL); 2 µL of 10× reaction buffer (250 m*M* Tris-HCl, pH 8.0, 50 m*M* $MgCl_2$); and 2 µL of *Taq* polymerase (3 U/µL).
2. For each sequencing reaction, transfer 4 µL of the above mixture to four thin-walled PCR tubes, each containing 2 µL of the respective termination mixture (ddATP, ddCTP, ddTTP, and ddGTP; *see* **Note 5**).
3. Overlay samples with 20 µL of mineral oil, and centrifuge for a few seconds in an Eppendorf tube, centrifuge at full speed.
4. Cycle sequencing is performed using a thermal cycling protocol empirically optimized for the T_m of the sequencing primer. When using sequencing primers of the same sequence and length as those used for PCR, a thermal cycling protocol identical to the one used for PCR usually gives good results (*see* **Note 6**).

3.3. Preparation of Sequencing Gel

For gel electrophoretic separation of sequencing reactions, an 8% denaturing polyacrylamide gel and standard sequencing equipment are used.

1. Clean glass plates thoroughly with 70% ethanol. Cover inner surface of one plate with a few drops or Sigmacote and let evaporate for 5 to 10 min. Set up glass plates with spacers and seal edges airtight using Scotch electrical tape. Put two strong metal clamps on each side.
2. Add 450 µL of 10% ammonium persulphate to one bottle (75 mL) of GelMix 8 and mix gently. Slowly fill the space between the glass plates (avoid air bubbles!). Let gel polymerize for at least 1 h at room temperature. (Wear protective gloves when handling unpolymerized acrylamide.)
3. After polymerization of gel, remove clamps and electrical tape from the lower end of the glass plates and transfer gel to sequencing apparatus. Fill the upper and lower buffer chambers with 500 mL of 1× TBE each. Rinse sample wells of gel with 1× TBE using a 10-mL disposable syringe with needle.

3.4. Gel Electrophoresis and Visualization of Sequencing Results

3.4.1. Electrophoresis and Contact Blotting

1. Add 3 µL of loading buffer to each tube containing the sequencing reactions with the respective "A," "C," "G," and "T" termination mixtures. Centrifuge for a few seconds

to mix, and denature samples for at least 2 min at 85°C using a heating block or thermal cycler.

2. Transfer 6 µL of each sample to the wells of the sequencing gel. Run gel at 35 mA for 2 to 4 h (until the upper dye front has reached the bottom end of the gel).
3. Remove glass plates from electrophoresis apparatus. Carefully take off the glass plate covered with Sigmacote on the inner side using a scalpel blade as a lever. Cover gel with nylon membrane (avoid air bubbles!) and cover with one layer of Whatman filter paper. Put on second glass plate and add approx 2 kg of weight. Leave for 30 min (*see* **Note 7**). Protect nylon membrane with one layer of plastic foil.
4. Crosslink nylon membrane (plastic wrapped, blotted side down) on an ultraviolet light box at 302 nm for 3 min (*see* **Note 8**). The membrane may now be air-dried, sealed in a plastic bag (*see* **Note 9**), and stored at 4°C until further use.

3.4.2. Visualization of Sequencing Bands

Visualization of sequencing results is performed in plastic hybridization bags (one blot per bag) at room temperature using the Digoxigenin Detection Kit (Boehringer). After crosslinking, wet blots may be used immediately for immunological detection. The following recipe applies to a 4∞4 lane gel (gently agitate membrane, except during the final step when incubating with substrate solution).

1. Incubate nylon membrane (in hybridization bag) for at least 30 min in approx 30 mL of blocking solution (0.5 g/100 mL TBS; *see* **Note 10**).
2. Wash blot three times in approx 100 mL of TBS (10 min each).
3. Incubate membrane with 50 mL of alkaline phosphatase-conjugated antidigoxigenin antibody for 1 h (50 µL antiserum/50 mL of TBS).
4. Wash three times in TBS (20 min each).
5. Prepare 30 mL of substrate solution (mix immediately before use): 150 µL of NBT (77 mg/mL in 70% *N,N*-dimethyl formamide [v/v]); 112 µL X-phosphate (50 mg/mL in *N,N*-dimethyl formamide); and 30 mL of alkaline phosphatase buffer, pH 9.5.

   Thoroughly remove washing solution from membrane. Add substrate solution. During this incubation step, cover plastic bag with black light protector and do not agitate (*see* **Note 11**).
6. Stop developing process when faint sequencing bands are beginning to appear. Add approx 1 L of deionized water to hybridization bag. Cut bag open and carefully remove membrane. Allow membrane to air-dry on sheets of Whatman filter paper or paper towels. Store stained, dried membranes in the dark (*see* **Note 12**). An example of a stained membrane is shown in **Fig. 2**.

4. Notes

1. Molecular biology-grade reagents are used for all experiments.
2. Oligonucleotide primers for PCR and sequencing may be designed by hand, but use of a computer program, such as Oligo (National Biosciences, Oslo, Norway), greatly facilitates this task. We have had good experience with primers ranging from 18 to 25 bp in length for both PCR and sequencing. The sequencing primer may be internally nested or identical in position and/or length to the primer used for PCR. Primers used in our laboratory were synthesized on an Applied Biosystems 394 DNA synthesizer at the Genzentrum of the University of Munich. Sequencing primers were custom-made and hapten-labeled using digoxigenin-3-*O*-methylcarbonylε-amino-caproic acid-*N*-hydroxy-succinimide ester and a 5′-oligopeptide linker (#1333054, Boehringer Mannheim).

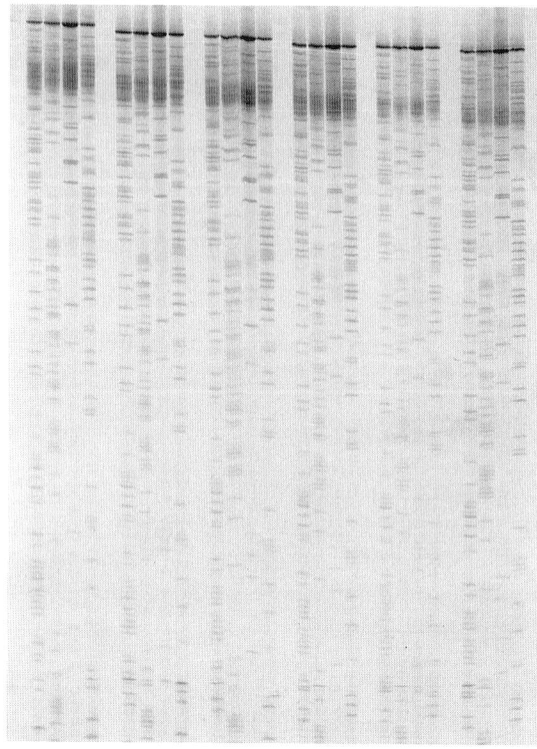

Fig. 2. Sequencing results obtained with primers *ND3* 5′-TCC CCA CCA TCA TAG CCA-3′ and *ND4* 5′-GGG TTT TGC AGT CCT TAG-3′, which amplify a 302-bp segment of the mitochondrial ND2 gene *(3)*. Sequencing results were visualized using alkaline phosphatase-conjugated antibodies directed against digoxigenin and NBT/X-phosphate as an enzymatic substrate. Contact blot of an 8% standard sequencing gel.

3. Purification of PCR products using the Wizard PCR Preps DNA Purification System: Add 50 to 100 µL of PCR product to 100 µL of purification buffer in an Eppendorf tube and mix. Then, add 1 mL of purification resin and mix thoroughly. Pipet mixture into a 2-mL syringe and push into a Wizard Prep column. Wash column with 2 mL of 80% isopropanol and spin for 20 s in an Eppendorf centrifuge at full speed. Let isopropanol evaporate (takes approx 5 min), and transfer column to new Eppendorf tube. Elute DNA by adding 50 µL of 1× TE to the column. After 1 min, spin column for 20 s to recover DNA completely. When ultrafiltration is used for purification of PCR products, 50 to 100 µL of PCR product are diluted with approx 400 µL of double-distilled water and transferred to a Microcon-30 concentrator. Centrifugation of the Microcon-30 column is done according to manufacturer's recommendation (e.g., 10 min at 12,000*g*) in an Eppendorf centrifuge. **Important:** Avoid contamination of both types of columns with mineral oil used for overlaying PCR.
4. A fluorometer is used to measure nanogram amounts of DNA. The TKO 100 (Hoefer) is a useful and reasonably priced alternative to full-sized spectrophotometers for measuring concentrations of double-stranded DNA. DNA concentrations as low as 10 ng/µL may be reliably determined using a standard setup of the TKO. For measuring the concentration of PCR products, the fluorometer is first calibrated with DNA standard (100 or 1000 ng/µL calf thymus DNA, depending on the DNA concentrations to be determined): 1 µL of

standard DNA is mixed with 1 mL of 1× TNE containing Hoechst dye No. 33258 and added to the quartz cuvet. Concentration of the Hoechst dye is 0.1 or 1 µL/mL in 1X TNE, depending on the concentration of the calibration standard.

5. Termination mixtures containing 7-deaza-dGTP are used to avoid band compression artifacts when sequencing guanine–cytosine-rich regions.
6. Similar to PCR, the success of *Taq* cycle sequencing depends on buffer conditions, especially the concentration of Mg^{2+} ions and pH. Buffer conditions for PCR and cycle sequencing may be optimized using commercial "optimizer" kits (e.g., K1220-01, Invitrogen, Leek, Netherlands). We have used *Taq* polymerases from Perkin–Elmer and Boehringer Mannheim with comparable success for both PCR and sequencing. Thermal cyclers from Perkin–Elmer (TC 480) and Biometra (Trio Thermo block) also gave comparable results in our hands. Sequencing protocols using smaller numbers of cycles, e.g., 20 to 25 may reduce the intensity of shadow bands.
7. Do not extend blotting time and do not use higher weights.
8. Protect eyes and skin against ultraviolet light (wear goggles, mask, coat, and gloves)!
9. To save on reagents, hybridization bags should be tightly sealed. However, leave one "long end," because bags will need to be reopened and resealed during subsequent incubation steps.
10. Blocking reagent (Boehringer Mannheim) should be prepared freshly before use. Dissolve blocking reagent at 50°C in TBS, and let cool down to room temperature. Blots can be stored in blocking solution at 4°C for a few days.
11. To avoid diffuse or spotty background, the volume of the substrate solution should be sufficiently large to cover the nylon membrane completely. Also, avoid folds in the hybridization bag. Sequencing bands should become visible within 15 min on incubation. Do not move the membrane during that time. Because the visualization process involves an enzymatic reaction, lowering the temperature of the incubation solutions may lengthen incubation times. We recommend developing at room temperature and checking the intensity of the stained sequencing bands by briefly lifting the dark cover from time to time.
12. The intensity of the sequencing bands may increase during the first few hours after the final incubation has been finished (if alkaline phosphatase and NBT/X-phosphate are used for visualization as in the present protocol). Therefore, the incubation with substrate solution should be stopped as soon as the first sequencing bands are clearly readable. **Important:** Stained, dried nylon membranes should not be exposed to sunlight because they will bleach rapidly. However, stained membranes can be stored safely for many months in the dark. For long-term documentation, we recommend taking photographs. Sequencing results may also be documented permanently by photocopying stained blots onto paper or overhead transparencies.

References

1. Kösel, S. and Graeber, M. B. (1993) Non-radioactive direct sequencing of PCR products amplified from neuropathological specimens. *Brain Pathol.* **3,** 421–424.
2. Kösel, S. and Graeber, M. B. (1994) Use of neuropathological tissue for molecular genetic studies: parameters affecting DNA extraction and polymerase chain reaction. *Acta Neuropathol.* **88,** 19–25.
3. Kösel, S., Egensperger, R., Mehraein, P., and Graeber, M. B. (1994) No association of mutations at nucleotide 5460 of mitochondrial NADH dehydrogenase with Alzheimer's disease. *Biochem. Biophys. Res. Commun.* **203,** 745–749.
4. Krishnan, B. R., Blakesley, R. W., and Berg, D. E. (1991) Linear amplification DNA sequencing directly from single phage plaques and bacterial colonies. *Nucleic Acids Res.* **19,** 1153.

5. Kwok, S. and Higuchi, R. (1989) Avoiding false positives with PCR. *Nature* **339,** 237–238.
6. Sambrook, J., Fritsch, E. F., and Maniatis, T. (1989) *Molecular Cloning: A Laboratory Manual,* 2nd ed., Cold Spring Harbor Laboratory Press, Cold Spring Harbor, New York.
7. Saiki, R. K., Gelfand, D. H., Stoffel, S., Scharf, S. J., Higuchi, R., Horn, G. T., et al. (1988) Primer-directed enzymatic amplification of DNA with a thermostable DNA polymerase. *Science* **239,** 487–491.

52

Direct Sequencing by Thermal Asymmetric PCR

Georges-Raoul Mazars and Charles Theillet

1. Introduction

Direct sequencing of polymerase chain reaction (PCR) products *(1)* has proven to be a powerful method in the generation of nucleic acid sequence data. Using these techniques, it is possible to produce microgram quantities of pure target DNA and subsequently its nucleotide sequence in a few hours, even, theoretically, from one single RNA or DNA molecule. However, problems have been encountered, and these have been attributed to the strong tendency of the short double-strand DNA templates to reanneal. In fact, compared with double-stranded plasmid DNA, which can be permanently denatured by alkali treatment and then form intermolecular interactions compatible with good sequencing efficiency, optimized conditions for direct sequencing are required before reannealing with short PCR product.

To obviate this, strategies have been developed, such as the generation of single-strand DNA template by asymmetric PCR *(2,3)*. Methods use either a disequilibrated concentration ratio between the two primers or a two-step amplification, both of which have their shortfalls. The first method is based on a large number of cycles, which is a potential source of misincorporation of errors, and optimized conditions enough to produce single-strand DNA that are strongly primer dependent. Moreover, it often has been the case that only one strand can easily be sequenced. The second case requires two physical separation steps, in which product contamination may occur.

Here, we propose a method combining the advantages of both symmetric and asymmetric PCR. It is based on a thermal asymmetry between the Tm of both primers. Annealing temperature of each primer is calculated with the formula: 69.3 + 0.41 (%GC) – 650/*L*, with *L* = primer length. PCR primers are designed to obtain a difference in T_m of at least 10°C. In the first step, double-stranded material is produced during 20 to 25 cycles (to minimize the yield of spurious products) using the lower Tm. During the second step, single-stranded DNA is generated using the higher T_m (**Fig. 1**).

Consequently, one primer is dropped out and linear amplification is obtained. The final quantity of single-stranded product is comparable with the one produced by Gyllensten and Erlich's method *(2)*. We applied thermal asymmetry to several sequences, which, in our hands, were difficult to sequence both from double-stranded DNA or with the Gyllensten and Erlich asymmetric PCR products (**Fig. 2**).

From: *Methods in Molecular Biology, Vol. 226: PCR Protocols, Second Edition*
Edited by: J. M. S. Bartlett and D. Stirling © Humana Press Inc., Totowa, NJ

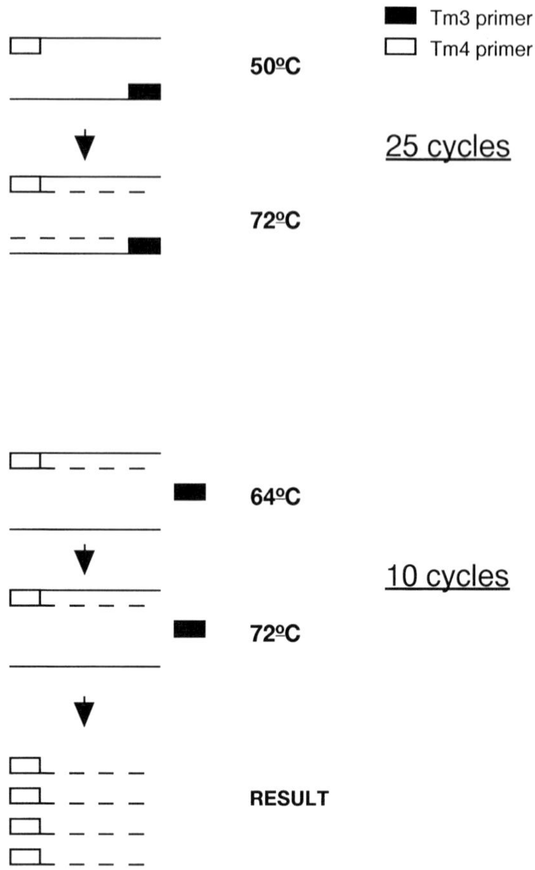

Fig. 1. Schematic representation of thermal asymmetric PCR.

These PCR fragments comprised the following: (1) exons 7 and 8 of the *p53* gene and (2) exon 1 of HRAS. This latter sequence is particularly GC-rich, and as part of another study, primer A was synthesized with a 40-bp GC stretch (to make a GC clamp). In conclusion, thermal asymmetric PCR allows direct sequencing of both strands with high reproducibility and reduced risk of contamination.

2. Materials

All solutions should be made according to the standard required for molecular biology, such as molecular biology-grade reagents and sterile distilled water. All reagents for sequencing are available commercially.

2.1. PCR Amplification

1. Primers were synthesized on Applied Amplifications, and PCR was performed on a Perkin–Elmer Cetus thermal cycler.
2. 1× *Taq* polymerase buffer: 10 m*M* Tris-HCl, pH 8.3, 2 m*M* $MgCl_2$, 50 m*M* KCl, 0.01% gelatin; 100 μ*M* of dNTPs.
3. *Taq* polymerase was purchased from Perkin–Elmer and used at 1 U/reaction.

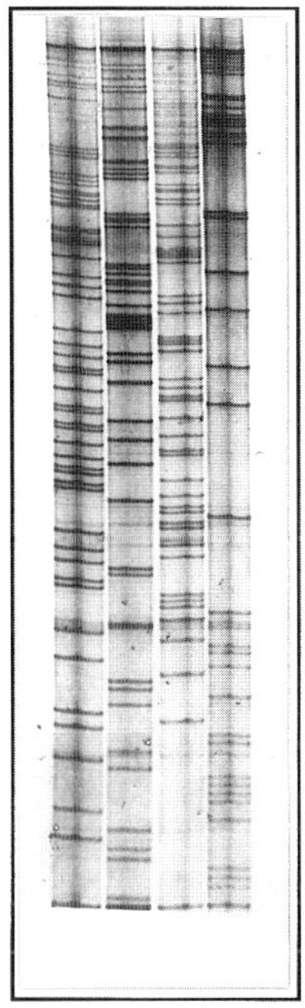

Fig. 2. Autoradiograph of a PCR product sequenced by thermal asymmetric PCR.

2.2. Purification and Sequencing of the PCR Product

2.2.1. Sequencing Reagents

1. Annealing buffer (5× concentrate): 200 m*M* Tris-HCl, pH 7.5, 100 m*M* $MgCl_2$, 250 m*M* NaCl, 0.1 *M* dithiothreitol (DTT).
2. Labeling nucleotide mixture (one for each dideoxy nucleotide): Each mixture contains 80 µ*M* dGTP, 80 µ*M* dATP, 80 µ*M* dTTP, 80 µ*M* dCTP, and 50 m*M* NaCl. In addition, the "G" mixture contains 8 µ*M* dideoxy-dGTP; the "A" mixture, 8 µ*M* ddATP; the "T," 8 µ*M* ddTTP; and the "C" 8 µ*M* ddCTP.
3. Stop solution: 95% formamide, 20 m*M* EDTA, 0.05% bromophenol blue, and 0.05% xylene cyanol FF.
4. Labeled dATP is [α-^{35}S] dATP from Amersham, and specific activity should be 1000 to 1500 Ci/mol.

2.2.2. Purification

Purification should be performed in Centricon 30 column (Amicon).

3. Methods

3.1. PCR Conditions

PCR conditions should be as follows: in a total volume of 25 µL, 20 pmol of each primer, 1× *Taq* polymerase buffer, and 1 U of *Taq* polymerase were incubated with 50 ng of genomic DNA. P53 primer A1 cttagtacctgaagggtgaaatattc ($T_m1 = 60°C$), P53 primer B1 gtagtggtaatctactgggacggaacagc ($T_m2 = 69°C$), P53 primer A2 taatctactgggacgga ($T_m3 = 50°C$), P53 primer B2 cccaagacttagtacctgaagggtg ($T_m4 = 64°C$). Cycling conditions were for 25 cycles: 92°C (30 s); T_m1 or 3 (30 s); 72°C (90 s), followed by 10 cycles: 92°C (30 s); T_m2 or 4 (30 s); 72°C (90 s).

3.2. Preparation of the PCR Product

1. Transfer PCR product directly by pipetting in a Centricon 30 and add 2 mL of water.
2. Spin at 5000*g* in a fixed-angled rotor in a Beckman-type centrifuge for 30 min at room temperature.
3. Add 2 mL of water. Spin again for 30 min and invert column and spin for 5 min at 1500*g*. This procedure efficiently removes the excess of dNTPs from the PCR. Volume recovered is typically 20 to 50 µL.
4. Typically, 7 µL of this purified product are used for single-strand sequencing according to the manufacturer's directions of United States Biochemicals (Cleveland, OH).

3.3. Sequencing Protocol

3.3.1. Annealing Template and Primer

1. For each template, a single annealing (and subsequent labeling) reaction is used. Combine the following:
   a. Primer 0.5 pmol (1 µL)
   b. DNA 7 µL
   c. Annealing buffer 2 µL
2. Warm the capped tube to 65°C for 2 min, and then allow the mixture to cool slowly to room temperature over a period of about 30 min.

3.3.2. Labeling Reaction

1. To the annealed template-primer add the following:
   a. DTT (0.1*M*) 1 µL
   b. Labeling nucleotide mix 2 µL
   c. [α-α-^{35}S] dATP 5 µCi (typically 0.5 µL)
   d. Sequenase 3 U from United States Biochemicals
2. Total volume should be approx 15 µL; mix thoroughly and incubate for 2 to 5 min at room temperature.

3.3.3. Termination Reactions

1. Label four tubes "A," "C," "G," and "T." Fill each with 2.5 µL of the appropriate dideoxy termination mixture.
2. When the labeling reaction is complete, transfer 3.5 µL of it to the tube (prewarmed to 37°C) labeled G. Similarly, transfer 3.5 µL of the labeling reaction to each of the other three tubes (A, T, and C).
3. After 2 to 5 min of incubation at 37°C, add 4 µL of stop solution to each termination reaction, mix, and store on ice.
4. To load the gel, heat the samples to 75 to 80°C for 2 min, and load 2 to 3 µL in each lane. Prerun a sequencing gel for 30 min, load, and run until bromophenol is just out of the gel.
5. Fix gel as usual and dry on Whatman 3MM paper. Correct sequencing yields a detectable signal using a bench-top Geiger counter.
6. Expose overnight without Saran™ paper at room temperature.

4. Notes

1. Instability of diluted solution of primers conserved at –20°C can sometimes be problematic. We recommend storing oligonucleotides as a dried powder and resuspending them in water prior to use.
2. Estimation of the yield of single-strand DNA produced can be achieved by Southern blotting: run a 2% agarose gel, blot following standard conditions, and probe with one of the PCR primers. Two bands should appear if you use higher T_m primer or only one band if you use lower T_m primer (corresponding to double-stranded DNA). An alternative strategy is to add α-dCTP[^{32}P] for the second step of amplification at high temperature: labeled single-stranded DNA should be exclusively produced.
3. We also sequenced PCR fragments after SSCP analysis. In this case, shifted SSCP bands were excised from the gel with a sterile razor blade and eluted in 50 µL of distilled water for 1 h at 65°C. A 1.5-µL aliquot of the eluate was subjected to thermal asymmetric PCR.
4. A good sequence can be obtained even if no primer is added for the sequencing reaction. The reason is that the low T_m primer from PCR is not completely removed by Centricon 30 purification.
5. The present protocol has been optimized for classical radioactivity labeled DNA sequencing, but should easily be adapted to automated fluorescent sequencing using fluorescent dye terminators.

References

1. Saiki, R. K., Gelfand, D. H., Stoffel, S., Scharf, S. F., Higuchi, R., Horn, R. T., et al. (1988) Primer-directed enzymatic amplification of DNA with a thermostable DNA polymerase. *Science* **239,** 487–491.
2. Gyllensten, U. B. and Erlich, H. A. (1988) Generation of single-stranded DNA by the polymerase chain reaction and its application to direct sequencing of the HLA-DQA locus. *Proc. Natl. Acad. Sci. USA* **85,** 7652–7656.
3. Wilson, R. K., Chen, C., and Hood, L. (1990) Optimization of asymmetric polymerase chain reaction for rapid fluorescent DNA sequencing. *BioTechniques* **8,** 184–189.

53

Analysis of Nucleotide Sequence Variations by Solid-Phase Minisequencing

Anu Suomalainen and Ann-Christine Syvänen

1. Introduction

The Sanger dideoxynucleotide sequencing method has been simplified by a number of methodological improvements, such as the use of the polymerase chain reaction (PCR) technique for generating DNA templates in sufficient quantities followed by affinity-capture techniques for convenient and efficient purification of the PCR fragments for sequencing, or altenatively the use of cyclic Sanger sequencing reactions that are easy to automate with laboratory robots, and the development of instruments for automatic on-line analysis of fluorescent products of the sequencing reactions (references to this book, solid phase sequencing, cyclic sequencing). Despite these technical improvements, the requirement for gel electrophoretic separation remains an obstacle when sequence analysis of large numbers of samples are needed, as in DNA diagnosis, or in the analysis of sequence variation for genetic, evolutionary, or epidemiological studies.

We have developed a method for analyzing DNA fragments that differ from each other in one or a few nucleotide positions *(1)* denoted solid-phase minisequencing, in which gel electrophoretic separation is avoided. Analogous to the methods for solid-phase sequencing of PCR products, the solid-phase minisequencing method is based on PCR amplification using one biotinylated and one unbiotinylated primer, followed by affinity-capture of the biotinylated PCR product on an avidin- or streptavidin-coated solid support. The nucleotide at the variable site is detected in the immobilized DNA fragment by a primer extension reaction: A detection step primer that anneals immediately adjacent to the nucleotide to be analyzed is extended by a DNA polymerase with a single labeled nucleotide complementary to the nucleotide at the variable site (**Fig. 1**). The amount of the incorporated label is measured, and it serves as a specific indicator of the nucleotide present at the variable site.

We have used the solid-phase minisequencing method for detecting numerous mutations causing human genetic disorders, for analyzing allelic variation in genetic linkage studies, and for identifying individuals *(2–4)*. The protocol described below is generally applicable for detecting any variable nucleotide. The method suits well for analyzing large numbers of samples because it comprises simple manipulations in

From: *Methods in Molecular Biology, Vol. 226: PCR Protocols, Second Edition*
Edited by: J. M. S. Bartlett and D. Stirling © Humana Press Inc., Totowa, NJ

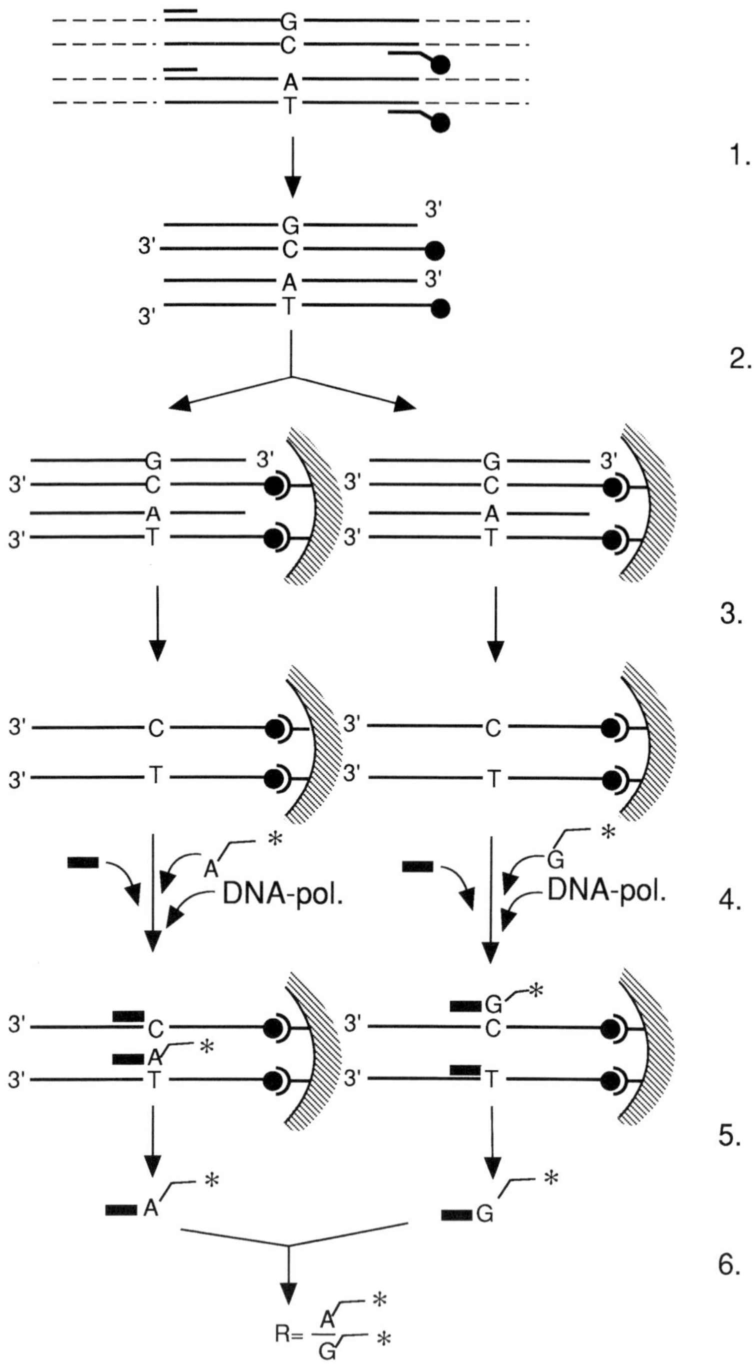

Fig. 1. Steps of the solid-phase minisequencing method. 1. PCR with one biotinylated (black ball) and one unbiotinylated primer. 2. Affinity-capture of the biotinylated PCR product in streptavidin-coated microtiter wells. 3. Washing and denaturation. 4. The minisequencing primer extension reaction. 5. Measurement of the incorporated label. 6. Calculation of the result.

a microtiter plate or test tube format and the result of the assay is obtained as an objective numeric value, which is easy to interpret. Furthermore, the solid-phase minisequencing method allows quantitative detection of a sequence variant present as a minority of less than 1% in a sample ***(2,3,5,6)***. We have used the sensitive quantitative analysis for detecting point mutations in malignant cells present as a minority in a cell population ***(5)*** and for analyzing heteroplasmic mutations of mitochondrial DNA ***(3,6)***. The high sensitivity is an advantage of the minisequencing method compared with dideoxynucleotide sequencing, in which a sequence variant must be present as 10 to 20% of a mixed sample to be detectable. A limitation of the solid-phase minisequencing method is that it is restricted to analyzing variable nucleotides only at positions predefined by the detection step primers used. The method is based on the use of equipment and reagents that are available from common suppliers of molecular biological products, facilitating easy setup. In the future, high-throughput analysis of nucleotide sequence variation will be performed by rapid, automatic methods based on homogeneous detection principles or alternatively using methods in microarray or chip formats. The minisequencing reaction principle is applicable for both types of assay formats ***(7,8)***.

2. Materials

2.1. Equipment

1. Programmable heat block, and facilities to avoid contamination in PCR.
2. Microtiter plates with streptavidin-coated wells (e.g., Combiplate 8, Labsystems, Helsinki, Finland; *see* **Note 1**).
3. Multichannel pipet and microtiter plate washer (optional).
4. Shaker at 37°C.
5. Water bath or incubator at 50°C.
6. Liquid scintillation counter.

2.2. Reagents

All the reagents should be of standard molecular biology grade. Use sterile distilled or deionized water.

1. Thermostable DNA polymerase. We use *Thermus aquaticus* (5 U/µL, Promega or Perkin–Elmer-ABI) or *Thermus brockianus* (Dynazyme™ II, 2 U/µL, Finnzymes, Espoo, Finland) DNA polymerase. Store at –20°C (*see* **Note 2**).
2. 10× concentrated DNA polymerase buffer: 500 m*M* m*M* $MgCl_2$, 1% (v/v) Triton X-100, and 0.1% (w/v) gelatin or 10× concentrated buffer supplied with the DNA polymerase enzyme. Store at –20°C.
3. dNTP mixture: 2 m*M* dATP, 2 m*M* dCTP, 2 m*M* dGTP, and 2 m*M* dTTP stored at –20°C.
4. PBS/Tween: 20 m*M* sodium phosphate buffer, pH 7.5, and 0.1% (v/v) Tween 20 store at 4°C. 50 mL is enough for several full-plate analyses.
5. TENT (washing solution): 40 m*M* Tris-HCl, pH 8.8, 1 m*M* EDTA, 50 m*M* NaCl, and 0.1% (v/v) Tween 20. Store at 4°C. Prepare 1 to 2 L at a time, which is enough for several full-plate analyses.
6. NaOH (50 m*M*; make fresh every 4 wk) stored in a plastic vial at room temperature ~20°C).
7. [^{3}H]-labeled deoxynucleotides (dNTPs): dATP to detect a T at the variant site, dCTP to

detect a G, etc. (Amersham-Pharmacia Biotech; [^{3}H]dATP, TRK 633; dCTP, TRK 625; dGTP, TRK 627; dTTP, TRK 576), stored at –20°C (*see* **Note 3**).

8. Scintillation fluid (for example Hi-Safe II, Wallac, Turku, Finland) stored at room temperature (~20°C).

2.3. Primer Design

1. PCR primers: One PCR primer of each pair is biotinylated at its 5′ end during the synthesis using a biotin-phosphoramidite reagent (for example Amersham–Pharmacia Biotech or Perkin-Elmer–ABI; *see* **Note 4**).
2. The detection step primer for the minisequencing analysis is a 20-mer oligonucleotide complementary to the biotinylated strand of the PCR product, designed to hybridize with the 3′ end immediately adjacent to the variant nucleotide to be detected (*see* **Fig. 1**) The minisequencing primer should be at least five nucleotides nested in relation to the unbiotinylated PCR primer.

3. Methods

3.1. PCR for Solid-Phase Minisequencing Analysis

The PCR is performed according to routine protocols, except that the amount of the biotin-labeled primer used is reduced not to exceed the biotin-binding capacity of the microtiter well (*see* **Note 1**). For a 50-µl PCR, we used 10 pmol of biotin-labeled primer and 50 pmol of the unbiotinylated primer. The PCR should be optimized (i.e., the annealing temperature and template amount) to be efficient and specific. To be able to use [^{3}H] dNTPs, which are low-energy β-emitters, for the minisequencing analysis, 1/10 of the PCR product should produce a single visible band after agarose gel electrophoresis, stained with ethidium bromide. There is no need for purification of the PCR product before the minisequencing analysis.

3.2. Solid-Phase Minisequencing Analysis

1. Affinity capture: Transfer 10-µL aliquots of the PCR product and 40 µL of the PBS/Tween solution to two streptavidin-coated microtiter wells (*see* **Note 5**). Include a control reaction, that is, a well with no PCR product. Seal the wells with a sticker and incubate the plate at 37°C for 1.5 h with gentle shaking.
2. Discard the liquid from the wells and tap the wells dry against a tissue paper.
3. Wash the wells three times at room temperature as follows: pipet 200 µL of TENT solution to each well, discard the washing solution and empty the wells thoroughly between the washings (*see* **Note 6**).
4. Denature the captured PCR product by adding 100 µL of 50 m*M* NaOH to each well, incubate at room temperature for 3 min. Discard the NaOH and wash the wells as in **step 3** above.
5. Prepare for each DNA fragment to be analyzed two 50-µL mixtures of nucleotide-specific minisequencing solution, one for detection of the normal and one for the mutant nucleotide (*see* **Note 7**). Mix 5 µL of 10× *Taq* DNA polymerase buffer, 10 pmol of detection step primer, 0.2 µCi (usually equals to 0.2 µL) of one [^{3}H] dNTP, 0.1 U of *Taq* DNA polymerase, and dH_2O to a total volume of 50 µL. It is obviously convenient to prepare master mixes for the desired number of analyses with each nucleotide.
6. Pipet 50 µL of one nucleotide-specific mixture per well, incubate the plate at 50°C for 10 min in a water bath or 20 min in an oven (*see* **Note 8**).
7. Discard the contents of the wells and wash them as in **step 3**.

8. Release the detection step primer from the template by adding 60 µL of 50 m*M* NaOH and incubating for 3 min at room temperature.
9. Transfer the eluted primer to the scintillation vials, add scintillation reagent, and measure the radioactivity, that is, the amount of incorporated label, in a liquid scintillation counter (*see* **Note 9**).
10. The result is obtained as counts per minute (cpm) values. The cpm value of each reaction expresses the amount of the incorporated [^{3}H] dNTP. Calculate the ratio (R) between the mutant and normal nucleotide cpms. In a sample of a subject homozygous for the mutant nucleotide the R will be >10, in a homozygote for the normal nucleotide R <0.1, and in the case of a heterozygote R varies between 0.5 and 2.0, depending on the specific activities of the [^{3}H] dNTPs (*see* **Note 10**).

4. Notes

1. The binding capacity of the streptavidin-coated microtiter well (Labsystems) is 2 to 5 pmol of biotinylated oligonucleotide. Other solid supports with higher biotin binding capacity, such as avidin-coated polystyrene beads (Fluoricon Assay Particles, IDEXX Corp., Portland ME; biotin-binding capacity over 2 nmol of oligonucleotide/ mg beads) ***(1)***, streptavidin-coated magnetic polystyrene beads (Dynabeads M-280, Dynal, Norway; biotin-binding capacity 300 pmol/mg) ***(5)***, or avidin-coated manifolds ***(9)*** can also be used. The biotin-binding capacity of a microtiter well allows reliable detection of up to 2% of a sequence variant present in the sample ***(6)***, whereas a detection sensitivity of less than 0.1% can be obtained with the bead-based format ***(5)***.
2. It is advantageous to use a thermostable DNA polymerase for the single-nucleotide primer extension reaction, as a high temperature, favorable for the simultaneous primer annealing reaction, can be applied. Thermostable DNA polymerases engineered for efficient incorporation of ddNTPs, such as the ThermoSequenase™ (Amersham-Pharmacia Biotech) enzyme, should preferably used for incorporating labeled ddNTPs.
3. Although the specific activities of the [^{3}H]dNTPs are low and their half lives are long (13 yr), the necessary precautions for working with [^{3}H] should be taken. Also, dNTPs or dideoxynucleotides labeled with other isotopes ([^{35}S] or [^{32}P]) ***(1)***, with haptens ***(1,10)*** or with fluorescent groups can be used ***(9)***.
4. If the biotin-labeled oligonucleotides are used without purification from unbiotinylated ones by high-performance liquid chromatography, polyacrylamide gel electrophoresis, or by ion exchange columns manufactured for this purpose (Perkin-Elmer-ABI), the level of biotinylation should be confirmed after the PCR. This can be performed by affinity capture of the biotinylated PCR product on an avidin matrix with high biotin-binding capacity, followed by detection of possible unbound products by agarose gel electrophoresis.
5. Each nucleotide to be detected at the variant site is analyzed in a separate well. Thus, at least two wells are needed per PCR product.
6. It is important that the PCR reagents are removed efficiently after the capturing reaction to avoid unspecific nucleotide incorporation in the minisequencing reaction. The washings can be performed in an automated microtiter plate washer or by manually pipetting the washing solution to the wells, discarding the liquid and tapping the plate against a tissue paper.
7. The minisequencing reaction mixture can be stored at room temperature for 1 to 2 h. It is convenient to prepare it during the affinity capture (**step 1** of **Subheading 3.2.**).
8. The conditions for hybridizing the detection step primer are not stringent, and the temperature of 50°C can be applied to analysis of most PCR products irrespectively of the sequence of the detection step primer. If the primer, however, is considerably shorter than a 20-mer or its guanine-cytosine content is low (melting temperature close to 50°C), lower

temperatures for the primer annealing may be required.

9. Streptavidin-coated microtiter plates made of scintillating polystyrene are available (Wallac, Finland). In this case the final washing, denaturation and transfer of the eluted detection primer can be omitted, but a scintillation counter for microtiter plates is needed *(11)*.
10. The ratio between the cpm values for the two nucleotides reflects the ratio between the two sequences in the original sample. Therefore, the solid-phase minisequencing method can be used for quantitative PCR analyses *(2–6)*. The R value is affected by the specific activities of the [^{3}H] dNTPs used, and if either the mutant or the normal sequence allows the detection step primer to be extended by more than one [^{3}H] dNTP, this will obviously also affect the R value. Both of these factors can easily be corrected for when calculating the ratio between the two sequences. Alternatively, a standard curve is constructed by mixing the two sequences in known ratios and plotting the obtained R-values as a function of the ratios to obtain a linear standard curve *(3,5,6)*. The test results can then be interpreted from the standard curve without the need of taking the specific activities of the number of [^{3}H] dNTPs incorporated into account.

References

1. Syvänen, A.-C., Aalto-Setälä, K., Harju, L., Kontula, K., and Söderlund, H. (1990) A primer-guided nucleotide incorporation assay in the genotyping of apolipoprotein E. *Genomics* **8,** 684–692.
2. Suomalainen, A. and Syvänen, A.-C. (2000) Quantitative analysis of human DNA sequences by PCR and solid-phase minisequencing. *Mol. Biotechnol.* **15,** 123–131.
3. Suomalainen, A., Kollmann, P., Octave, J.-N., Söderlund, H., and Syvänen, A.-C. (1993) Quantification of mitochondrial DNA carrying the $tRNA_{8344}^{Lys}$ point mutation in myoclonus epilepsy and ragged-red-fiber disease. *Eur. J. Hum. Genet.* **1,** 88–95.
4. Syvänen, A.-C. (1999) From gels to chips: "Minisequencing" primer extension for analysis of point mutations and single nucleotide polymorphisms. *Hum. Mutat.* **13,** 1–10.
5. Syvänen, A-C., Söderlund, H., Laaksonen, E., Bengtström, M., Turunen, M., and Palotie, A. (1992) N-ras gene mutations in acute myeloid leukemia: accurate detection by solid-phase minisequencing. *Int. J. Cancer* **50,** 713–718.
6. Suomalainen, A., Majander, A., Pihko, H., Peltonen, L., and Syvänen, A.-C. (1993) Quantification of $tRNA_{3243}^{Leu}$ point mutation of mitochondrial DNA in MELAS patients and its effects on mitochondrial transcription. *Hum. Mol. Genet.* **2,** 525–534.
7. Chen, X. and Kwok, P. Y. (1997) Template-directed dye-terminator incorporation (TDI) assay: a homogeneous DNA diagnostic method based on fluorescence resonance energy transfer. *Nucleic Acids Res.* **15,** 347–353.
8. Pastinen, T., Perola, M., Niini, P., Terwilliger, J., Salomaa, V., Vartiainen, E., et al. (1998) Array-based multiplex analysis of candidate genes reveals two independent and additive risk factors for myocardial infarction in the Finnish population. *Hum. Mol. Genet.* **7,** 1453–1462.
9. Pastinen, T., Partanen, J., and Syvänen, A.-C. (1996) Multiplex, fluorescent solid-phase minisequencing for efficient screening of DNA sequence variation. *Clin. Chem.* **42,** 1391–1397.
10. Sitbon, G., Hurtig, M., Palotie, A., Lönngren, J., and Syvänen, A.-C. (1997) A colorimetric mini-sequencing assay for the mutation in codon 506 of the coagulation factor V gene. *Thrombosis Haemostasis* **77,** 701–703.
11. Ihalainen, J., Siitari, H., Laine, S., Syvänen, A.-C., and Palotie, A. (1994) Towards automatic detection of point mutations: use of scintillating microplates in solid-phase minisequencing. *BioTechniques* **16,** 938–943.

54

Direct Sequencing with Highly Degenerate and Inosine-Containing Primers

Zhiyuan Shen, Jingmei Liu, Robert L. Wells, and Mortimer M. Elkind

1. Introduction

Among the many techniques of cloning new genes, one approach involves degenerate primers *(1–7)*. The approach usually requires the following three steps:

1. Using degenerate primers to amplify part of the gene of interest by PCR: The degenerate primers' sequences may be designed from known protein sequences or conserved regions of a gene family (e.g., **ref.** *2,4*). Because deoxyinosine can base pair with all of the four deoxyribonucleotides, it has been substituted for specific nucleic acids in degenerate primers to reduce the number of different primer sequences that would otherwise be needed in the reaction *(2,7,8)*.
2. A determination of which amplified PCR product(s) is from the gene of interest: If the target gene and the primers are only partially homologous, a moderate annealing stringency in PCR is usually necessary to obtain amplification. Moderate stringency may result in multiple PCR products. Although from the size of the PCR products it may be possible to predict which is from the gene of interest, sequencing analysis of the PCR products may be required.
3. The screening of a cDNA library using the correct PCR product as a probe and cloning the gene of interest.

Sequencing the amplified PCR product is one of the most important steps in this approach to gene cloning. To sequence the PCR fragment amplified by degenerate inosine-containing primers, the PCR fragment may be cloned into a sequencing vector, such as M13 bacteriophage. Sequencing is straightforward if primers specific to the vector are used. Theoretically, this method allows any unknown cloned DNA fragment to be sequenced. However, the *Taq* polymerase, which is used to amplify the target fragment, is thought to have relatively high misincorporation rates for dNTPs, or $\sim 10^{-4}$. Hence, it is possible that a copy of the product may contain one or more incorrect nucleotides. If such a copy has been cloned into the sequencing vector, the resulting sequencing data would be incorrect for that particular clone. Direct sequencing of PCR products can circumvent this problem because most of the fragments are exact replicas of the target molecule. Thus, the majority of the products used for sequencing would have the right nucleotide at a specified position and result in the correct sequencing

From: *Methods in Molecular Biology, Vol. 226: PCR Protocols, Second Edition*
Edited by: J. M. S. Bartlett and D. Stirling © Humana Press Inc., Totowa, NJ

ladder. Also, direct sequencing of PCR products avoids the time-consuming cloning of PCR products, and most of the available direct PCR sequencing protocols require relatively small amounts of template.

Many protocols are available for the direct sequencing of PCR products. Most of the protocols require specific sequencing primers. However, this means at least part of the specific base sequence of the template is needed. This requirement may not be met and, in many cases, information about the internal sequence of a gene may be lacking. This shortcoming may apply to the PCR products of degenerate inosine-containing primers of the cDNA of a new gene. Therefore, one may be forced to use the same degenerate inosine-containing primers that were used in the PCR step for direct sequencing. When primers have low degeneracy, they may be treated as sequence-specific primers, and some of the direct-sequencing protocols, such as those described in this volume, may be used with success. When only highly degenerate inosine-containing primers are available, these methods may not succeed.

To sequence a PCR product amplified via the use of a highly degenerate inosine-containing primers, several general factors must be kept in mind.

1. The sequencing primer(s) must anneal specifically to one site on the DNA fragment that is to be sequenced, that is, a secondary annealing site must be avoided. Therefore, stringent primer annealing temperatures are necessary.
2. A sufficient quantity of the specific primer should anneal to the correct site. Consequently, the primer annealing temperature cannot be too high.
3. Reassociation of the double-stranded DNA template should be minimized. This requirement generally can be met by using optimal PCR protocols for the sequencing reactions.

To perform the requirements above, a primer-labeling method, in which the primer is labeled at the 5′ end, may be worth considering. Linear PCR is used to generate the labeled dideoxynucleotide-terminated sequences ***(9,10)***. The use of this method minimizes problems of template reassociation and/or mismatching of the primer because the annealing time is relatively short. Also, the annealing temperature is higher than it would be in most protocols that use DNA polymerases other than *Taq*, such as T4 DNA polymerase, but the method requires a 5′ end-labeling step for which ^{35}S is generally not suitable compared with ^{32}P because of its lower specific activity and the lesser efficiency with which some enzymes label 5′ ends with α-^{35}S ATP versus α-^{32}P ATP. Only ^{32}P can be used, even though its greater radiation hazard owing to its higher β-particle emission and its shorter half-life make it less convenient. Furthermore, when highly degenerate primers are used, higher primer concentrations in the reaction mixture are needed to insure that sufficient specific priming will occur. The preceding increases the hazard as well as the cost.

To assure that sequencing primer(s) anneal to a DNA template specifically, to eliminate the need for 5′ end labeling, and to avoid reassociation of the double-stranded DNA template, a two-step cycle-sequencing protocol is described to sequence products amplified with degenerate inosine-containing primers. This method uses the same degenerate primers that were used in PCR amplification. The method can be broken down into two steps of linear PCR. The first step is for labeling the primers, and the second is for the random dideoxy termination. As shown in **Fig. 1**, in the first step, primers were extended and labeled with α-^{35}S -dATP. The extension is limited and performed under conditions of high stringency, low dNTP concentration, and a short

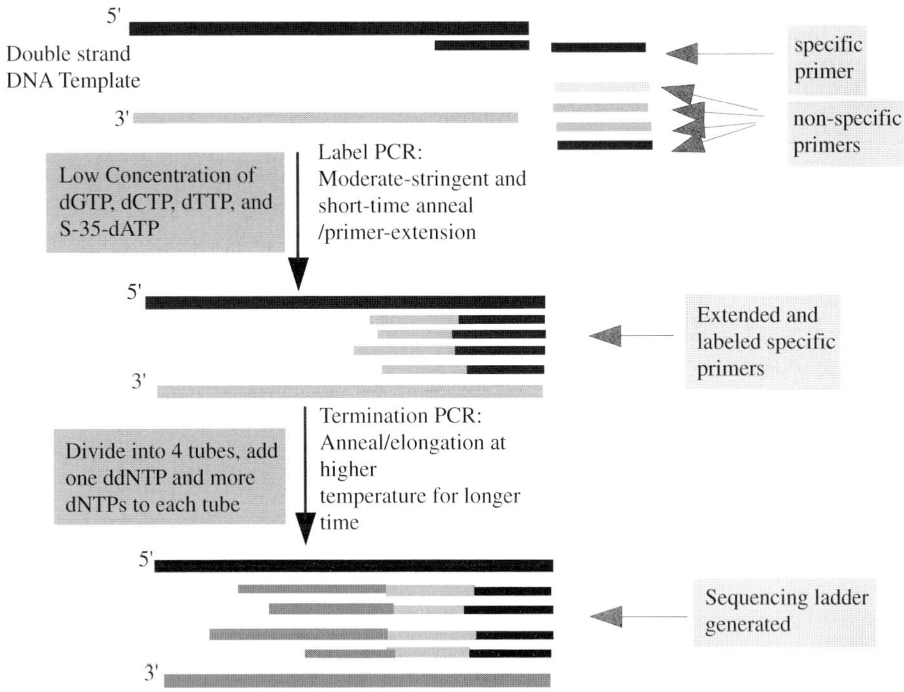

Fig. 1. Procedure of cycle sequencing with degenerate inosine-containing primers. Two linear PCR steps are involved. 1. Label PCR uses low dNTP concentrations, a low temperature, and short times for primer annealing/elongation to produce incomplete extension of specific primers. As a result, specific primers are labeled and extended. The extended and labeled primers have a higher melting temperature than the native printers. 2. Termination PCR using a higher annealing/elongation temperature and is performed with higher dNTP concentrations and in the presence of ddNTPs. Only the extended and labeled primers are involved in the termination reaction.

interval so that the specific primer in the mixture is favored and a limited length of primer extension is achieved. In the second step, dideoxynucleotide terminations are effected at a more stringent elevated annealing/elongation temperature. The result is that only the extended and labeled primers enter into the termination reactions.

We have used this method to sequence amplified cytochrome p450 cDNA fragments with a highly degenerate inosine-containing primer ***(1,2)***. In our case, a set of degenerate primers was used to amplify a presumably novel cytochrome *p450* gene(s). The upstream sense primer was a mix of 192, 20mer, containing three inosines that theoretically could anneal to 12,288 different sequences. The downstream antisense primer was a mix of 144, 23 mer, containing five inosines or 147,456 different possible sequences.

In what follows, we will only describe the sequencing reactions. Procedures for sequencing gel electrophoresis can be found in Chapter 51.

2. Materials

1. A thermal cycler: Cetus Perkin–Elmer Model 480 (*see* **Note 1**).
2. PCR tubes (0.5 mL).

3. Mineral oil.
4. Gel-purification kits/reagents, such as QIAEX Gel Extraction Kit (Qiagen #20020, Chatsworth, CA) or QiaOuick Gel Extraction Kit (Qiagen #28704).
5. All buffers and solutions must be free of DNase.
6. α-^{35}S-dATP (10 µCi/µL 1000 Ci/mmol; Amersham Corp).
7. Sequencing primers (degenerate primers) dissolved in H_2O, or 0.1× TE buffer.

   The sequencing reaction reagents can be homemade. However, we recommend purchasing them from a commercial company to ensure uniform performance. In the following materials, we include the catalog number for US Biochemicals (Cleveland, OH).
8. Reaction buffer (USB #71030): 260 m*M* Tris-HCl, pH 9.5; 65 m*M* $MgCl_2$.
9. Δ*Taq* DNA polymerase (USB #71059) or *Taq* DNA polymerase, (USB# 71057): 32 U/µL.
10. *Taq* DNA polymerase dilution buffer (USB #71051): 10 m*M* Tris-HCl, pH 8.0, 1 m*M* 2-mercaptoethanol, 0.5% Tween-20, and 0.5% Nonidet P-40.
11. Four separate primer label mixes: dGTP label mix: 3.0 µ*M* (USB #71034); dATP label mix: 3.0 µ*M* (USB #71036); dTTP label mix: 3.0 µ*M* (USB #71037); and dCTP label mix: 3.0 µ*M* (USB #71038).
12. Four separate termination mixes: ddG terminator mix: 15 µ*M* each dGTP, dATP, dTTP, dCTP, and 22.5 µ*M* ddGTP (USB #71020); ddA termination mix: 15 µ*M* each dGTP, dATP, dTTP, dCTP, and 300 µ*M* ddATP (USB #71035); ddT termination mix: 15 µ*M* each dGTP, dATP, dTTP, dCTP, and 450 µ*M* ddTTP (USB #71040); and ddC terminator mix: 15 µ*M* each dGTP, dATP, dTTP, dCTP, and 75 µ*M* ddCTP (USB #71025).
13. Stop/gel-loading solution (USB #70724): 95% formamide, 20 m*M* EDTA, 0.05% bromophenol blue, and 0.05% xylene cyanol FF.
14. 1× TE buffer: 10 m*M* Tris-HCl, pH 8.0, 1 m*M* EDTA.

3. Methods

3.1. Preparation of DNA as a Sequencing Template (see Note 2)

1. After gel electrophoresis, PCR fragment(s) of interest is cut out from the gel.
2. DNA in the gel is purified with the Qiagen gel-purification kit, and final PCR products are resuspended in a proper amount of 0.1× TE buffer.
3. To estimate the amount of PCR product, run an aliquot of the PCR products on an agarose gel. The amount of DNA may be estimated by a comparison with the amount of DNA that was used in the mol-wt ladder.

   In the following steps, always keep tubes on ice, unless otherwise indicated.
4. Prepare the following labeling PCR mix (*see* **Note 3**):

H_2O	0–8 µL
DNA (in 0.1× TE) (need total of 25–100 ng)	1–9 µL
Reaction buffer	2 µL
Degenerate primers (5–200 µ*M*)	1 µL
dGTP label mix	1 µL
dCTP label mix	1 µL
dTTP label mix	1 µL
α-^{35}*S*-dATP (10 µCi/µL >1000 Ci/mmol)	0.5 µL
Taq DNA polymerase (4 U/µL) (diluted in *Taq* dilution buffer)	2 µL
Total volume	17.5 µL

   Cover the label PCR mix with 10–20 µL of mineral oil.

3.2. Labeling PCR (see Note 4)

1. Run the following PCR program: presoak at 94°C for 3 to 5 min followed by 45 cycles of 95°C for 30 s and 52°C for 30 s.

2. Transfer 15 to 16 μL of the above labeled mixture to a new tube. Avoid carryover of any mineral oil. This can be done easily by putting the pipetting tip directly below the oil without touching the wall of the tube.
3. Optional (*see* **Note 5**): Load 1 to 2 μL with 1 μL of gel-loading buffer to a sequencing gel to check the labeling efficiency.
4. Termination PCR mix: For each of the labeled mixes, prepare four tubes labeled as "G," "A," "T," and "C." To each of the tubes, add 4 μL of termination mix G, A, T, or C (this can be done toward the end of label-PCR procedure). Add 3.5 μL of the label mix to each of the tubes. Cover the termination PCR mix with 8 to 10 μL of mineral oil.
5. Termination PCR: Cycle between 95°C for 30 s and 72°C for 90 s (*see* **Note 6**).
6. While the termination PCR is under way, prepare four clean 0.5-mL tubes labeled G, A, T, or C. To each of them add 4 μL of stop/gel-loading solution.
7. Transfer 6 to 7 μL of termination mix to these tubes with the stop/loading solution. Avoid carryover of mineral oil. Mix and spin down briefly and store at –20°C (good for up to 1 mo). These samples are ready for the sequencing gel (use 3 μL to load a gel). *See* Chapter 51.
8. Sequencing results: run sequencing gel, perform autoradiography, and read the sequence (*see* **Note 7**).

4. Notes

1. Cetus Perkin–Elmer thermal cycler Model 9600 also may be used. If it is, use 0.1-mL tubes; no mineral oil on the top of the reaction solution is needed. The PCR program should be adjusted accordingly in the procedure.
2. Other methods of DNA preparation are also acceptable as long as "clean" DNA is obtained.
3. Other {a-^{35}S-labeled nucleotides may also be used, but the label mix must be changed accordingly. The concentration of the stock of degenerate primers in the reaction is dependent on the degree of degeneracy. In our case, the stock concentration of our >100∞ degenerate primer was 200 μ*M*. Because radioactive ^{35}S is used for these experiments, always be careful and follow the safety operation procedure for your institute. Check with your radiation safety officer for the authorized amount of radioactivity that you can handle at any one time.
4. Depending on the sequencing primer, the annealing/elongation temperature or time may have to be optimized to give proper primer extension and labeling. The purposes of the label PCR is to have a sufficient amount of specific primer in the primer mix to anneal to a specific site on the DNA template and to extend the annealed primer for a limited nucleotide length with *Taq* DNA polymerase. The first purpose can be achieved by choice of an optimal annealing temperature and/or time. In our case (200 pmol of the 20 mer with inosine and a degeneracy of more than 100∞), we used 52°C and 30 s. Depending on circumstances, this temperature and the annealing time may need to be adjusted. The method of generating a limited elongated primer plus labeling of the primer is accomplished by using a shorter annealing/elongation time at a suboptimal temperature (for *Taq* activity), but still a stringent temperature for annealing and a low dNTP concentration. In this way, it is not necessary to know the sequence of the downstream flanking region of the sequencing primer.
5. Loading 1 to 2 μL of labeled mix to run a gel to check that the length of primer extension and label efficiency is optional. This can be run along with the sequencing sample after all the reactions are finished. Using our p450 degenerate sequencing primers under these labeling PCR conditions, we obtained an average primer extension of 15 to 25 bp.
6. The temperature used for both the annealing of labeled/extended primers to the DNA template for the elongation/termination reaction was 72°C. Only the prelabeled and pre-

extended primers, which are the specific primers in the primer mix, would be allowed during the termination/elongation because of the elevated temperature. If more template DNA is available, fewer cycles may be used.

7. Depending on the length of labeled primers, the readable sequence will vary. For our case of highly degenerate inosine-containing primers of p450 genes (*see* **Subheading 1.** for a description of our p450 primers) a ladder from 25 bp downstream of the primer was readable up to 300 bp.
8. A similar protocol of this method would be to omit one of the four dNTPs in the label step and use at least one α-^{35}S-labeled dNTP in the labeling mix. This will give an incomplete elongation of the sequencing primer during the labeling step because the primer extension will stop at the proper position when the omitted nucleotide is not present. The elongated primers may be labeled if the labeled nucleotide is by chance present between the sequence primer and the omitted nucleotide. This method is useful to sequence DNA when some sequence information immediately downstream from the sequencing primer is available. In such a case, one can decide which nucleotide to omit or to label in the label mix.

References

1. Shen, Z., Liu, J., Wells, R. L., and Elkind, M. M. (1993) Cycle sequencing using degenerate primers containing inosines. *BioTechniques* **15,** 82–89.
2. Shen, Z., Wells, R. L., Liu, J., and Elkind, M. M. (1993) Identification of a cytochrome p450 gene by reverse transcription-PCR using degenerate primes containing inosines. *Proc. Natl. Acad. Sci. USA* **90,** 11,483–11,487.
3. Shen, Z., Liu, J., Wells, R. L., and Elkind, M. M. (1994) cDNA cloning, sequence analysis, and induction by aryl hydrocarbons of a murine cytochrome p450 gene, Cyplbl. *DNA Cell Biol.* **13,** 763–769.
4. Shen, Z., Denison, K., Lobb, R., Gatewood, J., and Chen, D. J. (1995) The human and mouse homologs of yeast RAD52 genes: cDNA cloning, sequence analysis, assignment to human chromosome 12pl2.2-pl3, and mRNA expression in mouse tissues. *Genomics* **25,** 199–206.
5. Compton, T. (1990) Degenerate primers for DNA amplification, in *PCR Protocol, a Guide to Methods and Applications* (Innis, M. A., Gelfand, D. H., Sninsky, J. J., and White, T. J., eds.), Academic, San Diego, CA, pp. 39–45.
6. Lee, C. C. and Caskey, C. T. (1990) cDNA cloning using degenerate primers, in *PCR Protocol, a Guide to Methods and Applications* (Innis, M. A., Gelfand, D. H., Sninsky, J. J., and White, T. J., eds.), Academic, San Diego, CA, pp. 46–59.
7. Knoth, K. S., Roberds, S., Poteet, C., and Tamkun, M. (1988) Highly degenerate inosine-containing primers specifically amplify rare cDNA using the polymerase chain reaction. *Nucleic Acids Res.* **16,** 10932.
8. Erlich, H. A., Gelfand, D., and Sninsky, J. J. (1991) Recent advances in the polymerase chain reaction. *Sciences* **252,** 1643–1651.
9. Murray, V. (1989) Improved double strand DNA sequencing using the linear polymerase chain reaction. *Nucleic Acids Res.* **17,** 8889.
10. Smith, D. P., Jonstone, E. M., Little, S. P., and Hsiung, H. M. (1990) Direct DNA sequencing of cDNA inserts from plaques using the linear polymerase chain reaction. *Biotechniques* **9,** 48–52.

55

Determination of Unknown Genomic Sequences Without Cloning

Jean-Pierre Quivy and Peter B. Becker

1. Introduction

The inherent problems of sensitivity and specificity that one encounters when trying to determine a particular nucleotide sequence directly in its genomic context can be overcome by selective amplification of the region of interest. This amplification of the target DNA is usually achieved by one of two strategies: The relevant piece of DNA may be cloned and therefore amplified in a bacterial cell or, alternatively, the desired fragment may be amplified in vitro using PCR technology. Both strategies have drawbacks. The cloning of a specific genomic sequence is labor intensive, lengthy, and sometimes even difficult to achieve. The PCR amplification requires that enough sequence information is known to be able to design the two specific amplification primers and is therefore limited to sequencing alleles of already-known DNA. There are, however, many cases that would benefit from the determination of unknown genomic sequence close to a known piece of DNA. With a particular cDNA in hand, one may wish, for example, to determine genomic gene sequences, such as the promoter of the gene, its introns, or 5′- and 3′-nontranscribed regions. The protocol presented here uses ligation-mediated polymerase chain reaction (LM-PCR) to amplify unknown genomic DNA next to a short stretch (about 100 bp) of known sequence and details a convenient procedure to determine the new sequence by dideoxy sequencing *(1)*. The procedure may form the basis for "walking sequencing" strategies to determine large regions of continuous sequence information starting from a limited piece of known DNA.

The central feature of the LM-PCR technique is the ligation of a known short oligonucleotide, the "linker," to selected ends of genomic DNA fragments (**Fig. 1**). These generic linker sequences provide the second primer for amplification of linked fragments in combination with an oligonucleotide based on known sequences. LM-PCR was first introduced for genomic footprinting and chemical sequencing *(2,3)*. The disadvantages of chemical sequencing over the chain termination method *(1)* prompted us to adapt LM-PCR technology for direct dideoxy sequencing of genomic DNA *(4)*. The underlying procedure is derived from a variation of the original LM-PCR protocol called "Linker Tag Selection LM-PCR" *(5)*.

From: *Methods in Molecular Biology, Vol. 226: PCR Protocols, Second Edition*
Edited by: J. M. S. Bartlett and D. Stirling © Humana Press Inc., Totowa, NJ

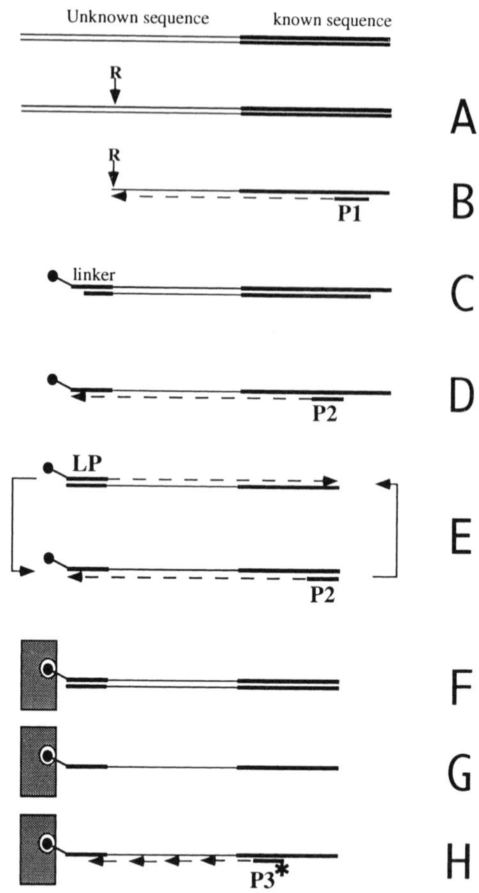

Fig. 1. The use of Linker Tag Selection LM-PCR for genomic dideoxy sequencing. Bold lines denote known sequences, and thin lines the unknown sequences to be determined. The arrows and dotted lines indicate primer extension reactions. R: Strategic restriction site. P1, P2, P3, and LP stand for the primers 1, 2, 3, and the linker primer, respectively. The biotin moiety on the linker is represented by a filled circle, and the streptavidin-coated paramagnetic beads by the shaded boxes. (H): Radiolabeled P3 is used to prime dideoxy sequencing reactions.

The steps of the reaction are outlined in **Fig. 1**. A restriction enzyme is selected that cleaves the genomic DNA somewhere within the unknown sequence but within 1 kb from the known sequence for which the specific primers have been designed. Cleavage creates a defined end (A). The cleaved genomic DNA is denatured, and the gene-specific primer 1 is annealed to the known sequence and extended by a polymerase until the end of the restriction fragment (B). This creates a blunt end to which a short double-stranded linker is ligated (C). The linker DNA consists of two complementary oligonucleotides, the longer one being biotinylated at its 5′-end. After denaturation, the specific primer 2, representing sequences more 3′ from primer 1 on the lower strand of the known sequence, is then annealed to the upper strand, which now carries the biotinylated linker oligonucleotide at its 5′-end. The primer is again extended to the end (D) creating a double-stranded fragment that contains a region of unknown DNA flanked by known sequences. The genomic sequences can now be

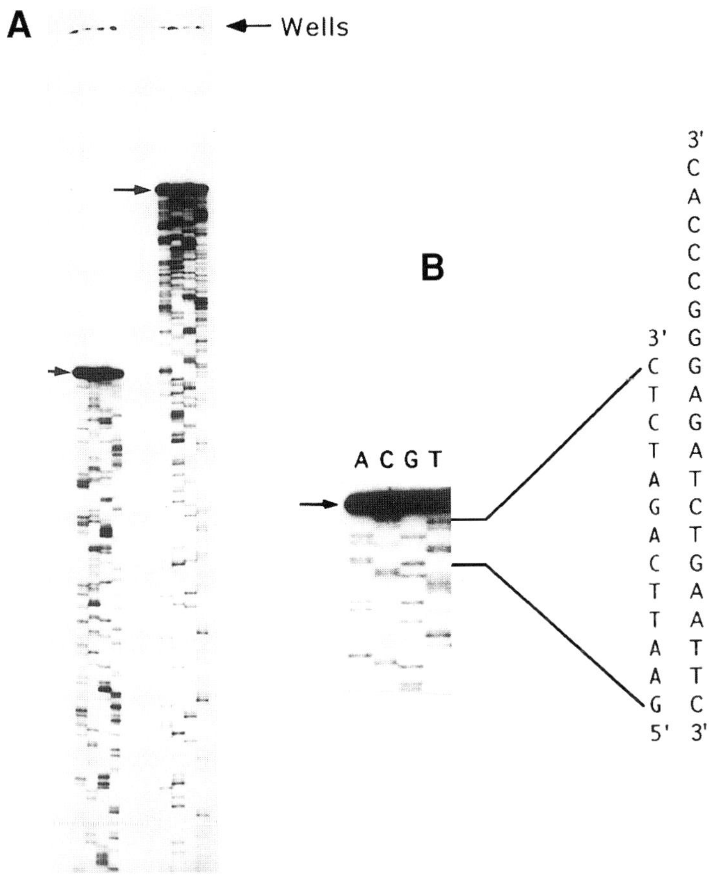

Fig. 2. Genomic sequences of the *Drosophila* hsp27 promoter region obtained following the outlined procedure. Arrows indicate the full-length restriction fragment. **(A)** Long and short gel runs of the same sequencing reaction obtained using the enzyme *Nru*I, which cut 821 bases upstream of the 5′-end of primer 1. **(B)** Upper part of a sequence obtained using the enzyme *Pst*I cutting 332 bases away from the 5′-end of primer 1. The sequence of the linker oligonucleotide at the end of the genomic sequence can be unambiguously identified.

amplified by PCR using a combination of the biotinylated linker primer and primer 2 (E). The biotinylated amplification products are then immobilized on streptavidin-coated paramagnetic beads and purified from the PCR in a magnetic field (F). This step efficiently removes unincorporated primers, which interfere with the subsequent sequencing reaction. The immobilization also facilitates the handling of the template fragments during the subsequent steps and specifically allows the sequencing of a single-stranded template. The immobilized fragments are denatured, and the complementary strand is removed by washes (G). Radioactively labeled specific primer 3, again located 3′ to primer 2 on the lower strand, now serves to prime a standard chain termination sequencing reaction that finally generates the sequencing ladder (H). Sequences start to be readable about 25 bases 3′ from primer 3, and may extend for up to 1 kb depending on the location of the restriction site and the efficiency of the overall process (**Fig. 2A**).

Important features of the procedure are the use of a proofreading polymerase for the PCR amplification (the Vent DNA polymerase possesses a 3′-5′ exonuclease activity) that decreases the error rate during the amplification, and the use of an enzyme without the 3′-5′ exonuclease activity (e.g., Vent Exo) for efficient single primer extensions. The immobilization of the amplified fragments on paramagnetic beads exploiting the strong streptavidin/biotin interaction is crucial for the efficiency of the sequencing reaction because it allows the efficient removal of interfering primers from the PCR and permits the use of a single-strand template ***(4,5)***. The attachment of the sequenced DNA fragment to the solid support does not create a steric hindrance for the polymerase. Frequently, the sequence of the linker oligonucleotide itself can be determined at the very end of the genomic sequence (*see* **Fig. 2B**).

The length of sequence determined critically depends on the ability to resolve long fragments with single nucleotide resolution, provided that the restriction site used for linker ligation is not too close to the known sequence. Fragments of over 800 bp have yielded reliable sequence information (**Fig. 2A**). Longer sequences can be obtained in walking strategies where the newly determined sequence is in turn used to design further reaching sets of primers. The presented strategy critically relies on the previous identification of a suitable restriction site, ideally between 0.5 and 1 kb away from the known sequence. Too short fragments will yield little new sequence information, whereas large fragments (exceeding 1 kb) do not work, presumably because of decreasing efficiencies in DNA denaturation and primer extensions reactions. Because any kind of restriction enzyme will work, a site can be conveniently identified on a Southern blot testing a small selection of enzymes that cut the genome at a reasonable frequency. Alternatively, a selection of enzymes can simply be tried at random in an LM-PCR sequencing reaction. To increase the chances that the reaction will work, the genomic DNA may be cleaved with a whole cocktail of enzymes that collectively have a high likelihood to produce a suitable restriction fragment.

2. Materials

2.1. Purification and Restriction of Genomic DNA

1. Suspension of nuclei from desired organism (*see* **Note 1**).
2. 0.5 *M* EDTA, pH 8.0.
3. RNase A, DNase free, 10 mg/mL (Boehringer, Mannheim).
4. Aqueous solution of *N*-lauroylsarkosine (sarkosyl), 20% (P/V) (Sigma).
5. Proteinase K, 10 mg/mL (Merck).
6. Phenol, highest quality, neutralized, and equilibrated with TE (10 m*M* Tris-HCl, pH 7.5; 1 m*M* EDTA; Aurresco).
7. Phenol/chloroform/isoamyl alcohol mixture (25:24:1; Aurresco).
8. Chloroform:isoamyl alcohol (24:1; Merck).
9. 0.3 *M* sodium acetate, pH 5.2.
10. Ethanol 100%.
11. Ethanol 80%.
12. TE: 10 m*M* Tris-HCl, pH 7.5, 1 m*M* EDTA.
13. Restriction enzyme with suitable 10× reaction buffer (*see* **Note 2**).

2.2. Primer Design, Primer Purification, and Annealing of the Linker Primer (see Note 3)

The primers were synthesized on an ABI 394 DNA synthesizer and gel purified (*see* **Subheading 3.2.**).

1. Primer 1 (P1): a 18-22 mer with a calculated T_m around 45 to 50°C. Working concentration: 0.5 pmol/µL.
2. Primer 2 (P2) should have the same melting temperature as the long-linker primer (*see* **Note 4**) used for the PCR amplification, in our case a 25 to 27 mer with a calculated T_m of 60 to 65°C. The guanine-cytosine content is usually between 45 and 55%. It does not need to overlap with P1, but a 5-bp overlap has worked. It should be internal to primer 2 to increase the specificity of the overall reaction. Working concentration: 10 pmol/µL.
3. Primer 3 (P3): an 18 to 22 mer with a calculated T_m around 45 to 50°C, but longer oligos with higher GC contents will also work. It should be internal to P2 to increase the specificity. It will be ^{32}P kinased with an SA of 10^7 cpm/pmol (*see* **Subheading 3.3.**). Working concentration: 0.2 pmol/µL.
4. Long-linker oligonucleotide: 5′ CACCCGGGAGATCTGAATTC 3′ (*see* **Note 4**). It is biotinylated at its 5′-end during synthesis by incorporation of a biotin-2-*o*-propylphosphoramidite. It should be unphosphorylated.
5. Short-linker oligonucleotide: 5′ GAATTCAGATC 3′, dephosphorylated.
6. Oligo loading mix: 10% glycerol in formamide.
7. Denaturing polyacrylamide gel: 14.5% acrylamide, 0.5% bis-acrylamide, 7 *M* urea, 1× TBE. Size: 25 × 25 × 0.1 cm.
8. Formamide loading buffer: 96% formamide, 0.05% xylene cyanol, 0.05% bromophenol blue, 10 m*M* EDTA.
9. PE buffer: 50 m*M* Tris-HCl, pH 7.5, 100 m*M* NaCl, 1 m*M* EDTA, 1% phenol (v/v).
10. Chloroform : isoamyl alcohol (24 : 1; Merck).
11. 5 *M* LiCl.
12. 1 *M* $MgCl_2$.
13. Ethanol 100%.
14. Ethanol 80%.
15. TBE: 90 m*M* Tris-borate, 1 m*M* EDTA, pH 8.3.

2.3. Kinasing of Primer 3

1. P3 at 10 pmol/µL.
2. Polynucleotide kinase buffer 10×: 700 m*M* Tris-HCl, pH 7.6, 100 m*M* $MgCl_2$, 50 m*M* DTT.
3. γ^{32}PATP, 5000 Ci/mmol (Redivue, Amersham).
4. Polynucleotide kinase 10 U/µL (NE Biolabs).
5. 50 m*M* EDTA, pH 8.0.
6. G-25 fine spin columns (Boehringer, Mannheim).

2.4. First Primer Extension

1. P1 at 0.5 pmol/µL in TE, pH 7.5.
2. 1 *N* NaOH.
3. TES buffer: 560 m*M* TES, free acid (Sigma), 240 m*M* HCl, 100 m*M* $MgCl_2$.
4. Vent buffer (10×): 100 m*M* KCl; 100 m*M* $(NH_4)_2SO_4$, 200 m*M* Tris-HCl, pH 8.8, 20 m*M* $MgSO_4$, 0.1% Triton X-100 (NE Biolabs).
5. dNTP solution: 10 m*M* dNTPs, (Boehringer, Mannheim); keep in small aliquots at –20°C. Do not freeze/thaw more than three times.
6. Vent Exo DNA polymerase, 2 U/µL (NE Biolabs).

2.5. Ligation

1. Ligase buffer (10×): 500 m*M* Tris-HCl, pH 7.5, 100 m*M* $MgCl_2$, 100 m*M* DTT, 10 m*M* ATP, 250 µg/mL BSA (NE Biolabs).

2. Solution of 40% PEG 8000 (Sigma), filtered through a 0.22-μm filter. It will take some time to dissolve the PEG in water. Incubate at room temperature on a rotating wheel for several hours. It also takes some force to filter the solution using a syringe.
3. T4 DNA ligase, 400 U/μL (NE Biolabs).
4. Annealed linker (*see* **Subheading 3.4.**).

2.6. PCR

1. TE: 10 m*M* Tris-HCl, pH 8.5; 1 m*M* EDTA, pH 8.5.
2. Phenol:chloroform:isoamyl alcohol (25:24:1; Aurresco).
3. Solution of 7.5 *M* ammonium acetate containing 25 μg/mL yeast tRNA (Boehringer Mannheim). Crude yeast tRNA has to be cleaned by multiple organic extractions and ethanol precipitation.
4. Vent buffer (*see* **Subheading 2.4.**).
5. dNTP solution (*see* **Subheading 2.4.**).
6. 100 m*M* $MgSO_4$.
7. P2 solution, 10 pmol/μL (*see* **Subheading 2.2.**).
8. Long-linker primer, 10 pmol/μL (*see* **Subheading 2.2.**).
9. Vent DNA polymerase 2 U/μL (NE Biolabs).
10. Perkin–Elmer thermal cycler.
11. PCR tubes (Perkin–Elmer).
12. Mineral oil (PCR-grade).

2.7. Tag Selection of the PCR Products

1. Dynabeads M-280 streptavidin (Dynal, Oslo, 10 mg/mL).
2. Magnetic particle concentrator (MPC, Dynal).
3. Phosphate-buffered saline (PBS), pH 7.4.
4. PBS, pH 7.4; 0.01% BSA (molecular biology grade).
5. BW solution: a 1:1 mixture of TE and 5 *M* NaCl.

2.8. Sequencing Reaction (see Note 7)

1. MPC.
2. 150 m*M* NaOH, freshly prepared.
3. TE, pH 7.5.
4. Vent buffer (*see* **Subheading 2.4.**).
5. Termination mixes made up in 1× vent buffer:
   A-mix: 900 μ*M* ddATP, 30 μ*M* dATP, 100 μ*M* dCTP, 100 μ*M* dGTP, 100 μ*M* dTTP.
   C-mix: 480 μ*M* ddCTP, 30 μ*M* dATP, 37 μ*M* dCTP, 100 μ*M* dGTP, 100 μ*M* dTTP.
   G-mix: 400 μ*M* ddGTP, 30 μ*M* dATP, 100 μ*M* dCTP, 37 μ*M* dGTP, 100 μ*M* dTTP.
   T-mix: 720 μ*M* ddTTP, 30 μ*M* dATP, 100 μ*M* dCTP, 100 μ*M* dGTP, 33 μ*M* dTTP.
6. Labeled P3 (*see* **Subheading 3.3.**).
7. Circumvent sequencing buffer 10×: 100 m*M* KCl, 100 m*M* $(NH_4)_2SO_4$, 200 m*M* Tris-HCl, pH 8.8, 50 m*M* $MgSO_4$ (NE Biolabs).
8. Triton X-100 solution, 3% (v/v) in water.
9. Vent Exo DNA polymerase, 2 U/μL (NE Biolabs).
10. Formamide loading buffer (*see* **Subheading 2.2.**).
11. Sequencing gel.
12. Fixing solution: 10% acetic acid, 10% methanol.
13. Dupont NEN reflection™ films and corresponding cassettes.

3. Methods

3.1. Purification of Genomic DNA and Restriction (see Note 1)

1. Isolate nuclei from cells of interest by suitable methods (*see* **Note 1**) and spin them down to obtain the nuclear pellet.
2. Resuspend the pellet in 1 mL of 0.5 *M* EDTA. Avoid harsh vortexing and vigorous pipetting to prevent shearing (*see* **Note 1**).
3. Add 25 µL of RNase A and 25 µL of Sarkosyl, mix by inverting the tube, and incubate for 3 h at 37°C on a rotating wheel.
4. Add 25 µL of proteinase K, mix by inverting the tube, and incubate overnight at 37°C on a rotating wheel.
5. Add 1 mL of phenol and mix by inverting the tube several times. Spin to separate the phases and collect the lower phase and interphase (the lower phase is the aqueous phase owing to the high density of 0.5 *M* EDTA).
6. Repeat **step 5**, but do not take the interphase.
7. Add 1 mL of phenol:chloroform and collect the upper phase (which is now the aqueous phase).
8. Dialyze overnight (or longer) against TE, pH 7.5, with at least four changes of TE. Avoid a large increase in volume by keeping the dialysis bag tight.
9. Precipitate DNA with one-tenth vol of 0.3 *M* Na acetate and 2.5 vol of cold absolute ethanol.
10. Spin at 4°C to collect DNA pellet, wash with 80% ethanol, remove residual ethanol, but do not dry too long since the DNA will be difficult to redissolve.
11. Dissolve DNA in TE, pH 7.5, and store at 4°C.
12. Restriction digest of genomic DNA: Digest 10 µg of DNA with 1 U/µg of restriction enzyme according to the manufacturer's recommendation (*see* **Notes 2** and **4**) for at least 3 h (up to overnight).
13. Extract DNA once with phenol, once with phenol:chloroform, once with chloroform, and precipitate with one-tenth vol of 0.3 *M* Na acetate and 2.5 vol of cold 100% ethanol.
14. Spin at 4°C, wash DNA pellet with 80% ethanol, and remove residual ethanol in the SpeedVac. Again, do not dry too long. Resuspend DNA in TE, pH 7.5, and adjust concentration to 1 µg/µL (OD at Aµ260).
15. Extract once more with chloroform, and remove traces of chloroform in the SpeedVac.

3.2. Purification of Oligonucleotide Primers and Annealing of the Linker Fragment

1. Dry down 75 nmol of the oligonucleotide in the SpeedVac concentrator and dissolve in 75 µL of oligo loading mix. Heat for 5 min at 75°C and load 5∞15 µL onto a prerun denaturing polyacrylamide gel. In a separate slot load some formamide-loading buffer to monitor the run. Electrophorese until the bromophenol marker dye migrates to two thirds of the gel.
2. By ultraviolet shadowing ***(6)***, locate the band corresponding to the full-length oligonucleotide and excise from the gel using a razor blade.
3. Transfer the polyacrylamide gel slice to a 1.5-mL reaction tube containing 1 mL of PE buffer and incubate overnight at 37°C.
4. Filter the supernatant through a 0.22-µm filter, prewetted with PE, with the help of a 2-mL syringe.
5. Wash another 100 µL of PE buffer through the filter.
6. Extract the pooled PE solutions with chloroform and dispense 2∞450 µL of the upper phase into fresh tubes.

7. Precipitate the oligonucleotides by the addition of 36 µL of 5 *M* LiCl, 4.5 µL of 1 *M* $MgCl_2$, and 1 mL cold 100% ethanol.
8. Mix by vortexing and let precipitate at –70°C for 15 min.
9. Spin at 4°C for 20 min in a tabletop centrifuge, wash the pellet with 80% ethanol, dry in the SpeedVac, and resuspend the oligonucleotide in TE, pH 7.5.
10. Determine the concentration (OD at A_{260}) and dilute some aliquots to the working concentration.
11. Annealing of linker oligos: combine in a 1.5-mL reaction tube: 20 pmol/µL of each linker oligonucleotide in 250 m*M* Tris (pH 7.5), 5 m*M* $MgCl_2$. Heat at 95°C for 5 min, transfer to a beaker containing boiling water, and allow to cool slowly at room temperature for 5 h (up to overnight in the cold room). Aliquot the linker solution and store at –20°C. Aliquots are only used once and never refrozen.

3.3. Kinasing of Primer 3

1. Combine in a tube 1 µL of P3 (10 pmol), 3 µL of 10× T4 polynucleotide kinase, and 10.5 µL of water.
2. Add 15 µL of γ^{32}PATP and 0.5 µL of polynucleotide kinase.
3. Incubate for 30 min at 37°C.
4. Add 20 µL of 50 m*M* EDTA, pH 8.0, and heat at 65°C for 10 min.
5. Purify the labeled oligonucleotide from the unincorporated label by a G-25 spin column.

3.4. First Primer Extension

1. Combine in a tube: 0.5 to 1 µg of restricted genomic DNA (*see* **Subheading 3.1.** and **Note 4**), 1 µL P1 (0.5 pmol), 1 µL 1 *N* NaOH, and water to 8 µL.
2. Incubate at 65°C for 5 min.
3. Immediately add 2 µL of TES buffer and mix.
4. Spin to collect and incubate for 10 min at room temperature.
5. Add 9 µL of a mix containing 2 µL of 10× Vent buffer, 0.4 µL of 10 m*M* dNTPs, 6.6 µL of H_2O.
6. Incubate at 50°C for 10 to 20 min.
7. Add 1 µL of the Vent Exo (2 U).
8. Incubate for 10 min at 76°C.
9. Chill on ice, spin to collect liquid, and proceed immediately to ligation.

3.5. Ligation

1. Prepare a premix containing 5 µL of 10× ligase buffer, 19 µL of 40% PEG 8000, and 5 µL of annealed linker (20 pmol/µL). Mix well by pipetting since the resulting solution is very viscous.
2. Add 29 µL of this premix to the first primer extension reaction (*see* **Section 3.4.**).
3. Add 1 µL of T4 DNA ligase (400 U), mix well by pipetting, and incubate overnight at 17°C.

3.6. PCR

1. To the ligation reaction, add 150 µL of TE, pH 8.5, and mix by vortexing.
2. Add 150 µL of phenol:chloroform:isoamyl alcohol (25:24:1) mix by vortexing, and then spin for 5 min.
3. Collect the upper aqueous phase and transfer to a tube containing 10 µL of 7.5 *M* NH_4Ac/yeast tRNA. Add 750 µL of cold 100% ethanol, mix by vortexing, and let precipitate for 15 min at –20°C.

4. Centrifuge for 20 min at 4°C, remove the supernatant, wash the pellet with 500 µL of 80% ethanol, and dry in the SpeedVac.
5. Dissolve DNA in 20 µL of water, transfer to a 500-µL PCR tube, and keep on ice.
6. Prepare a premix containing 5 µL of 10× Vent buffer, 1 µL of 10 m*M* dNTPs, 1 µL of 100 m*M* $MgSO_4$ (*see* **Note 5**), 1 µL of P2 solution, 1 µL of long-linker primer, and 19.5 µL of water.
7. Immediately before use, add 1.5 µL of Vent DNA polymerase (3 U) to the premix and mix by pipetting. Add premix to the PCR tube containing the DNA, then add two drops of mineral oil, and keep on ice.
8. Transfer the tube to the thermal cycler (one droplet of mineral oil in sample holders) preheated to 95°C, incubate for 2.5 min at 95°C, and subject to 18 cycles as follows: 95°C for 1 min, 60°C for 2 min (*see* **Note 5**), and 76°C for 3 min. Allow an increase of 5 s/cycle for the extension step, and end the cycling by a 10-min incubation at 76°C.
9. The PCR can be used immediately for the labeling step or stored frozen at –20°C.

3.7. Linker Tag Selection of the PCR Product

1. Resuspend the Dynabeads M-280 streptavidin well and take 50 µL (500 µg beads) into a 1.5-mL reaction tube.
2. Concentrate in the magnetic rack (MPC, Dynal), and remove the supernatant (0.01% BSA; *see* **Note 6**).
3. Wash the beads with 100 µL of PBS (pH 7.4), concentrate, and remove the supernatant. Repeat this step.
4. Resuspend beads in 100 µL of PBS/0.01% BSA by pipetting up and down until a homogenous suspension is achieved. Concentrate and remove supernatant.
5. Wash the beads with 100 µL of BW solution by pipetting up and down until no aggregates are seen. Concentrate again. Repeat this step.
6. Finally resuspend the beads in 100 µL of BW solution. The beads are now ready to be used.
7. Add the PCR to the beads and mix well. Avoid mineral oil, which spoils the magnetic separation of beads (we do not recommend a chloroform extraction because traces of this solvent adversely affects later steps in the reaction).
8. Incubate the tube on a rotating wheel at room temperature for 30 min. Assure that the beads do not sediment in the tube but also avoid a spreading of the liquid over the entire tube wall. A suitable agitation can be obtained by adjusting the rotation angle.
9. Concentrate beads and discard supernatant.
10. Wash the beads with 100 µL of BW solution.

3.8. Template Denaturation and Sequencing Reaction (see Note 7)

1. Resuspend beads well in 100 µL 150 m*M* NaOH by pipetting up and down. Incubate at room temperature for 5 min with occasional gentle agitation, and then for 2 min at 50°C.
2. Spin shortly to collect condensed liquid and beads trapped in the lid. Concentrate beads.
3. Remove the supernatant and resuspend the beads in 100 µL of 150 m*M* NaOH. Spin shortly to collect all the NaOH solution and concentrate beads.
4. Discard supernatant, resuspend the beads in 100 µL of TE, pH 8.5, and concentrate again. Repeat this wash with TE.
5. Resuspend the beads in 50 µL of Vent buffer and keep on ice.
6. Prepare four tubes labeled A, C, G, and T containing 3 µL of the respective termination mixes (*see* **Subheading 2.8.**).

7. Prepare a premix containing 0.2 pmol labeled P3, 1.5 µL of 10× circumvent sequencing buffer, 1 µL of Triton X-100, and add water to 15 µL.
8. Concentrate beads and then remove supernatant.
9. Resuspend the beads in 15 µL of premix.
10. Transfer 3.5 µL of this bead suspension into each tube containing a termination mix and mix by pipetting.
11. Heat at 95°C for 10 s and then transfer to 50°C. Incubate for 10 to 20 min with occasional gentle resuspension.
12. Add 1 µL of Vent Exo (2 U) to each tube, mix well, and immediately transfer to 76°C.
13. Incubate for 10 min and then chill on ice.
14. Spin to collect liquid, concentrate beads, and discard supernatant.
15. Resuspend beads in 50 µL of water, spin shortly to collect liquid, and concentrate beads.
16. Carefully remove all liquid and resuspend the beads in 4 µL of formamide loading buffer:0.15 NaOH (2:1). Mix well to dissolve all aggregates (*see* **Note 8**).
17. Leave for 5 min at room temperature.
18. Incubate 3 min at 76°C, spin to collect liquid, and chill on ice.
19. Concentrate beads and then transfer supernatant into a fresh tube on ice.
20. Check with a hand monitor that most of the radioactivity is in the supernatant. The bead pellet always contains radioactivity owing to trapped P3. Usually, we do not re-extract the beads.
21. Load on a prerun sequencing gel (*see* Chapter 51); reactions can be stored at –20°C.
22. Fix gel in fixing solution, dry onto blotting paper, and expose the dried gel to X-ray film in the presence of an intensifying screen. Readable sequences are usually obtained after overnight exposure.

4. Notes

1. The genomic DNA used for genomic sequencing needs to be clean and undegraded. Any shearing of the DNA during preparation and handling before the first primer extension must be avoided. Nicks between the restriction site and the P1 priming site will be converted to blunt ends during the first primer extension and will give rise to a background of dominant bands in all four sequencing lanes. We detail here one particular protocol that consistently yields high molecular weight genomic DNA of good quality. In principle, other methods can be followed. In any case, we advise to start a DNA prep with a nuclei isolation. Some suggestions for how to prepare nuclei have been described *(7,8)*.
2. Enzymes that produce blunt ends, with 5′- or 3′-overhangs can be used because the fragment is anyway converted to a blunt end after the first primer extension.
3. All oligonucleotides should be gel-purified (*see* **Subheading 3.2.**) and stored in water or TE, pH 7.5, at two concentrations: a stock solution (determined after the purification) and a diluted working solution that should not be frozen and thawed more than five times. The sequence-specific primers 1–3 need to be designed for each new sequencing project. The long and short linker oligonucleotides are constant. The distances between primers 1/2 and 2/3 should not be longer than 10 bases; overlaps of up to 5 bp are tolerated, but not required.
4. This protocol was used to sequence in the context of the *Drosophila* genome, which required changes from the original protocol described for mammalian DNA. If sequences are to be determined in the context of the 10 times more complex mammalian genome, a few parameters have to be adjusted: The amount of starting DNA used should be increased about five times, and the longer-linker oligos should be taken from the original procedure *(3)*.

5. The annealing temperature and the optimal magnesium concentration may be optimized specifically for each primer, but the given conditions worked reasonably well for most of the primers we tested.
6. In general, beads are concentrated by placing the 1.5-mL reaction tube on the magnetic rack for 10 to 20 s. If left for too long, the bead pellets become too tight and therefore difficult to resuspend. The supernatant is removed with a pipet tip with the tube still in the rack. Great care is taken at every step to resuspend the beads well by pipetting the suspension up and down. A drying of the bead pellets should also be avoided.
7. The reagents used for sequencing (*see* **Subheadings 2.8.** and **3.8.**) are available as a kit (Circumvent sequencing kit, NE Biolabs).
8. **Subheading 3.8., step 19** allows the separation of the labeled fragments from the beads, which facilitates gel loading. However, the presence of the beads in the sample load does not adversely affect the migration of DNA fragments.

References

1. Sanger, F., Nicklen, S., and Coulson, A. R. (1977) DNA sequencing with chain-termination inhibitors. *Proc. Natl. Acad. Sci. USA* **74,** 5463–5467.
2. Pfeifer, G. P., Steigerwald, S. D., Mueller, P. R., Wold, B., and Riggs, A. (1989) Genomic sequencing and methylation analysis by ligation mediated PCR. *Science* **246,** 810–813.
3. Mueller, P. R., Garrity, P. A., and Wold, B. (1993) Ligation mediated PCR for genomic sequencing and footprinting, in *Current Protocols in Molecular Biology* vol. 2 (Ausubel, F. M., Brent, R., Kingston, R. E., Moore, D. D., Seidman, J. G., Smith, J. A., et al, eds.), Current Protocols, New York, pp. 15.5.1–15.5.26.
4. Quivy, J.-P. and Becker, P. B. (1994) Direct dideoxy sequencing of genomic DNA by ligation-mediated PCR. *Biotechniques* **16,** 238–241.
5. Quivy, J.-P. and Becker, P. B. (1993) An improved protocol for genomic sequencing and footprinting by ligation-mediated PCR. *Nucleic Acids Res.* **21,** 2779–2781.
6. Sambrook, J., Fritsch, E. F., and Maniatis, T. (1989) *Molecular Cloning: A Laboratory Manual*, Cold Spring Harbor Laboratory Press, Cold Spring Harbor, NY.
7. Wu, C. (1989) Analysis of hypersensitive sites in chromatin. *Methods Enzymol.* **170,** 269–289.
8. Bellard, M., Dretzen, G., Giangrande, A., and Ramain, P. (1989) Nuclease digestion of transcriptionally active chromatin. *Methods Enzymol.* **170,** 317–346.

56

Cloning PCR Products for Sequencing in M13 Vectors

David Walsh

1. Introduction

Although numerous methods are now available for direct sequencing of polymerase chain reaction (PCR) products, cloning of amplified DNA for sequencing in M13 vectors remains an attractive approach because of the high quality of sequence information generated from single-stranded bacteriophage DNA templates.

Cloning of PCR products is in theory straightforward but in practice is often problematical, as widely reported *(1–4)*. Difficulties are generally ascribed to modifications of the DNA termini by *Taq* DNA polymerase. After completion of thermal cycling, the enzyme may remain associated with DNA ends and thus interfere with subsequent ligations, unless specific steps are included for its removal or inactivation. Carryover of *Taq* DNA polymerase and residual dNTPs into restriction digests can also result in end filling of 5′-overhangs *(5)*, severely reducing the efficiency of cohesive-ended cloning strategies using restriction sites within PCR primers. Removal of *Taq* DNA polymerase by proteinase K digestion *(6)* or repeated phenol/chloroform extractions *(5)* circumvents these problems.

Furthermore, the terminal transferase activity of *Taq* DNA polymerase catalyzes the nontemplate-directed addition of a single nucleotide, almost invariably deoxyadenosine (dA), to the 3′ ends of amplified DNA molecules *(7)*. The resulting "ragged ends" must be removed if blunt-end ligation to *Sma*I-cut vector is required. This is best achieved by utilizing the strong 3′ to 5′ exonuclease activity of T4 DNA polymerase, which in the presence of low concentrations of dNTPs removes 3′ overhangs from double-stranded DNA. Cloning of amplified DNA into linearized 5′-dephosphorylated vector also necessitates the presence of 5′ phosphate groups on the PCR products, achieved by kinasing either the primers before amplification or the PCR product itself. Conveniently, 3′-dA removal by T4 DNA polymerase and 5′-phosphorylation by T4 polynucleotide kinase can be performed simultaneously *(8)*.

Difficulties in cloning PCR products as blunt-ended molecules may be avoided by incorporating restriction sites into the PCR primers and cloning products more efficiently as cohesive-ended molecules. The major problem encountered here is the failure of some restriction endonucleases to cleave toward the ends of DNA fragments. The presence of 4 bp 5′ to the recognition sequence is sufficient for efficient cutting

From: *Methods in Molecular Biology, Vol. 226: PCR Protocols, Second Edition*
Edited by: J. M. S. Bartlett and D. Stirling © Humana Press Inc., Totowa, NJ

by most, but not all M13 polylinker enzymes (>75% digestion in 2 h by *Eco*RI, *Kpn*I, *Ava*I, *Xma*I, *Pst*I, *Bam*HI, *Sac*I, and *Xba*I, but <10% digestion in 2 h by *Acc*I, *Sph*I, *Sal*I, and *Hind*III) ***(9,10)***. Cutting efficiency may be improved by use of primers with longer regions 5′ to the recognition site, but this solution is not recommended. Rather, inefficiently cut terminal sites should be converted to internal sites by concatamerizing the PCR product with T4 DNA ligase in the presence of PEG ***(10,11,12)***. Internalized restriction sites are cut with greatly improved efficiency, facilitating a corresponding increase in cloning efficiency of amplified DNA.

In addition to the required product, the completed PCR mixture contains residual dNTPs and primers and often artifactual short amplification products and primer dimers. These unwanted species may be present in molar excess and, if carried forward into DNA-modifying reactions, are likely to result in reduced cloning efficiency of the required product. Thus, it is beneficial to purify the required PCR product before modification, either by means of a spin filtration device, such as Microcon™, or by glass bead isolation from agarose using Geneclean™.

This chapter describes the manipulations required for the efficient blunt- and cohesive-ended cloning of PCR products into M13 vectors.

2. Materials

1. Phenol:chloroform: Mix equal volumes of Tris-buffered phenol, pH >7.5 (nucleic-acid-grade) and chloroform (AR-grade). Store at 4°C in a dark glass bottle.
2. Chloroform/isoamyl alcohol (24:1): Store at 4°C.
3. Spin filtration device, such as Amicon Microcon™ (Beverly, MA) or Promega Wizard™ PCR purification unit (Madison, WI).
4. Variable-speed microcentrifuge.
5. TE buffer: 10 m*M* Tris-HCl, 1 m*M* EDTA, pH 8.0.
6. T4 DNA polymerase.
7. 10× T4 DNA polymerase buffer: 500 m*M* NaCl, 100 m*M* Tris-HCl, pH 7.9, 100 m*M* $MgCl_2$, and 10 m*M* DTT, 500 μg/mL BSA.
8. T4 polynucleotide kinase.
9. 10× polynucleotide kinase (PNK) forward reaction buffer: 500 m*M* Tris-HCl, pH 7.5, 100 m*M* $MgCl_2$, 50 m*M* DTT, 500 μg/mL BSA.
10. T4 DNA ligase.
11. 10× T4 DNA ligase buffer: 500 m*M* Tris-HCl, pH 7.8, 100 m*M* $MgCl_2$, 100 m*M* DTT, 250 μg/mL BSA.
12. Calf intestinal alkaline phosphatase (CIP).
13. 10× CIP buffer: 200 m*M* Tris-HCl, pH 8.0, 10 m*M* $MgCl_2$, 10 m*M* $ZnCl_2$.
14. Restriction enzymes at 10 U/μL and 10× reaction buffers.
15. 0.5 *M* EDTA, pH 8.0.
16. 3 *M* sodium acetate, pH 5.3.
17. Absolute ethanol, stored at –20°C.
18. Ultrapure stock of all four dNTPs: Make 20 m*M* stock in sterile water and store at –20°C. Avoid freeze/thaw.
19. 10 m*M* ATP: Store in aliquots at –20°C. Avoid freeze/thaw.
20. 60% (w/v) PEG 8000 in water, filter-sterilized. Store at room temperature away from direct sunlight.
21. Ultrapure low-melting-point agarose.
22. Glass bead DNA isolation kit, such as Geneclean (Bio101, Madison, WI).

23. M13 RF DNA purchased from supplier or prepared in-house by CsCl density gradient centrifugation.
24. Isopropyl-β-thiogalactopyranoside (IPTG): 20 mg/mL stock in sterile water, stored at –20°C.
25. 5-Bromo-4-chloro-3-indolyl-βD-galactoside (X-gal): 20 mg/mL stock in dimethyl formamide, stored at –20°C in glass vial.
26. *Escherichia coli* strain containing the F plasmid (e.g., JM101, JM107) and maintained on M9 minimal agar.

3. Methods

3.1. Purification of PCR Products from Taq *DNA Polymerase, Primers, and dNTPs (see Notes 1 and 2)*

1. Extract the completed PCR with an equal volume of chloroform. Spin in a microfuge at 13,000*g* for 2 min to separate the aqueous and organic phases.
2. Extract the upper aqueous phase twice with an equal volume of phenol:chloroform and once with an equal volume of chloroform/isoamyl alcohol.
3. Transfer the upper aqueous layer (up to 100 µL) into a Microcon unit housed in a 1.5-mL Eppendorf tube and add 400 µL of TE buffer. Spin in a microfuge at 500*g* for 15 min (Microcon-100) or at 14,000*g* for 6 min (Microcon-50). Add a further 400 µL of TE to the sample reservoir, and spin as before. Volume retained will now be 50 to 100 µL. If required (PCR product present in low yield), concentration down to a volume of 10 to 20 µL is achieved by a further spin cycle. Each cycle reduces the concentration of salts, PCR primers, and dNTPs by approx 95%.
4. Invert the unit in a fresh tube and spin at 500*g* for 2 min to recover purified PCR product.
5. Check recovery by agarose gel electrophoresis of an aliquot of the concentrated product.

If amplified DNA is to be cloned by cohesive-end ligation via restriction sites incorporated into PCR primers, and these sites are known to cut efficiently, the purified product can now be digested with restriction endonucleases without further processing. PCR products to be cloned by blunt-end ligation or via digestion with restriction enzymes that cut inefficiently at DNA termini should be processed as follows.

3.2. Simultaneous End Repair and Phosphorylation of PCR Products (see Notes 3 and 4)

1. Set up a reaction containing: 100 ng to 1 µg of purified PCR product, 3 µL of 10× T4 DNA polymerase buffer, 100 µ*M* each dNTP, 1 m*M* ATP, 0.5 U T4 DNA polymerase, 5 U T4 polynucleotide kinase, and H_2O to 30 µL. Incubate at 25°C for 20 min.
2. Stop the reaction by incubating at 75°C for 10 min in the presence of 5 m*M* EDTA, pH 8.0.
3. Increase the volume to 100 µL with H_2O, and perform one extraction with phenol/chloroform and one with chloroform/isoamyl alcohol.
4. Remove the aqueous phase to a fresh tube, and add 0.1 vol 3 *M* sodium acetate, pH 5.3, and 2.5 vol cold absolute ethanol. Mix well and store at –20°C for 1 h or at –70°C for 20 min. Precipitate DNA by centrifugation at 13,000*g* for 10 min in a microfuge.
5. Remove the supernatant carefully and add 0.5 mL cold 70% ethanol to the pellet. Spin again at 13,000*g* for 2 min. Discard the supernatant as before, vacuum dry the pellet (2–5 min), and finally dissolve DNA in 10 µL of TE.

PCR products are now flush-ended and phosphorylated at their 5′ ends, ready for direct cloning into *Sma*I-cut, dephosphorylated M13 vector or for concatamerization, as required.

3.3. Concatamerization/Digestion of PCR Products (see Notes 5–8)

1. To 10 µL PCR product, add 2 µL of 10× T4 DNA ligase buffer, 7 µL 60% PEG 8000 (20% final), and 2 U T4 DNA ligase. Incubate at room temperature overnight.
2. Increase the volume to 100 µL with water, and perform one extraction with phenol:chloroform. Avoiding the white PEG precipitate at the interface, remove 10 µL of the aqueous phase to check extent of concatamerization by electrophoresis through 0.8% agarose. Run out alongside an aliquot of the original PCR product and DNA size markers (*see* **Note 8**).
3. If concatamerization is judged to be successful (PCR product present as trimers and larger species), extract the remaining 90 µL once with chloroform/isoamyl alcohol and precipitate with sodium acetate/ethanol as above. Dissolve in 20 µL TE.
4. Add 10 U of appropriate restriction enzyme(s), 3 µL 10× reaction buffer, and H_2O to 30 µL. Incubate at the required temperature for 1 h.

   If cutting with two enzymes that have different salt requirements, digest with the low-salt enzyme first, heat-inactivate at 65°C for 20 min (or phenol-extract heat-stable enzymes), then adjust salt concentration with 1 *M* NaCl, and add 10 U of the second enzyme. Where two enzymes require completely different buffers, phenol/chloroform-extract and sodium acetate/ethanol-precipitate the DNA in between each digest.
5. Check to ensure the digest now contains only monomer-size PCR product by agarose gel electrophoresis. If larger species remain, indicating incomplete digestion of concatamers, add more enzyme and continue digestion.
6. When digestion is complete, electrophorese the digest mixture through 0.8% low-melting-point agarose and recover the PCR product by glass bead isolation using Geneclean.

3.4. Preparation of M13 Vector DNA for Ligation

3.4.1. Digestion with Restriction Endonucleases

Where digestion of the vector with two enzymes is required, the ability of each enzyme to cleave toward the end of linear DNA molecules should be considered to determine the preferred order of sequential addition New England Biolabs catalog, Reference Appendix; (*see* **Note 9**). Enzymes that cut less efficiently toward DNA termini should be used first. In this situation where directional cloning is required, digest M13 derivatives containing the polylinker in both orientations, for example, M13mp18 and M13mp19.

1. Set up the following digest: 1 µg M13 RF DNA, 5 U restriction enzyme, 2 µL 10× reaction buffer, and H_2O to 20 µL.
2. Incubate for 1 h at the appropriate temperature (25°C for *Sma*I and 37°C for all other polylinker enzymes). Remove 1 µL to analyze extent of digestion by electrophoresis through 0.8% agarose. If digestion is not complete, add more enzyme and continue incubation.
3. When complete, extract the digest once with phenol:chloroform and precipitate DNA with sodium acetate/ethanol as in **Subheading 3.2., steps 4–6**. Dissolve the DNA pellet in 10 µL of TE and digest with a second enzyme if required.

M13 DNA cut with a single enzyme should now be treated with calf intestinal alkaline phosphatase to reduce recircularization during ligation.

3.4.2. Dephosphorylation of Vector DNA

1. To 1 μg linearized M13 DNA in 10 μL TE, add CIP 0.05 U for 5′ overhangs, 0.5 U for 3′ overhangs or blunt ends, 5 μL of 10× CIP buffer, and H_2O to 50 μL of total volume.
2. Incubate at 37°C for 60 min.
3. Inactivate CIP by heating the reaction to 75°C for 10 min in the presence of 5 m*M* EDTA, pH 8.0.
4. Extract the reaction once with phenol:chloroform and recover DNA by sodium acetate/ethanol precipitation as in **Subheading 3.2., steps 4–6**. Dissolve the DNA pellet in 20 μL of TE.,
5. Check recovery of M13 vector and insert DNA by electrophoresing an aliquot of each through 0.8% agarose.

3.5. Ligation of PCR Products into M13 Vectors (see Notes 10 and 11)

1. Set up the ligation reaction and add components in the following order: 50 ng of M13 vector DNA, 1 μL of 10 m*M* ATP (1 m*M* final), 1 μL of 10× ligation buffer, 1–4 μL of DNA insert (3- to 5-fold molar excess), 5 U T4 DNA ligase for blunt termini, 1 U T4 DNA ligase for cohesive termini, and H_2O to 10 μL of total volume.
2. Set up a negative control ligation in which an equal volume of water is substituted for the PCR DNA insert and a positive control ligation containing an appropriate blunt- or cohesive-ended restriction fragment, preferably of a similar size to the PCR product.
3. Incubate overnight at 14°C for cohesive-end ligations or at room temperature for blunt-end ligations.
4. Transform 2.5 to 5 μL of the ligation reaction into *E. coli*-competent cells and plate in a soft agar overlay containing 0.33 m*M* IPTG and 0.03% X-gal. Identify recombinant phage clones by blue/white selection (*see* **Note 11**).

4. Notes

1. The use of a spin filtration unit to purify the PCR product from unincorporated nucleotides and primers is only recommended if the PCR mixture does not contain unwanted species larger than the nucleotide cutoff value of the unit. For the Microcon-100, this corresponds to 300 bases (single-stranded) or 125 bp (double-stranded). If larger unwanted products are present, the target product should be purified by electrophoresis through low-melting-point agarose followed by adsorption to glass beads.
2. If a spin filtration device is not available, PCR products can be partially purified from residual primers and dNTPs by precipitation with sodium acetate/ethanol, followed by a 70% ethanol wash. Better removal of such reactants can be achieved by adjusting to 2 *M* ammonium acetate and adding 2 vol of ethanol, although it should be noted that ammonium ions are a strong inhibitor of T4 DNA polymerase and must be thoroughly removed by extensive washing in 70% ethanol before the end-repair step.
3. Occasionally, PCR products end-repaired and kinased as described may fail to clone as blunt-ended molecules, most probably because of the persistence of *Taq* DNA polymerase bound at DNA termini. In this situation, residual enzyme can be removed by adding to the sample 50 μg/mL proteinase K in 10 m*M* Tris-HCl, pH 7.8, 5 m*M* EDTA, 0.5% (v/v) SDS, and incubating at 37°C for 30 min. Extract with phenol/chloroform, and precipitate PCR products with sodium acetate/ethanol.
4. It is important not to exceed the recommended amount of T4 DNA polymerase enzyme or the incubation time of 20 to 30 min for the end-repair reaction, since both may result in excessive exonuclease activity and nonblunt "nibbled ends." T4 DNA polymerase also has excessive exonuclease activity at higher temperatures (37°C).

5. For palindromic restriction enzyme sites, concatamerization of PCR products containing terminal half-sites reconstitutes the site. For example, end-to-end ligation of DNA molecules with terminal sequences GGA-3′ and 5′-TCC reconstitutes the *BamH*I recognition sequence. This allows the use of shorter PCR primers containing fewer extraneous nucleotides that do not hybridize to the target sequence.
6. Intermolecular joining of PCR products is stimulated by macromolecular exclusion molecules, such as PEG 8000, with maximal stimulation occurring in the range 15 to 25% (w/v).
7. Concatamerization and digestion of PCR products containing nucleotides 5′ to the restriction site generates a small cohesive-ended fragment that can coprecipitate with the full-length product and give rise to false positives on ligation into M13. Purification of the required product by glass bead isolation from low-melting-point agarose before cloning is therefore recommended. Eighty to ninety percent of clear plaques examined should then be found to contain the required insert. This figure falls to 10 to 20% if purification is not performed.
8. To allow analysis of concatamerized PCR products by gel electrophoresis, PEG must first be removed by extracting with phenol/chloroform because the presence of the polymer prevents entry of DNA into agarose.
9. When digesting M13 RF DNA with two enzymes that cut at closely spaced sites in the polylinker, it is preferable to perform the digests sequentially, rather than simultaneously, even when both enzymes are active in the same buffer, in order to maximize the cutting efficiency of each enzyme.
10. The number of clear plaques that can be expected depends on the strategy being followed. Cloning via direct cutting of efficiently recognized terminal restriction sites should yield 50 to 100 clear plaques per transformation. Rather fewer, typically 10 to 20, are produced by the concatamerization/digestion and blunt-end approaches.
11. M13 clones containing the required insert can quickly be identified by PCR using primers used for the original amplification. Use sterile toothpicks to transfer phage particles from individual clear plaques into 0.5-mL microcentrifuge tubes containing all components of the original PCR mixture minus the template. Vortex lightly and add mineral oil. Heat to 95°C for 2 min to lyse phage and then perform 25 cycles of the original temperature regimen. Analyze by agarose gel electrophoresis.

References

1. Buchman, G. W., Schuster, D. M., and Rashtchian, A. (1992) Rapid and efficient cloning of PCR products using the CloneAmp system. *Focus* **14,** 41–45.
2. Crowe, J. S., Cooper, H. J., Smith, M. A., Sims, M. J., Parker, D., and Gewert, D. (1991) Improved cloning efficiency of polymerase chain reaction (PCR) products after proteinase K digestion. *Nucleic Acids Res.* **19,** 184.
3. Krowczynska, A. M. and Henderson, M. B. (1992) Efficient purification of PCR products using ultrafiltration. *Biotechniques* **13,** 286–289.
4. Liu, Z. and Schwartz, L. M. (1992) An efficient method for blunt-end ligation of PCR products. *Biotechniques* **12,** 28–30.
5. Bennett, B. L. and Molenaar, A. J. (1994) Cloning of PCR products can be inhibited by *Taq* polymerase carryover. *BioTechniques* **16,** 32.
6. Hitti, Y. S. and Bertino, A. M. (1994) Proteinase K and T4 DNA polymerase facilitate the blunt-end subcloning of PCR products. *Biotechniques* **16,** 802.
7. Clarke, J. M. (1988) Novel non-templated nucleotide addition reactions catalysed by prokaryotic and eukaryotic DNA polymerases. *Nucleic Acids Res.* **16,** 9677–9686.

8. Wang, K., Koop, B. F., and Hood, L. (1994) A simple method using T4 DNA polymerase to clone polymerase chain reaction products. *Biotechniques* **17,** 236.
9. New England Biolabs catalog (1994) Reference Appendix, pp. 180,181.
10. Kaufman, D. L. and Evans, G. A. (1990) Restriction endonuclease cleavage at the termini of PCR products. *Biotechniques* **9,** 304.
11. Jung, V., Pestka, S. B., and Pestka, S. (1993) Cloning of polymerase chain reaction-generated DNA containing terminal restriction endonuclease recognition sites, in *Methods in Enzymology*, vol. 218 (Wu, R., ed.), Academic, London, pp. 357–362.
12. Pfeiffer, B. H. and Zimmerman, S. B. (1983) Polymer-stimulated ligation: Enhanced blunt- or cohesive-end ligation of DNA or deoxyribonucleotides by T4 DNA ligase in polymer solutions. *Nucleic Acids Res.* **11,** 7853–7871.

57

DNA Rescue by the Vectorette Method

Marcia A. McAleer, Alison J. Coffey, and Ian Dunham

1. Introduction

A major advance in physical mapping of the human genome was the development of yeast artificial chromosome (YAC) vectors *(1)*. This has enabled the cloning of pieces of DNA several hundred kilobases in length *(2)*. The availability of such large cloned genomic DNA fragments means that by ordering a series of overlapping YAC clones, a contiguous stretch of DNA, several megabases in length, can be isolated around a genomic region of interest (e.g., the region of a chromosome linked to a particular disease gene). The successful isolation of terminal sequences of a given YAC can be very useful in assembling an ordered "contig" of YAC clones. Such terminal clones may be used directly as hybridization probes or sequenced and used to generate sequence tagged sites (STSs) to identify overlaps between, and isolate other, members of the contig. Several methods have been used to this end, including PCR with vector-specific primers in combination with primers designed either for repetitive elements, such as *Alu* sequences *(3)*, or in combination with random nonspecific primers *(4)*. However, these techniques rely on a suitable repetitive element or random primer sequence occurring close enough to the end of the YAC so as to be amplified by PCR. Furthermore, probes isolated in this manner may well contain highly repetitive sequences that, if unsuccessfully blocked, will increase nonspecific signal in any subsequent hybridization procedures *(5)*.

The vectorette method was originally described by Riley et al. *(6)*. YAC DNA is digested with a restriction enzyme, and the resulting fragments are ligated to a linker molecule to create a vectorette "library," that is, a complex mixture of restriction fragments with linker ligated to each end. Within this library are fragments that contain the YAC vector/genomic DNA junction, which includes the terminal sequences of the YAC (**Fig. 1**).

The linker molecule consists of two long (>50 nucleotides) preannealed oligonucleotides incorporating a suitable 5′-overhang corresponding to the restriction enzyme used in the initial YAC digest. Blunt-ended linkers may also be used. Although the oligonucleotides comprising the linker are complementary at the 5′- and 3′-ends, there is a region of noncomplementarity in the middle where the two strands are unable to pair and a vectorette "bubble" is formed. The PCR is then performed on this mixture using one of two vector-specific primers (designed either for the centric or acentric

From: *Methods in Molecular Biology, Vol. 226: PCR Protocols, Second Edition*
Edited by: J. M. S. Bartlett and D. Stirling © Humana Press Inc., Totowa, NJ

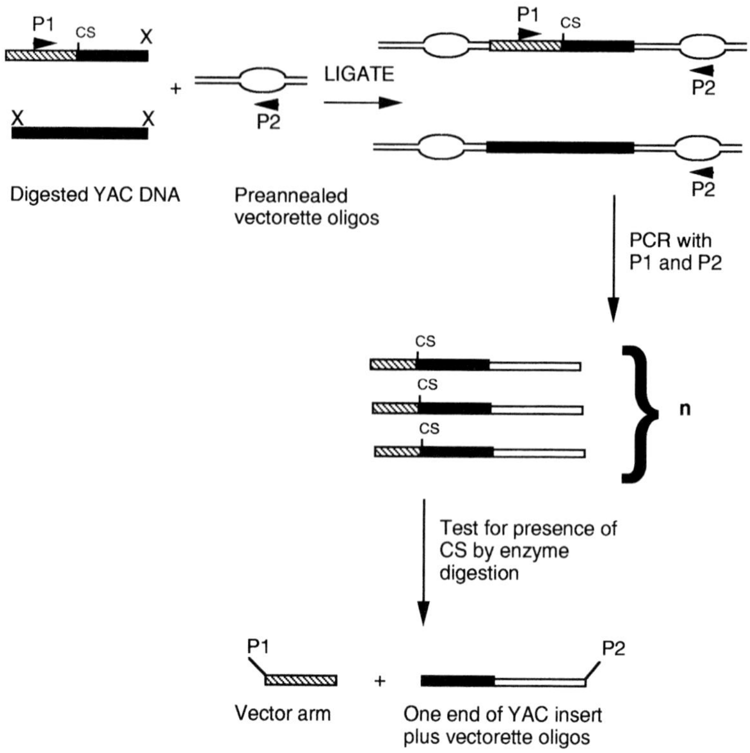

Fig. 1. A schematic representation of the vectorette method. Solid boxes represent genomic DNA, and the hatched boxes represent YAC vector sequence. YAC DNA is digested with a restriction enzyme, X. After ligation to annealed vectorette oligos, products are amplified with a vectorette-specific primer (P2) and a primer specific for one or other of the YAC vector arms (P1). Only fragments containing vector/insert junction are amplified. Confirmation of the presence of the cloning site (CS) within the amplified fragment can be obtained by digestion of the hybrid fragment with the enzyme that cuts at the cloning site, releasing a fragment diagnostic of the vector arm (**Table 2**) together with one or more fragments corresponding to the genomic DNA insert. CS = cloning site.

vector arms) in combination with a linker-specific primer. The linker-specific primer corresponds to the sequence of the linker ligated to the 5′-end of each DNA strand and has no complement on the other strand of the "bubble." It is therefore unable to anneal to template until the complementary sequence has been generated by priming off the vector-specific sequence. Thus, only those fragments containing binding sites for the vector-specific primer (i.e., DNA including and immediately adjacent to the cloning site of the YAC vector) will be successfully amplified by the PCR. The amplification products may then be used as DNA probes, for DNA sequencing, or may be cloned into a suitable vector.

A recent adaptation of the vectorette method has been used to isolate possible gene fragments from selected regions of the genome without prior knowledge of gene sequence *(7)*. This method is termed Island Rescue PCR (IRP), and relies on the fact that nearly all housekeeping genes and over 40% of tissue-specific genes have a CpG island in or near the 5′-end of the gene *(8)*. Such CpG islands have a significantly increased C + G content compared with the bulk of genomic DNA. These CpG islands can be detected in native human genomic DNA, by rare-cutting restriction endonucleases that recognize unmethylated CpG-containing sequences. The principles of the vectorette method described above are used except the YAC DNA in this instance is digested with restriction endonucleases that specifically recognize CpG-containing sequences, for example, *Sac*II, *Eag*I. Therefore, YAC DNA will be cut at CpG-rich sites, which may be associated with a gene. The mixture is then ligated to the preannealed vectorette oligos, and PCR in this instance is driven by an *Alu*-specific primer together with the vectorette oligo described above. Northern blot analysis may then be used to test that amplified sequences are associated with expressed mRNAs. There are two main drawbacks to this method. First, because DNA in yeast is not differentially methylated, all CpG-containing restriction sites will be cut whether or not they are associated with an unmethylated island in native genomic DNA. Therefore, a portion of the amplified fragments may not be associated with an expressed mRNA. Second, as with all *Alu*-PCR based methods, there is a requirement for an *Alu* sequence close enough to the restriction site to allow amplification by the *Taq* polymerase. However, in terms of transcript mapping, where the previously described methods, for example, direct selection/cDNA enrichment ***(9)***, exon trapping ***(10)***, probing cDNA libraries directly with radiolabeled YAC DNA ***(11)***, all have limitations, IRP may prove to be a rapid and useful technique for the identification of transcriptional units within complex sources of DNA.

Although the vectorette method was originally developed for rescuing the vector-insert junctions of YACs, it may be used to isolate sequences adjacent to any known sequence, for example, the identification of intron/exon boundaries in a specified gene ***(12)***. This chapter describes in detail the application of the vectorette method to isolating terminal sequences from YACs.

2. Materials

All solutions should be made to the standard required for molecular biology, that is, using sterile distilled water and molecular-biology-grade reagents.

1. T4 DNA ligase, 1 U/µL and 5X T4 DNA ligase buffer (0.25 *M* Tris-HCl, pH 7.6, 50 m*M* $MgCl_2$, 5 m*M* ATP, 5 m*M* DTT, 25% [w/v] polyethylene glycol-8000; Gibco BRL, Paisley, Scotland).
2. The sequences of the oligonucleotides used in this chapter are given in **Table 1** and are taken from **ref.** ***(13)***. The vector-specific primers are designed against pYAC4 (these can be replaced with appropriate vector primers or *Alu*-specific primers if performing IRP). The vectorette oligonucleotides described are suitable for blunt-ended ligations. If desired, a suitable overhang at the 5′-end of the "top" strand oligonucleotide may be incorporated to facilitate "sticky ended" ligations.

   Oligonucleotides were synthesized by phosphoramidite chemistry on an Applied Biosystems 392 DNA/RNA synthesizer. After deprotection (7 h at 55°C), oligonucleotides are dried in a centrifugal evaporator (alternatively, the standard ethanol precipitation procedure may be used). Oligonucleotides used in PCR are resuspended in H_2O to a concentration of 20 µ*M*. Vectorette oligonucleotides are purified by high-performance liquid chromatography (12% polyacrylamide gel electrophoresis may also be used). Before use, equimolar quantities of the "top" and "bottom" oligonucleotides are preannealed in 25 m*M* NaCl by heating at 65°C for 5 min and left to cool to room temperature. A working concentration of 1 µ*M* is used in ligations. All oligonucleotides are stored at –20°C.

Table 1
Oligonucleotide Sequences for Vectorette PCR

Vectorette oligonucleotides (for blunt-ended ligations)	
"Top" strand	
CAAGGAGAGGACGCTGTCTGTCGAAGGTAAGGAACGGACGAGAGA AGGGAGAG	
"Bottom" strand	
CTCTCCCTTCTCGAATCGTAACCGTTCGTACGAGAATCGCTGTCCTC TCCTTG	
Universal vectorette primer 224	
CGAATCGTAACCGTTCGTACGAGAATCGCT	
pYAC4-specific primers	
Centric ("left") arm	
1089	CACCCGTTCTCGGAGCACTGTCCGACCGC
Sup4-2	GTTGGTTTAAGGCGCAAGAC
pYACL *(13)*	AATTTATCACTACGGAATTC
Acentric ("right") arm	
1091	ATATAGGCGCCAGCAACCGCACCTGTGGCG
Sup4-3	GTCGAACGCCCGATCTCAAG
pYACR *(13)*	CCGATCTCAAGATTACGGAATTC

All oligonucleotide sequences are written in the 5′→3′ direction.

3. PCR is performed using a GeneAmp PCR reagent kit (Perkin–Elmer, Warrington, UK) in 10 m*M* Tris-HCl (pH 8.3), 50 m*M* KCl, 1.5 m*M* $MgCl_2$, 0.01% (w/v) gelatin containing 200 µ*M* of each dNTP and 1.0 µ*M* of each primer. Amplitaq is added to a concentration of 1.25 U/50 µL reaction and Perfect Match (Stratagene, Cambridge, UK) to a concentration of 5 U/50 µL reaction and overlaid with mineral oil (Sigma, Poole, UK). DNA amplification is performed in an Omnigene thermocycler (Hybaid, Teddington, UK).

3. Methods

1. Take half an agarose plug (approx 50–100 µL containing 1–2 µg DNA) of miniprep YAC DNA (DNA in solution may also be used; *see* **Note 1**) and wash as follows: 3 × 20 min in 10 m*M* Tris-HCl, 0.1 m*M* EDTA, pH 7.4 (1 mL/plug) at 50°C. 1 × 20 min in 10 m*M* Tris-HCl, 0.1 m*M* EDTA, pH 7.4 (1 mL/plug) at room temperature.
2. Preincubate plugs for 30 min at 37°C in 100 µL of the appropriate enzyme buffer (*see* manufacturer's recommendation).
3. Remove buffer, and replace with 100 µL of fresh enzyme buffer containing 20 to 30 U of restriction enzyme (*see* **Note 2**), and incubate overnight at the recommended temperature (usually 37°C). After digestion, the plug may be cut into three, and one portion electrophoresed through a 1.0% agarose mini gel alongside a similar amount of untreated YAC DNA to test for complete digestion. One slice may be stored dry at 4°C and redigested if incomplete digestion has occurred.
4. Incubate one third of the agarose plug from **step 3** in 1 mL of 1× ligation buffer for 1 h on ice.
5. Replace with 100 µL of fresh 1× ligation buffer. To this add 10 µL of preannealed blunt-ended vectorette linker (at 1 µ*M*; *see* **Subheading 2., item 2**), that is, 10 pmol of linker.
6. Heat to 65°C for 15 min to melt the agarose plug, and then equilibrate at 37°C (approx 5 min).

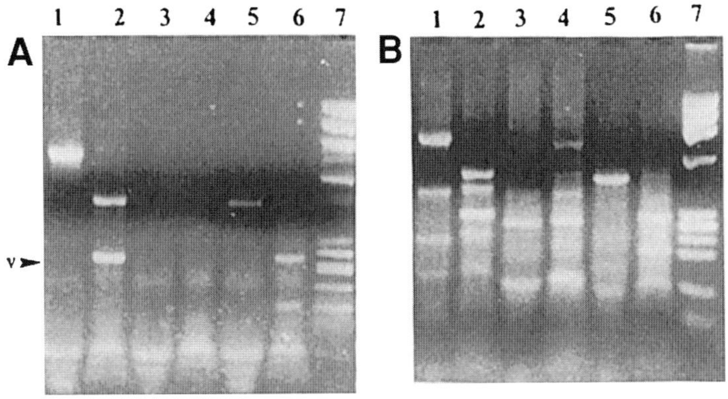

Fig. 2. Three vectorette "libraries" were created using the blunt-ended restriction enzymes: *Pvu*II (lanes 1 and 2), *Stu*I (lanes 3 and 4), and *Rsa*I (lanes 5 and 6). PCR was performed using oligos specific for the centric arm of the pYAC4, 1089, and the universal vectorette oligo, 224. In **A**, 5 µL of Perfect Match have been added to each PCR, whereas in **B**, this has been omitted. Ten microliters of untreated product were loaded on a 2.5% agarose minigel in lanes 1, 3, and 5, whereas samples in lanes 2, 4, and 6 were first digested with *Eco*RI to release the vector arm from the genomic fragment. Lane 7 contains *Hae*III fragments of ΦX RF DNA (Gibco BRL, Paisley, Scotland). *Stu*I-digested YAC has failed to produce a PCR product (**A**, lanes 3 and 4), probably through the lack of an enzyme site close to the vector/insert junction. *Pvu*II- and *Rsa*I-digested YAC yields products of approximately 800 and 500 bp (lanes 1 and 5), respectively, which on digestion with *Eco*RI release vector fragments (V) of the predicted size 287 bp together with the terminal *Pvu*II and *Rsa*I fragments of the YAC insert (500 and 200 bp).

7. When the reaction mix is equilibrated, add 1 µL of T4 DNA ligase (1 U/µL) and incubate at 37°C. After 1 h, add 400 µL of 10 m*M* Tris-HCl, 0.1 m*M* EDTA, pH 8.0, and mix thoroughly. The vectorette library may now be stored in aliquots at –20°C.
8. Two sets of PCR mixes need to be prepared for each vectorette library constructed. The first contains a primer, 1091, specific for the "right" arm of the YAC vector (i.e., the acentric arm encoding the *URA3* gene) together with the vectorette-specific oligo (224), whereas the second contains a primer, 1089, specific for the "left" arm of the YAC vector (i.e., the centric arm, which contains the *CEN4* gene) together with 224 (*see* **Table 1**). Each reaction is carried out in 50 µL buffer described in **Subheading 2., item 3**, including 5 µL of Perfect Match (*see* **Note 3** and **Fig. 2**) using the following cycling conditions: 94°C for 1 min, 1 cycle, followed by 93°C for 1 min, 65°C for 1 min, and 72°C for 3 min, 38 cycles, and followed by 72°C for 5 min, 1 cycle. For IRP, *see* **Note 4** for suggested primer sequences.
9. Confirmation that PCR products originate from the terminal sequences of YAC clones can be obtained by demonstrating the presence of the YAC vector cloning site in the hybrid fragment. This is done by digesting the PCR product with a restriction enzyme that cleaves within the cloning site. To 9 µL of PCR product, add 1 µL of 10× restriction enzyme buffer (*see* manufacturer's recommendation), 10 U of enzyme, and incubate for 1 h at 37°C. When the vector is pYAC4, 10 U of *Eco*RI may be added directly to 9 µL of PCR product, without addition of enzyme buffer. Restriction fragments can be visualized on a 2.5% agarose minigel containing ethidium bromide (0.5 µg/mL; **Fig. 2**). Note: ethidium bromide is a powerful mutagen and gloves should be worn at all times. The distances from the primer sequences described in **Subheading 2., item 3** to the *Eco*RI cloning site of pYAC4 are given in **Table 2**.

Table 2
Positions of Primer Sequences Described in Table 1 with Respect to the *Eco*RI Sequence in the Cloning Site of pYAC4

Centric arm		Acentric arm	
1089→*Eco*RI	287 bp	1091→*Eco*RI	172 bp
Sup4-2→*Eco*RI	40 bp	Sup4-3→*Eco*RI	29 bp
pYACL→*Eco*RI	17 bp	pYACR→*Eco*RI	20 bp

10. A second PCR may be performed to reduce the amount of vector DNA contained in the amplified product. A nested vector-specific primer that anneals closer to the cloning site (**Tables 1** and **2**) is used in combination with the vectorette-specific oligo. Either use 1 µL of the primary PCR or toothpick the fragment found to cut with *Eco*RI in **step 9** (not the restriction digestion product) directly from the agarose gel into a PCR containing: for "left" arm products: Sup4-2 + 224 or pYACL + 224, and for "right" arm products: Sup4-3 + 224 or pYACR + 224. The same cycling conditions as those described in **step 8** are used, but the annealing temperature is reduced to 59°C and only 20 cycles are performed. Ten microliters may be visualized on a 2.5% agarose minigel.
11. PCR products may now either be sequenced directly, radiolabeled and used as a hybridization probe (*see* **Note 5**), or subcloned using a suitable cloning system, such as pCR-Script™ SK (Stratagene) or TA-cloning™ system (Invitrogen, Leek, The Netherlands).

4. Notes

1. Use approx 1 µg of solution DNA for each restriction enzyme digest. Reactions should be performed in the buffers recommended by the manufacturers for 4 h at the specified temperature. Before ligation (**Subheading 3., step 5**), enzymes should be heat-inactivated (65°C for 15 min is usually sufficient), extracted with phenol : chloroform (equal volume of ratio 1 : 1), ethanol-precipitated by standard methods (2 vol 95% ethanol with one-tenth vol 3 *M* sodium acetate, pH 5.6) and resuspended to a concentration of 250 ng/µL. Ligations can be performed in a volume of 10 µL with 1 µL preannealed vectorette oligos.
2. It is important to check that there are no recognition sites for a given restriction endonuclease between the sequences corresponding to the vector-specific primers and the cloning site. If such a site were present, it would be cleaved in the initial digest and a vector-only fragment would be amplified. Suitable enzymes for pYAC4 are *Rsa*I, *Pvu*II, and *Stu*I.
3. The addition of Perfect Match to the PCR reduces the number of nonspecific bands generated (compare **Fig. 2A** with **B**), although some laboratories have found little difference on its omission.
4. IRP is a variant of *Alu*-vectorette PCR that can be used to generate probes from YACs as an alternative to *Alu*-PCR. For IRP, the universal vectorette primer 224 is used together with primer sequences that recognize a human *Alu* repeat. For example: 5′-GGATTACAGGCGTGAGCCAC-3′ and 5′-GATCGCGCCACTGCAC TCC-3′ (both sequences taken from **ref.** *7*). The thermocycling conditions described in **Subheading 3., step 8** may also be used for these two sets of primers.
5. Probes generated by this method may contain highly repetitive sequences. Therefore, it is advisable to pre-reassociate the labeled probe with total human genomic DNA prior to any hybridization procedure. Make probe up to 250 µL with H_2O. Add 125 µL of 10 mg/mL sonicated total human DNA (Sigma) and boil for 5 min. Snap-chill on ice for 5 min, and then add probe to hybridization as normal.

References

1. Burke, D. T., Carle, G. F., and Olson, M. V. (1987) Cloning of large segments of exogenous DNA into yeast by means of artificial chromosome vectors. *Science* **236,** 806–812.
2. Cohen, D., Chumakov, I., and Weissenbach, J. (1993) A first-generation physical map of the human genome. *Nature* **366,** 698–701.
3. Nelson, D. L., Ledbetter, S. A., Corbo, L., Victoria, M. F., Ramirez-Soli, R., Webster, T. D., et al. (1989) *Alu* polymerase chain reaction: A method for rapid isolation of human-specific sequences from complex DNA sources. *Proc. Natl. Acad. Sci. USA* **86,** 6686–6690.
4. Wesley, C. S., Myers, M. P., and Young, M. W. (1994) Rapid sequential walking from termini of cosmid, P1 and YAC inserts. *Nucleic Acids Res.* **22,** 538–539.
5. Cole, C. G., Patel, K, Shipley, J., Sheer, D., Bobrow, M., Bentley, D. R., et al. (1992) Identification of region-specific yeast artificial chromosomes using pools of *Alu* element-mediated polymerase chain reaction probes labelled via linear amplification. *Genomics* **14,** 931–938.
6. Riley, J., Ogilvie, D., Finniear, R., Jenner, D., Powell, S., Anand, R., et al. (1990) A novel, rapid method for the isolation of terminal sequences from yeast artificial chromosome (YAC) clones. *Nucleic Acids Res.* **18,** 2887–2890.
7. Valdes, J. M., Tagle, D. A., and Collins, F. S. (1994) Island rescue PCR: A rapid and efficient method for isolating transcribed sequences from yeast artificial chromosomes and cosmids. *Proc. Natl. Acad. Sci. USA* **91,** 5377–5381.
8. Larsen, F., Gundersen, G., Lopez, R., and Prydz, H. (1992) CpG islands as gene markers in the human genome. *Genomics* **13,** 1095–1107.
9. Lovett, M., Kere, J., and Hinton, L. (1991) Direct selection: A method for isolation of cDNAs encoded by large genomic regions. *Proc. Natl. Acad. Sci. USA* **88,** 9628–9632.
10. Buckler, A. J., Chang, D. D., Graw, S. L., Brook, J. D., Haber, D. A., Sharp, P. A., et al. (1991) Exon amplification: A strategy to isolate mammalian genes based on RNA splicing. *Proc. Natl. Acad. Sci. USA* **88,** 4005–4009.
11. Elvin, P., Slynn, G., Black, D., Graham, A., Butler, R., Riley, J., et al. (1990) Isolation of cDNA clones using yeast artificial chromosome probes. *Nucleic Acids Res.* **18,** 3913–3917.
12. Roberts, R. G., Coffey, A. J., Bobrow, M., and Bentley, D. R. (1993) Exon structure of the human dystrophin gene. *Genomics* **16,** 536–538.
13. Coulson, A., Kozono, Y., Lutterbach, B., Shownkeen, R., Sulston, J., and Waterston, R. (1991) YACs and the *C. elegans* genome. *Bioessays* **13,** 413–417.

58

Technical Notes for Sequencing Difficult Templates

David Stirling

1. Introduction

There are a number of template types that are generally recognized as being difficult to sequence. These can include sequences with a high guanine–cytosine (G/C) content, sequences that are very rich in adenine/thymine (A/T), sequences with a marked secondary structure, or large regions of homopolymer. There is no one solution to these difficulties; however, there are a number of approaches that can be used to improve the quality of sequencing data obtained from each type of problem sequence.

2. High G/C Content

Sequences with high G/C content have higher melting temperatures, and the incorporation of the dye-labeled terminators is also less efficient. The addition of DMSO to a final concentration (v/v) of 5% or betaine (final concentration of 1 *M*) to cycle sequencing reactions can greatly improve results. Changing the cycling conditions to use a higher melting temperature can help, as can the use of altered sequencing chemistry. Applied Biosystems produce a dGTP Big Dye kit where dITP normally used in big dye chemistry is replaced with dGTP. Adding more *Taq* and dNTPs can also help.

3. A/T-Rich Sequences

A/T-rich sequences are generally not as difficult as G/C-rich templates. A/T-rich primers will have low melting temperatures and so may need to be longer (24–26 bases) to increase the melting temperature closer to 55°C. Dye primer sequencing is generally more even through A/T regions.

4. Secondary Structure

Sudden loss of sequencing data generally indicates a problem with secondary structure. Where there are runs of Gs followed by runs of Cs, for instance, hairpin loops can form, impeding the progress of the polymerase and resulting in a stop in the data. Any of the approaches suggested for high G/C content sequences can be used.

From: *Methods in Molecular Biology, Vol. 226: PCR Protocols, Second Edition*
Edited by: J. M. S. Bartlett and D. Stirling © Humana Press Inc., Totowa, NJ

5. Homopolymer Regions

Enzyme slippage can occur when a run of the same bases occur consecutively (sometimes less than 10 for G and C). One solution is to use an anchored primer. This consists of a string of the repeated base, followed by a degenerate 3′ position of the other three bases. This allows the sequence to be resumed after the homopolymer but gives no information regarding the number of bases in the repeat. Designing a primer 30 to 40 bases from the start of the homopolymer and increasing the *Taq* concentration can improve the chances of reading through the homopolymer.

VIII

In Situ PCR and PRINS

59

PCR-Based Detection of Nucleic Acids in Chromosomes, Cells, and Tissues

Technical Considerations on PRINS and In Situ *PCR and Comparison with* In Situ *Hybridization*

Ernst J. M. Speel, Frans C. S. Ramaekers, and Anton H. N. Hopman

1. Introduction

The polymerase chain reaction (PCR) is an extremely sensitive technique allowing the detection of rare and low copy nucleic acid sequences (up to 1–10 copies in DNA or mRNA extracts from 1 million cells) by solution-phase amplification using specific primer sets and *Taq* DNA polymerase and visualization of the resulting PCR products by gel electrophoresis and blotting techniques (see chapters in this book). However, the obligatory cell and tissue destruction required for nucleic acid extraction does not permit the correlation of results with histopathological features or the localization of targets in specific cell types. This may now be overcome by combining recently developed micromanipulation systems, such as laser-assisted microdissection, to isolate the cells of interest (up to the level of a single cell) from a large population of cells or from the tissue, and to apply single-cell PCR on the extracted nucleic acids (for a review, see **ref. *1***). Alternatively, and in case it is unknown in which cell a certain target nucleic acid, for example, a virus particle, may be present, cellular localization of DNA and RNA can be accomplished by *in situ* hybridization (ISH). This procedure has a history of more than 30 years and has been improved continuously. In particular, the development of nonradioactive approaches and the recent implementation of tyramide signal amplification have made ISH a powerful technique for use in many applications *(2)*. Some 10 to 15 years ago, however, ISH detection limits were only in the range of 10 to 20 copies of mRNA or viral DNA per cell, and probe detection periods could be very long when using radioactive procedures *(3–5)*. Hence, in the end of the 1980s and 1990s, several strategies had been developed to improve the threshold levels as well as the efficiency of nucleic acid detection *in situ*, such as target and signal amplification methods (**Table 1**). In this chapter, we focus on the technical aspects of the primed *in situ* labeling (PRINS) and *in situ* PCR procedures, originally introduced by Koch et al. *(6)* and Haase et al. *(7)*, respectively, as examples of nucleic acid target amplification methods. The pros and cons of both techniques will be discussed and compared with

From: *Methods in Molecular Biology, Vol. 226: PCR Protocols, Second Edition*
Edited by: J. M. S. Bartlett and D. Stirling © Humana Press Inc., Totowa, NJ

Table 1
Approaches to Amplify Nucleic Acid Target Sequences and (Immuno) Cytochemical Detection Signals *in situ* (Adapted from ref. *2*)

	Target	Reference
Nucleic acid target amplification		
In situ polymerase chain reaction (*in situ* PCR)	DNA	(This chapter)
Primed *in situ* labeling (PRINS) and repeated/cycling PRINS	DNA	(This chapter)
Rolling circle amplification	Hybridized oligonucleotide (Padlock probe)	***(55)***
In situ reverse transcriptase (RT) PCR	RNA	***(8–10,13)***
In situ self-sustained sequence replication (3SR)	RNA	***(56)***
In situ transcription/PRINS	RNA	***(57,58)***
Detection signal amplification		
Branched DNA amplification		***(59)***
Catalyzed reporter deposition/tyramide signal amplification (CARD/TSA)		***(2,14)***
Mirror image complementary antibodies (MICA)		***(60)***
Enzyme antibody polymer system (EPOS/EnVision)		***(61,62)***
Enzyme-labeled antibody–avidin conjugates		***(63)***
End Product Amplification (anti-DAB antibody strategy)		***(64)***

the current protocols for ISH using tyramide signal amplification. Special emphasis will be on the conditions needed to achieve an optimal balance between nucleic acid detection *in situ* and preservation of cell and tissue morphology, including discussions on sample fixation and pretreatment, oligonucleotide/probe hybridization, *in situ* primer extension and amplification, and detection of incorporated reporter molecules. For more comprehensive reviews and applications of PRINS, *in situ* PCR, and ISH, we refer to the literature ***(2,8–17)***.

2. PRINS

2.1. PRINS vs ISH

Particularly in the field of cytogenetics, the PRINS labeling technique has become an alternative to ISH for the localization of nucleic acid sequences in chromosome and cell preparations ***(6,8,12,17)***. Occasionally, its application to the detection of chromosome copy numbers and viral DNA in frozen and formaldehyde-fixed, paraffin-embedded tissue sections have been reported ***(12,18,19)***. Whereas in an ISH approach a nucleic acid probe with incorporated reporter molecules is hybridized to its cellular target, the PRINS method is based on the use of high concentrations of unlabeled primers (restriction fragment, PCR product, or oligonucleotide) that allow a very fast hybridization (annealing) to denatured, complementary target sequences *in situ* (**Fig. 1**).

These primers serve as initiation sites for *in situ* chain elongation catalyzed by *Taq* DNA polymerase (in the appropriate buffer containing 1.5 m*M* $MgCl_2$) using the target DNA as a template. DNA labeling takes place during the elongation step when labeled nucleotides are incorporated. Fluorochrome-labeled nucleotides can be detected directly by fluorescence microscopy, while haptenized (e.g., biotin,

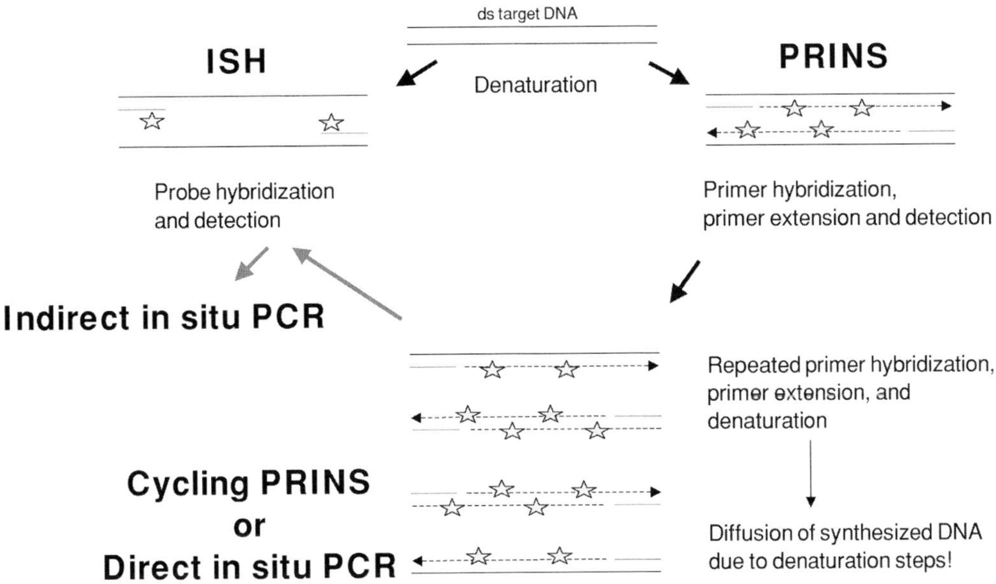

Fig. 1. Basic differences between in ISH, PRINS, and cycling PRINS DNA labeling and direct and indirect *in situ* PCR. *Incorporated reporter molecule (fluorochrome or hapten [biotin, digoxigenin, dinitrophenyl]), which can be detected as described in **Subheading 2.1.**

digoxigenin, dinitrophenyl) nucleotides can be visualized by the additional application of fluorochrome- or enzyme-conjugated avidin or antibody molecules, followed by fluorescence microscopy or brightfield visualization of enzyme reaction products *(2)*. Image recording and analysis is usually performed by using a CCD camera or confocal scanning laser microscope. In this overview, the use of radioactivity as reporter molecule will not be considered because of disadvantages related to cost, instability, biohazard potential, the time required for autoradiographic detection, and the poor resolution of the final *in situ* signals.

2.2. Sample Fixation and Processing

In methanol:acetic acid (3:1)-fixed metaphase spreads (adhered to noncoated glass slides), a PRINS reaction with oligonucleotides specific for centromeric or telomeric repeats usually runs for only 5 to 30 min, resulting in bright and easily localized signals and a high signal-to-noise ratio (**Fig. 2A**).

Nevertheless, background staining might occur because of nicks in the chromosomal DNA that may act as primers for unspecific labeling. Thus, it is recommended to use freshly prepared chromosome preparations or optionally perform a DNA ligase reaction to close the nicks before PRINS labeling *(17,20)*. Also, in interphase cells and frozen tissue sections, these repetitive target sequences can be efficiently detected provided that the specimens are adhered to coated (e.g., organosilane) glass slides, fixed in methanol:acetic acid (3:1), and pretreated with a mild protease digestion (e.g., 100 µg/mL pepsin or 0.025% proteinase K; **Fig. 2B** and **refs.** ***12,18***). Because of the thorough fixation, short denaturation temperature, and speed of the reaction, the morphology of chromosomes and cell nuclei in methanol:acetic acid (3:1)-fixed cell preparations is usually well preserved. On frozen and formaldehyde-fixed, paraffin-

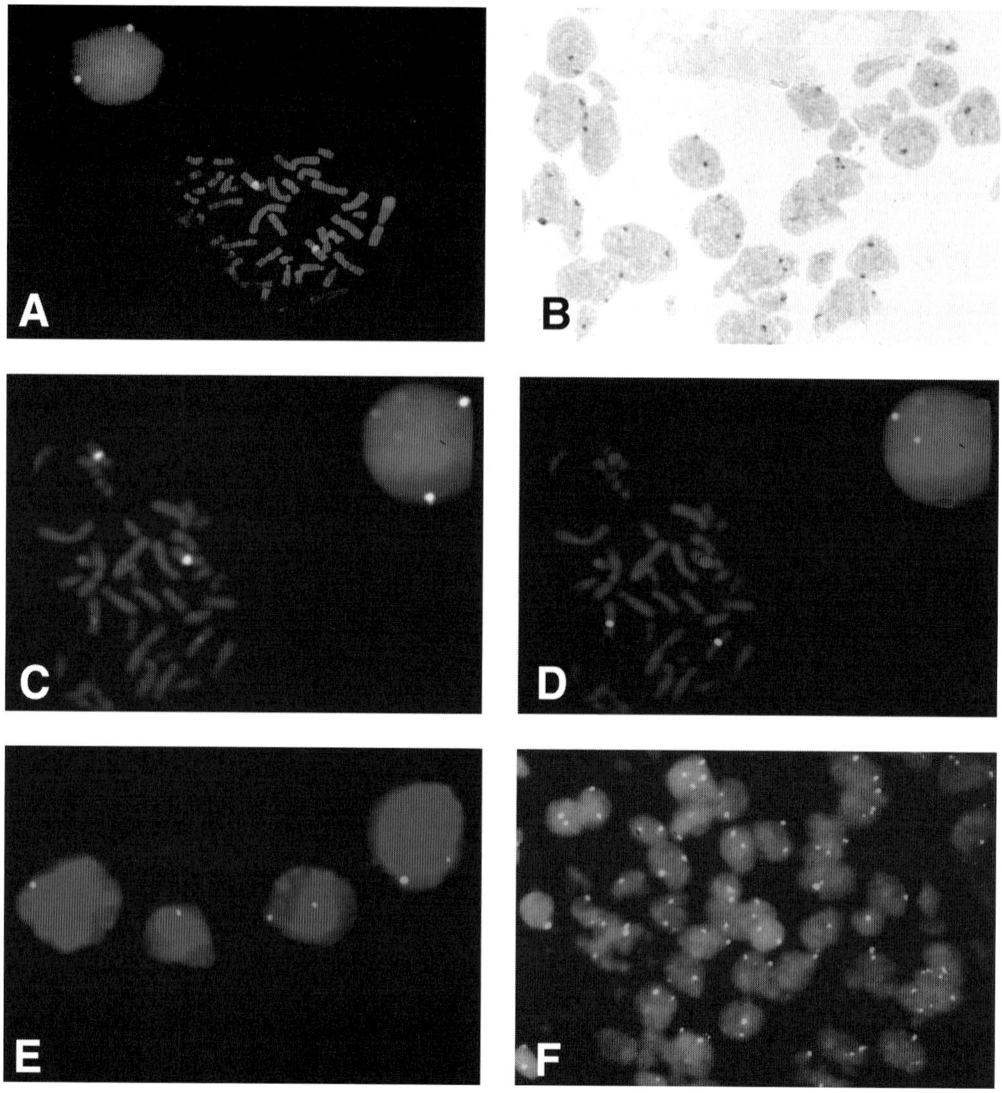

Fig. 2. **(A)** Fluorescence PRINS labeling of chromosome 9 centromeres with digoxigenin/sheep anti-digoxigeninFITC in human metaphase spreads and Vectashield embedding with propidium iodide (PI) counterstaining. **(B)** Brightfield diaminobenzidine visualization of chromosome 9 centromeres labeled with biotin by PRINS and detected with AvidinPeroxidase in hematoxylin counterstained cell nuclei in the urothelium. (**C** and **D**) Fluorescence double-color PRINS labeling of chromosome 7 (**C**) and 9 (**D**) centromeres with, respectively, biotin/AvidinTexasRed and digoxigenin/sheep anti-digoxigeninFITC in human metaphase spreads and Vectashield embedding with diamidinophenylindole (DAPI) counterstaining. **(E)** Fluorescence ISH detection of 1 to 2 integrated HPV 16 genomes (8 kb) in the SiHa cell line using digoxigenin-labeled HPV 16 genomic DNA and rhodamin-labeled tyramide signal amplification, and Vectashield embedding with DAPI counterstaining. **(F)** Fluorescence PRINS labeling after pepsin pretreatment and thermal cycling in buffer without PCR reagents, of chromosome 9 centromeres with digoxigenin/sheep anti-digoxigeninFITC in urothelium and Vectashield embedding with DAPI counterstaining.

embedded tissue sections, however, nuclear morphology may be less well preserved after PRINS DNA labeling because in our experience more powerful tissue pretreatment steps (extensive protein removal) are required than used for ISH to allow for an efficient *in situ* primer elongation step *(18)*. In combination with the high denaturation temperature, as a result, discrete nuclear morphology is more difficult to maintain. Although the small size of oligonucleotides used for PRINS (usually 18 to 35 nucleotides) greatly facilitates their accessibility to genomic target sequences in comparison with the larger probes used for ISH, the subsequent *in situ* primer elongation reaction, thus, appears to be the success-determining step in the PRINS procedure. Particularly on tissue sections, this requires more extensive pretreatment steps than necessary for optimal ISH, which is a perhaps unexpected but important finding to realize.

2.3. Primers and Hybridization Conditions

The specificity of the PRINS reaction is dependent on the choice of the primer sequences as well as the conditions of primer annealing and extension used. Both single and paired primers have been described for use in PRINS. On basis of interchromosomal differences in the α-satellite and other satellite DNA repeats, human chromosome-specific PRINS primers have been designed for all human chromosomes, except for chromosome 14 and 22, which are detected simultaneously with a single 14/22 primer *(21,22)*. Chromosome-specific PRINS primers have also been constructed for other species, such as the pig *(23)*, as well as for other human DNA repeat regions, including telomeres and ALU sequences *(24–26)* and for specific single-copy genes *(27,28)*. Optimization of the stringency of primer annealing (in sufficient reaction volume to prevent evaporation leading to concentration and temperature shifts), which reduces mispriming during the PRINS reaction, has been established by substituting the initially used thermoblocks or waterbaths by programmable thermal cyclers equipped with a flat plate block, which allow for precise and durable temperature control (up to 0.2°C accuracy). As a consequence, semi-automated PRINS protocols are now available, which offer a high reproducibility of nucleic acid labeling *(8,9,29,30)*.

2.4. Multicolor PRINS and Combination with Immunocytochemistry

Multicolor detection of up to three DNA target sequences *in situ* have been performed using subsequent PRINS reactions with different primers and labeled oligonucleotides *(30–33)*. This enables several targets to be analyzed simultaneously (**Fig. 2C,D**), which, for example, can make the evaluation of chromosome aberrations in clinical samples more robust. In addition, multicolor PRINS might be especially valuable in cases where only one or a few cells are available for analysis, for example, preimplantation genetic diagnosis. A prerequisite of performing multiple PRINS reactions in sequence is to stop the first PRINS reaction adequately and to avoid the further labeling of the first produced DNA strand with differently labeled nucleotides used in the subsequent PRINS reaction. This can be achieved by incubation with ddNTPs and Klenow DNA polymerase to block the free 3′ ends of the produced DNA strand *(16,33)*, although others have reported that this step can be omitted, probably because of either complex chromatin conformations or incomplete DNA denaturation after relatively long extension reactions in the first PRINS reaction that hamper the access of *Taq* polymerase ***(31,32)***. In principle, the number of *in situ* DNA targets to be detected simultaneously can be extended further by increasing the number of subsequent PRINS reactions applying different reporters/fluorochromes or combinations of two or more reporters in different ratios. However, the efficiency of this procedure is expected to decrease after multiple sequential PRINS reactions and, therefore, ISH is the preferred technique to use for the localization of more than 3 targets (**Table 2** and **refs.** ***34,35***). PRINS has also been successfully combined with the immunocytochemical detection of proteins in multicolor approaches to, for example, immunophenotype cells harboring a specific chromosomal aberration or viral infection, to investigate chromosome distribution and segregation in cells during processes such as polyploidization and aneuploidization, and to identify possible relationships of different families of DNA sequences with, for example, proteins associated with different chromosome-specific structures, such as the kinetochore complex (*see* Chapter 63).

Particularly, the rapidity, improved probe accessibility and lack of formamide for hybridization, thereby preventing the destruction of protein epitopes, are advantages of applying PRINS instead of ISH in these procedures.

2.5. Improvement of the Detection Sensitivity

The major drawback of PRINS for a long time proved to be its inability to convincingly detect single-copy gene sequences ***(27)***. This is caused by the fact that the *in situ* primer extension by *Taq* DNA polymerase in the biological material (adhered to glass slides) is limited to relatively short lengths (in the range of maximum a few hundreds of basepairs), probably caused by (1) the local chromatin organization of the target sequence; (2) the binding of (part of) the target to remnant proteins or the glass slide; and/or (3) the presence of nicks in the DNA where the polymerase reaction will stop ***(12)***. As a consequence, a single PRINS reaction to localize a single copy gene sequence with one primer (pair) will hardly result in a positive signal in the microscope, as the current detection sensitivity with ISH is approx 1 to 5 kb ***(2,36)***. The problem has now been overcome by using either multiple target-specific primers in a single PRINS reaction combined with reporter detection by the tyramide signal amplification procedure ***(28)*** or repeated PRINS reactions with the same reaction mixture and primers on specimen preparations (also called cycling PRINS) (***8,20,29,37,38*** and Chapter 60). Essential improvements of the first approach consisted of (1) the treatment of 1-d-old metaphase slides with 0.02 *N* HCl to remove loosely bound protein and thereby to render the DNA more accessible to the primer; (2) the use of multiple (four to five) primers for one locus; (3) one PRINS reaction and stringent washings in SSC to achieve optimal specificity; (4) the use of TaqStart, a monoclonal antibody against *Taq* DNA polymerase, which prevents nonspecific amplification and formation of primer-dimers; and (5) the use of biotin incorporation combined with biotin-labeled tyramide signal amplification ***(28)***. With this procedure a couple of single-copy genes have been detected in chromosome preparations with high efficiency ***(39)***. It will be interesting to see whether this approach can be applied to clinical cell and tissue specimens as well, for example, for rapid and reliable detection of microorganisms and chromosomal alterations in cancer specimens. Alternatively, cycling PRINS has been used on metaphase spreads and blood smears to detect low and single-copy DNA sequences resulting in more intense (up to 15×) fluorescence *in situ* signals than seen after a single

PRINS reaction (**Fig. 1**, **refs.** ***20,29,37,38,40***). The repeated PRINS reactions appear to label not only the target sequence (as observed after the last PRINS cycle) but also the DNA that has been synthesized during the preceding cycles. Several protocols have been described with suggestions for optimal gene localization, including (1) the use of single or multiple primers generating DNA products of 250 to 550 bp in length; (2) alternative fixation and pretreatment protocols (e.g., ethanol and microwave treatment instead of methanol:acetic acid [3:1] fixation); 3) the use of a hot start (applying the reaction mixture and/or *Taq* polymerase at the annealing or a higher temperature to the slide to avoid mispriming or primer oligomerization during PRINS/PCR) before 1 to 20 PRINS cycles, eventually preceding by PCR with the same primers but without labeled nucleotides; and (4) adaptation of the PRINS reaction volume and conditions on the glass slide during cycling. However, despite these recommendations, a main concern comprises the reproducibility of cycling PRINS, because in most studies variable and relatively low frequencies (10–70%) of chromosomes or cells harboring PRINS signals were reported. Besides the discussed possibilities that may result in background (nicks, mispriming), the major disadvantage of applying multiple cycles of PRINS (or PCR, see later) on cell and tissue preparations is the inevitable diffusion of newly synthesized DNA products, inextricably bound up with the denaturation steps, from the site of synthesis inside and/or outside the cells, followed by possible extracellular generation of amplificants (**Fig. 1. refs.** ***9,20,29,41–43***). Several studies have provided evidence for this phenomenon by demonstrating the expected DNA products using gel electrophoresis of the reaction mixture after cycling PRINS. The observation of stronger but often more diffuse and less discretely localized PRINS signals in only a low percentage of chromosomes or cell nuclei fits in with the view that labeled DNA in these cases is retained at or near the site of synthesis by possible entrapment in the chromatin or nonspecific binding to the surrounding cellular structures, whereas diffusion of DNA away from the site of synthesis has been taken place in the remaining negative chromosomes or nuclei. That diffusion occurs may be explained by the fact that methanol:acetic acid (3:1) fixation is an effective procedure to extract proteins, thus limiting the possibillities to entrap synthesized DNA molecules efficiently in the chromosome structures. Although postfixation in paraformaldehyde after cycling PRINS has been suggested to reduce diffusion of produced DNA ***(17)***, this will of course only be of help in the chromosomes and nuclei harboring signals.

Thus, because of several drawbacks concerning the efficiency and reproducibility of cycling PRINS in localizing either low or single-copy DNA sequences *in situ*, a single PRINS reaction with multiple primers combined with tyramide signal amplification is recommended for this purpose (as a rapid alternative to ISH with locus-specific probes).

3. *In Situ* PCR

3.1. *In Situ PCR vs Cycling PRINS*

In line with the development of PRINS, several groups working in the fields of pathology and microbiology have introduced the successful combination of PCR and ISH to visualize specific amplified nucleic acid sequences in cell and particularly formaldehyde-fixed, paraffin-embedded tissue preparations. Most studies have focused on the detection of (pro)viral (foreign) nucleic acid sequences, but in addition the

Table 2
Comparison of PRINS, *in Situ* PCR, and ISH for Localization of DNA Target Sequences *in Situ*

	PRINS	*In situ* PCR	ISH
Main application	Chromosome and gene identification and quantification in cells and chromosomes for use in clinical and cancer genetics and preimplantation genetic diagnosis	Detection of altered genes or foreign (e.g., viral) DNA in paraffin-embedded tissue section for use in molecular pathology	As for PRINS and *in situ* PCR, and localization of DNA sequences to study 3D organization of the interphase nucleus
Crucial steps (*see* **Fig. 1**)	Primer diffusion to DNA target Primer hybridization Primer elongation and DNA labeling by *Taq* polymerase	Primer diffusion to DNA target Primer hybridization Primer elongation (and DNA labeling) by *Taq* polymerase PCR while limiting diffusion of newly synthesized DNA products	Probe diffusion to DNA target Probe hybridization
Procedure			
Specimen, fixative, and preferred protein removal procedure (**Ch** = chromosomes, **Ce** = cells, **Ti** = tissues)	Ch: Methanol : acetic acid (3 : 1) Ce, Ti: Methanol : acetic acid (3 : 1) and *mild* pepsin digestion but to limit diffusion of PCR products	Ce, Ti: 4–10% (para)formaldehyde and *tuned* proteinase K digestion to allow reagents to access the target DNA	**Ch, Ce:** Methanol : acetic acid (3 : 1), **Ce:** 70% ethanol and *mild* pepsin **Ti:** 4–10% (para)formaldehyde combined with *strong* pepsin and 1M sodium thiocyanate or microwave
Main probe type	Single or multiple Oligonucleotide(s) (20–30 mer)	Oligonucleotide primer pair (20–30 mer)	Cloned genomic or cDNA (1–100 kb) in plasmid, cosmid, PAC or BAC vector
DNA labeling *in situ*	Specimen denaturation Primer annealing Incorporation of labeled nucleotides by DNA polymerase	Incorporation of labeled nucleotides during PCR (10–30 cycles) on the pretreated specimen (= *direct in situ PCR*), or PCR followed by FISH detection of amplified DNA (= *indirect in situ PCR*)	Hybridization of a denatured, labeled DNA probe on the pretreated, denatured specimen
Specificity	Primer annealing	Primer annealing and/or ISH after PCR	Probe hybridization and stringent washes
Label detection	Direct (fluorochromes) or indirect (antibody detection and signal amplification)	Idem	Idem
Advantages	Good accessibility of reagents by optimal protein removal	High sensitivity (target few hundred bp) optimal protein removal	Good accessibility of reagents by

	Direct labeling during 1 PCR cycle Multiple-target detection (up to 3) Use of primers with different annealing temperatures Relative accurate DNA localization Short turnaround time (few hours)	Relative high sensitivity (target 1–5 kb) High specificity and efficiency Accurate DNA localization Well-preserved morphology Multi-target detection (up to 32)	
Disadvantages	Limited primer elongation due to nicks and target DNA bound to the glass slide Relative low sensitivity (single-copy DNA) Relative poor morphology on tissue Subsequent PRINS reactions needed for multiple-target PRINS Nonspecific incorporation of labeled nucleotides (nicks, mispriming) leading to limited signal-to-noise ratio	Only reasonable accessibility of PCR reagents and poor primer elongation due to suboptimal protein removal at the start Increasing number of PCR cycles improves primer elongation but also diffusion of PCR products. This results in: Low amplification efficiency, variable reproducibility and poor DNA localization Relative poor morphology due to heating steps in the PCR procedure Further disadvantages: Nonspecific incorporation of labeled nucleotides at nicks (direct in situ PCR) and amplification due to mispriming Only single-target detection Relative long turnaround time (1–2 days)	Relative long turnaround time (1–2 days)
Comments	Muliple oligonuclotides and signal amplification have improved signal-to-noise ratio and enabled single copy gene detection	Signal amplification may improve signal-to-noise ratio	Signal amplification have improved signal-to-noise ratio and evaluation of results

technique has also been applied to identify endogenous DNA and RNA sequences in human cells (for reviews, *see* **refs. *8–10,13***). *In situ* PCR techniques are theoretically straightforward and comprise (1) sample fixation and pretreatment to improve the accessibility of the target nucleic acid sequences by the PCR primers, nucleotides, and *Taq* polymerase enzymes and to avoid diffusion of PCR-generated amplificants; (2) PCR amplification in the cell by *Taq* polymerase using the target DNA as a template; and (3) direct (by incorporation of labeled nucleotides during PCR) or indirect (by ISH with labeled nucleic acid probes) detection of the amplified nucleic acid molecules. Visualization of labeled target or probe DNA can be performed as described in **Subheading 2.1.** Thus, in principle direct *in situ* PCR is identical to the cycling PRINS procedure described above when cellular DNA is directly labeled during PCR cycling (**Fig. 1**). Consequently, the conditions for optimal chromosome labeling by (cycling) PRINS will also apply for the direct *in situ* PCR approach, although they need to be adapted for the specimen of interest and the fixative used to process cells and tissues.

3.2. Sample Fixation and Processing

It has been reported that 4 to 10% buffered formaldehyde is the fixative of choice for successful *in situ* PCR on cells and (frozen) tissues and that these specimens should be adhered to coated (e.g., organosilane) glass slides to prevent loss of tissue adherence during the in situ PCR procedure ***(5,9)***. In addition, cell and tissue preparations need to be subjected to a protease (pepsin, trypsin, or protease K) treatment to create holes in cellular membranes and remove cross-linked DNA-binding proteins from nuclear DNA, thereby facilitating the accessibility of the nucleic acids *in situ* by the PCR and detection reagents, as well as the ISH probes (in the indirect *in situ* PCR procedure). Importantly, for every protease and tissue, the optimal balance between time and concentration should be determined (usually 5 to 30 min of 2 mg/mL protease at room temperature or 37°C) to avoid overdigestion, leading to poor tissue morphology and possible leakage (diffusion) of amplified products from the cell in which they were generated, or insufficient protein removal, resulting in a decreased or completely absent in situ signal caused by very inefficient or failure of amplification ***(5,9)***. A striking point to notice here is that the generally used protease pretreatment for *in situ* PCR is less powerful than the one applied usually for optimal ISH on formaldehyde-fixed, paraffin-embedded preparations, in which prior to the protease digestion treatments with 1 *M* sodium thiocyanate and, optionally, 85% formic acid/0.3% H_2O_2 are performed to achieve optimal conditions for ISH while preserving nuclear morphology ***(44)***. Although this may still be sufficient to allow the PCR reagents to reach the target DNA, we know from our own results that efficient *in situ* primer extension rather requires a more powerful tissue pretreatment step than used for ISH (*see* **Subheading 2.2.** and **3.3.2.**). Indeed, we have not been able to detect human centromere repeats by PRINS labeling using tissue pretreatment conditions optimal for ISH with centromere-specific plasmid probes (unpublished results), indicating that *in situ* primer extension during the first couple of cycles of PCR on the slide is most likely impossible or very inefficient. This might in part explain the low amplification efficiency (restricted sensitivity) often obtained by *in situ* PCR, which furthermore can be caused by diffusion of the synthesized DNA amplificants outside the cell (*see* **Fig. 1** and below).

3.3. In Situ Amplification and Detection of Intracellular PCR Products

PCR protocols with optimal stringency of primer annealing and using generally longer extension times, increasing concentrations of *Taq* polymerase and $MgCl_2$, and/or the addition of bovine serum albumin in the reaction mixture on the cell and tissue preparations as compared with solution-phase PCR have been established by using the newly designed programmable thermal cyclers as described for (cycling) PRINS (*see* **Subheading 2.3.** and **refs.** ***5,9,30***). After PCR amplification, visualization of intracellular PCR products is achieved either directly through immunohistochemical detection of labeled nucleotides (*see* **Subheading 2.1.**) that have been incorporated into PCR products during thermal cycling (direct *in situ* PCR) or indirectly by ISH with a labeled probe and subsequent immunohistochemical detection (*see* Chapter 63 and **refs.** ***5,9***).

3.3.1 *Direct* In Situ *PCR*

Although direct in situ PCR is more rapid than indirect *in situ* PCR by eliminating the need for subsequent ISH, this procedure has proved unreliable with respect to the specificity of the results obtained ***(9,43,45)***. Even when the hot start procedure is performed (*see* **Subheading 2.4.**), the direct detection approach yields significant false-positive results, especially when working with tissue sections that have been dried at 56 to 65°C for several hours (introduction of nicks in the DNA) ***(5,9)***. This is the result of a number of artifacts, including incorporation of labeled nucleotides into (1) nonspecific PCR products resulting from mispriming ("endogenous priming" artifacts) and primer oligomerization, which seems to occur less likely inside nuclei ***(5)*** but may play a role during extracellular amplification of diffused DNA products), (2) single- and double-stranded nicks introduced by tissue fixation, cutting, and drying; and (3) fragmented DNA undergoing "repair" by DNA polymerase ("repair" artifacts). Repair artifacts may particularly be evident in apoptotic cells or samples that have been pretreated with DNase before *in situ* reverse transcriptase PCR for mRNA detection ***(9,45)***. These artifacts may only slightly be reduced by using an exonuclease-free DNA polymerase or by carrying out a DNA ligase reaction to close the nicks or a ddNTP reaction to prevent a subsequent extension reaction ***(9,17,33)***. Thus, caution and adequate use of appropriate controls are recommended in the interpretation of data produced by direct in situ PCR (**Table 3** and **Subheading 3.4.**).

3.3.2. *Indirect* In Situ *PCR*

The indirect in situ PCR technique, therefore, is the preferred approach to use, because the intracellular target-specific PCR products are identified by ISH, whereas the nonspecific amplificants will not be detected. Probes targeted to regions in between the primers used for PCR, usually oligonucleotide probes of 20 to 40 bases, represent the ideal ISH probes for reasons of specificity because they assure that the detected signal is the PCR product and not the result of for example primer oligomerization. ISH with these small probes, however, is usually less sensitive than with cDNA or genomic probes. Because primer oligomerization seems not to occur inside nuclei during PCR ***(10)***, the latter probes are recommended because of the increased number of reporter molecules they carry, leading to a higher detection sensitivity. Immunohistochemical

Table 3
Summary of Control Experiments Required for DNA Detection by PRINS, *in situ* PCR, and ISH (Modified from ref. *9*).

Method	Control experiment	Purpose
General	Use of known positive and negative control samples as well as mixtures of these (cells) in different ratios	Control for specificity and sensitivity
	Immunophenotpying the cell types of interest	Control for specificity and sensitivity
	Solution-phase PCR on extracted DNA of sample under study or directly on mildly fixed cell suspension	Control for sensitivity, false negative results and DNA quality
	Omission of primary antibody in immunohistochemical detection	Control for background induced by endogenous enzyme activity (colorimetric detection) or nonspecific sticking of secondary and/or tertiary detection reagents to sample and/or glass slide
	Use of no, random, or irrelevant (labeled) primers/probes, or vector sequences without a target-specific insert and/or detection reagents to sample and/or glass slide	Control for background induced by endogenous enzyme activity (colorimetric detection) or non-specific sticking of probe
(Cycling) PRINS and Direct *in situ* PCR	Omission of DNA polymerase	Control for background induced by endogenous enzyme activity (colorimetric detection) or nonspecific sticking of detection reagents to sample and/or glass slide
	Omission of primers	Control for artifacts related to endogenous priming, DNA repair and extension of nicks present in the DNA
Indirect *in situ* PCR	Omission of DNA polymerase	Control for effect of amplification and sensitivity of ISH

detection of labeled nucleotides within target or probe DNA can be performed as described previously (**Subheading 2.1.** and **refs.** ***2,46***). Because of the autofluorescence often present in tissue preparations, the limited stability of fluorochromes, and the preference of histopathologists to analyze permanent preparations by brightfield microscopy, colorimetric detection systems have been most frequently used ***(5,9)***. In these procedures, the activity of alkaline phosphatase or horseradish peroxidase, coupled to (immuno)histochemical reagents (e.g., antibodies or (strept)avidin molecules), is visualized by enzyme precipitation reactions (NBT/BCIP and DAB are most often used, respectively). Strikingly, in most cases suboptimal detection formats have been used, combining, for example, only a single antibody layer (alkaline phosphatase-conjugated anti-digoxigenin Fab fragments) with a relatively poor localizing enzyme reaction (NBT/BCIP) and omitting the possibility to apply the sensitive tyramide signal amplification procedure ***(2,14)***. As a consequence, relatively poor signal-to-noise ratios have often been obtained, which may further contribute to the restricted increase in detection sensitivity obtained by (in)direct *in situ* PCR when compared with ISH. Moreover, this might also explain in part the relatively low sensitivity reported with the ISH procedure when the PCR step is omitted, resulting in a detection limit of only 20 to 40 viral copies per cell ***(4,5)***. As can be seen in **Fig. 2E**, current optimal ISH protocols ***(44)*** are able to identify the 1 to 2 copies of HPV 16 DNA integrated in the genome in the often used SiHa control cell line without any prior PCR step.

Because the combination of PCR and ISH in the indirect *in situ* PCR method is essential to guarantee that the obtained signal is specific, one may consider this approach also as a rather cumbersome ISH method, in which sample pretreatment consists of fixation and protease digestion in combination with heating (thermal cycling) during nucleic acid amplification (by PCR). Moreover, the increase in detection sensitivity as compared with conventional ISH is rather limited even after optimization of the procedures, and cell morphology and nucleic acid localization are often poor (*see* **Subheading 3.4.** and **Table 2**). Furthermore, the question arises if the increase in the final ISH signal intensity is really caused by the *in situ* amplification of target sequences alone, or is rather the result of heating the specimen by thermal cycling, because it has been shown that microwaving or thermal cycling without *in situ* PCR can also result in increased sensitivity of ISH ***(47,48)***. In this respect, we have reported a similar effect localizing human centromere 9-specific DNA in routinely-fixed, paraffin-embedded tissue sections, where only after pepsin digestion in combination with thermal cycling in buffer without PCR reagents it proved to be possible to perform a primed in situ labeling reaction with a chromosome 9-specific primer with positive outcome (**Fig. 2F** and **ref.** ***14***).

Indirect *in situ* PCR protocols are available for the simultaneous detection of two DNA targets in formaldehyde-fixed, paraffin-embedded tissue sections as well as for the detection of DNA in combination with a protein detected subsequently by immunohistochemistry ***(49)***. However, without using appropriate controls (**Table 3**) to adequately address the possible leakage of synthesized DNA molecules to target-negative sites, as is the principal drawback of indirect *in situ* PCR, caution is recommended by interpretation of the data.

3.4 Sensitivity and Evaluation of In Situ *PCR Results*

Only a few studies have tried to give an indication of the increase in *in situ* signal intensity by comparing indirect *in situ* PCR with ISH. In the most optimal situation, a 50-fold increase in sensitivity has been reported when cell preparations were used ***(7,50)***. This relatively poor result after PCR amplification thus lies far beneath the amplification efficiency that can be achieved by solution-phase PCR. As has been discussed above, this is the result of many factors, including suboptimal sample pretreatment, inefficient and nonspecific *in situ* primer extension, as well as diffusion of PCR products because of the denaturation steps during PCR. In addition, background signals introduced by the ISH procedure, as a result of nonspecific binding of probe and detection reagents to the sample and the coated glass slide as the result of suboptimal stringent washings after probe hybridization and application of too less diluted probe and/or detection conjugates, may even further contribute to a decrease of the difference in signal intensity between in situ PCR and ISH. Because so many factors may influence the final signal intensity observed, it is not advisable to use *in situ* PCR results for quantification purposes.

Nevertheless, an amplification factor of 10- to 50-fold would still be a very acceptable increase in sensitivity, provided that reproducible and specific results would be obtained by indirect *in situ* PCR. However, despite the fact that the detection of one target copy per cell have been reported by using a single primer pair that amplifies a sequence of a few hundred basepairs ***(5,10)***, indirect *in situ* PCR appears also to be hampered by restricted specificity of results and, moreover, is unable to distinctly localize nucleic acid sequences in cell and tissue preparations ***(9,51)***. Even when assuming that PCR conditions are optimal and artifacts caused by mispriming and nicks are eliminated, the diffusion of generated PCR products from the site of synthesis inside and/or outside the cell is an almost impossible process to control. Therefore, many creative approaches intended to minimize the impact of diffusion have been described, such as optimal sample fixation and pretreatment, reduction of PCR cycle numbers, generation of longer or more complex PCR products, incorporation of labeled nucleotides to make bulkier PCR products, and the embedding of samples in agarose or protein matrices ***(9,43,50,52).*** Nevertheless, in most cases only a discrimination between positive and negative nuclei can be determined, whereas for example the number and site of viral integration in the nucleus, as shown in **Fig. 2E** by ISH, as well as the discrimination between viral integration and replication, are almost impossible to identify.

Thus, appropriate controls at each step in the (in)direct *in situ* PCR procedure are essential to demonstrate specificity and to correctly interpret the results (**Table 3** and **refs.** *5,9,42*). Control experiments should include (1) samples that either harbor or lack the target of interest (or a known mixture of these cells to verify the expected result) or use of irrelevant primers that cannot find targets in the cells under study (e.g., viral-specific primers in uninfected cells); (2) omission of the DNA polymerase from the PCR mixture to detect nonspecific sticking of ISH probes and detection reagents to the slide; (3) omission of primers to detect artifacts related to endogenous priming (nicks in the DNA) and DNA repair in the direct *in situ* PCR approaches. In case of reverse transcription *in situ* PCR to detect (m)RNA sequences *in situ*, RNase pretreatment of samples and the omission of the reverse transcription step are important controls.

4. Conclusion

During the past 10 to 15 yr, several strategies have been developed to improve the threshold levels of nucleic acid detection *in situ* by either labeling and/or amplification of target nucleic acid sequences (prior to ISH), for example, (cycling) PRINS and *in situ* PCR, or by amplification of the detection signals after the hybridization procedure, for example, PRINS or ISH followed by tyramide signal amplification (**Table 1**). Although PRINS and *in situ* PCR have been developed and applied in different research disciplines and only sporadically have been compared with each other and/or with ISH in the same study, all three techniques have been extensively studied and optimized to cope with the central question how to achieve the most optimal balance between specific in situ nucleic acid detection with a high signal-to-noise ratio and preservation of cell and tissue morphology. As a result, all techniques are capable to detect repetitive as well as single copy DNA sequences to date. However, because of evident differences in specificity, reproducibility, and detection sensitivity between PRINS, *in situ* PCR, and ISH, in our opinion the most suitable technique to be used for nucleic acid detection in biological samples is ISH, optionally combined with tyramide signal amplification to achieve the highest sensitivity.

In methanol:acetic acid (3:1)-fixed chromosome and cell preparations, a single PRINS reaction is a suitable alternative to ISH allowing for the rapid, specific, and reliable localization of repetitive and single copy DNA sequences ***(28,39)***. In the latter case, PRINS should be conducted with multiple carefully selected and target specific oligonucleotide primers and be combined with tyramide signal amplification. The exact detection limit, however, is unknown, and will be dependent on the ratio between the obtained specific PRINS signal and the background noise resulting from mispriming and labeling of nicks in the DNA, for which control experiments should be performed. Cycling PRINS is not recommended because until now this technique has not proved to be reliable with respect to efficiency and specificity because of problems of DNA diffusion and nonspecific labeling of DNA ***(29,38,40,51)***. Although initially the turn-around time (usually 16 h) and the generation and expensive purchase of probes have been considered as real disadvantages of ISH as compared with PRINS ***(12)***, the current situation is clearly a different one. ISH can be performed in very short turnaround times if repetitive and directly labeled probes are used. Furthermore, ISH probes are well available now because of the complete sequencing of the human and other genomes, and although still expensive can be used for the specific localization of DNA probes even under 1 kb ***(36)*** as well as for the simultaneous localization of 24 or more DNA targets in chromosomes and cells ***(34,35)***. In contrast, in a single PRINS reaction a huge amount of primer(s) is used together with *Taq* DNA polymerase and the amount of hapten-labeled nucleotides that is normally used for labeling of 1 to 2 µg of ISH probe. In addition, the use of tyramide signal amplification required for the localization of single copy genes will lead to high costs if applied for clinical diagnostics as a result of the fact that it is a patented trademark. Moreover, only a limited number of targets can be detected simultaneously by subsequent PRINS reactions, and additional advantages of PRINS over ISH, for example, the ability to discriminate between human centromeres 13 and 21 for chromosome enumeration analysis in prenatal diagnosis, have been counteracted by the development of specific probe sets that combine chromosome-specific single-copy or chromosome painting probes for discrimination.

In methanol:acetic acid (3:1)-fixed and protease-pretreated frozen tissue sections and cell preparations (touch preparations, blood smears, smears of bone marrow aspirates) so far only repetitive sequences have been visualized efficiently by a single PRINS reaction ***(8,12,18)***. Single-copy DNA sequences have been successfully detected by cycling PRINS and ISH ***(2,40)***, but because of the drawbacks involved with the use of the cycling PRINS procedure as described above an ISH approach is preferred for this purpose. The most optimal results have been obtained by applying (single or multiple-target) ISH to 70% ethanol-fixed samples that have been pretreated with pepsin ***(44,53)***.

A couple of studies have compared DNA detection, particularly genomic human papillomavirus DNA, in formaldehyde-fixed, paraffin-embedded cell and tissue preparations by *in situ* PCR, PRINS, and ISH ***(5,19,51,54)***. With all techniques HPV-specific sequences were efficiently detected, but only the one or two HPV 16 DNA sequences, known to be integrated in the genome of the human SiHa cell line, could be visualized by ISH or indirect *in situ* PCR (*see* also **Fig. 2E**). Interestingly, two of these studies ***(51,54)*** reported the use of ISH in combination with tyramide signal amplification as the method of choice, since it provided the same sensitivity and a much better reproducibility and reliability than the more cumbersome and poorly reproducible indirect in situ PCR method. Apparently, under optimal conditions ISH combined with the high amplification power of the signal amplification system can result in the same detection limits as can be reached by the relatively low amplification efficiency of indirect in situ PCR (as compared with solution-phase PCR). Furthermore, it provides optimal localization of signals and discrimination between viral integration and replication within well-preserved cells and nuclei (Hopman et al., manuscript in preparation). This is almost impossible to achieve with *in situ* PCR because of diffusion of PCR products in the cell nucleus as well as leakage out of the nucleus to target-negative cells, which is still the most important disadvantage of an *in situ* PCR approach.

References

1. Hahn, S., Zhong, X. Y., Troeger, C., Burgemeister, R., Gloning, K., and Holzgreve, W. (2000) Current applications of single-cell PCR. *Cell. Mol. Life Sci.* **57,** 96–105.
2. Speel, E. J. M. (1999) Detection and amplification systems for sensitive, multiple-target DNA and RNA in situ hybridization: looking inside cells with a spectrum of colors. *Histochem. Cell Biol.* **112,** 89–113.
3. Höfler, H., Childers, H., Montminy, M. R., Lechan, R. M., Goodmann, R. H., and Wolfe, H. J. (1986) In situ hybridization methods for the detection of somatostatin mRNA in tissue sections using antisense RNA probes. *Histochem J.* **18,** 597–604.
4. Burns, J., Graham, A. K., Frank, C., Fleming, K. A., Evans, M. F., and McGee, J. O'. D. (1987) Detection of low copy human papillomavirus DNA and mRNA in routine paraffin sections of cervix by non-isotopic in situ hybridization. *J. Clin. Pathol.* **40,** 858–864.
5. Nuovo, G. J. (2000) In situ localization of PCR-amplified DNA and cDNA, in In situ hybridization protocols. *Methods in Molecular Biology* (Darby, I. A., ed.), Humana Press, Totowa, NJ, pp. 217–238.

6. Koch, J., Kölvraa, S., Petersen, K. B., Gregersen, N., and Bolund, L. (1989) Oligonucleotide-priming methods for the chromosome-specific labelling of alpha satellite DNA in situ. *Chromosoma* **98,** 259–265.
7. Haase, A. T., Retzel, E. F., and Staskus, K. A. (1990) Amplification and detection of lentiviral DNA inside cells. *Proc. Natl. Acad. Sci. USA* **87,** 4971–4975.
8. Gosden, J. R., ed. (1997) *PRINS and In Situ PCR protocols. Methods in Molecular Biology*, Vol. 71, Humana Press, Totowa, NJ.
9. Long, A. A. and Komminoth, P. (1997) In situ PCR, in *PRINS and in situ PCR protocols. Methods in Molecular Biology*, Vol. 71 (Gosden, J. R., ed.), Humana Press, Totowa, NJ, pp. 141–161.
10. Nuovo, G. J. (1997) PCR in situ hybridization, protocols and applications, 3rd ed., Lippincott-Raven, New York.
11. Totos, G., Tbakhi, A., Hauser-Kronberger, C., and Tubbs, R. R. (1997) Catalyzed reporter deposition: a new era in molecular and immunomorphology—Nanogold-silver staining and colorimetric detection and protocols. *Cell Vision* **4,** 433–442.
12. Wilkens, L., Tchinda, J., Komminoth, P., and Werner, M. (1997) Single- and double-color oligonucleotide primed in situ labeling (PRINS): applications in pathology. *Histochem. Cell Biol.* **108,** 439–446.
13. Herrington, C. S. and O'Leary, J. J. (1998) *PCR In Situ Hybridization: A Practical Approach*, Oxford University Press, Oxford, U.K.
14. Speel, E. J. M., Hopman, A. H. N., and Komminoth, P. (1999) Amplification methods to increase the sensitivity of in situ hybridization. Play card(s). *J. Histochem. Cytochem.* **47,** 281–288.
15. Darby, I. A. (2000) *In Situ Hybridization Protocols. Methods in Molecular Biology, Vol. 123*, Humana Press, Totowa, NJ.
16. Coullin, P., Roy, L., Pellestor, F., Candelier, J.-J., Bed"hom, B., Guillier-Gencik, Z., et al. (2002) PRINS, the other in situ DNA labeling method that is useful in cellular biology. *Am. J. Med. Genet.*, in press.
17. Hindkjœr, J., Bolund, L., and Kolvraa, S. (2001) Primed in situ labeling, in *Methods in Cell Biology*, Vol. 63–64, Academic Press, San Diego, pp. 55–68.
18. Speel, E. J. M., Lawson, D., Ramaekers, F. C. S., Gosden, J. R., and Hopman, A. H. N. (1996) Rapid brightfield detection of oligonucleotide primed in situ (PRINS)-labeled DNA in chromosome preparations and frozen tissue sections. *BioTechniques* **20,** 226–234.
19. Ramael, M., Van Steelandt, H., Stuyven, G., Van Steenkiste, M., and Degroote, J. (1999) Detection of human papilloma virus (HPV) genomes by the primed in situ labelling technique. *Pathol. Res. Pract.* **195,** 801–807.
20. Gosden, J. and Lawson, D. (1994) Rapid chromosome identification by oligonucleotide-primed in situ DNA synthesis (PRINS). *Hum. Mol. Genet.* **3,** 931–936.
21. Koch, J., Hindkjœr, J., Kolvraa, S., and Bolund, L. (1995) Construction of a panel of chromosome-specific oligonucleotide probes (PRINS-primers) useful for the identification of individual human chromosomes in situ. *Cytogenet. Cell. Genet.* **71,** 142–147.
22. Pellestor, F., Girardet, A., Lefort, G., Andréo, B., and Charlieu, J.-P. (1995) Selection of chromosome specific primers and their use in simple and double PRINS technique for rapid in situ identification of human chromosomes. *Cytogenet. Cell. Genet.* **70,** 138–142.
23. Roguel-Gaillard, C., Hayes, H., Coullin, P., Chardon, P., and Vaiman, M. (1997) Swine centromeric DNA repeats revealed by primed in situ (PRINS) labeling. *Cytogenet. Cell Genet.* **79,** 79–84.
24. Gosden, J., Hanratty, D., Starling, J., Fantes, J., Mitchell, A., and Porteous, D. (1991) Oligonucleotide-primed in situ DNA synthesis (PRINS): A method for chromosome mapping, banding, and investigation of sequence organization. *Cytogenet. Cell. Genet.* **57,** 100–104.

25. Krejci, K. and Koch, J. (1999) An in situ study of variant telomeric repeats in human chromosomes. *Genomics* **58,** 202–206.
26. Go, Y., Rakotoarisoa, G., Kawamoto, Y., Randrianjafy, A., Koyama, N., and Hirai, H. (2000) PRINS analysis of telomeric sequence in seven lemurs. *Chrom. Res.* **8,** 57–65.
27. Cinti, C., Santi, S., and Maraldi, N. M. (1993) Localization of a single copy gene by PRINS technique. *Nucleic Acids Res.* **21,** 5799–5800.
28. Kadandale, J. S., Tunca, Y., and Tharapel, A. T. (2000) Chromosomal localization of single copy genes SRY and SOX3 by primed in situ labeling (PRINS). *Microbial. Comparative Genomics* **5,** 71–74.
29. Harrer, T., Schwinger, E., and Mennicke, K. (2001) A new technique for cyclic in situ amplification and a case report about amplification of a single copy gene sequence in human metaphase chromosomes through PCR-PRINS. *Hum. Mutat.* **17,** 131–140.
30. Pellestor, F. (2002) PRINS and cycling PRINS on chromosomes and sperm cells, in *Chromosome Analysis Protocols*, 2nd ed, Methods in Molecular Biology (Speel, E. J. M., and Hopman, A. H. N., eds.), Humana Press, Totowa, NJ, in press.
31. Volpi, E. V. and Baldini, A. (1993) MultiPRINS: A method for multicolor primed in situ labeling. *Chrom. Res.* **1,** 257–260.
32. Hindkjœr, J., Koch, J., Terkelsen, C., Brandt, C. A., Kolvraa, S., and Bolund, L. (1994) Fast, sensitive multicolor detection of nucleic acids in situ by primed in situ labeling (PRINS). *Cytogenet. Cell. Genet.* **66,** 152–154.
33. Speel, E. J. M., Lawson, D., Hopman, A. H. N., and Gosden, J. (1995) Multi-PRINS: multiple sequential oligonucleotide primed in situ DNA synthesis reactions label specific chromosomes and produce bands. *Hum. Genet.* **95,** 29–33.
34. Schröck, E., Du Manoir, S., Veldman, T., Schoell, B., Wienberg, J., Ferguson-Smith, M. A., et al. (1996) Multicolor spectral karyotyping of human chromosomes. *Science* **273,** 494–497.
35. Speicher, M. R., Ballard, S. G., and Ward, D. C. (1996) Karyotyping human chromosomes by combinatorial multi-fluor FISH. *Nat. Genet.* **12,** 368–375.
36. Schriml, L. M., Padilla-Nash, H. M., Coleman, A., Moen, P., Nash, W. G., Menninger, J., et al. (1999) Tyramide signal amplification (TSA)-FISH applied to mapping PCR-labeled probes less than 1 kb in size. *BioTechniques* **27,** 608–613.
37. Terkelsen, C., Koch, J., Kolvraa, S., Hindkjœr, J., Pedersen, S., and Bolund, L. (1993) Repeated primed in situ labeling: formation and labeling of specific DNA sequences in chromosomes and nuclei. *Cytogenet. Cell Genet.* **63,** 235–237.
38. Troyer, D. L., Goad, D. W., Xie, H., Rohrer, G. A., Alexander, L. J., and Beattle, C. W. (1994) Use of direct in situ single-copy (DISC) PCR to physically map five porcine microsatellites. *Cytogenet. Cell Genet.* **67,** 199–204.
39. Tharapel, S. A. and Kadandale, J. S. (2002) Primed in situ labeling (PRINS) for evaluation of gene deletions in cancer. *Am. J. Med. Genet.*, in press.
40. Paskins, L., Brownie, J., and Bull, J. (1999) In situ polymerase chain reaction and cycling primed in situ amplification: improvements and adaptations. *Histochem. Cell Biol.* **111,** 411–416.
41. Jonveaux, J. (1991) PCR amplification of specific DNA sequences from routinely fixed chromosomal spreads. *Nucleic Acids Res.* **19,** 1946.
42. Höfler, H. (1993) In situ polymerase chain reaction: toy or tool? *Histochemistry* **99,** 103–104.
43. Teo, I. A. and Shaunak, S. (1995) Polymerase chain reaction in situ: An appraisal of an emerging technique. *Histochem. J.* **27,** 647–659.
44. Hopman, A. H. N. and Ramaekers, F. C. S. (1998) Processing and staining of cell and tissue material for interphase cytogenetics, in *Current Protocols in Cytometry* (Robinson, J. P., ed.), John Wiley & Sons, New York, pp. 8.5.1–8.5.22.

45. Sällström, J. F., Zehbe, I., Alemi, M., and Wilander, E. (1993) Pitfalls of in situ polymerase chain reaction (PCR) using direct incorporation of labelled nucleotides. *Anticancer Res.* **13,** 1153.
46. Speel, E. J. M., Hopman, A. H. N., and Komminoth, P. (2000) Signal amplification for DNA and mRNA in situ hybridization, in *In Situ Hybridization Protocols. Methods in Molecular Biology* (Darby, I. A., ed.), Humana Press, Totowa, NJ, pp. 195–216.
47. Sperry, A., Jin, L., and Lloyd, R. V. (1996) Microwave treatment enhances detection of RNA and DNA by in situ hybridization. *Diagn. Mol. Pathol.* **5,** 291–296.
48. Pilling, A., Endersby-Wood, H., and Jones, S. (1997) Increased sensitivity of in situ hybridization: implications for in situ amplification. *Diagn. Mol. Pathol.* **6,** 174.
49. Nuovo, G. J. (2001) Co-labeling using in situ PCR: A review. *J. Histochem. Cytochem.* **49,** 1329–1339.
50. Komminoth, P., Long, A. A., Ray, R., and Wolfe, H. J. (1992) In situ polymerase chain reaction detection of viral DNA, single copy genes and gene rearrangements in cell suspensions and cytospins. *Diagn. Mol. Pathol.* **1,** 85–97.
51. Wiedorn, K. H., Kühl, H., Galle, J., Caselitz, J., and Vollmer, E. (1999) Comparison of in situ hybridization, direct and indirect in situ PCR as well as tyramide signal amplification for the detection of HPV. *Histochem. Cell Biol.* **111,** 89–95.
52. Patterson, B., Till, M., Otto, P., Goolsby, C., Furtado, M., McBride, L., and Wolinsky, S. (1993) Detection of HIV-1 DNA and messenger RNA in individual cells by PCR-driven in situ hybridization and flow cytometry. *Science* **260,** 976–979.
53. Speel, E. J. M., Richter, J., Moch, H., Egenter, C., Saremaslani, P., Rütimann, K., et al. (1999) Genetic differences in endocrine pancreatic tumor subtypes detected by comparative genomic hybridization. *Am. J. Pathol.* **155,** 1787–1794.
54. Zehbe, I., Hacker, G. W., Su, H., Hauser-Kronberger, C., Hainfeld, J. F., and Tubbs, R. (1997) Sensitive in situ hybridization with catalyzed reporter deposition, streptavidin-nanogold, and silver acetate autometallography. *Am. J. Pathol.* **150,** 1553–1561.
55. Lizardi, P. M., Huang, X., Zhu, Z., Bray-Ward, P., Thomas, D. C., and Ward, D. C. (1998) Mutation detection and single molecule counting using isothermal rolling circle amplification. *Nat. Genet.* **19,** 225–232.
56. Höfler, H., Pütz, B., Mueller, J., Neubert, W., Sutter, G., and Gais, P. (1995) In situ amplification of measles virus RNA by the self-sustained sequence replication reaction. *Lab. Invest.* **73,** 577–585.
57. Tecott, L. H., Barchas, J. D., and Eberwine, J. H. (1988) In situ transcription: specific synthesis of complementary DNA in fixed tissue sections. *Science* **240,** 1661–1664.
58. Mogensen, J., Kolvraa, S., Hindkjœr, J., Petersen, S., Koch, J., Nygård, M., et al. (1991) Nonradioactive, sequence-specific detection of RNA in situ by primed in situ labeling (PRINS). *Exp. Cell Res.* **196,** 92–98.
59. Player, A. N., Shen, L.-P., Kenny, D., Antao, V. P., and Kolberg, J. A. (2001) Single-copy gene detection using branched DNA (bDNA) in situ hybridization. *J. Histochem. Cytochem.* **49,** 603–611.
60. Mangham, D. C. and Isaacson, P. G. (1999) A novel immunohistochemical detection system using mirror image complementary antibodies (MICA). *Histopathology* **35,** 129–133.
61. Pastore, J. N., Clampett, C., Miller, J., Porter, K., and Miller, D. (1995) A rapid immunoenzyme double labeling technique using enzyme polymer system (EPOS) reagents. *J. Histotechnol.* **18,** 35–40.
62. Kämmerer, U., Kapp, M., Gassel, A. M., Richter, T., Tank, C., Dietl, J., and Ruck, P. (2001) A new rapid immunohistochemical staining technique using the EnVision antibody complex. *J. Histochem. Cytochem.* **49,** 623–630.

63. Van Gijlswijk, R. P. M., Van Gijlswijk-Janssen, D. J., Raap, A. K., Daha, M. R., and Tanke, H. J. (1996) Enzyme-labeled antibody-avidin conjugates: new flexible and sensitive immunochemical reagents. *J. Immunol. Methods* **189,** 117–127.
64. Chen, B.-X., Szabolcs, M. J., Matsushima, A. Y., and Erlanger, B. F. (1996) A strategy for immunohistochemical signal enhancement by end-product amplification. *J. Histochem. Cytochem.* **44,** 819–824.

60

Cycling Primed *In Situ* Amplification

John H. Bull and Lynn Paskins

1. Introduction

Primed *in situ* amplification (PRINS) is a technique for the visualization of specific sequences, usually repeat sequences, in fixed cell nuclei. When viewed on the microscope, the resulting signals can be *see*n as spots within nuclei, providing a means to visualize telomeres, centromeric regions, Alu repeats, or other sequences.

At its simplest, the PRINS reaction is a primer extension conducted under a sealed coverslip on a microscope slide. A mix of oligonucleotide primer, dNTPs (including a labeled dNTP), and *Taq* DNA polymerase is applied to a preparation of fixed cell nuclei on the slide, which is placed on a heating block and subjected to a round of denaturation, annealing, and extension. During this process, the nuclei are held in place while the DNA strands are made available for oligonucleotide annealing and extension by the polymerase. Unincorporated nucleotides are washed off, and the incorporated labeled nucleotide is detected typically by fluorescence *(1,2)*.

PRINS has many applications in common with the widely used fluorescence *in situ* hybridization (FISH) technique *(3)*. However, PRINS has significant advantages in the speed of the reaction, avoidance of toxic chemicals, lack of dependence on a carefully controlled stringency wash step, and in the use of easily synthesized oligonucleotides rather than expensive probes. For example, specific primers can be used for most human chromosomes or pairs of chromosomes *(4)*, giving discrete subnuclear spots that allow chromosome enumeration. The target sequence is typically satellite repeat (e.g., α-satellite, a family of 171-bp repeat units present at the centromeres of human chromosomes) and the strength of the signal is a function of the number of repeat units at the target site.

Cycling PRINS represents a modification where the primer extension is repeated several times (*see* **Fig. 1**), with the result that multiple labeled DNA strands are synthesized with a concomitant increase in signal *(5)*. This has obvious advantages in sensitivity, but in practice there are constraints to the signal build-up in cycling PRINS, which center around the fixation method used to prepare the nuclei or cells. PRINS depends on unimpeded access of reagents to the nuclear DNA. As the newly primed strand is synthesized, it is held in place by base pairing to the nuclear template DNA. When the reaction is cycled, there may be nothing to stop the newly synthesized strand diffusing away from the site of synthesis. In our laboratory, this problem appears

From: *Methods in Molecular Biology, Vol. 226: PCR Protocols, Second Edition*
Edited by: J. M. S. Bartlett and D. Stirling © Humana Press Inc., Totowa, NJ

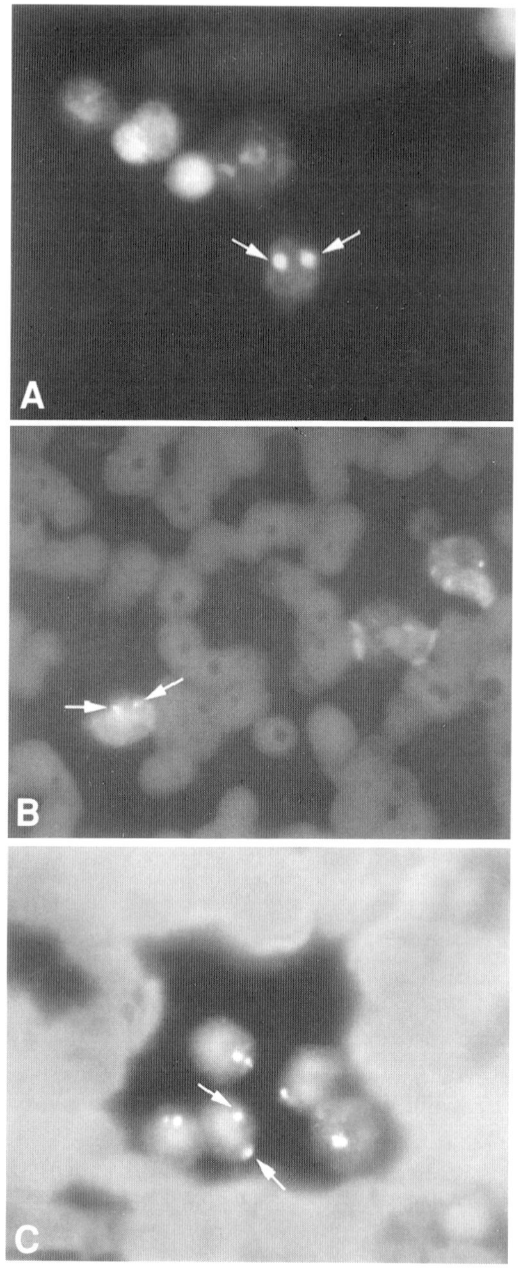

Fig. 1. Results of cycling PRINS using α-satellite primers for specific chromosomes (*see* **Subheading 3.1.**), DIG-dUTP incorporation and FITC detection. **(A)** cultured lymphoblast cells settled onto a slide labeled for chromosome X, **(B)** peripheral blood smear also labeled for chromosome X, and **(C)** smear of bone marrow aspirate labeled for chromosome 7. *See* **Note 4** for details of cell preparation.

most severe when the cytoplasm has been stripped away, as in typical methanol/acetic acid fixed metaphase spread preparations used in cytogenetic analysis *(6)*. The naked nucleus provides no barrier to diffusion, and although PRINS is very efficient, cycling PRINS confers no advantage as a steady state of signal is reached after the first cycle.

Many elaborate schemes can be imagined to overcome this, involving crosslinking agents, looped or circular structures, and photoactivatable anchors. The method described here simply uses ethanol fixation of whole cells. When ethanol is used, signal retention in cycling PRINS is markedly improved *(7)*. Background labeling of the nuclear DNA is increased with this fixative, but this is easily counteracted by microwave boiling of the slides before the PRINS reaction is set up. Using an extended initial denaturation step also helps reduce background.

The method described here is applicable to cell preparations, such as blood smears, cytospin, or gravity-settled cell preparations. Once dry, the slides can be fixed and stored for several months before cycling PRINS analysis. Although it is presently limited to whole cell preparations and *see*ms incompatible with the analysis of metaphase spreads, it is simple, robust, and effective.

2. Materials

2.1. Slide Preparation

1. Glass slides and coverslips: high-grade, dust-free slides are a sound investment and require no further preparation (e.g., Menzel Glaser Super Premium, Fisher Scientific, Loughborough, UK). Coated slides can be useful if there are doubts around loss of valuable cells (*see* **Note 1**). Coverslips can be 22 × 22 mm or bigger.
2. Cell preparation (*see* **Subheading 3.2.**).
3. Ethanol (100%).
4. Slide staining jar, for example, a Coplin jar.
5. 10 m*M* Tris-HCl, 5 m*M* EDTA, pH 7.0.
6. Anti-bumping granules (Merck).
7. 800-W Microwave oven (e.g., Matsui M162TC).
8. Graded ethanol solutions: 100%, 90%, and 70% v/v with water.

2.2. Cycling PRINS

1. AmpliTaq Gold DNA polymerase (PE Biosystems).
2. 10× PCR buffer: 500 m*M* KCl, 100 m*M* Tris-HCl, pH 8.3, 200 m*M* MgCl2 (*see* **Note 2**).
3. dNTPs (Amersham Pharmacia Biotech Inc., Uppsala, Sweden). Dilute to make a 50× stock solution of 10 m*M* each dATP, dCTP, dGTP; and 1.8 m*M* dTTP.
4. Digoxygenin-11-dUTP (DIG-dUTP; 10 nmol/µL, Roche Molecular Biochemicals, Lewes, UK) (*see* **Note 3**).
5. Oligonucleotide primer (50 µ*M*, *see* **Subheading 3.1.**).
6. Double distilled water.
7. High-grade clean coverslips and rubber solution (*see* **Note 4**) or Amplicovers and Ampliclips (PE Biosystems).
8. Flat-bed or slide-adapted thermal cycling ("PCR") machine (e.g., Hybaid, Ashford, UK or PE Biosytems).
9. Stop buffer: 500m*M* NaCl, 50 m*M* EDTA, pH 7.0.
10. Water bath at 65°C.

2.3. Detection

1. Water bath at 45°C, and staining jar (e.g., a Coplin jar).
2. Incubator set at 37°C and sandwich box.
3. 20× SSC: 3.0 *M* NaCl, 0.30 *M* tri-sodium citrate, pH 7.3.
4. Wash buffer: 4× SSC (diluted from 20× stock), 0.05% Triton×-100.
5. Dried skimmed milk powder.
6. Blocking buffer: Wash buffer with the addition of 5% skimmed milk powder (which can be stored for a few days at 4°C).
7. Anti DIG-FITC (Roche Molecular Biochemicals). Store at –20°C in 5-µL aliquots.
8. Antifade mountant (e.g., Vectashield, Vector Laboratories, Burlingame CA).
9. Counterstain: propidium iodide solution (20 µg/mL, Sigma, Dorset, UK) (*see* **Note 5**).
10. Fluorescence microscope with appropriate filters for FITC and PI (e.g., Zeiss Axioskop®, Carl Zeiss, Thornwood, NY).

3. Methods

3.1. Primer Design

These protocols were tested using α-satellite-specific primers for DYZ1 (D599: TGGGCTGGAATGGAAAGGAATCGAAAC), DXZ1 (E563: ATAATTTCCCATA-ACTAAACACA), D17Z1 (E571: AATTTCAGCTGACTAAACA), D7Z1 (E528: AGCGATTTGAGGACAATTGC), and D3Z1 (E570: TCTGCAAGTGGATATTTAAA) *(1)*. It has been suggested that one or more of these are used to set up the technique when using human cells (*see* **Note 5**).

3.2. Cell Preparation

The type of cells you use will of course depend on your experimental system. Aim to prepare cells in a monolayer, preferably slightly separated from each other, with several hundred cells in an area of 20 to 30 mm^2 for ease of location at the microscopy stage (*see* **Note 5**). In practice, 10 to 20 cells may be all that is needed for good results.

Human peripheral blood leukocytes provide a robust positive control and, satisfyingly, are easily available in plentiful and cheap supply (*see* **Note 6**). The majority red cells do not interfere with the PRINS reaction.

1. Obtain a few drops of blood from your finger using a suitable puncture device (e.g.. Haemolance, HaeMedic AB, Munka Ljungby, Sweden).
2. Spot approx 5 µL whole blood onto one end of a slide. Use a second microscope slide to spread the blood. This is done by holding the end of the second slide at an angle of roughly 45 degrees onto the blood spot, allowing it to spread under the edge, then pushing with some force to smear the cells. When done properly, a monolayer or "feather" of cells a few mm wide will form at the end of the smear (*see* **Note 5**). With practice, it is easily possible to make 2 or 3 small smears per slide, each using 1 to 2 µL of blood. This is useful if space on the flat-block PCR machine is limiting.
3. Allow to air dry.

3.3. Fixation

1. Fix by immersion in ethanol (*see* **Note 7**) for 5 min at room temperature (1 min is sufficient for blood smears).
2. Remove slides, drain, and air dry. Slides can be used immediately or stored at 4°C for several months.

3.4. Microwave Pretreatment

1. Place about 10 anti-bumping granules in a 50-mL Coplin jar and add the microscope slides. Slides can be placed back to back so that a jar contains 10 slides.
2. Fill the jar with 10 m*M* Tris-HCl, 5 m*M* EDTA (40–50 mL).
3. Place the jar in the center of the microwave oven, switch on full power until boiling, and boil for a further 50 s.
4. Quickly transfer the slides to a staining jar containing 70% ethanol at room temperature. Leave for 1 min.
5. Pass through 90% and 100% ethanol for 1 min each. Slides can either be air dried for immediate use or stored at 4°C in 100% ethanol for several months if desired.

3.5. Cycling PRINS

Reaction volumes will vary with your system (*see* **Note 8**). As a rough guide, 22- × 22-mm coverslips require about 15 µL, and 50- × 22-mm coverslips or Amplicovers require about 40 µL.

1. Prepare 100-µL reaction mix as follows: 2 µL of dNTP mix, 0.4 µL of DIG-dUTP, 10 µL of 10× PCR buffer, 2 µL (10 units) of AmpliTaq Gold DNA polymerase, 2 µL oligonucleotide primer, and 83.6 µL water.
2. Place 15 to 40 µL on the slide area, according to the cell preparation area and size of your coverslip, or 40 µL for Amplicovers.
3. For coverslips, seal with rubber solution and allow this to dry (a fan or laminar flow cabinet speed this up). For Amplicovers, follow manufacturers' instructions.
4. Transfer the slides to a flat block thermal cycler. A suitable program for the primers described is: 18 min at 95°C (to activate the AmpliTaq Gold DNA polymerase: reduce to 5 min for standard *Taq*), then 94°C for 1 min, 60°C for 1 min, and 72°C for 1 min for 15 cycles (*see* **Note 8**).
5. Peel off rubber glues, or take off Amplicovers, and immerse slides in a Coplin jar containing stop buffer at 65°C. Coverslips may come off with the glue, if not they will fall off readily in the stop bath. After 1 min, transfer slides to a jar containing wash buffer (we have also used 2× SSC, 45°C for 5 min to stop the reaction).

3.6. Detection

Do not allow slides to dry during this process.

1. Prepare blocking buffer, and pipet 40 µL on to a clean cover slip. Shake the slide free of excess wash buffer, then pick up the cover slip with slide. Leave the slide at room temperature for 5 min.
2. Dilute the anti-DIG FITC 1/100 in blocking buffer (e.g., add 500 µL of blocking buffer to a 5-µL aliquot. This is usually a suitable dilution, although batch-to-batch variation may be encountered).
3. Remove the coverslip from the slide and drain the excess fluid by shaking or briefly blotting one edge against an absorbent paper towel. Pipet 40 µL of diluted anti-DIG FITC on to the coverslip and replace the slide. Incubate in a moist chamber (e.g., a sandwich box lined with damp filter paper) at 37°C for 30 min.
4. Wash the slides 3× for 2 min in prewarmed wash buffer (42°C) in a Coplin jar.
5. Prepare mountant: add 3.75 µL of propidium iodide (*see* **Note 9**) stock to 100 µL of Vectashield.

6. Shake the slide free of excess liquid and mount as follows: pipet 40 µL of mountant on to a clean coverslip and pick up the coverslip with the slide. Place between two layers of tissue and press to spread the mountant and expel the excess. Seal with rubber solution (for long term storage) and allow to dry.
7. Slides can be viewed at once or stored for up to a few months in the dark at 4°C without significant loss of signal. Typical results are shown in **Fig. 1**.

4. Notes

1. Coated slides are often used in histological procedures where harsh treatments can cause loss of sample from the slide surface. They can be bought from Fisher or PE Biosystems or prepared by using poly L-Lysine or aminopropyltriethoxysilane ***(8)***. The Amplicover® and Ampliclip® system, together with slides, supplied by PE Biosystems (Foster City, CA) provides an alternative option for sealing the reaction solution onto the slide surface, but also requires use of their In Situ PCR 1000 thermal cycler and tool for assembling the slide chamber. Although relatively expensive, this is the best system for more than five cycles of PRINS.
2. The relatively high concentration of $MgCl_2$ is to counteract loss of solution phase Mg thought to occur during cycling.
3. Alternatively, biotin-labeled dUTP can be used at the same concentration, in which case a suitable detection reagent is avidin-FITC (Vector Labs). This can be substituted for anti-DIG FITC, at a 1:500 dilution.
4. Most commercially available brands for cycle puncture repair perform adequately. Rema Tip Top (Munich, Germany) is recommended.
5. Hematological smears work well, as do cytocentrifuge preparations of lymphocytes or neutrophils or cultured cells. Cells prepared by these methods are flattened and have a distinct morphological appearance, which is an advantage in interpreting and recording results because subnuclear spots tend to be in the same focal plane. We have used standard hematological preparations, such as bone marrow or peripheral blood smears, Giemsa-stained slides, and archival slides stored at room temperature, for over a year without problems. For cultured cells, or leukocyte preparations made from fresh blood (e.g., Lymphoprep, Sigma), cytocentrifuge ("cytospin") preparations are ideal (e.g., Shandon or Hiraeus cytocentrifuges and equipment). Alternatively, cells can be allowed to settle on coated slides for 10 to 20 min then allowed to dry after draining off liquid. These methods are easy to perform, but molecular biologists unfamiliar with cytological preparations might benefit from a tutorial in a hospital hematology or histology laboratory. Crystallized solids from support medium or isotonic solutions are not usually problematic because they are removed in the fixation and microwaving procedure. For extra phenotypic information, immunocytochemical staining can be done, and signals can be visualized alongside PRINS signal ***(7,9)***.
6. Hygiene precautions (i.e., hand washing) should be taken. The implications of sampling one's own blood might merit consideration: An inadvertent diagnosis of aneuploidy may cause distress! Chromosomes 3, 7, and 17 should be safe in this regard.
7. Ethanol fixation works by precipitating proteins irreversibly from solution. Other precipitating fixatives we have successfully used include methanol and acetone.
8. Reactions can be performed under coverslips or using the PE Applied Biosystems system (Amplicovers and Ampliclips). If using coverslips, some slides may start to dry out after around 10 cycles, but this will depend on the size of your coverslip and the type of rubber solution used to seal it to the slide. We find that 5 to 7 cycles is a good compromise. Using the PE system, we have cycled up to 70 times.

9. Propidium iodide binds specifically to DNA and emits a red fluorescence distinct from FITC. If desired, DAPI (4′,6-Diamidino-2-phenylindole 2 HCl; Sigma) can also be used at the same concentration. This emits a strong blue fluorescence, which sometimes aids in locating the cells on the slide. As an alternative detection system, Texas Red-conjugated reagents are available for detection of DIG or biotin. This can be combined with the DAPI counterstain, leaving the green channel free for detection of a third label, for example, via immunostaining *(7,9)*.

References

1. Gosden, J. and Lawson, D. (1994) Rapid chromosome identification by oligonucleotide-primed in situ synthesis (PRINS). *Hum. Mol. Genet.* **3,** 931–936.
2. Hindkjaer, J., Koch, J., Terkelsen, C., Brandt, C. A., Kolvraa, S., and Bolund, S. (1994) Fast, sensitive multicolour detection of nucleic acids in situ by primed in situ labelling (PRINS). *Cytogenet. Cell Genet.* **66,** 152–154.
3. Trask, B. J. (1991) Fluorescence in situ hybridization: Applications in cytogenetics and gene mapping. *Trends Genet.* **7,** 149–54.
4. Koch, J., Hindkjaer, J., Kolvraa, S., and Bolund, L. (1995) Construction of a panel of chromosome-specific oligonucleotide probes (PRINS-primers) useful for the identification of individual human chromosomes in situ. *Cytogenet. Cell Genet.* **71,** 142–147.
5. Gosden, J. and Hanratty, D. (1993) PCR in situ: A rapid alternative to in situ hybridisation for mapping short, low-copy number sequences without isotopes. *BioTechniques* **15,** 78–83.
6. Spowart, G. (1994) Mitotic metaphase chromosome preparation from peripheral blood for high resolution, in *Methods in Molecular Biology, Vol. 29: Chromosome Analysis Protocols* (Gosden, J. R., ed.), Humana, Totowa, NJ, pp. 1–10.
7. Paskins, L., Brownie, J., and Bull, J. (1999) In situ polymerase chain reaction and cycling primed in situ amplification: improvements and adaptations. *Histochem. Cell Biol.* **111,** 411–416.
8. Jackson, P. and Blyth, D. (1993) Immunolabelling techniques for light microscopy, in *Immuncytochemistry, a Practical Approach* (Beesley, J. E., ed.), IRL, Oxford, UK. pp. 15–42.
9. Speel, J. M., Lawson, D., Ramaekers, C. S., Gosden, J. R., and Hopman, A. H. N. (1997) Combined immunocytochemistry and PRINS DNA synthesis for simultaneous detection of phenotypic and genomic parameters in cells, in *Methods in Molecular Biology, Vol. 71: PRINS and in Situ PCR Protocols* (Gosden, J.R., ed.), Humana, Totowa, NJ, pp. 53–60.

61

Direct and Indirect *In Situ* PCR

Klaus Hermann Wiedorn and Torsten Goldmann

1. Introduction

In recent years, the development of *in situ* technologies has made good progress. *In situ* hybridization (ISH) has become an important tool and has enabled the pathologist to demonstrate infectious pathogens or mRNAs in tissue sections or cytospins without destruction of morphology, thus enabling the assignment of signals to individual cells or cell compartments ***(1–9)***.

Although ISH has contributed substantially to the diagnosis and understanding of neoplastic and infectious diseases, the detection of low copy DNA and RNA sequences by conventional ISH remained difficult in the past because of the relatively low sensitivity of ISH, irrespective of whether the investigation was performed using radioactive or nonradioactive hybridization probes ***(8,9a,10–15)***. Nonradioactive probes, especially biotin and digoxigenin (DIG), which are less hazardous to work with, can be much more quickly developed, allow a much higher spatial resolution, and have been shown to be at least as sensitive as radioactive probes ***(10–14)***. Several different detection systems ***(14,16–20)*** have been used to enhance signal intensities. Nevertheless, conventional ISH usually will enable detection of high to medium copy number nucleic acids only.

However, target amplification by *in situ* PCR (IS-PCR) or reverse transcription *in situ* PCR (RT-IS-PCR) have been shown to even allow the detection of low copy number DNAs or RNAs ***(1,4–6,8,15,21–23)***. There exist two different approaches to IS-PCR: direct and indirect IS-PCR (**Fig. 1**).

Indirect IS-PCR requires an additional ISH step and is more cumbersome but will usually yield reliable results. Direct IS-PCR is often hampered by nonspecific products, especially when performed on paraffin-embedded tissue sections. These false positives in direct IS-PCR are predominantly primer-independent artifacts resulting from DNA repair and endogenous priming ***(5,21,22,24,25)***. These pathways are also operative in indirect IS-PCR but will not produce false positives because no labeled nucleotides are incorporated during the amplification step. Primer-dependent artifacts like mispriming can be controlled by hot start maneuvers, although this is in general not quite as easy as in solution-phase PCRs. In addition, diffusion artifacts may be involved in the generation of false-positive cells in paraffin-embedded tissues undergoing direct as well

From: *Methods in Molecular Biology, Vol. 226: PCR Protocols, Second Edition*
Edited by: J. M. S. Bartlett and D. Stirling © Humana Press Inc., Totowa, NJ

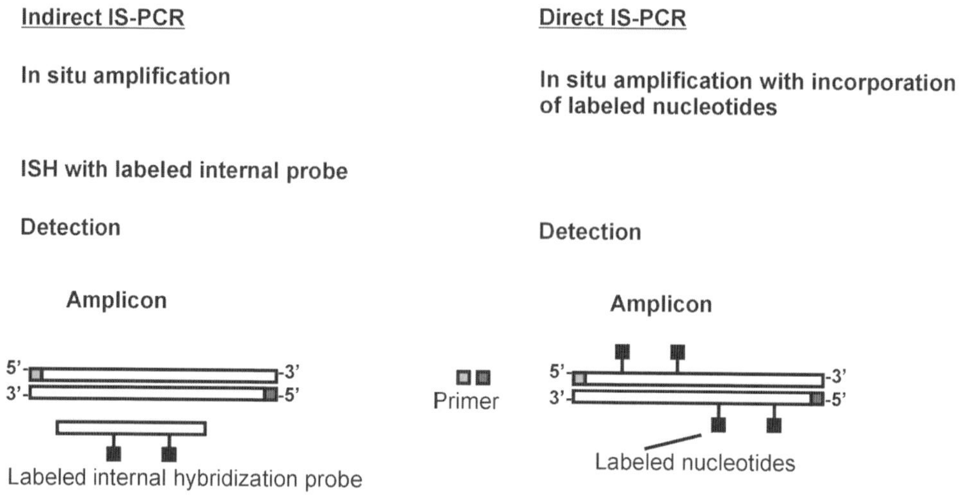

Fig. 1. IS-PCR can be performed with indirect and direct approach. With direct IS-PCR during amplification, unlabeled and labeled nucleotides are used in the reaction mix. Thus, labeled nucleotides are incorporated into the PCR products. Therefore, subsequent to the amplification, direct detection of the labeled PCR products can be performed. In contrast with indirect IS-PCR, only unlabeled nucleotides are used with the reaction mix. Therefore, the amplicons are unlabeled, and for detection of the PCR products subsequent to the amplification, an ISH with a labeled internal probe (a probe that does not contain the primer sites) has to be performed. Thus, the detection cannot be performed until after ISH is finished.

as indirect IS-PCR, although they are predominantly associated with cell suspensions *(4,22)*. Optimized fixation and permeabilization as well as reduction of PCR cycle numbers to less than 30 cycles are likely to minimize but not totally exclude this phenomenon. With RT-IS-PCR, in contrast to solution-phase PCR, an additional problem will arise. Depending on the primers chosen, false-positive nuclear signals may arise because usually the primers will anneal to cDNA as well as to genomic DNA in the nucleus (which is eliminated during RNA extraction and additional DNAse digestion for removing residual DNA in approaches using solution phase RT-PCR) and will therefore not only generate PCR-products in the cytoplasm but also in the nucleus. To circumvent this problem, cDNA-specific primers, which will only anneal to the cDNA, may be designed for RT-IS-PCR *(**15**)*. In addition, a DNAse pretreatment may be performed before RT-IS-PCR to destroy the genomic DNA. However, the general potential of this technique is controversially discussed *(5,**15**,26)*.

Nonetheless, although direct IS-PCR will contain more risks for false-positive results, both methods are described in this chapter especially because direct IS-PCR offers a more convenient way and will yield reliable results when performed on nonparaffin-embedded samples like cytospins.

2. Materials

1. *In situ* Thermal cycling machine (Hybaid AGS, Heidelberg, Germany; Shandon, Frankfurt, Germany) (*see* **Note 1**).
2. Wash Module (Hybaid AGS, Heidelberg, Germany; Shandon, Frankfurt, Germany).
3. SuperFrostPlus slides (Menzel Gläser, Braunschweig, Germany) (*see* **Note 2**).

4. Cover Slips (Omnilab, Germany).
5. Microtome (Leica, Germany).
6. Heating Oven for 220°C (WTB Binder, Germany).
7. Water bath (e.g., GFL 1083, GFL, Germany).
8. Pattex Supermatic 200plus (Henkel, Düsseldorf, Germany).
9. Forceps.
10. Laboratory gas burner.
11. DEPC (Sigma, Deisenhofen, Germany, Cat-No. D5758).
12. 5 *N* NaOH (Merck, Germany).
13. 1 *N* HCl (Merck, Germany).
14. NaCl (Merck, Germany).
15. $NaH_2PO_4H_2O$ (Merck, Germany).
16. $Na_2HPO_42H_2O$ (Merck, Germany).
17. Tris-HCl (Sigma, Deisenhofen, Germany).
18. EDTA (Sigma, Deisenhofen, Germany).
19. 10× phosphate-buffered saline (PBS): 1.3 *M* NaCl, 5 m*M* NaH_2PO_4, 95 m*M* Na_2HPO_4. Adjust to pH 7.2 with orthophosphoric acid.
20. 5% buffered formalin pH 7.2 (Frontell, Germany).
21. 4% buffered paraformaldehyde, pH 7.2. Dissolve 40 g of paraformaldehyde (Merck ultra pure 4005, Germany) in 1× PBS. Adjust to pH 7.2 with NaOH and fill up to 1000 mL with 1× PBS. Filter through Wathman No 1.
22. 100% ETOH (Merck, Germany).
23. Xylene (Merck, Germany).
24. *Taq* Polymerase 5 U/µL (Roche Diagnostic GmbH, Mannheim, Germany, Cat No. 1146173).
25. 0.1 *M* Tris-HCl, 0.1 *M* NaCl, pH 7.5.
26. 1 *M* $MgCl_2$ (Sigma, Deisenhofen, Germany).
27. $CaCl_2$ (C3306, Sigma, Deisenhofen, Germany).
28. 20× SSC: 3 *M* NaCl and 0.3 *M* Na_3 Citrate. Adjust to pH 7.0 with HCl.
29. Bovine serum albumin (BSA) 20 mg/ mL (Roche Diagnostic GmbH, Mannheim, Germany, Cat-No. 711454).
30. 10% SDS (Sigma, Deisenhofen, Germany, Cat-No. L 4522).
31. 100% deionized formamide (Sigma, Deisenhofen, Germany, Cat-No. F9037).
32. Dextran sulfate (Sigma, Deisenhofen, Germany, Cat-No. D8906): 50% dextran sulfate. Dissolve 50 mg in 100 mL of DEPC-treated distilled water. Heat to 60°C and shake to speed up the dissolving.
33. DIG Wash and Block Buffer Set (Roche Diagnostic GmbH, Mannheim, Germany, Cat-No. 1585762) containing 10× washing buffer, 10× blocking solution (=10% blocking solution), 10× detection buffer, 10× maleic acid buffer.
34. NBT 100 mg/mL (Nitroblue tetrazolium chloride; Roche Diagnostic GmbH, Mannheim, Germany Cat-No. 1383213).
35. BCIP 50 mg/mL (5-Brom-4-chlor-3-indoyl-phosphate) (Roche Diagnostic GmbH, Mannheim, Germany, Cat-No. 138221).
36. Mounting media (as supplied; Permount SP15-500, Fisher Scientific, Wiesbaden, Germany).
37. Counterstain (Nuclear Fast Red and Eosin, Merck, Germany).
38. Anti-Dig or Anti-biotin antibody conjugates (Roche Diagnostic GmbH, Mannheim, Germany Cat-No. 1426303, 1426311, 1093274, 1207733).
39. Hybridization probes (Eurogentec, Belgium or MWG Biotech, Ebersberg, Germany).
40. DNAse (Roche Diagnostic GmbH, Mannheim, Germany, Cat-No. 776785).

41. Proteinase K solution: Proteinase K (Roche Diagnostic GmbH, Mannheim, Germany; 250 µg/mL), 100 m*M* Tris-HCl and 50 m*M* EDTA.
42. 10× Target Retrieval (Dako, Hamburg, Germany, Cat-No. S1700).
43. DNAse solution: 40 m*M* Tris-HCl, pH 7.4, 6 m*M* $MgCl_2$, 2 m*M* $CaCl_2$, and 1 U/µL RNAse free DNAse (Roche Diagnostic GmbH, Mannheim, Germany).
44. IL6 PCR mix: 10 m*M* Tris-HCl, pH 9.2, 50 m*M* KCl, 3 m*M* $MgCl_2$, 0.2 m*M* dNTPs (Roche Diagnostic GmbH, Mannheim, Germany); 0.1% BSA (Roche Diagnostic GmbH, Mannheim, Germany); 10 µ*M* DIG-11-dUTP (Roche Diagnostic GmbH, Mannheim, Germany) (*see* **Note 3**); 5 U/sample *Taq*-Polymerase (Roche Diagnostic GmbH, Mannheim, Germany); and 0.4 µ*M* of sense and antisense primer.
45. Sequence of IL6 primers (*see* **Note 4**): sense primer: 5′ CTTCTCCACAAGCGCCTTC-3′; antisense primer: 3′ CTAAGTTACTCCTCTGAACGG-5′.
46. A typical hybridization mix is composed of (exact compositions depends on the probe in use): 2× up to 5× SSC; up to 50% formamide; 5% up to 10% dextran sulfate (*see* **Note 5**); 0.1% SDS; 0.1% BSA; 1% up to 2% blocking solution; 250 µg/ mL fish sperm DNA (Roche Diagnostic GmbH, Mannheim, Germany); 1–10 ng/µL hybridization probe.

Additionally for RT-IS-PCR:

47. DNAse/RNAse solution: 40 m*M* Tris-HCl, pH 7.4, 6 m*M* $MgCl_2$, 2 m*M* $CaCl_2$, 1 U/µL RNAse free DNAse (Roche Diagnostic GmbH, Mannheim, Germany), and500 µg/mL RNAse (Roche Diagnostic GmbH, Mannheim, Germany).
48. Reverse transcription buffer (RT-buffer): 50 m*M* Tris-HCl, pH 8.3, 75 m*M* KCl, 3 m*M* $MgCl_2$, 0.5 m*M* dNTPs (Roche Diagnostic GmbH, Mannheim, Germany), 10 m*M* DTT (GibcoBRL, Karlsruhe, Germany), 1 U/µL RNAsin (Promega, Mannheim, Germany), 2 U/µL M-MLV-RT (GibcoBRL, Karlsruhe, Germany) (*see* **Note 6**), 0.1 µg/µL Random Hexamers (Roche Diagnostic GmbH, Mannheim, Germany).

3. Methods

If not otherwise mentioned, all incubations are performed at ambient temperature (20°C). Incubations on the thermocycler are performed using the humidity chamber. Samples are processed without coverslips and sealing if not otherwise mentioned.

3.1. Prevention of DNAse and RNAse Contamination

1. Incubate all slides, coverslips and microtome blades for 12 h at 220°C to inactivate DNAse and RNAse (*see* **Notes 7** and **8**).
2. If RNA detection is required, include 0.1% DEPC in all solutions and incubate overnight.
3. Autoclave solutions for 20 min at 1.2 bar (15 psi) at 121°C.

3.2. Fixation

Incubate tissues for 24 h in 5% buffered formalin (for detection of DNA) or 4% buffered paraformaldehyde (for detection of mRNA) (*see* **Note 9**). With cell suspensions and cytospins a 30 min incubation is usually sufficient.

3.3. Paraffin Embedding

The distilled water used to dilute ETOH should be autoclaved and prepared with DEPC-treated water to ensure the absence of DNAse and RNAse.

1. Incubate tissue for 30–60 min in 70% ETOH.
2. Incubate tissue for 30–60 min in 80% ETOH.

3. Incubate tissue for 30–60 min in 90% ETOH.
4. Incubate tissue for 30–60 min in 96% ETOH.
5. Incubate tissue for 30–60 min in 100% ETOH.
6. Incubate tissue for 30–60 min in 100% ETOH.
7. Incubate tissue for 60 min in xylene: 100% ETOH (1 : 1).
8. Incubate tissue 2 × 60 min in xylene.
9. Incubate tissue 3 × 60 min in paraffin.
10. Allow to cool and harden the paraffin for several hours.

3.4. Preparation of Samples

3.4.1. Tissue Sections

1. Use Teflon coated SuperFrostPlus slides (*see* **Note 2**).
2. Cut section of 2- to 5-μm thickness (*see* **Note 10**).
3. Change the microtome blade frequently and clean thoroughly with xylene after each block to prevent crosscontamination between samples.
4. Incubate slides overnight at 50°C.

3.4.2. Cytospins

1. Centrifuge cells on slides.
2. Incubate slides for at least 30 min at 50°C.

3.5. Deparaffinization

1. Incubate slides for 10 min at 70°C.
2. Incubate slides for 10 min in fresh xylene (*see* **Note 11**).
3. Incubate slides for 2 min in fresh xylene.
4. Incubate 2 × 1 min in 100% ETOH.
5. Incubate 1 min in 90% ETOH.
6. Incubate 1 min in 70% ETOH.
7. Incubate 1 min in 50% ETOH.
8. Incubate 1 min in distilled water.
9. Dry Slides for up to 10 min at 37°C (on the thermocycler).

3.6. Permeabilization

The following permeabilization protocols are for optimized fixation conditions and may vary with fixation conditions and tissues (*see* **Notes 9** and **12**).

3.6.1. By Protease

1. Incubate 15 min at 37°C (on the thermocycler) with 100 μL of Proteinase K solution.
2. Incubate 2 × 5 min in 0.1 *M* Tris-HCl, 0.1 *M* NaCl, pH 7.5.

3.6.2. By Target Retrieval

1. Put slides in a Coplin jar filled with 1 × Target Retrieval (Dako). Place Coplin jar in a waterbath and incubate 35 min at 95°C.
2. Place Coplin jar outside the water bath and allow to cool for 20 min.
3. Incubate 2 × 5 min in 0.1 *M* Tris-HCl, 0.1 *M* NaCl, pH 7.5.

3.7. DNAse Treatment

1. Incubate in 100 μL of DNAse solution at 37°C for 12 h on the thermocycler. Cover slides with coverslips.

2. Incubate 4 min at 95°C.
3. Incubate 2 × 5 min in 0.1 *M* Tris-HCl, 0.1 *M* NaCl, pH 7.5, in a Coplin jar.

3.8. DNAse/RNAse Treatment

If destruction of RNAse is necessary too (for example to generate negative controls), a combined DNAse/RNAse treatment is recommended. Proceed as mentioned under **Subheading 3.7.**, but use the DNAse/RNAse solution.

3.9. Quenching

If Peroxidase/AEC or -/DAB systems are used during the detection then quenching is required to inactivate endogenous peroxidase.

1. Incubate 30 to 45 min in 100 µL of 0.6% hydrogen peroxide (*see* **Note 13**).
2. Incubate 1 min in 50% ETOH.
3. Incubate 1 min in 70% ETOH.
4. Incubate 1 min in 90% ETOH.
5. Dry Slides for 5 min at 37°C (on the thermocycler).

3.10. Reverse Transcription

For all RT and IS-PCR, the Omnislide and MISHA thermocyclers have to be set to simulated slide control and calibration factor 100.

If you are investigating DNA targets, then proceed to **Subheading 3.11.**

1. Incubate for 60 min at 39°C in 15 to 20 µL of RT-buffer per sample on the thermocycler. Cover sample with coverslips and seal with PattexSupermatic200Plus (*see* **Note 14**).
2. Stop reaction by incubation at 92°C for 10 min.
3. Remove coverslips (*see* **Note 15**) and discard reaction mixture.

3.11. IS-PCR

Use aerosol resistant pipet tips for preparing the PCR.

1. Prepare PCR mix without *Taq*-Polymerase in Eppendorf tubes.
2. Heat PCR mix to 80°C.
3. Heat slides to 80°C on the thermocycler.
4. Add *Taq*-Polymerase to PCR mix and vortex briefly.
5. Pipet 16 µL of the PCR mix onto the specimen.
6. Immediately cover with a coverslip using a forceps and seal with PattexSupermatic 200Plus.
7. Heat forceps in a laboratory gas burner before proceeding to the next sample to avoid cross contamination.
8. Repeat **steps 5–7** for each specimen.
9. Start the PCR program.
10. Remove coverslips (*see* **Note 15**) and discard reaction mixture.

The PCR protocol for IL6 is composed of one cycle at 92°C for 5 min, 32 cycles each with 1 min denaturation step at 92°C, 1 min annealing step at 56°C, 2 min extension step at 72°C. The program is finished by an additional extension for 7 min at 72°C.

3.12. ISH

This step is only necessary if indirect IS-PCR is performed. Otherwise proceed to **Subheading 3.13.** All reactions are performed on the thermocycler.

1. Prepare fresh hybridization mix without fish sperm DNA.
2. Denature fish sperm DNA for 10 min at 100°C in a water bath.
3. Cool fish sperm DNA on ice for 2 min.
4. Add fish sperm DNA to the hybridization mix.
5. Pipet 16 μL of the hybridization mix onto each sample.
6. Immediately cover with a coverslip using a forceps and seal with PattexSupermatic200Plus (*see* **Note 14**).
7. Heat forceps in a laboratory gas burner before proceeding to the next sample to avoid cross contamination.
8. Incubate 10 min at 95°C.
9. Incubate 1 to 16 h between 30 to 65°C (*see* **Note 16**).
10. Remove coverslips (*see* **Note 15**) and discard reaction mixture.

3.13. Washing and Detection

1. Incubate 2 × 5 min in 0.1× SSC at 20°C (*see* **Note 17**).
2. Incubate 10 min in 0.1× SSC at 45°C in the Wash Module (*see* **Note 17**).
3. Incubate 1 min in 1× washing buffer (DIG Wash and Block buffer Set, Roche Diagnostic GmbH, Mannheim, Germany).
4. Incubate 60 min in 1× blocking solution at 37°C.
5. Add 100 μL of 1:250 diluted Anti-DIG-AP-conjugate per sample and incubate 60 min at 37°C on the thermocycler.
6. Incubate 2 × 5 min in 1× washing buffer at 20°C.
7. Incubate 2 × 5 min in 1× detection buffer at 20°C.
8. Incubate in 100 μL per sample freshly made substrate solution (up to 12 h) in the dark. Substrate solution is composed of (per mL): 4.5 μL NBT, 3.5 μL BCIP, 992 μL of detection buffer (all reagents Roche Diagnostic GmbH, Mannheim, Germany) (*see* **Note 18**).
9. Counterstain 10 min with nuclear fast red or eosin at 20°C (intensity of counterstain may be adjusted by reducing or prolonging incubation period).
10. Mount with Permount SP15-500 (Fisher Scientific).

3.14. Controls

Several controls have to be performed to ensure the specificity of the results.

1. Omission of RT (should result in cells showing no signal).
2. Omission of *Taq*-Polymerase (should result in cells showing no signal).
3. Omission of primers (should result in cells showing no signal).
4. RT IS-PCR of a housekeeping gene, such as GAPDH, or actin to ensure that the reverse transcription and PCR was successful because GAPDH should be demonstrable in each cell. Furthermore, this should indicate if the permeabilization was successful.
5. PCR of Alu-repeats to ensure that PCR and permeabilization was successful.
6. Omission of the Anti-Dig-POD or Anti-DIG-AP to detect endogenous enzyme activity.
7. Tissues that are definitively negative for the sequence under investigation. If such tissues are not available, the control samples may be treated either with DNAse, RNAse or both.
8. Tissues that are definitively positive for the sequence under investigation.

Furthermore, at least in the phase of establishing a new PCR protocol for new primers, all IS-PCRs should be controlled by simultaneous solution-phase PCRs on serial sections of the same samples.

3.15. Primer Design

Primers should be designed to be cDNA specific (if RNA targets are under investigation), that is, to span an intron *(15)* to disable amplification of genomic sequences during RT-IS-PCR. Furthermore, primers should be chosen to give small amplification products because the efficiency of IS-PCR will be reduced with longer products.

3.16. Closing Remarks

PCR has become an important diagnostic as well as research tool in molecular biology, clinical chemistry, and pathology. With the invention of IS-PCR, the amplification power of solution-phase PCR with no limitations in the amount of template was hoped to be transferred to the *in situ* techniques. Unfortunately, this has become reality only partly because IS-PCR is often hampered by poor reproducibility, specificity, and reliability *(8,9)* and by the cumbersome protocol. For semiquantitative *in situ* studies of gene expression in combination with image analysis, RT-IS-PCR seems to be of little value because of the tremendously varying amplification efficiency of IS-PCR *(15,27)*. For these applications, ISH with subsequent signal amplification by biotinyl tyramide proved to be the method of choice. This approach has been shown to be an excellent alternative for IS-PCR. With respect to most applications, generally signal amplification procedures are more suitable than target amplification by direct or indirect IS-PCR and exhibit a sensitivity similar to that of IS-PCR *(2,8,9,25,28)*. Adequate choice of hybridization probes provided signal amplification allows even the detection of single-copy virus sequences *(28)*.

4. Notes

1. We have made good experience with the Omnislide *in situ* Thermocycler (Hybaid AGS Germany) and the MISHA (Shandon Germany) because the slides are fitted horizontally onto the blocks in the humidity chamber thus requiring sealing only during the PCR whereas those apparatus where slides have to be fitted vertically onto the blocks require sealing during each step of the procedure resulting in a much more cumbersome protocol. For achieving reliable results, it is of utmost importance to use thermocyclers specially designed for *in situ* PCR procedures. For detailed instructions regarding the setup of the thermocycler and protocols, *see* instruction manual and *(29)*.
2. Although expensive (E 0.50 per slide) we recommend the use of Teflon coated SuperFrost-Plus slides (Menzel-Gläser, Braunschweig, Germany). These behaved well with respect to adhesion of tissue even after prolonged cycling protocols and because of the hydrophobic Teflon coating around the well do not require the cumbersome use of hydrophobic pens to outline the reaction area. Hydrophobic pens usually have to be used several times during a PCR protocol to guarantee a closed border around the tissue sample. However, repeated application of the hydrophobic pens requires drying of the slide and the tissue, which can produce strong background staining.

   Slides with at least two wells should be used so that controls can be run on the same slide simultaneously. In this case, Teflon coating will effectively prevent contamination between the two samples. We used slides with wells of 17-mm diameter each. The volumes indicated during the protocol proofed well in covering samples of this size but have to be adjusted for wells with different diameters.
3. DIG-dUTP is omitted if indirect IS-PCR is performed. Instead of DIG-dUTP, biotin-labeled nucleotides can be used too.
4. Unlike in solution-phase, PCR primers for IS-PCR should be designed to give amplicons

of 200 to 800 bp because with larger amplicons the amplification efficiency will be reduced dramatically and will often result in false negative samples.

5. Because dextran sulfate may produce background staining, especially if higher concentrations are used, reduce the amount of dextran sulfate to diminish the background staining. For ISH in contrast to filter hybridization, usually 10% dextran sulfate is the upper limit.
6. The reverse transcription can also be performed with other transcriptases, such as Superscript (GibcoBRL, Karlsruhe, Germany). In this case, the protocol has to be adjusted accordingly.
7. For SuperFrostPlus slides, this is only a precaution as when stored (dust free) and handled under appropriate conditions, we have encountered neither DNAse nor RNAse contamination.
8. Coverslips may be siliconized prior to the incubation at 220°C if adherence of tissue sample to the coverslips is observed during the PCR procedure.
9. To obtain reproducible results, it is essential that fixation is always performed at the same temperature and for the same time. Otherwise, the permeabilization conditions have to be adapted and optimized for each PCR. A difference of 5°C during fixation may give false negative results because of changes in cell permeability. Unfortunately, tissue in molecular pathology is usually not fixed under standardized conditions. Therefore, several permeabilization conditions have to be tested for each sample.

   To improve fixation and therefore DNA and RNA preservation, tissues should be cut to the minimum size prior to fixation.

   For mRNA detection paraformaldehyde will yield a better preservation of RNA than formalin. Perform fixation at 4°C where possible. The extraction of RNA from formalin- or paraformaldehyde-fixed tissues for doing a solution-phase PCR to verify the *in situ* results especially when establishing a new protocol is usually very ineffective. For these applications, we recommend the use of the Hope fixative (patent pending, Dr. Olert, Institut für Kinderpathologie, Universität Mainz, Germany available at DCS, Hamburg, Germany) which additionally yields a much better preservation of RNA compared to other fixatives.
10. Thinner sections show better adhesion to slides during IS-PCR, although this reduces the number of target sequences, which may result in some negative cells.
11. Use fresh reagents for each incubation. Because xylenes saturate rapidly with paraffin use 200 mL of fresh xylene for each batch of 15 to 20 slides.
12. Although fixed under the same conditions, different tissues may require adapted permeabilization parameters. When optimizing permeabilization parameters with respect to morphology, it is usually better to prolong Proteinase K incubation than to increase the concentration of Proteinase K ***(15)***. If diffusion artifacts are a major problem the reduction of the concentration of Proteinase K with simultaneous prolongation of incubation turned out to be advantageous ***(15)***. If you encounter background staining this may be reduced by the use of target retrieval instead of Proteinase K ***(15)***. Proteinase K (250 µg/ mL) is optimal for portio and condylomata biopsies fixed for 24 h in buffered formalin whereas for leukocytes 10 to 30 µg/mL Proteinase K and for bronchial epithelial cells 30 to 50 µg/mL Proteinase K performed well. Leukocytes and bronchial epithelial cells were fixed with 4% buffered paraformaldehyde for 30 min.
13. The concentration of endogenous peroxidase varies significantly between tissues. Leukocytes exhibiting high endogenous peroxidase levels usually require longer incubation (45 min) than other tissues. If you encounter signals in your negative controls try to diminish those signals either by prolongation of the quenching reaction or by increasing the concentration of H_2O_2 up to 3%.
14. Evaporation control is of the utmost importance for reliable and reproducible results as unspecific signals because of the evaporation of the reaction mixture may result from insuf-

ficient sealing. From the many methods for sealing, the use of Pattex Supermatic200Plus offers a convenient and reliable way to prevent leakage ***(15,29,30)***. It is easier to handle than nail polish with the same good sealing capacity. Furthermore, it seems to represent a biocompatible adhesive as it does not interfere with enzyme activity during PCR.

15. Coverslips can easily be removed by an initial scalpel cut along the sealed edge, which is to be lifted first. To facilitate the removal of the coverslips, the slides can be stored at 4°C for 1 to 2 min to completely harden the gluten if it is still viscous after the incubation at 92°C *(8,**15,29**)*.
16. The lower the temperature and the shorter the incubation period, the better will be morphology which, nevertheless, is negatively affected by the IS-PCR procedure.
17. Temperatures, concentration of SSC, and incubation period have to be adjusted for the probe in use.
18. Antibody concentrations have to be adapted for different probes and according to the efficiency of the PCR. One can choose from a wide variety of chromogens. Furthermore, a peroxidase-based system can be used using AEC+ or DAB+ (Dako) as chromogens, which usually will result in more localized signals *(8,**28**)* than those achieved with NBT/BCIP. Detection solutions and antibody concentrations have to be adjusted.

Acknowledgments

The authors are indebted to Prof. Dr. Dr. E. Vollmer for making it possible to participate in the development of this technology and H. Kühl for excellent technical assistance.

References

1. Chen, R. H. and Fuggle, S. V. (1993) In situ cDNA polymerase chain reaction. A novel technique for detecting mRNA expression. *Am. J. Pathol.* **143,** 1527–1534.
2. Goldmann, T., Becher, B., Wiedorn, K. H., Pirow, R., Deutschbein, M. E., Vollmer, E., et al. (1999) Epipodite and fat cells as sites of hemoglobin synthesis in the branchiopod crustacean Daphnia magna. *Histochem. Cell Biol.* **112,** 335–339.
3. Heniford, B. W., Shum-Siu, A., Leonberger, M., and Hendler, F. J. (1993) Variation in cellular, E. G.F receptor mRNA expression demonstrated by in situ reverse transcriptase polymerase chain reaction. *Nucleic Acids Res.* **21,** 3159–3166.
4. Long, A. A., Komminoth, P., and Wolfe, H. J. (1992) Detection of HI.V provirus by in situ polymerase chain reaction. *N. Engl. J. Med.* **327,** 1529.
5. Nuovo, G. J. (1994) *PCR in Situ Hybridization Protocols and Applications*, Second Ed., Raven Press, New York.
6. Patel, V. G., Shum-Siu, A., Heniford, B. W., Wieman, T. J., and Hendler, F. J. (1994) Detection of epidermal growth factor receptor mRNA in tissue sections from biopsy specimens using in situ polymerase chain reaction. *Am. J. Pathol.* **144,** 7–14.
7. Snijders, P. J. F., van den Brule, A. J. C., Schrijnemakers, H. F. J., Snow, G., Meijer, C. J. L. M., and Walboomers, J. M. M. (1990) The use of general primers in the polymerase chain reaction permits detection of a broad spectrum of human papillomavirus genotypes. *J. Gen. Virol.* **71,** 173–181.
8. Wiedorn, K. H., Kühl, H., Galle, J., Caselitz, J., and Vollmer, E. (1999) Comparison of in-situ hybridization, direct and indirect in-situ PCR as well as tyramide signal amplification for the detection of HPV. *Histochem. Cell Biol.* **111,** 89–95.
9. Wiedorn, K. H., Goldmann, T., Kühl, H., Galle, J., and Vollmer, E. (1999) GenPoint™: A new biotinyl tyramide based signal amplification system for significantly increasing sensitivity of in situ hybridization (ISH). *Elec. J. Pathol. Histol.* 994–999.

9a. Furuta, Y., Shinohara, T., Sano, K., Meguro, M., and Nagashima, K. (1990) In situ hybridization with digoxigenin-labelled DNA probes for detection of viral genomes. *J. Clin. Pathol.* **43,** 806–809.
10. Allan, G. M., Todd, D., Smyth, J. A., Mackie, D. P., Burns, J., and McNulty, M. S. (1989) In situ hybridization: an optimised detection protocol for a biotinylated DNA probe renders it more sensitive than a comparable ^{35}S-labeled probe. *J. Virol. Methods* **24,** 181–190.
11. Burns, J., Graham, A., Frak, C., Fleming, F., Evans, M., and McGee, J. (1987) Detection of low copy human papilloma virus DNA and mRNA in routine paraffin sections of cervix by non-isotopic in situ hybridization. *J. Clin. Pathol.* **40,** 858–864.
12. Furuta, Y., Shinohara, T., Sano, K., Meguro, M., and Nagashima, K. (1990) In situ hybridization with digoxigenin-labelled DNA probes for detection of viral genomes. *J. Clin. Pathol.* **43,** 806–809.
13. Nuovo, G. J. and Richart, R. M. (1989) A comparison of biotin and ^{35}S-based in situ hybridization methodologies for detection of human papilloma-virus DNA. *Lab. Invest.* **61,** 471–476.
14. Syrjänen, S., Partanen, P., Mäntyjärvi, R., and Syrjanen K. (1988) Sensitivity of in situ hybridization techniques using biotin- and 35S-labeled human papillomavirus (HPV) DNA probes. *J. Virol. Methods* **19,** 225–238.
15. Wiedorn, K. H., Lang, D., Köhl, H., Galle, J., and Vollmer, E. (1999) Sensitive in situ detection of modulated expression levels of interleukin 6 (IL6) and interleukin 8 (IL8) in bronchial epithelium and leucocytes using cDNA specific primers and direct RT in-Situ-PCR. *Elec. J. Pathol. Histol.* 993–996.
16. Herrington, C. S., Burns, J., Graham, A. K., Evans, M., and McGee, J. O. (1989) Interphase cytogenetics using biotin and digoxigenin labeled probes I: relative sensitivity of both reporter molecules for detection of HPV 16 in CaSki cells. *J. Clin. Pathol.* **42,** 592–600.
17. Holm, R., Karlsen, F., and Nesland, J. M. (1992) In situ hybridization with nonisotopic probes using different detection systems. *Mod. Pathol.* **5,** 315–319.
18. Lassus, J., Niemi, K.-M., Marjamaki, A., Syrjanen, S., Kartamaa, M., Lehmus, A., et al. (1992) Comparison of four in situ hybridization methods, based on digoxigenin and biotin-labelled probes, in detecting HPV DNA in male condylomata acuminata. *Int. J. STD AIDS* **3,** 196–203.
19. Mifflin, T. E., Bowdwn, J., Lovell, M. A., Burns, D. E., Hayden, F. G., Graschel, D. H. M, et al. (1987) Comparison of radioactive and biotinylated DNA probes for the detection of cytomegalovirus. *Clin. Biochem.* **20,** 231–235.
20. Niedobitek, G., Finn, T., Herbst, H., and Stein, H. (1989) In situ hybridization using biotinylated probes. An evaluation of different detection systems. *Pathol. Res. Pract.* **184,** 343–348.
21. Komminoth, P. and Long, A. A. (1993) In situ polymerase chain reaction. An overview of methods, applications and limitations of a new molecular technique. *Virchows Archiv.* **64,** 67–73.
22. Komminoth, P. and Long, A. A. (1995) In situ polymerase chain reaction—methodology, applications and nonspecific pathways, in *PCR Applications Manual*, Boehringer Mannheim, Mannheim, pp. 97–106.
23. Long, A. A., Komminoth, P., and Wolfe, H. J. (1993) Comparison of indirect and direct in situ polymerase chain reaction in cell preparations and tissue sections. Detection of viral DNA gene rearrangements and chromosomal translocations. *Histochemistry* **99,** 51–162.
24. Söllström, J. F., Zehbe, I., Alemi, M., and Wilander, E. (1993) Pitfalls of in situ polymerase chain reaction (PCR) using direct incorporation of labelled nucleotides. *Anticancer Res.* **13,** 1153–1154.

25. Komminoth, P. and Werner, M. (1997) Target and signal amplification: approaches to increase the sensitivity of in situ hybridization. *Histochem. Cell Biol.* **108,** 325–333.
26. Komminoth, P., Adams, V., Long, A. A., Roth, J., Saremaslani, P, Flury, R., et al. (1994) Evaluation of methods for hepatitis C virus (HCV) detection in liver biopsies: Comparison of histology, immunohistochemistry, in situ hybridization, reverse (RT) PCR and in situ RT PCR. *Pathol. Res. Pract.* **190,** 1017–1025.
27. Wiedorn, K. H., Schaaf, B., Kühl, H., Spuck, S., Galle, J., Dalhoff, K., et al. (1999) Granulocyte Colony Stimulating Factor (G-CSF) induced modulation of cytokine expression: A semiquantitative in situ analysis using target amplification by reverse transcriptase in situ PCR (RT IS PCR) as well as signal amplification by in situ hybridization (ISH) with catalyzed reporter deposition (CARD). *Verhandlungen der Deutschen Gesellschaft für Zytologie* **21,** 63–65.
28. Wiedorn, K. H., Goldmann, T., Kühl, H., Rohrer, Ch., and Vollmer, E. (2000) Single-copy virus detection by GenPoint in situ hybridisation on the Eppendorf Mastercycler. *BioTech Int.* **12,** 12–13.
29. Wiedorn, K. H. (1996) In situ Hybridisierungshandbuch zum MISHA in situ Thermocycler, in *Misha Multipler In-Situ-Hybridisierungsautomat für die In-Situ-Hybridisierung und Polymerase-Kettenreaktion (PCR)*, Shandon, Frankfurt.
30. Stapleton, M. J., Levin, M. C., and Jacobson, S. (1994) DNA Amplification within cells on slides: Advances in evaporation and temperature control. *Cell Vision* **1,** 177–181.

62

Reverse Transcriptase *In Situ* PCR

New Methods in Cellular Interrogation

Mark Gilchrist and A. Dean Befus

1. Introduction

The advent of the reverse transcriptase polymerase chain reaction (RT-PCR) technique represents a quantum leap in sensitivity over preceding methods of detecting mRNA transcripts, such as Northern blotting. With the arrival of such sensitive techniques, it has become possible to amplify RNA transcripts from very small amounts of template nucleic acid, thus opening new avenues of research that were previously off limits because of difficulties in obtaining adequate quantities and quality of RNA *(1,2)*. However, RT-PCR suffers from the same limitations as its predecessor because the isolation of RNA necessitates the destruction of the cells/tissue involved, thus preventing the identification of the specific cell source of the mRNA *(3)*. Conversely, *in situ* hybridization allows the specific localization of mRNA to the cells of origin, but the methodology is much less sensitive than RT PCR *(3)*. A methodology that combines the best attributes of *in situ* hybridization (specific cellular localization) and RT-PCR (high sensitivity) would be desirable. RT *in situ* PCR provides these attributes, allowing for the location and detection of low copy RNA species, amplified within individual intact cells *(4)*.

The method (**Fig. 1**) involves fixation of cells in suspension, followed by controlled digestion of the crosslinked cellular proteins with a proteolytic enzyme. This allows the entry of the RT and PCR reagents into the cell. Next, genomic DNA is removed by a DNase digestion step, thus ensuring that only mRNA is amplified. Subsequent steps of RT and PCR are then undertaken, using a "labeled" reporter nucleotide, which results in its direct incorporation into the PCR product. This is followed by a detection step that visualizes the amplified mRNA, either by chromogenic or radioactive means (**Fig. 1**). The specificity of the reaction can be then established by several methods (*see* **Note 1**).

Methods of RT *in situ* PCR, although sharing fundamental steps, have varied greatly between laboratories *(5–7)*. On the basis of our experience, RT *in situ* PCR seems to be best suited for the detection of mRNA in single-cell suspensions, in which fixation and pretreatments can be optimally controlled *(8)*. The method outlined below is similar to that developed by Nuovo et al. *4,5)* and has been adopted and built upon through our

From: *Methods in Molecular Biology, Vol. 226: PCR Protocols, Second Edition*
Edited by: J. M. S. Bartlett and D. Stirling © Humana Press Inc., Totowa, NJ

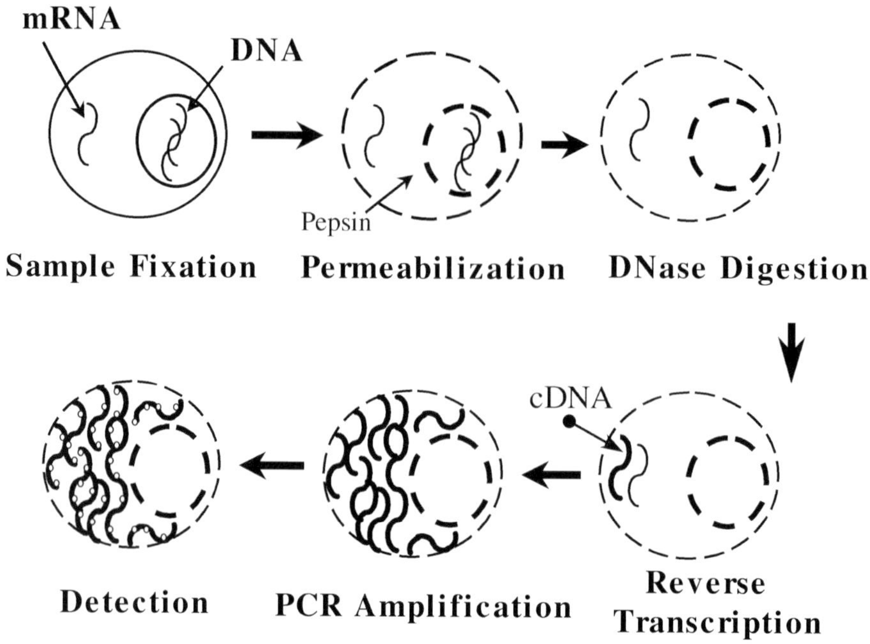

Fig. 1. Schematic representation of the steps involved in the general protocol outlined in the text. After fixation of the cells in suspension, the crosslinks that formed during fixation are reduced by limited protease digestion. To remove the possibility of amplifying DNA during the PCR, a DNase step is then performed to remove contaminating genomic DNA. A reverse transcription reaction is then performed to convert the mRNA to cDNA. PCR amplification with gene specific primers is then completed. The PCR product is then detected by chromogenic, fluorescent or radioactive means. Many commonly used RT *in situ* PCR procedures follow similar guidelines, differing in specific details, such as fixative used, protease used, digestion time, and method of detection.

experiences. The inclusion of appropriate controls in every run helps insure accurate results. Controls to indicate that genomic DNA has been removed as a source of false-positive signals (negative control), or failure of PCR amplification because of inadequate protease digestion or flaws in the RT-PCR protocol resulting in false-negative results (positive control) must be tested on every slide. Therefore, a slide schematic of the necessary controls and test spots is included to aid in the elimination of spurious results (**Fig. 2**).

2. Materials

1. Phosphate-buffered saline (PBS): 130 m*M* NaCl, 10 m*M* sodium phosphate, pH 7.4. Store at room temperature.
2. 10% neutral buffered formalin (BDH). Store at room temperature.
3. Heparinase I (for heparin containing mast cell populations; Sigma).
4. Heparinase buffer: 5 m*M* Tris-HCl, pH 7.5, 1 m*M* $CaCl_2$, 7.5 U RNasin.
5. Pepsin (5000 U/mL), made fresh in 0.01 *N* HCl for immediate use (Boehringer Mannheim).
6. RNase-free DNase I (Boehringer Mannheim).
7. DNase solution: DNase I (10 U/μL) in 0.1 *M* sodium acetate, pH 5.0, 5 m*M* $MgSO_4$.

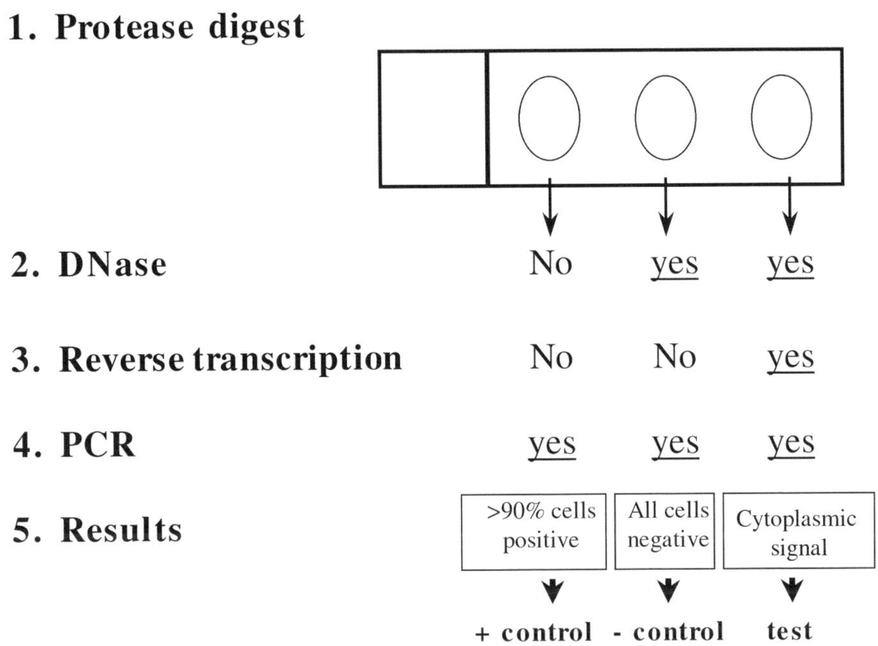

Fig. 2. Schematic of the appropriate controls, how they are achieved, and the expected results that indicate a successful reaction.

8. Primers (*see* **Subheading 4.6.**). The following primer pair yields a 295-bp amplicon for rat TNF mRNA: 5-TACTGAACTTCGGGGTGATCGGTCC-3 and 5-CAGCCTTGTCCCTT GAAGAGAACC-3.
9. M-MLV RT enzyme (Gibco/BRL).
10. Reverse transcription buffer: 75 m*M* KCl, 3 m*M* $MgCl_2$, 50 m*M* Tris-HCl, 0.1 *M* DTT (pH 8.3), 1 µ*M* antisense primer or 25 µg/mL oligo-$(dT)_{12-18}$ primer, and 2500 U/mL M-MLV RT enzyme.
11. *Taq* DNA polymerase (Gibco/BRL).
12. PCR buffer: 50 m*M* KCl, 10 m*M* Tris-HCl (pH 8.3), 4.5 m*M* $MgCl_2$, 80 µ*M* mixed dNTP, 16 µ*M* digoxigenin-11-dUTP, 1.2 µ*M* each primer, and 120 U/mL *Taq* polymerase.
13. Digoxigenin-11-dUTP (Boehringer Mannheim).
14. Anti-Digoxigenin antibody, alkaline phosphatase conjugated (Boehringer Mannheim).
15. Wash buffer 1: 0.1 *M* TRIS, 0.15 *M* NaCl, pH 7.5.
16. Wash buffer 2: 0.1 *M* TRIS, 0.15 *M* NaCl, pH 9.5.
17. Chromagen: 4-nitro blue tetrazolium chloride (NBT)/5-bromo-4-chloro-3-indoyl phosphate (BCIP) (Boehringer Mannheim).
18. Substrate solution : 10µL of NBT, 7.5 µL of BCIP in 2 mL of buffer 2.
19. Water, di-ethyl pyrocarbonate (DEPC) treated, RNase free.
20. Silane-coated glass slides.
21. Thermal cycling PCR machine or dedicated *in situ* PCR machine.

3. Methods

3.1. Cell Fixation

1. Cells collected from tissue culture flasks, blood, or other sources are collected in a 15-mL polypropylene tube.

2. Cells are pelleted by centrifugation in a tabletop centrifuge at 800*g* for 5 min at 4°C.
3. Cells should be washed once by resuspending in 10 mL of PBS and centrifuged again at 800*g* for 5 min at 4°C.
4. The supernatant is discarded, and the pelleted cells are then fixed in 10 mL of 10% neutral buffered formalin for 16 h at room temperature (*see* **Note 2**).
5. After fixation the cells are twice washed in DEPC water, recovering the cells each time by centrifugation at 800*g* for 5 min at 4°C.
6. The cells are then resuspended in DEPC water.
7. Cells are then dropped (75 µL) onto slides in three discrete areas and allowed to air dry at room temperature. Cell concentration should be between 5000 to 10,000 cells per droplet/spot. (**Fig. 2**) (*see* **Note 3**).

3.2. Protease Digestion

1. Dissolve 20 mg of pepsin in 9.5 mL of DEPC-treated water and then add 0.5 mL of 0.2 *N* HCl. The final concentration of pepsin should be 5000 U/mL.
2. Add 200 µL of this mix to each spot on the slide (*see* **Note 4**). The digestion time has to be standardized for each cell type (*see* **Note 5**). For mast cell lines we use a digestion time of 45 min at 37°C.
3. After the digestion is completed, the pepsin is inactivated with a 1-min wash in DEPC water in a Coplin jar.
4. Then, wash the slides in 100% ethanol for 1 min.
5. Allow to air dry at room temperature (*see* **Note 6**).

3.3. Pretreatments (Optional)

In our studies of mast cells, the presence of granule-associated heparin in most preparations is a major obstacle ***(9)***. Heparin is a potent inhibitor of RT and *Taq* polymerase enzymes, and therefore a heparinase digestion step must be included to remove heparin before proceeding.

1. 50 µL of heparinase digest buffer containing 333 U/mL of heparinase I is added to the slide and incubated for 2 h at room temperature.
2. Wash the slides in DEPC water for 1 min.
3. Then wash the slides in 100% ethanol for 1 min.
4. Allow the slides to air dry at room temperature.

3.4. DNase Treatment

This step is used to remove chromosomal DNA. It is added to the negative control and test cell spots only. Complete removal of genomic DNA is essential for correct interpretation of the test cells.

1. The negative control and the test cell spots are treated with 50 µL of DNase solution per spot, and incubated overnight at 37°C.
2. After the overnight digestion, wash the slides in DEPC water for 1 min.
3. Then wash the slides in 100% ethanol for 1 min.
4. Allow the slides to air dry at room temperature.

3.5. Reverse Transcription

1. Add 50 µL of the master mix to the test spot ONLY (*see* **Note 7**).
2. Incubate 1 h at 37°C in a moist chamber. Transcripts of low abundance may require longer incubation of the RT step (*see* **Note 8**).

3. Wash slides in DEPC treated water for 1 min.
4. Then wash the slides in 100% ethanol for 1 min.
5. Allow the slides to air dry at room temperature.

3.6. PCR

1. Place 50 μL of the PCR master mix on EACH spot (*see* **Note 9**).
2. A good starting program for the TNF primers is outlined: 94°C for 2 min, followed by 30 cycles of: 94°C for 1 min , 45°C for 2 min, 72°C for 2 min, and a final step at 4°C until ready to develop the slides. However, this may require optimization for each gene and cell type studied (*see* **Note 10**).
3. Once the slides have finished cycling, remove the coverslip and transfer the slides to wash buffer 1 for 5 min.
4. Do not allow the slides to air dry from this point on.

3.7. Digoxigenin Detection and Color Development

This step uses an alkaline phosphatase (AP)-labeled anti-digoxigenin antibody to detect the incorporated digoxigenin-dUTP. Color development is accomplished with NBT/BCIP Chromagen, which is oxidized to a purple/blue color by AP.

1. Prepare a 1:300 dilution (*see* **Note 11**) of anti-DIG antibody using wash buffer 1 as the diluent. Add 150 μL of the diluted antibody to the slide. Incubate the slides in a moist chamber for 30 min at room temperature.
2. Wash the slides with wash buffer 2 for 5 min.
3. Prepare the substrate solution.
4. Add the substrate solution to the slides and develop for 5 min to 1 h at room temperature. Monitor the purple/blue color development under the microscope.
5. Stop the color reaction by washing the slides in water.
6. Wash the slides for 1 min in 100% ethanol.
7. Wash in xylene, 1 min.
8. Mount using Permount and a coverslip.

3.8. Controls and Expected Results

The positive control (–DNAse, –RT, +PCR) should show intense nuclear staining in >90% of the cell population. This indicates incorporation of digoxigenin into strand breaks of genomic DNA by *Taq* polymerase and ensures that the protease digestion and PCR were adequate (**Fig. 3b**).

The negative control (+ DNAse, –RT, +PCR) should show no staining. This indicates that all genomic DNA has been digested and thus any staining in the test is caused by mRNA expression (**Fig. 3c**).

The test (+ DNAse, +RT, +PCR), if positive, should show staining predominantly over the cytoplasm and not in the nucleus. The cytoplasm is the correct cellular compartment for most mRNA (**Fig. 3a**), although it may have an intense perinuclear localization, especially in highly granulated cells *(8)*.

3.9. Future Directions

Initial development of RT *in situ* PCR lent itself immediately to areas of pathology and detection of viruses important in human disease. Publications in the field provided new insights into the pathogenesis and involvement of viruses, such as human papilloma virus (HPV) in cervical carcinoma *(10)* and Human Immunodeficiency Virus (HIV)

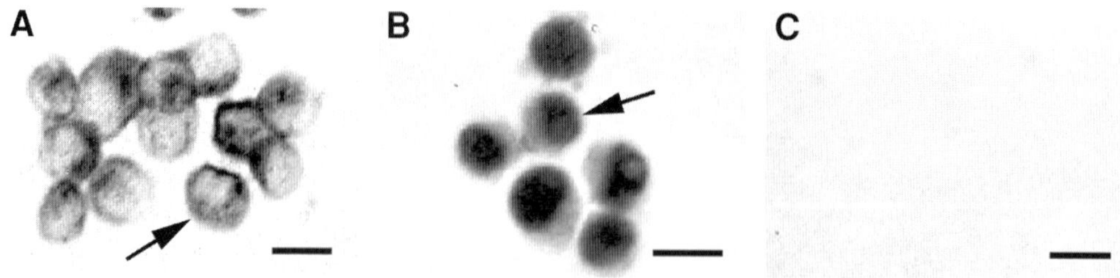

Fig. 3. RT *in situ* PCR detection of mRNA in a rat mast cell line (RCMC 1.11.2) to show representative results. **(A)** Positive signal localized in the cytoplasm of RCMC indicating the presence of TNF mRNA. **(B)** Positive control showing nuclear staining only and indicating that protease digestion, PCR, and detection steps are optimized. **(C)** Negative control showing no staining, indicating that genomic DNA is not causing a nonspecific signal. Original magnification ×1000; bar = 10 μm.

(4,11). In both cases, viral nucleic acid was found to be latently expressed in many cells, a result that was only obtainable with RT *in situ* PCR.

This methodology is now progressing into the area of cytokine research. Cytokine production and regulation has been difficult to study because of their low levels of expression and rapid turnover. Previously, our knowledge of cytokine expression and interaction was obtained from studies performed in cell culture systems, far removed from a true in vivo environment. With recent progresses in this method, it has become feasible to identify the source, kinetics and tissue distribution patterns of cytokines *in situ*, and their patterns of distribution in both health and disease ***(12,13)***.

The future of RT *in situ* PCR looks promising. Growth in the area will progress with the ability to colocalize both mRNA signals by RT *in situ* PCR, combined with a technique such as immunohistochemistry for protein detection. Such a protocol will allow researchers the power to identify both protein and gene expression in the same cell, contributing greatly to the understanding of molecular interactions at the cellular level.

4. Notes

1. Several methods have been used to confirm the specificity of the products generated by this methodology ***(8)***. In our experience, the labeled PCR product can be directly isolated from the cells on the slide itself. Once isolated the product can be confirmed either by Southern blot analysis or by cloning and sequencing of the generated product.
2. Fixation is a critical parameter in this protocol. A wide variety of fixatives have been used with success by other investigators, including 10% formalin, 4% paraformaldehyde, ethanol, and methanol/acetic acid ***(14,15)***. We have used 10% neutral-buffered formalin as our fixative of choice because of its good cell structural preservation qualities and minimal reactivity with RNA ***(16)***, thus facilitating good cDNA amplification.
3. Because cells/sections must remain on the slide through the rigors of digestion, pretreatments, and thermocycling, they must be placed on silane-coated slides. We use slides available from Perkin–Elmer for our *in situ* studies. Other sources of slides, including those produced "in-house" as part of a regular regime for use in immunohistochemistry can also be used with excellent results.
4. To keep the RT-PCR master mixes from evaporating during cycling, some method of covering the sections must be devised. We have used several techniques with similar success, including dedicated neoprene covers produced by Perkin–Elmer, as well as glass

coverslips anchored with nail polish, and polyethylene bags, cut to size and held in place with nail polish *(5,17)*. It is important to remove ALL visible air bubbles because these will be expanded as a result of the cycling process, which can result in some parts of sections being inaccessible to the PCR reagents, resulting in false negatives. Different means of covering the cell spots will result in different volumes of reagent required for each step, and they should be adjusted accordingly to save costs.

5. Protease digestion reduces the protein cross-links that form during the fixation step, thus allowing reagents to enter the cell. We currently use pepsin (5000 U/mL) at 37°C. The duration of pepsin digestion depends on the type of fixative used and the extent of fixation, with mast cell, macrophage, and lymphocyte cell lines taking ~45 min of digestion, whereas other cells such as *in vivo* derived mast cells requiring much less digestion (~15 min). However, other temperatures (22°C) and enzymes (trypsin, proteinase K) have been used by various groups *(7)*. Both these enzymes can give excellent results as well. We have settled on pepsin as our enzyme of choice, as it works at low pH (3.0), and is thus easily inactivated by a wash in water (~pH 7.5).
6. As a first step in the process of optimizing the digestion time and to become familiar with the methodology, we use a shortened protocol *(5)*. This involves using a single slide, with each individual spot being subjected to a different digestion time (e.g., 20 min, 40 and 60 min). After washing off the protease, the slide is then subjected to PCR and development, with no DNase or RT steps. By looking at the nuclear staining obtained with the different times, the investigator will be able to pick the time that gives >90% of the nuclei positive but maintains the cellular architecture. This optimal time can then be used for complete RT *in situ* PCR studies on this same batch of fixed cells. Some investigators have noted that they use >50% of the nuclei positive as an indication of optimal digestion *(4,5)*. We generally see >90% positive nuclei and have chosen this as our optimal indicator of digestion.
7. Recommended primer lengths are between 20 to 30 nucleotides. The PCR product should be at least 300 bp in length to assure that the amplicons remain inside the cell. The RT solution should contain an antisense primer capable of initiating first strand cDNA production. Oligo-dT can be used, although we have found that a gene specific primer works best.
8. Numerous RT enzymes are commercially available. We have used both Maloney Murine Leukemia Virus (MMLV) and Avian Myeloblastosis Virus (AMV) RT enzymes with identical results. We do our RT steps at 37°C, and vary the time of incubation from 1 to 3 h, depending on the signal seen with a specific cell type.
9. The critical parameter in the composition of the PCR solution is the $MgCl_2$ concentration. We found that Mg^{2+} concentration in the range of 3.0 to 5.0 m*M* works best, which is about four times higher than that used in solution-based PCR. This necessary increase in Mg^{2+} concentration is thought to be caused by binding of Mg^{2+} to the glass slides *(5,7)*.
10. If no product is detected with the cycling program given, then the program will have to be optimized. Take into account the T_m of the primers being used, and vary the number of cycles. In general terms, we begin with an annealing temperature that is 5°C below that of the primer Tm. Furthermore, depending on the abundance of the mRNA being amplified, upwards of 35 cycles may be required to obtain a detectable signal.
11. Transposed onto the technique of *in situ* RT-PCR is the necessity to detect the amplified product using immunohistochemical techniques. This introduces numerous other variables into the protocol. We have found that antibody concentrations of 0.75 to 3.75 μg/mL provide optimal signal with little background staining.

Acknowledgments

The authors recognize Dr. Osamu Nohara for his contribution to the development of our understanding of this procedure.

References

1. Mullis, K. B. and Faloona, F. A. (1987) Specific synthesis of DNA in vitro via a polymerase-catalysed chain reaction. *Methods Enzymol.* **155,** 335–350.
2. Veres, G., Gibbs, R. A., Scherer, S. E., and Caskey, C. T. (1987) The molecular basis of the sparse fur mouse mutation. *Science* **237,** 415–417.
3. Long, A. A. (1998) In situ polymerase chain reaction: foundation of the technology and today's options. *Eur. J. Histochem.* **42,** 101–109.
4. Nuovo, G. J., Forde, A., MacConnell, P., and Fahrenwald, R. (1993) In situ detection of PCR-amplified HIV-1 nucleic acids and tumor necrosis factor cDNA in cervical tissues. *Am. J. Pathol.* **143,** 40–48.
5. Nuovo, G. J. (1992) *PCR in situ Hybridization: Protocols and Applications.* Raven Press, New York.
6. Chen, R. H. and Fuggle, S. V. (1993) In situ cDNA polymerase chain reaction. *Am. J. Pathol.* **143,** 1527–1534.
7. Teo, I. A. and Shaunak, S. (1995) Polymerase chain reaction in situ: An appraisal of an emerging technique. *Histochem. J.* **27,** 647–659.
8. Nohara, O., Gilchrist, M., Dery, R. E., Stenton, G. R., Hirji, N. S., and Befus, A. D. (1999) Reverse transcriptase in situ polymerase chain reaction for gene expression in rat mast cells and macrophages. *J. Immunol. Methods* **226,** 147–158.
9. Gilchrist, M., MacDonald, A. J., Neverova, I., Ritchie, B., and Befus, A. D. (1997) Optimization of the isolation and effective use of mRNA from rat mast cells. *J. Immunol. Methods* **201,** 207–214.
10. Chiu, K. P., Cohen, S. H., Morris, D. W., and Jordan, G. W. (1992) Intracellular amplification of proviral DNA in tissue sections using the polymerase chain reaction. *J. Histochem. Cytochem.* **40,** 333–341.
11. Patterson, B. K., Till, M., Otto, P., Goolsby, C., Furtado, M. R., McBride, L. J., et al. (1993) Detection of HIV-1 DNA and messenger RNA in individual cells by PCR-driven in situ hybridization and flow cytometry. *Science* **230,** 1350–1354.
12. Hagimoto, N., Kuwano, K., Nomoto, Y., Kunitake, R., and Hara, N. (1997) Apoptosis and expression of Fas/Fas ligand mRNA in bleomycin-induced pulmonary fibrosis in mice. *Am. J. Respir. Cell Mol. Biol.* **16,** 91–101.
13. Peters, J., Krams, M., Wacker, H. H., Carstens, A., Weisner, D., Hamann, K., et al. (1997) Detection of rare RNA sequences by single-enzyme in situ reverse transcription polymerase chain reaction: High-resolution analysis of interleukin-6 mRNA in paraffin sections of lymph nodes. *Am. J. Pathol.* **150,** 469–476.
14. O'Leary, J. J. (1994) Importance of fixation procedures on DNA template and its suitability for solution phase PCR and PCR in situ hybridization. *Histochem. J.* **26,** 337–346.
15. Long, A. A., Komminoth, P., and Wolf, H. J. (1993) Comparison of indirect and direct in situ polymerase chain reaction in cell preparations and tissue sections. *Histochemistry* **99,** 151–162.
16. Teo, I. A. and Shaunak, S. (1995) PCR in situ: aspects which reduce amplification and generate false-positive results. *Histochem. J.* **27,** 660–669.
17. Patel, V. G., Shum-Siu, A., Heniford, B. W., Weiman, T. J., and Hendler, F. J. (1994) Detection of epidermal growth factor receptor mRNA in tissue sections from biopsy specimens using in situ polymerase chain reaction. *Am. J. Pathol.* **144,** 7–14.

63

Primed *In Situ* Nucleic Acid Labeling Combined with Immunocytochemistry to Simultaneously Localize DNA and Proteins in Cells and Chromosomes

Ernst J. M. Speel, Frans C. S. Ramaekers, and Anton H. N. Hopman

1. Introduction

In the past decade, the primed in situ (PRINS) labeling technique has become an alternative to fluorescence *in situ* hybridization (ISH) for the localization of nucleic acid sequences in chromosome, cell, and tissue preparations ***(1–8)***. The PRINS method is based on the rapid annealing of unlabeled primers (restriction fragment, PCR product, or oligonucleotide) to complementary target sequences *in situ*. These primers serve as initiation sites for *in situ* chain elongation using *Taq* DNA polymerase and labeled nucleotides. Incorporated fluorochrome-labeled nucleotides can be detected directly by fluorescence microscopy, whereas haptenized (e.g., biotin, digoxigenin, dinitrophenyl) nucleotides can be visualized by the additional application of fluorochrome- or enzyme-conjugated avidin or antibody molecules ***(4,5,9,10)***, followed by fluorescence microscopy or brightfield visualization of enzyme reaction products. Particularly, rapidity, simplicity, and cost-effectiveness have made the PRINS technique a useful tool in cytogenetics and cell biology. Its detection sensitivity, however, seems to be limited to repetitive targets for a long time. Only recently, the combined use of multiple oligonucleotides for 1 locus together with tyramide signal amplification have shown the first reproducible results demonstrating single-copy gene detection by PRINS ***(11)***.

With immunocytochemistry (ICC), specific information can be obtained regarding the presence or absence of proteins or antigens in chromosomes, cells, and tissue sections, thus allowing one to characterize the function of structural proteins in chromosomes or to phenotype cells (for example, their type of differentiation and proliferative activity). In addition, protocols have been developed to efficiently combine protein and nucleic acid detection in the same biological material to, for example, immunophenotype cells harboring a specific chromosomal aberration or viral infection, determine cytokinetic parameters of tumor cell populations that are genetically of phenotypically aberrant, and study the structural organization of chromosomes and the cell nucleus ***(10,12)***. The success and sensitivity of such a combined procedure depends on factors, such as preservation of cell morphology and protein epitopes, accessibility

From: *Methods in Molecular Biology, Vol. 226: PCR Protocols, Second Edition*
Edited by: J. M. S. Bartlett and D. Stirling © Humana Press Inc., Totowa, NJ

of nucleic acid targets, lack of crossreaction between the different protein and nucleic acid detection procedures, and good color contrast and stability of the fluorochromes and enzyme cytochemical precipitates applied. A variety of procedures have been reported that combine ICC and ISH (for a review, *see* **ref.** ***10,12***), in most cases applying ICC before ISH to prevent the destruction of antigenic determinants by the ISH procedure because of enzymatic digestion, postfixation, denaturation at high temperatures, and hybridization in formamide. For reasons of rapidity, probe accessibility, and lack of formamide for hybridization, the substitution of ISH by PRINS may be an extra advantage in such a combined procedure.

Here, we present three protocols for combined ICC and PRINS DNA labeling. In the first procedure, a sensitive, high-resolution fluorescence alkaline phosphatase (APase)-Fast Red ICC staining method ***(13,14)*** is performed before subsequent PRINS labeling of DNA target sequences to enable the simultaneous detection of surface antigens (EGF receptor, neural cell adhesion molecule) and repeated chromosome-specific DNA sequences in somatic cell hybrid and tumor cell lines. The fact that the Apase-Fast Red precipitate withstand subsequent proteolytic digestion and denaturation steps guarantees the most optimal conditions for efficient PRINS labeling ***(15)***. The second procedure has been recently described to investigate chromosome distribution and segregation in cells during processes, such as polyploidization and aneuploidization. This protocol starts with the PRINS labeling of chromosome centromeres followed by the staining of the mitotic spindle by tubulin ICC, because the authors found the antibody epitope to be better preserved when ICC came after DNA labeling ***(16,17)***. The third protocol has been used on metaphase chromosomes to identify possible relationships of different families of DNA sequences with, for example, proteins associated with different chromosome-specific structures, such as the kinetochore complex. This approach again starts with ICC staining followed by PRINS DNA labeling ***(18,19)***.

2. Materials

2.1. Protocol 1

2.1.1. Fluorescence Immunophenotyping by Alkaline Phosphatase Cytochemistry

1. Cold methanol (–20°C), cold acetone (4 and –20°C), and cold 70% ethanol (–20°C).
2. Normal goat serum (NGS).
3. Monoclonal antibody EGFR1, directed against the epidermal growth factor receptor (a kind gift of V. van Heyningen, Edinburgh, UK).
4. Monoclonal antibody 163A5, directed against a cell-surface marker of J1-C14 cells ***(20)***.
5. Monoclonal antibody RNL1, directed against the neural cell adhesion molecule (N-CAM) ***(21)***.
6. Alkaline phosphatase-conjugated goat anti-mouse IgG (GAMAPase) (Dako, Glostrup, Denmark).
7. Naphthol-ASMX-phosphate (Sigma, St. Louis, MO).
8. Fast Red TR (Sigma).
9. Polyvinylalcohol (PVA), MW 40,000 (Sigma).
10. 10× phosphate-buffered saline (PBS): 1.37 *M* NaCl, 30 m*M* KH_2PO_4, 130 m*M* Na_2HPO_4.
11. APase buffer: 0.2 *M* Tris-HCl, pH 8.5, 10 m*M* $MgCl_2$, 5% PVA. Dilute from stock solutions 1 *M* Tris-HCl, pH 8.5 and 1 *M* $MgCl_2$ and add 5% (w/v) PVA. Dissolve PVA by using a microwave.
12. Blocking buffer: 1× PBS (diluted from stock 10× PBS), 0.05% Triton X-100, 2–5% NGS.

Table 1
Sequences of Oligonucleotide Primers Used in PRINS

Name	Human origin	DNA sequence
E528	Chromosome 7 centromere	AGCGATTTGAGGACAATTGC
G33	Chromosome 9 centromere	AATCAACCCGAGTGCAATC
G35	Chromosome 11 centromere	GAGGGTTTCAGAGCTGCTC
D600	Chromosome X centromere	TCCATTCGATTCCATTTTTTTCGAGAA

13. Washing buffer: 1× PBS, 0.05% Triton X-100.
14. Glass coverslips 50 × 24 mm (Menzel Gläzer, Braunschweig, Germany).
15. Humid chamber.
16. Coplin jars (50 or 100 mL).

2.1.2 PRINS DNA Labeling

1. Pepsin from porcine stomach mucosa (2500–3500 U/mg; Sigma).
2. Ultrapure dNTP set (Amaersham Pharmacia Biotech, Little Chalfont, UK): 100 m*M* solutions of dATP, dCTP, dGTP, and dTTP.
3. Biotin-16-dUTP, Digoxigenin-11-dUTP, and Fluorescein-12-dUTP (Roche Molecular Biochemicals, Basel, Switzerland).
4. Oligonucleotide primer (*see* **Table 1** and **Note 1**).
5. *Taq* 1 DNA polymerase (Roche) or AmpliTaq (Perkin–Elmer).
6. Bovine serum albumin (BSA; Sigma).
7. Dried skimmed milk powder.
8. FITC-conjugated avidin (AvFITC; Vector, Burlingame, CA).
9. FITC-conjugated sheep anti digoxigenin Fab fragments (SHADigFITC; Roche).
10. Vectashield (Vector).
11. 4′,6-diamidino-2-phenyl indole (DAPI; Sigma).
12. 0.01 *M* HCl.
13. 1× PBS: diluted from 10× PBS (*see* **Subheading 2.1.1., item 10**).
14. Postfixation buffer: 1% formaldehyde (diluted from 37% formaldehyde (Merck, Darmstadt, Germany) in 1× PBS.
15. 20 × SSC: 3 *M* NaCl, 300 m*M* trisodium citrate, pH 7.0.
16. 10× *Taq* buffer: 500 m*M* KCl; 100 m*M* Tris-HCl, pH 8.3, 15 m*M* $MgCl_2$, 0.1% BSA.
17. PRINS stop buffer: 500 m*M* NaCl, 50 m*M* EDTA, pH 8.0.
18. Blocking buffer: 4 × SSC (diluted from stock 20 × SSC), 0.05% Triton X-100, 5% skimmed milk powder.
19. Washing buffer: 4 × SSC, 0.05% Triton X-100.
20. Coplin jars (50 or 100 mL).
21. Ethanol/37% HCl (100 : 1)-cleaned microscope slides and coverslips (Menzel).
22. Rubber cement.
23. Water bath at 65°C.
24. Thermal cycler (Hybaid Omnigene Flatbed; Hybaid, Ashford, UK; *see* **Note 2**).
25. Humid chamber.
26. Incubator set at 37°C.
27. Leica DM fluorescence microscope (Leica, Wetzlar, Germany) equipped with filter sets for DAPI, FITC, and TRITC.
28. Kodak 400 ASA film.
29. Black and white CCD camera and Isis 4 analysis software (Metasystems, Sandhausen, Germany).

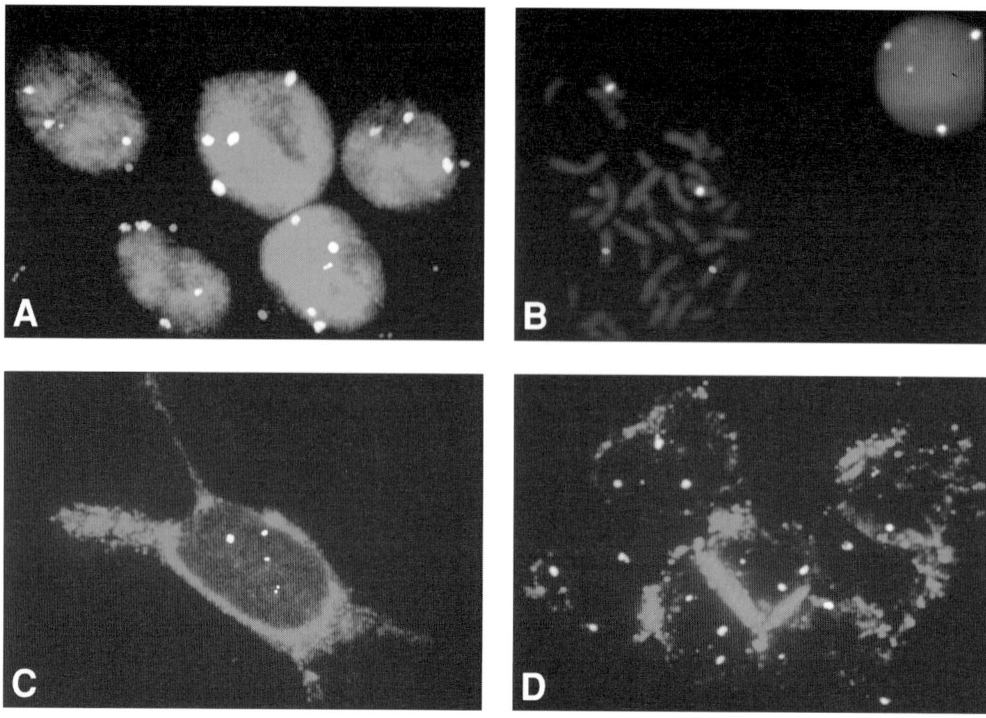

Fig. 1. Fluorescence detection of **(A)** chromosome 9 centromeres with digoxigenin/SHADigFITC in T24 cells after PRINS and Vectashield embedding with PI counterstaining. **(B)** chromosome 7 and 9 centromeres with, respectively, biotin/AvidinTexasRed anf digoxigenin/SHADigFITC after double-target PRINS and Vectashield embedding with DAPI counterstaining. **(C)** EGF receptor with alkaline phosphatase-Fast Red and four chromosome 7 centromeres with biotin/avidinFITC in C121-TN6 cells after immunocytochemistry followed by PRINS and Vectashield embedding. **(D)** NCAM antigen with alkaline phosphatase-Fast Red and three chromosome 9 centromeres with digoxigenein/SHADigFITC in H460 cells after immunocytochemistry followed by PRINS and Vectashield embedding.

30. Bio–Rad MRC 600 confocal scanning laser microsope equipped with the laser lines 488, 568, and 648 nm (Bio–Rad Laboratories, Veenendaal, The Netherlands).

2.2. Protocol 2

2.2.1. PRINS DNA Labeling

1. Dried skimmed milk powder.
2. PRINS reaction mixture components: dNTPs, labeled dUTPs, oligonucleotide primers, and *Taq* DNA polymerases as described in **Subheading 2.1.2., items 2–6** and **16**.
3. 1× PBS (diluted from 10× PBS, *see* **Subheading 2.1.1., ref. *10***) containing 0.1% EDTA and 0.2% BSA.
4. Methanol: acetic acid (9:1), freshly prepared.
5. 1× PBS (diluted from 10× PBS, *see* **Subheading 2.1.1., ref. *10***).
6. Permeabilization buffer: 1× PBS containing 0.1% Triton X 100.
7. A series of 70, 96, and 100% ethanol.
8. Denaturation buffer: 70% formamide/2 × SSC pH 7.0.

9. Washing buffer: 4 × SSC (diluted from 20 × SSC, *see* **Subheading 2.1.2., item 15**), 0.05% Triton X 100.
10. Blocking buffer: 4 × SSC (diluted from 20 × SSC, *see* **Subheading 2.1.2., item 15**), 0.05% Triton X 100, 5% dried skimmed milk powder.
11. Glass slides (Menzel).
12. Coplin jars.
13. Cytocentrifuge (Shandon, Astmoor, UK).
14. Waterbath at 70°C.
15. Thermal cycler (Hybaid Omnigene Flatbed; Hybaid).
16. Incubator at 37°C.

2.2.2. Immunofluorescence Detection of Mitotic Spindle

1. Mouse monoclonal anti-β-tubulin antibody (Sigma).
2. TRITC-conjugated donkey anti-mouse IgG F(ab′)2 fragments (DAMTRITC; Jackson Immunoresearch, West Grove, PE).
3. Normal goat serum (NGS).
4. Vectashield (Vector).
5. DAPI (*see* **Subheading 2.1.2., item 11**).
6. TOTO-3 iodide (Molecular Probes, Eugene, OR).
7. Blocking buffer: 1× PBS (diluted from stock 10× PBS, *see* **Subheading 2.1.1., item 9**), 0.05% Triton X-100, 2-5% NGS.
8. Washing buffer: 1× PBS (diluted from stock 10× PBS, *see* **Subheading 2.1.1., item 9**), 0.05% Triton X-100.
9. 1× PBS.
10. Vectashield (Vector) containing 0.5 μg/mL DAPI or 3000× diluted TOTO-3 iodide (from 1 m*M* stock in DMSO).
11. Humid chamber.
12. Coplin jar (50 or 100 mL).
13. Fluorescence microscope, CCD camera, and confocal microscope (*see* **Subheading 2.1.2., items 27–30**).

2.3. Protocol 3

2.3.1. Immunofluorescence Detection of Structural Proteins in Chromosomes

1. Normal goat serum (NGS).
2. Primary antibody.
3. FITC-conjugated secondary antibody.
4. Alcohol-cleaned glass slides and coverslips (Menzel).
5. Hypotonic solution: 75 m*M* KCl.
6. Potassium chromosome medium (KCM) solution: 120 m*M* KCl, 20 m*M* NaCl, 10 m*M* Tris-HCl, pH 8.0, 0.5 m*M* EDTA, and 0.1% (v/v) Triton X-100.
7. Blocking buffer: KCM containing 2 to 5% NGS.
8. Fixation solution: KCM containing 10% formalin (from 37% stock solution; Merck).
9. Cytocentrifuge (Shandon).
10. Coplin jar (50 or 100 mL).
11. Humid chamber.

2.3.2. PRINS DNA Labeling

1. Immersion solution: 0.1 *M* NaOH.
2. Neutralization solution: 0.01 *M* Tris-HCl, pH 7.4.

3. Methanol:acetic acid (3 : 1).
4. A series of cold (4°C) 70, 96, and 100% ethanol.
5. Denaturation solution: 30 m*M* NaOH/1 *M* NaCl (pH >12).
6. PRINS reaction, detection, and mounting components, as described in **Subheading 2.1.2., items 2–26**.
7. Slide evaluation equipment, as described in **Subheading 2.1.2., items 27–30**.

3. Methods

3.1. Protocol 1

3.1.1. Fluorescence Immunophenotyping by Alkaline Phosphatase Cytochemistry

1. Hybrid (C121-TN6, J1-C14) and tumor (H460) cell lines are cultured on glass slides by standard methods ***(20,22,23)***, fixed in either cold methanol (–20°C) for 5 s and cold acetone (4°C) for 3 × 5 s, cold acetone (–20°C) for 10 min, or 70% ethanol (–20°C) for 10 min, air-dried, and stored at –20°C until use (*see* **Note 3**).
2. Incubate slides for 10 min at room temperature with 100 µL of blocking buffer under a coverslip in a humid chamber.
3. Remove the coverslip, discard the blocking buffer, and incubate the slides for 30 to 45 min at room temperature with 50 µL of undiluted culture supernatant of the appropriate antigen-specific monoclonal antibody containing 2% NGS under a coverslip in a humid chamber.
4. Wash slides for 2 × 5 min with washing buffer in a Coplin jar.
5. Incubate slides for 30 to 45 min at room temperature with 50 µL of GAMAPase, diluted 1:50 in blocking buffer (*see* **Note 4**), under a coverslip in a humid chamber.
6. Wash slides for 5 min with washing buffer, and for 5 min with 1× PBS.
7. Visualize the antigen with the alkaline phosphatase-Fast Red (APase-Fast Red) reaction: Mix 4 mL of APase buffer, 1 mg of naphthol-ASMX-phosphate in 250 µL of buffer without PVA and 5 mg of Fast Red TR in 750 µL of buffer without PVA just before use and overlay each sample with 100 µL under a coverslip. Incubate the slides for 5 to 15 min at 37°C and wash 3× 5 min with 1× PBS (*see* **Notes 5–7**).

3.1.2 PRINS DNA Labeling

1. Process cells for PRINS as follows: wash slides for 2 min at 37°C with 0.01 *M* HCl, incubate the samples with 100 µg/mLpepsin in 0.01 *M* HCl for 20 min at 37°C, wash again with 0.01 *M* HCl for 2 min, and post-fix the slides in 1% formaldehyde in 1× PBS for 10 min at room temperature. Wash cells in 1× PBS for 5 min at room temperature, followed by a wash step in 1× *Taq* buffer for 5 min at room temperature (all steps in a Coplin jar).
2. Prepare the PRINS reaction mix on ice as follows: Dilute 100 m*M* dATP, dGTP, and dCTP 1:10 with distilled water. Dilute 100 m*M* dTTP 1:100. Put together in a microcentriphuge tube: 1 µL of each of the diluted dNTPs, 1 µL of either 1m*M* Biotin-16-dUTP, Digoxigenin-11-dUTP, or Fluorescein-12-dUTP (*see* **Note 8**), 5 µL of 10× *Taq* buffer, 250 ng of oligonucleotide (*see* **Note 9**), 1 U *Taq* polymerase, and distilled water to 50 µL.
3. Place 40 µL of this mixture under a coverslip on the slide, seal with rubber cement, air-dry the rubber cement, and transfer to the heating block of the thermal cycler.
4. Each PRINS reaction cycle consists of 2 min at 94°C (denaturation of cellular DNA, *see* **Note 10**), 5 min at the appropriate annealing temperature (*see* **Note 11**) and 15 min at 72°C for *in situ* primer extension.
5. Stop the PRINS reaction by transferring the slides (after removal of the rubber solution seal) to 50 mLof PRINS stop buffer in a Coplin jar at 65°C for 1 min (*see* **Note 12**).
6. Transfer the slides to washing buffer at room temperature and wash 5 min.

7. Place 40 µL of blocking buffer under a coverslip on the slide and leave for 5 min at room temperature in a humid chamber to reduce background staining in the detection procedures.
8. Wash slides for 1 × 5 min in washing buffer in a Coplin jar.
9a. For reactions using Biotin-16-dUTP: Dilute AvFITC 1:100 in blocking buffer and apply 50 µL under a coverslip. Incubate slides for 30 min at 37°C in a humid chamber (*see* **Note 13**).
9b. For reactions using Digoxigenin-11-dUTP: Dilute SHADigFITC 1 : 100 in blocking buffer and treat as in 9a (*see* **Note 13**).
9c. Fluorescein-12-dUTP needs no additional reporter and is simply mounted as described in 11 (*see* **Note 13**).
10. Wash slides for 2 × 5 min in washing buffer in a Coplin jar. Optionally, you may wash the slides for 5 min in 1× PBS and dehydrate them.
11. Mount the slides in Vectashield containing 0.5 µg/mLDAPI.
12. Examine the slides under a fluorescence microscope. Selected cells can be either directly photographed using Kodak 400 ASA film, visualized with a charge-coupled device (CCD) camera, or scanned with a confocal scanning laser microscope (CSLM).

3.2. Protocol 2

3.2.1 PRINS DNA Labeling

1. Hybrid and tumor cell lines are cultured on glass slides, as described in **Subheading 3.1.1., step 1**), or in suspension according to standard methods ***(16,17,20,22,23)***. Cells are rinsed in 1× PBS/0.1% EDTA/0.2% BSA, cytocentrifuged on glass slides at 65*g* for 4 min (in case of growth in suspension), fixed in methanol : acetic acid (9 : 1) for 10 min at room temperature in a Coplin jar, and airdried.
2. Rehydrate cells in 1× PBS for 5 min, permeabilize them with 0.1% Triton X 100 in 1× PBS for 5 to 10 min (both in a Coplin jar), dehydrate in a series of 70, 96, and 100% ethanol and airdry.
3. Denature slides in 70% formamide/2 × SSC, pH 7.0, for 2 min at 70°C in a Coplin jar, dehydrate in a series of 70, 96 (both at 4°C), and 100% ethanol, and airdry.
4. Prepare the PRINS reaction mix and apply it on the prewarmed (annealing temperature) slides (on the block of the thermal cycler) as described in **Subheading 3.1.2., steps 2** and **3**).
5. Perform the PRINS reaction for 5 min at the appropriate annealing temperature (*see* **Note 11**) and for 15 min at 72°C for chain elongation on a thermal cycler.
6. Stop the PRINS reaction by removing the coverslips (*see* **Note 12**) and wash the slides in washing buffer for 3 × 5 min at room temperature followed by a 5-min wash step in 1× PBS (all steps in a Coplin jar).
7. Apply a blocking and wash step as described in **Subheading 3.1.2., steps 7** and **8**.
8. In case of PRINS labeling with Biotin-16-dUTP or Digoxigenin-11-dUTP, detect the haptens as described in **Subheading 3.1.2., steps 9** and **10**).

3.2.2. Immunofluorescence Detection of the Mitotic Spindle

1. Incubate the slides for 30 to 45 min at room temperature with 50 µL of mouse anti-β-tubulin primary antibody, diluted 1 : 50 in blocking buffer, under a coverslip in a humid chamber.
2. Wash the slides for 2 × 5 min with washing buffer in a Coplin jar.
3. Incubate slides for 30 to 45 min at room temperature with 50 µL of DAMTRITC, diluted 1 : 100 in blocking buffer (*see* **Note 4**), under a coverslip in a humid chamber.

4. Wash the slides for 2 × 5 min with washing buffer and for 5 min with 1× PBS (all steps in a Coplin jar).
5. Mount slides in Vectashield containing 0.5 µg/mLDAPI or 3000× diluted TOTO-3 iodide.
6. Examine the slides as described in **Subheading 3.1.2., step 12**.

3.3. Protocol 3

3.3.1. Immunofluorescence Detection of Structural Proteins in Chromosomes

1. Prepare metaphase spreads from human peripheral blood lymphocytes as described previously ***(18,19)***.
2. After hypotonic treatment of the cell suspension for 10 min at 37°C in 75 m*M* KCl, cytocentrifuge the cells onto alcohol-cleaned glass slides with 65–275*g* for 4 min and air dry for 2 min.
3. Place slides in a Coplin jar containing KCM solution for 10 min at room temperature, followed by incubation for 10 min at room temperature with 100 µL of blocking buffer under a coverslip in a humid chamber.
4. Incubate slides for 30 to 45 min at room temperature with 50 µL of the primary antibody, diluted in blocking buffer, under a coverslip in a humid chamber.
5. Wash slides for 2 × 5 min with KCM in a Coplin jar.
6. Incubate slides for 30 to 45 min at room temperature with 50 µL of FITC-conjugated secondary antibody, diluted in blocking buffer, under a coverslip in a humid chamber (*see* **Note 4**).
7. Wash slides for 2 × 5 min with KCM, and fix the chromosomes in fixation solution for 5 to 15 min at room temperature in a Coplin jar.
8. Wash slides in destilled water, airdry and store in the dark at room temperature until use.

3.2.3. PRINS DNA Labeling

1. To improve the signal intensity after PRINS labeling, the slides are immersed in 0.1 *M* NaOH for 10 to 40 s (*see* **Note 14**) followed by neutralization with 2 × 5-min washes in 0.01 *M* Tris-HCl, pH 7.4, and a wash in destilled water (all steps in a Coplin jar) before airdrying.
2. Fix slides in methanol : acetic acid (3 : 1) for 2 × 2 min in a Coplin jar (*see* **Note 15**), wash in 0.01 *M* Tris-HCl, pH 7.4, for 2 × 5 min in a Coplin jar, dehydrate slides in a series of cold (4°C) 70, 96, and 100% ethanol, and airdry (*see* **Note 16**).
3. Denature chromosomal DNA in denaturation solution for 45 min at 4°C and neutralize in 0.01 *M* Tris-HCl, pH 7.4, for 2 × 5 min at room temperature (all steps in a Coplin jar).
4. Remove excess fluid by draining and blow slides dry with a jet of air.
5. Perform PRINS reaction, mounting and evaluation of the slides as described in **Subheading 3.1.2., steps 2–12**).

4. Notes

1. So far, specific oligonucleotide primers have been defined for the centromere regions of up to 20 chromosomes ***(7,24,25)***. Particularly, the ability of primers to differentiate between closely related sequences has made it possible to define α-satellite primers for some chromosomes indistinguishable by FISH with centromeric probes, such as for chromosomes 13 and 21, that only exhibit a one-base differerence at their 3′ end. Furthermore, a number of chromosome-specific telomere primers have been generated, as well as HPV-specific and single-gene-specific oligonucleotides ***(11,26,27)***.
2. Several companies commercialize specialized thermal cyclers with a flat block, including Hybaid, Perkin–Elmer, Techne Corporation (Cambridge, UK), and MJ Research Inc. (Watertown, MA). Because of differences in the design of the heating block, PRINS

conditions always need to be optimized for the respective instrument used. Because an accurate temperature at the top surface of the slide is crucial for a successful PRINS reaction, some cyclers, such as the Hybaid Omnigene, possess incorporated software to compensate for the temperature difference between the block and the surface of the slide.

3. Because on methanol-acetone fixed cells we frequently observed a poor preservation of cell morphology as well as fluorescent staining of the entire nucleus (by PRINS labeling), probably caused by nuclease activities that survive this mild type of fixation, other fixatives should be and have been successfully tested that are compatible with antigen detection and result in a better cell morphology and specific PRINS labeling. For example, a fixation with cold 70% ethanol (–20°C) for 10 min proved to be a valid alternative on H460 cells.
4. If further amplification of the (immuno)cytochemical signal is needed, a third detection step may be added after this second incubation step. Alternatively, the avidin biotinylated enzyme (e.g., alkaline phosphatase) complex system may be applied as well as the tyramide signal amplification system (in combination with peroxidase conjugates). For details of possible reagents to use, *see* **Note 13** ***(7,10,12)***.
5. It is recommended to monitor the enzyme reaction under the microscope to adjust the reaction time to ensure the precipitate becoming discretely localized and not so dense that it shields nucleic acid sequences in the PRINS reaction.
6. To ensure the specificity of the APase-Fast Red staining, a control slide with FITC-conjugated secondary antibodies is recommended for comparison. Staining specificity can be lost if cells contain endogenous APase activity. This endogenous enzyme activity can be inhibited by the addition of levamisole (Sigma) to the reaction medium to a final concentration of 1–5 m*M*.
7. Do not dehydrate the slides after the APase reaction because the precipitate dissolves in organic solvents. Optionally, you may air dry the slides after rinsing in distilled water.
8. In the case of labeling with biotin-16-dUTP or fluorescein-12-dUTP a 4× decrease of the concentration of dTTP in the PRINS reaction mix resulted in significant stronger labeling of DNA sequences. Under the described standard conditions, digoxigenin-11-dUTP provides the highest sensitivity. However, all the modified nucleotides are suited for detection of one or multiple repeated sequences *in situ*. In this respect, the number of different fluorochrome- and hapten-labeled nucleotides is still increasing to date, which can be obtained from a number of companies, such as, e.g., Perkin–Elmer Life Science, Roche, Amersham, Dako, and Molecular Probes.
9. The concentration of the appropriate oligonucleotide resulting in positive signals need to be determined by experiment. Generally, 250 ng/slide (50–250 pmol) in 40 µL is used for primers of 16 to 35 bases complementary to repeated sequences.
10. Seperate denaturation of cellular DNA in 70% formamide/2xSSC, pH 7.0, for 2 min at 70°C before the PRINS reaction as is usually performed for chromosome preparations, resulted in no or only weak PRINS labeling of DNA sequences in the interphase nuclei of the cell preparations. Whether this is caused by inefficient primer annealing or extension is not clear. The same phenomenon is also observed for PRINS on frozen tissue sections ***(6,7)***.
11. The optimum primer annealing temperature is only determined empirically. We usually try a series from 45 to 70°C in 5°C steps.
12. For detection of multiple DNA targets by sequential PRINS reactions (MULTIPRINS), it was found essential to prevent the free 3′ ends of the newly synthesized DNA from being used as primers for subsequent reactions. This can be achieved by incubating the slides with Klenow DNA polymerase together with ddNTPs. The reaction mix is made up as follows: Dilute 5 m*M* of all four ddNTPs 1:10 with distilled water. Put together in a centrifuge tube 2.5 µL of each of the ddNTPs (Amersham Pharmacia), 5 µL of 10× Klenow

buffer (500 m*M* Tris-HCl, pH 7.2; 100 m*M* $MgSO_4$; 100 m*M* DTT; 1.5 mg.mLBSA), 1 U Klenow DNA polymerase (Roche), and distilled water to 50 µL. Incubate slide with 40 µL under a coverslip for 1 h at 37°C in a humid chamber, followed by dehydration and airdrying before running the next PRINS reaction ***(5)***.

13. Amplification of PRINS signals can be achieved as follows:
    a. AvFITC detection of Biotin-16-dUTP may be followed by incubation with biotinylated goat anti-avidin (Vector), 1:100 diluted in blocking buffer, and again AvFITC.
    b. SHADigFITC detection of digoxigenin-11-dUTP may be followed by incubation with FITC-conjugated anti-sheep IgG (Roche), or as described for FITC-12-dUTP amplification.
    c. Fluorescein-12-dUTP signals may be amplified by incubation with monoclonal mouse or polyclonal rabbit anti-FITC (Dako) and FITC-conjugated rabbit anti-mouse IgG or swine anti-rabbit IgG (Dako).
    d. Amplification of PRINS signals may also be achieved by using peroxidase-mediated deposition of hapten- or fluorochrome-labeled tyramides ***(28–30)***.
14. The time required will vary according to the repeated DNA family of sequences under study (size of target) and the cell type from which the chromosome preparations are made.
15. This again may vary according to the cell type and target size under study.
16. The slides can be stored in the dark for several weeks at room temperature at this stage.

References

1. Koch, J., Kölvraa, S., Petersen, K. B., Gregersen, N., and Bolund, L. (1989) Oligonucleotide-priming methods for the chromosome-specific labelling of alpha satellite DNA *in situ*. *Chromosoma* **98,** 259–265.
2. Koch, J., Mogensen, J., Pedersen, S., Fischer, H., Hindkjder, J., Kölvraa, S., et al. (1992) Fast one-step procedure for the detection of nucleic acids *in situ* by primer-induced sequence-specific labeling with fluorescein-12-dUTP. *Cytogenet. Cell. Genet.* **60,** 1–3.
3. Gosden, J., Hanratty, D., Starling, J., Fantes, J., Mitchell, A., and Porteous, D. (1991) Oligonucleotide-primed *in situ* DNA synthesis (PRINS): A method for chromosome mapping, banding, and investigation of sequence organization. *Cytogenet. Cell. Genet.* **57,** 100–104.
4. Gosden, J. and Lawson, D. (1994) Rapid chromosome identification by oligonucleotide-primed *in situ* DNA synthesis (PRINS). *Hum. Mol. Genet.* **3,** 931–936.
5. Speel, E. J. M., Lawson, D., Hopman, A. H. N., and Gosden, J. (1995) Multi-PRINS: multiple sequential oligonucleotide primed *in situ* DNA synthesis reactions label specific chromosomes and produce bands. *Hum. Genet.* **95,** 29–33.
6. Speel, E. J. M., Lawson, D., Ramaekers, F. C. S., Gosden, J. R., and Hopman, A. H. N. (1996) Rapid brightfield detection of oligonucleotide primed in situ (PRINS)-labeled DNA in chromosome preparations and frozen tissue sections. *BioTechniques* **20,** 226–234.
7. Gosden, J. R., ed. (1997) *PRINS and In Situ PCR Protocols.* Methods in Molecular Biology, Vol. 71, Humana Press, Totowa, NJ.
8. Wilkens, L., Tchinda, J., Komminoth, P., and Werner, M. (1997) Single- and double-color oligonucleotide primed in situ labeling (PRINS): Applications in pathology. *Histochem. Cell Biol.* **108,** 439–446.
9. Hindkjder, J., Koch, J., Terkelsen, C., Brandt, C. A., Kölvraa, S., and Bolund, L. (1994) Fast, sensitive multicolor detection of nucleic acids in situ by primed in situ labeling (PRINS). *Cytogenet. Cell. Genet.* **66,** 152–154.
10. Speel, E. J. M. (1999) Detection and amplification systems for sensitive, multiple-target DNA and RNA in situ hybridization: Looking inside cells with a spectrum of colors. *Histochem. Cell Biol.* **112,** 89–113.

11. Kadandale, J. S., Wachtel, S. S., Tunca, Y., Sid Wilroy Jr., R., Martens, P. R., and Tharapel, A. T. (2000) Localization of SRY by primed in situ labeling in XX and XY sex reversal. *Am. J. Med. Genet.* **95,** 71–74.
12. Speel, E. J. M., Ramaekers, F. C. S., and Hopman, A. H. N. (1995) Cytochemical detection systems for in situ hybridization, and the combination with immunocytochemistry. Who is still afraid of red, green and blue? *Histochem. J.* **27,** 833–858.
13. Speel, E. J. M., Herbergs, J., Ramaekers, F. C. S., and Hopman, A. H. N. (1994) Combined immunocytochemistry and fluorescence in situ hybridization for simultaneous tricolor detection of cell cycle, genomic, and phenotypic parameters of tumor cells. *J. Histochem. Cytochem.* **42,** 961–966.
14. Herbergs, J., Speel, E. J. M., Ramaekers, F. C. S., De Bruïne, A. P., Arends, J. W., and Hopman, A. H. N. (1996) Combination of lamin immunocytochemistry and in situ hybridization for the analysis of chromosome copy numbers in tumor cell areas with high nuclear density. *Cytometry* **23,** 1–7.
15. Speel, E. J. M., Lawson, D., Ramaekers, F. C. S., Gosden, J. R., and Hopman, A. H. N. (1997) *Combined immunocytochemistry and PRINS DNA synthesis for simultaneous detection of phenotypic and genomic parameters in cells, in PRINS and in situ PCR protocols*, Methods in Molecular Biology, Vol. 71 (Gosden, J. R., ed.), Humana Press, Totowa, NJ, pp. 53–60.
16. Roy, L., Coullin, P., Vitrat, N., Hellio, R., Debili, N., Weinstein, J., et al. (2001) Asymmetrical segregation of chromosomes with a normal metaphase/anaphase checkpoint in polyploid megakaryocytes. *Blood* **97,** 2238–2247.
17. Coullin, P., Roy, L., Pellestor, F., Candelier, J.-J., Bed"hom, B., Guillier-Gencik, Z., et al. (2002) PRINS, the other in situ DNA labeling method that is useful in cellular biology. *Am. J. Med. Genet.* **107,** 127–135.
18. Jeppesen, P. (1994) Immunofluorescence techniques applied to mitotic chromosome preparations, in *Chromosome Analysis Protocols,* Methods in Molecular Biology (Gosden, J. R., ed.), Humana Press, Totowa, NJ, pp. 253–286.
19. Mitchell, A. R. (1997) Chromosomal PRINS DNA labeling combined with indirect immunocytochemistry, in *PRINS and in situ PCR protocols*, Methods in Molecular Biology (Gosden, J. R., ed.), Humana Press, Totowa, NJ, pp. 61–70.
20. Glaser, T., Housman, D., Lewis, W. H., Gerhard, D., and Jones, C. (1989) A fine-structure deletion map of chromosome 11p: analysis of J1 series hybrids. *Somat. Cell Mol. Genet.* **15,** 477–501.
21. Boerman, O. C., Mijnheere, E. P., Broers, J. L. V., Vooijs, G. P., and Ramaekers, F. C. S. (1991) Biodistribution of a monoclonal antibody (RNL-1) against the neural cell adhesion molecule (NCAM) in athymic mice bearing human small-cell lung-cancer xenografts. *Int. J. Cancer* **48,** 457–462.
22. Dorin, J. R., Inglis, J. D., and Porteous, D. J. (1989) Selection for precise chromosomal targeting of a dominant marker by homologous recombination. *Science* **243,** 1357–1360.
23. Carney, D. N., Gazdar, A. F., Bepler, G., Guccion, J. G., Marangos, P. J., Moody, T. W., et al. (1985) Establishment and identification of small cell lung cancer cell lines having classic and variant features. *Cancer Res.* **45,** 2913–2923.
24. Koch, J., Hindkjder, J., Kölvraa, S., and Bolund, L. (1995) Construction of a panel of chromosome-specific oligonucleotide probes (PRINS-primers) useful for the identification of individual human chromosomes in situ. *Cytogenet. Cell. Genet.* **71,** 142–147.
25. Pellestor, F., Girardet, A., Lefort, G., Andréo, B., and Charlieu, J.-P. (1995) Selection of chromosome specific primers and their use in simple and double PRINS technique for rapid in situ identification of human chromosomes. *Cytogenet. Cell. Genet.* **70,** 138–142.

26. Krejci, K. and Koch, J. (1999) An in situ study of variant telomeric repeats in human chromosomes. *Genomics* **58,** 202–206.
27. Ramael, M., Van Steelandt, H., Stuyven, G., Van Steenkiste, M., and Degroote, J. (1999) Detection of human papilloma virus (HPV) genomes by the primed in situ labelling technique. *Pathol. Res. Pract.* **195,** 801–807.
28. Bobrow, M. N., Harris, T. D., Shaughnessy, K. J., and Litt, G. J. (1989) Catalyzed reporter deposition, a novel method of signal amplification. Application to immunoassays. *J. Immunol. Methods* **125,** 279–285.
29. Speel, E. J. M., Hopman, A. H. N., and Komminoth, P. (1999) Amplification methods to increase the sensitivity of in situ hybridization. Play card(s). *J. Histochem. Cytochem.* **47,** 281–288.
30. Speel, E. J. M., Hopman, A. H. N., and Komminoth, P. (2000) Signal amplification for DNA and mRNA in situ hybridization, in *In Situ Hybridization Protocols*, Methods in Molecular Biology (Darby, I. A., ed.), Humana Press, Totowa, NJ, pp. 195–216.

IX

Cloning and Mutagenesis

64

Cloning and Mutagenesis

A Technical Overview

Helen Pearson and David Stirling

1. Introduction

Strategies for cloning polymerase chain reaction (PCR) products and performing in vitro site-directed mutagenesis are legion, and the following chapters outline five robust and reliable protocols. Before embarking on such a strategy, however, it is worth considering if it is entirely necessary. Even high-fidelity, proofreading *Taq* polymerases carry the risk of misincorporated bases being included, especially late in the PCR, when dNTP concentrations may become limiting. Thus, if the product is cloned, there is a chance that the clone selected contains a misincorporated base. If the purpose of the exercise is simply to determine the sequence of the original template, it may be more appropriate to sequence the PCR product directly and so avoid such cloning bias. If the object is to produce a clone that can be further used in expression studies for instance, it is imperative that all cloned material is sequenced to verify its integrity, prior to expression.

2. Cloning

2.1. T Overhangs

Taq polymerases lacking 3′ to 5′ exonuclease activity tends to add nontemplate-directed nucleotides to the ends of double-stranded DNA fragments. If blunt end cloning is to be used, these overhangs need to be "polished," and Chapter 65 by Kai Wang provides an optimized method using T4 DNA polymerase. Because the predominant nucleotide added in this nontemplate-directed manner is adenosine, many successful protocols have used vectors with a thymidine overhang to direct the cloning. This allows rapid cloning into a shuttle vector, from which the fragment can then be restricted and cloned further. Alternatively, Horton and colleagues present an elegant protocol (Chapter 66) using T-linkers to enable cloning into specific sites.

2.2. Restriction Site Primers

Neither blunt-end cloning nor the use of T-overhang vectors allows directional cloning. If the region of template DNA to be amplified contains suitable restriction sites, the product can simple be digested and cloned in the same way as any other DNA fragment. If this is not feasible, it is possible to introduce restriction site sequences

From: *Methods in Molecular Biology, Vol. 226: PCR Protocols, Second Edition*
Edited by: J. M. S. Bartlett and D. Stirling © Humana Press Inc., Totowa, NJ

into PCR products by having these sequences incorporated into the 5′ end of the PCR primer(s). The short restriction site sequence on the 5′ end of the PCR primer will not hybridize, but as long as the 3′ hybridizing region is long enough (i.e., its T_m is high enough; ~20 mer), the primer will specifically bind to the appropriate site. The PCR product will thus have an additional DNA sequence at the 5′ end that will contain the endonuclease restriction site. A similar or different restriction site sequence can be added via the other PCR primer. If the other primer has a different restriction sequence, then the PCR fragment can be inserted in a directional-dependent manner in a host plasmid. There are a number of potential problems with this method that should be considered. There is no easy way to prevent internal sites containing similar restriction sequences from being cut when the end of the PCR product are cut. Care should therefore be taken to use restriction sites that are not present in the fragment to be amplified. Restriction sequences are inverse repeat sequences, thus the potential exists for primer dimer association and resultant non-productive annealing. Finally, when restriction sites are located very close to the end of an amplified fragment, the efficiency of cleavage of those sites can be markedly impaired. It may therefore be necessary to include not just the restriction site but an additional 5 to 10 bases to avoid this problem.

3. Mutagenesis

In vitro site-directed mutagenesis is an invaluable technique for studying protein structure–function relationships, gene expression, and vector modification. Several methods have appeared in the literature, but many of these methods require single-stranded DNA as the template. The reason for this, historically, has been the need for separating the complementary strands to prevent reannealing. Use of PCR in site-directed mutagenesis accomplishes strand separation by using a denaturing step to separate the complementing strands and allowing efficient polymerization of the PCR primers. PCR site-directed methods thus allow site-specific mutations to be incorporated in virtually any double-stranded plasmid, eliminating the need for M13-based vectors or single-stranded rescue.

Three divergent strategies for mutagenesis are outlined in the following chapters; however, several points applicable to all three should be here. First, it is often desirable to reduce the number of cycles during PCR when performing PCR-based site-directed mutagenesis to prevent clonal expansion of any (undesired) second-site mutations. Limited cycling, which would result in reduced product yield, can be offset by increasing the starting template concentration. Second, a selection must be used to reduce the number of parental molecules coming through the reaction. This is of particular importance when the parental molecules are used in high concentrations. Third, because of the tendency of some thermostable polymerases to add nontemplate-directed nucleotides to the ends of double-stranded DNA fragments, it is often necessary to incorporate an end-polishing step into the procedure prior to end-to-end ligation of the PCR-generated product containing the incorporated mutations in one or both PCR primers (*see* **Subheading 2.1.**).

Finally, even if the presence of the desired mutation is confirmed by restriction digest or sequencing, it is essential to sequence the entire region of manipulated DNA to ensure that there has been no undesirable mutation introduced by the PCR processes.

65

Using T4 DNA Polymerase to Generate Clonable PCR Products

Kai Wang

1. Introduction

Polymerase chain reaction (PCR) mediated through *Taq* DNA polymerase has become a simple and routine method for cloning, sequencing, and analyzing genetic information from very small amounts of materials *(1)*. *Taq* DNA polymerase, like some other DNA polymerases, lacks 3′ to 5′ exonuclease activity and will add nontemplate-directed nucleotides to the ends of double-stranded DNA fragments. Because of the strong preference of the *Taq* polymerase for dATP, the nucleotide added is almost exclusively an adenosine *(2)*. This results in generating "ragged" unclonable amplification products *(2,3)*. Restriction endonuclease sites are often incorporated into the amplification primers so that clonable PCR products can be generated by restriction enzyme cleavage *(4)*. However, the possible secondary sites located within amplified products often complicate the cloning and interpretation of PCR results. A cloning system exploiting the template-independent terminal transferase activity of *Taq* polymerase has been reported *(5–7)*. However, a special vector with thymidine (T) overhanging ends has to be used in the process.

T4 DNA polymerase has very strong exonuclease and polymerase activities in a broad range of reaction conditions *(8)*. By adapting its strong enzymatic activities, a simple and efficient method to generate clonable PCR fragments with T4 DNA polymerase has been developed *(9)*. The T4 DNA polymerase not only repairs the ends of the PCR products, but it also removes the remaining primers in the reaction with its strong single-stranded exonuclease activity. Therefore, this method usually does not require multiple sample handling, buffer changes, or gel purification steps. Instead, a simple alcohol precipitation step is used to purify the PCR products.

The blunt-end cloning protocol can be modified for sticky-end cloning. Even though this may increase cloning efficiency to a certain extent, a purification step, to remove excess deoxynucleotides from PCRs, has to be added before adding T4 DNA polymerase.

From: *Methods in Molecular Biology, Vol. 226: PCR Protocols, Second Edition*
Edited by: J. M. S. Bartlett and D. Stirling © Humana Press Inc., Totowa, NJ

2. Materials

2.1. PCR

1. DNA template containing the sequence of interest.
2. Oligonucleotide primers.
3. *Taq* DNA polymerase.
4. 10× PCR and enzymatic repair buffer: 500 m*M* Tris-HCl, pH 9.0, 25 m*M* $MgCl_2$, 500 m*M* NaCl, and 5 m*M* DTT. Commercial 10× PCR buffer also works well.
5. 1.5 m*M* 10× deoxynucleotides (dNTP) solution. Concentrated stock solution (100 m*M*) can be obtained from Amersham Biosciences (Piscataway, NJ) or Boehringer Mannheim (Indianapolis, IN).
6. Gel electrophoresis and PCR equipment.

2.2. End Repair and Blunt-End Cloning

1. Enzymes (T4 DNA polymerase, T4 polynucleotide kinase, and T4 DNA ligase). Enzymes can be purchased from Boehringer Mannheim, Invitrogen (Carlsbad, CA) or any other provider.
2. 1 m*M* of ATP solution. Concentrated stock solution (100 m*M*) can be obtained from Boehringer Mannheim.
3. Isopropyl alcohol.
4. Vector (blunt end and dephosphorylated).
5. 10× Ligase buffer: 660 m*M* Tris-HCl, pH 7.6, 66 m*M* $MgCl_2$, 10 m*M* ATP, 1 m*M* spermidine, and 10 m*M* DTT. Commercially available 10× ligase buffers also works well.
6. TE buffer: 10 m*M* Tris-HCl, pH 7.5, 1 m*M* EDTA.
7. 5 *M* NaCl.

2.3. End Repair and Sticky-End Cloning

1. Sephacryl S-400 spin column. A commercial spin column (MicroSpin S-400HR) can be obtained from Pharmacia.
2. Enzymes (T4 DNA polymerase, T4 polynucleotide kinase, and T4 DNA ligase). Enzymes can be purchased from Boehringer Mannheim, Invitrogen, or any other provider.
3. ATP and dNTPs.
4. Isopropyl alcohol.
5. Vector (digested and dephosphorylated).
6. 10× Ligase buffer: 660 m*M* Tris-HCl, pH 7.6, 66 m*M* $MgCl_2$, 10 m*M* ATP, 1 m*M* spermidine, and 10 m*M* DTT. Commercially available 10× ligase buffers also works well.
7. 0.5 *M* EDTA, pH 8.0.
8. 5 *M* NaCl.

3. Method

3.1. Primer Design

For blunt-end cloning, no special primer is needed. However, secondary structure and stretch of homopolymer should be avoided. For sticky-end cloning, depending on restriction site selected, specific sequences should be included so that compatible ends can be generated after T4 DNA polymerase treatment (*see* **Subheading 3.4.**).

3.2. PCR

1. Prepare the following in a PCR tube: 5 μL of 10× PCR buffer; 0.25 μg of genomic DNA; 1 μ*M* of each primer; 0.15 m*M* of each deoxynucleotide (5 μL of 1.5 m*M* stock dNTP solution), 1 U *Taq* polymerase, and deionized H_2O to a final volume of 50 μL.

2. Amplification conditions largely depend on the specific applications. However, a general cycling profile can be used in most experiments: 94°C for 7 min (initial denaturation); 94°C for 30 s (amplification), 55°C for 45 s, 72°C for 90 s; and 72°C for 10 min (extension).
3. Examine the PCR amplification results with agarose gel electrophoresis (*see* **Note 1**).

3.3. End Repair and Blunt-End Cloning

1. Add the following to PCR tubes directly to repair the PCR products (*see* **Note 2**): 1 U of T4 DNA polymerase; 1 UL of 4 m*M* dNTP solution (optional; *see* **Note 3**); 5 U of T4 polynucleotide kinase (*see* **Note 4**); and 1 µL of 1 M*M* ATP.
2. Incubate the reaction tubes at 25°C (room temperature) for 20 min (*see* **Notes 5** and **6**). Stop the reactions by adding 3 mL of 0.5 *M* EDTA, pH 8.0.
3. Incubate the reaction tubes at 70°C for 10 min to inactive the enzymes.
4. Precipitate the PCR products by adding 5 µL of 5 *M* NaCl and 60 µL of isopropyl alcohol (*see* **Notes 7**, **ref.** ***8***).
5. Resuspend the DNA fragments in 20 µL of TE or water.
6. Take 2 µL of DNA (containing approx 20–50 ng of PCR product) and mix with ligase and vector for ligation (*see* **Note 8**): 1 µL of 10× ligase buffer; 1 µL of T4 ligase; 2 µL of repaired DNA; dephosphorylated vector (60–150 ng); and add deionized H_2O to a final volume of 10 µL.
7. Incubate at 16°C overnight.
8. Dilute the ligation reaction fivefold in TE buffer. Use 2 µL of the diluted ligation reaction for transformation (*see* **Note 9**).

3.4. End Repair and Sticky-End Cloning

An *Eco*RI site is used as an example in the following protocol **Fig. 1).** However, depending on the desired cloning site, a different combination of dNTP should be added in the "repair" reaction (*see* **Subheading 3.4., step 2**).

1. Spin through the PCR mixture (40 µL) in a pre-equilibrated Sephacryl S-400HR spin column (*see* **Note 10**) to remove excess dNTP.
2. Add the following to the column-purified PCR fragments to generate sticky ends: 5 µL of 10× PCR buffer; 1 U of T4 DNA polymerase; 5 U of T4 polynucleotide kinase; 1 µL of 4 M*M* dCTP and dGTP; 1 µL of 1 M*M* ATP; and add deionized H_2O to a final volume of 50 µL.
3. Incubate the reaction tubes at 25°C (room temperature) for 20 min. Stop the reaction by adding 3 µL of 0.5 *M* EDTA, pH 8.0.
4. Heat inactivate the enzymes by placing the reaction tubes at 70°C for 10 min.
5. Precipitate the PCR products by adding 5 µL of 5 *M* NaCl and 60 µL of isopropyl alcohol ***(8)***.
6. Resuspend the DNA fragments in 20 µL of TE or water.
7. Take 2 µL of DNA (containing approx 20–50 ng of DNA) and mix with ligase and *Eco*RI digested, dephosphorylated vector for cloning: 1 µL of 10× ligase buffer; 1 µL of T4 ligase; 2 µL of repaired DNA; dephosphorylated vector (60–150 ng); and add deionized H_2O to a final volume of 10 µL
8. Incubate at 16°C overnight.
9. Dilute the ligation reaction fivefold in TE buffer. Use 2 µL of the diluted ligation reaction for transformation (*see* **Note 9**).

4. Notes

1. In case of multiple PCR products from a single reaction, the specific products should be purified by gel electrophoresis based on size after repair reaction. Several different

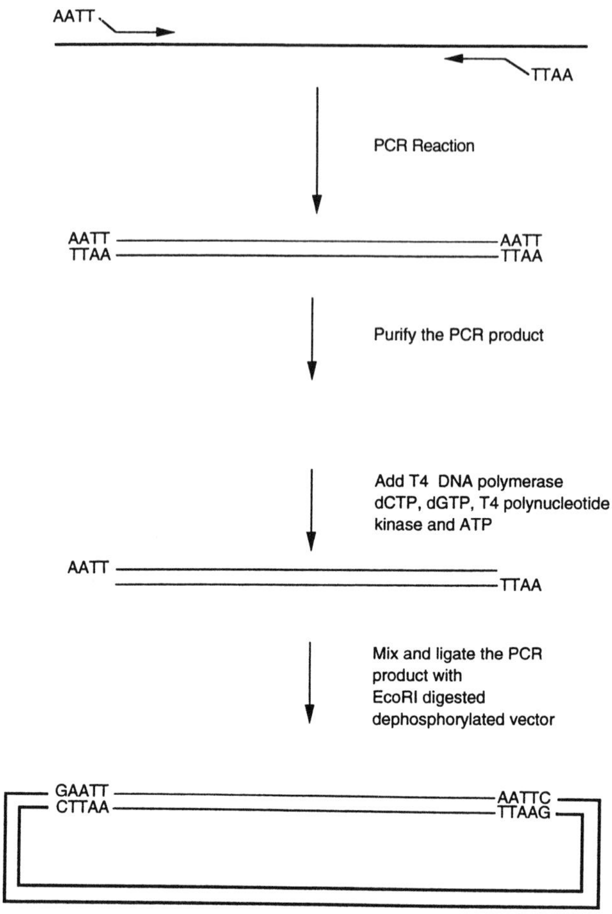

Fig. 1. A brief outline of the strategy used to generate sticky-end PCR products with T4 DNA polymerase.

methods can be used to purify DNA fragments from agarose gel, such as phenol extraction from low-melting gel ***(8)***, the "glassmilk" method, or simple low-speed centrifugation ***(10)***. The phenol extraction method has been found to be less expensive and able to recover a sufficient amount of clean DNA for cloning.

2. This protocol uses one buffer for all the enzymes that include *Taq* polymerase in PCR, T4 polymerase, and T4 polynucleotide kinase in end-repair reaction. Therefore, a slightly higher concentration of reagents and enzymes can be added in the reaction.
3. T4 DNA polymerase can be added directly into the PCR tube without providing additional nucleotides. However, T4 DNA polymerase balances its exonuclease and polymerase activities based on the concentration of available deoxynucleotides. Depending on the length of amplification products, number of amplification cycles, and nucleotide sequence composition of amplified region, the remaining nucleotide concentration after PCR amplification may be different from experiment to experiment. To avoid unnecessary confusion, supplemental nucleotides are routinely added for end-repair reaction.
4. T4 polynucleotide kinase is not needed when vector used has not been treated with phosphatase previously. However, dephosphorylated vector should be used to increase the cloning efficiency.

5. Room temperature (25°C) was chosen for the reaction since T4 DNA polymerase has excessive exonuclease activity at 37°C.
6. The T4 polynucleotide kinase works well at room temperature as opposed to the higher reaction temperature (37°C) regularly used *(8)*.
7. Although the PCR products purified directly by alcohol precipitation after end repairing are sufficient for routine cloning, passing the repaired PCR product mixtures through a gel filtration column prior to the alcohol precipitation can greatly enhance the cloning efficiency.
8. In the ligation reaction, we routinely used 1:1 molar ratio between vector (dephosphorylated) and insert. Generally more than 200 recombinant clones can be obtained with 0.4 µL of the ligation reaction. Therefore, a single ligation reaction for each PCR product is sufficient for most applications.
9. Because of its reliability and high transformation efficiency, commercial $CaCl_2$-treated competent cells are used for the transformation step. The bacteria strain we routinely used is DH10B (BRL; MAX efficiency DH10B competent cells). However, various competent cells can be purchased from companies, such as Stratagene and Invitrogen.
10. Prepacked Sephacryl S-400HR spin columns (MicroSpin S-400HR) can be purchased from Pharmacia. Alternatively, the spin columns can be prepared from bulk gel filtration matrix (Sephacryl S-400HR, Pharmacia) as described in other protocol books *(8)*. The filtration medium (prepacked or bulk filtration matrix) contains 20% alcohol; therefore, the spin columns should be washed and equilibrated with TE.

References

1. Saiki, R. K., Scharf, S., Faloona, F., Mullis, K. B., Horn, G. T., Erlich, H. A., et al. (1985) Primer-directed enzymatic amplification of DNA with a thermostable DNA polymerase. *Science* **230,** 1350–1354.
2. Clark, J. M. (1988) Novel non-templated nucleotide addition reactions catalyzed by procaryotic and eucaryotic DNA polymerases. *Nucleic Acids Res.* **16,** 9677–9686.
3. Scharf, S. (1990) *PCR Protocols: A Guide to Methods and Applications*, Academic Press, San Diego, CA.
4. Jung, V., Pestka, S. B., and Pestka, S. (1990) Efficient cloning of PCR generated DNA containing terminal restriction endonuclease recognition sites. *Nucleic Acids Res.* **18,** 6156.
5. Mead, D. A., Pey, N. K., Herrnstadt, C., Marcil, R. A., and Smith, L. M. (1991) A universal method for the direct cloning of PCR amplified nucleic acid. *Bio/Technology* **9,** 657.
6. Kovalic, D., Kwak, J., and Weisblum, B. (1991) General method for direct cloning of DNA fragments generated by the polymerase chain reaction. *Nucleic Acids Res.* **19,** 4560.
7. Marchuk, D., Drumm, M., Saulino, A., and Collins, F. (1991) Construction of T-vectors, a rapid and general system for direct cloning of unmodified PCR products. *Nucleic Acids Res.* **19,** 1154.
8. Sambrook, J., Fritsch, E. F., and Maniatis, T. (1989) *Molecular Cloning: A Laboratory Manual*, 2nd ed., Cold Spring Harbor Laboratory, Cold Spring Harbor, NY.
9. Wang, K., Koop, B. F., and Hood, L. (1994) A simple method using T4 DNA polymerase to clone polymerase chain reaction products. *Biotechniques* **17,** 236–238.
10. Heery, D. M., Gannon, F., and Powell, R. (1990) A simple method for subcloning DNA fragments from gel slices. *Trends Genet.* **6,** 173.

66

A T-Linker Strategy for Modification and Directional Cloning of PCR Products

Robert M. Horton, Raghavanpillai Raju, and Bianca M. Conti-Fine

1. Introduction

The propensity of *Taq* polymerase to add 3′-A overhangs *(1,2)* to polymerase chain reaction (PCR)-amplified DNA has made possible a simple method for cloning PCR products into a T-vector (Invitrogen, San Diego, CA) *(3–5)*. Here, we present a related strategy that uses T-linkers to add sequences, such as restriction sites, to the ends of PCR products (*see* **Note 1**). A single-base T overhang at the end of a synthetic double-stranded oligonucleotide linker allows ligation of the linker to the unpolished ends of a PCR product. This avoids the expense of adding the "extra" sequences to sequence-specific primers.

1.1. Examples

Two T-linker designs are presented here. In each case, the T-linker is a double-stranded synthetic oligonucleotide composed of complementary oligos (either TL-A and TL-B for the NdeT linker or HisTL-A and HisTL-B for the HisT-linker) with a single 3′ overhanging t at one end

1.1.1. NdeTL

The basic principles involved in using a T-linker are shown using the Nde-T-linker in **Fig. 1**. This T-linker contains complete *Eco*RI and *Not*I sites, and a third site (*Nde*I) is partially present, except for its final g; the overhanging t is part of this site. The 5′ end of TL-B is phosphorylated (indicated by an asterisk) so that it can be ligated. The other end of the linker contains a sticky *Hind*III-compatible end, which was not used in the approach described here. However, because this 5′ overhang is filled in by the polymerase during PCR, these extra bases serve as a "clamp" or spacer, which permits the *Eco*RI site to be cut.

The overhanging t of the linker matches the a added by the polymerase and directs the ligation of TL-B, allowing reamplification of the sequence using TL-A as a primer. The resulting molecule contains the original PCR-amplified sequence flanked by inverted repeats of the T-linker.

If one of the sequence-specific primers has a g at its 5′ end, an *Nde*I site will be formed. This site is "split" between the T-linker and the PCR product to be cloned. In

From: *Methods in Molecular Biology, Vol. 226: PCR Protocols, Second Edition*
Edited by: J. M. S. Bartlett and D. Stirling © Humana Press Inc., Totowa, NJ

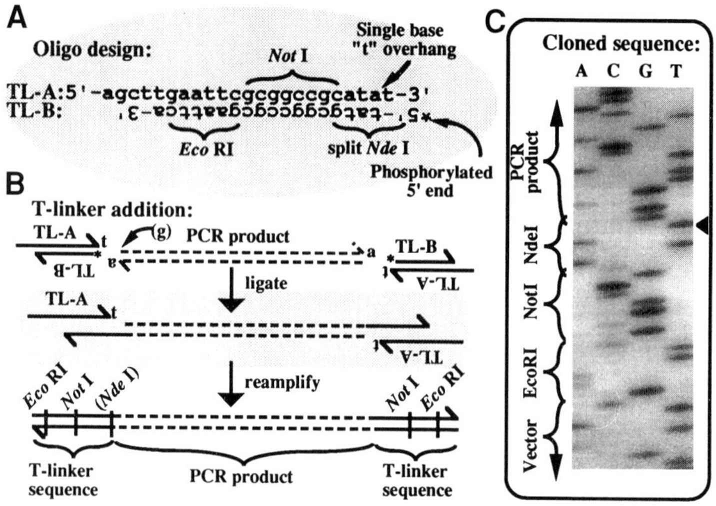

Fig. 1. Cloning of PCR products with a T-linker. **(A)** Oligonucleotide design. **(B)** Addition of the T-linker to a PCR product. The parenthetical g and *Nde*I indicate that if a g is the first base in the PCR product, the *Nde*I site will be completed. **(C)** Sequence from a pBluescript KS II+ plasmid containing a T-linker-reamplified insert cloned into its *Eco*RI site. This sequence was obtained using an fmol cycle sequencing kit (Promega, Madison, WI). The base representing the overhanging t of the T-linker is indicated by the solid triangle to the right of the sequence. Since the PCR product begins with a g, a complete *Nde*I site is formed.

this way, very minor restraints in the PCR primer sequences (having one start with g, but not the other) can be used to complete the site at only one end of the reamplified product. One simple application of this idea is directional cloning (*see* **Notes 2** and **3**).

The use of this T-linker is illustrated in **Fig. 1**. A portion of a T-cell receptor V region was amplified from Jurkat tumor line cDNA using primers Vb8-1cpe5 and hpVbe3. The ligated products were reamplfied with TL-A and cloned into the *Eco*RI site of pBluescript KS II$^+$. The sequence of the resulting product is shown in **Fig. 1C**. At the bottom of the gel is sequence from the polylinker of the vector to the *Eco*RI site, followed by the T-linker. The overhanging t is marked. The g completing the *Nde*I site is the first base of Vb81cp.5.

1.1.2. HisTL

Our second example adds another level of sophistication, as shown in **Fig. 2**. Here an additional ability of the T-linker is brought into use. A new and useful sequence is added in addition to the restriction sites, namely a "histidine tag" sequence, which will be used for affinity purification of the expressed recombinant α3 protein on a Ni^{2+} column (Qiagen, Chatsworth, CA). Note that the *Hind*III site is created only at the 3′ end of the product in this case, because only the 3′ primer begins with t. The histidine tag portion is removed from the 3′ end by cutting with *Hind*III. No special sequences have been added to the sequence-specific primers hacpe5 and hacpe3, but their positions have been chosen to take advantage of the design of the HisT-linkers. The initial PCR-amplified sequence begins with a full codon to put the sequence in

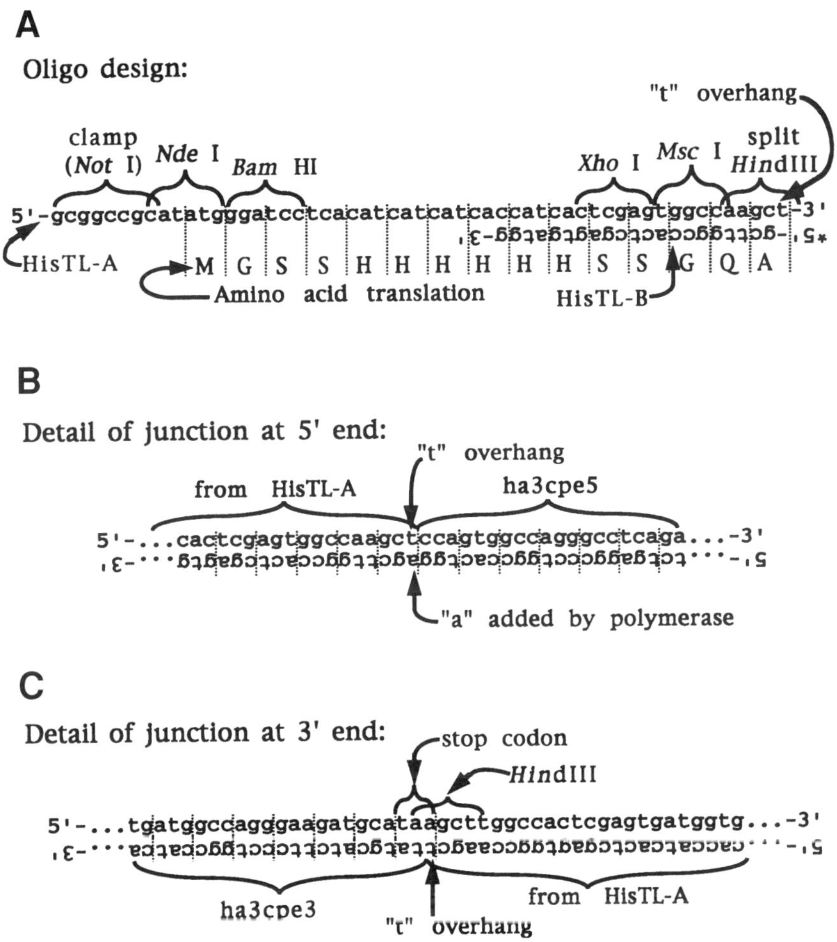

Fig. 2. Using a T-linker to add a "histidine tag" to one end of a PCR product. **(A)** Oligonucleotide design. The *Nde*I site allows the product to be cloned into an appropriate expression vector, in which translation begins at the ATG codon included in this site. The amino acids that make up the histidine tag are shown, and the reading frame is indicated by vertical bars between codons. The *Not*I site in HisTL-A is in parentheses because it would not be expected to cut so close to the end of a DNA molecule (although it should be possible to use it if the primer were phosphorylated and the products ligated to form concatamers). Its main purpose is to serve as a clamp to allow efficient cutting at the *Nde*I site. **(B)** Sequence at the junction between the HisT-linker and the 5′ end of the PCR-amplified sequence. The 5′ primer begins at the first base in a codon, to put the sequence in frame with the histidine tag. **(C)** Sequence at the 3′ junction. The 3′ primer begins with ta; the complementary ta in the other strand is converted to a stop codon "taa" when the extra a is added by *Taq* polymerase. Because the 5′ primer begins with c, the *Hind*III site is completed only at the 3′ end of the molecule. This allows directional cloning of the product, and removal of the T-linker sequences from the 3′ end.

the proper reading frame with the His tag. It ends with "ta"; the "a" added by the polymerase finishes the termination codon (taa) and is included in the *Hind*III site.

2. Materials

1. cDNA template: This was reverse transcribed from total RNA using Superscript RNase H-reverse transcriptase (Life Sciences) and an oligo (dT) primer using manufacturers instructions.
2. Reagents for PCR: 10× buffer: 500 m*M* KCl, 100 m*M* Tris-HCl, pH 8.3; red sucrose is a PCR-compatible gel-loading dye consisting of ~1 m*M* cresol red in 60% sucrose ***(6)***; *Taq* DNA polymerase, 10 m*M* dNTP mix, 10 m*M* $MgCl_2$ solution (Perkin–Elmer Cetus, Norwalk, CT).
3. Reagents for agarose gel electrophoresis.
4. GeneClean (Bio 101, La Jolla, CA).
5. Ligation reagents: T4 DNA ligase (Stratagene, La Jolla, CA) 8 U/µL, ligase buffer supplied by manufacturer, 10 m*M* ATP (Pharmacia, Piscataway, NJ), and PEG 8000 (Aldrich, Milwaukee, WI). Recently, we have used T4 ligase buffer from Life Technologies (Gibco-BRL, Gaithersburg, MD), which already contains ATP and PEG.
6. T-linker oligonucleotides:
   a. Nde-T-linkers: TL-A: (23-mer) 5′-agcttgaattcgcggccgcatat-3′; TL-B: (18-mer)* 5′-tatgcggccgcgaattca-3′;
   b. His-T-linkers: HisTL-A: (55-mer) 5′-gcggccgcatatgggatcctcacatcatcatcaccatcactcgagt-ggccaagct-3′; HisTL-B: (21-mer) *5′-gcttggccactcgagtgatgg-3′

      The 5′ end of each "B" oligonucleotide is phosphorylated (represented by the *) so that it can be ligated to the (nonphosphorylated) end of the PCR product. The 5′ end of the "A" oligonucleotide is not phosphorylated. The "B" oligo only needs to be long enough to bind the "A" oligo during the ligation and in the annealing steps of the early rounds of reamplification. A 5′ overhang on oligo "A" is not a problem, because this is filled in by the polymerase during reamplification.
   c. Dissolve primers at a stock concentration of 10 µ*M*, which is generally considered 20× for PCR. After mixing T-linker primers, they are at a final concentration of approx 5 µ*M* for ligation reactions.
7. Ampligrease (*see* **ref. 7**): plain petroleum jelly, Vaseline brand, or generic is suitable. Apply quality control checks as described *(7)*.
8. Reagents for bacterial culture, including competent *Escherichia coli*, appropriate antibiotic, LB agar.
9. 95% ethanol containing 2% (w/v) potassium acetate.
10. 75% ethanol.

3. Methods

3.1. PCR

The HisT-linker was used to clone the coding region of the α3 subunit of the nicotinic acetylcholine receptor from human bronchial epithelium (manuscript in preparation; *see* **Note 2**) into the *E. coli* expression vector pT7-7 ***(8)***. The following sequence-specific primers were used:

Vb8-1cpe5:	5′-ggagttatccagtcacc-3′
hpVbe3:	5′-gggaattcgtcgactgctggcrcagarrta-3′
ha3cpe5:	5′-ccagtggccagggcctcaga-3′
ha3cpe3:	5′-tatgcatcttccctggccatca-3′

1. Perform PCRs under fairly standard conditions. For example, for cloning the α3 subunit of the nicotinic receptor, the conditions for amplification were as follows: 3 μL of 10× PCR buffer, 4.5 μL of 10 m*M* $MgCl_2$, 3 μL of dNTPs (2 m*M* each), 6 μL of red sucrose, 16.5 μL of H_2O, 0.75 μL of ha3cpe5 primer (10 μ*M*), 0.75 μL of ha3cpe3 primer (10 μ*M*), 1 μL of cDNA template, and 0.25 μL of *Taq* polymerase.
2. Perform PCR using 40 cycles at 94, 52, and 72°C, each for 0.5 min, in a programmable circulating air oven (ProOven, Integrated Separation Systems).
3. Products were purified by agarose gel electrophoresis and GeneClean (Bio 101; *see* **Chapter 18**).

3.2. Ligation (see Note 4)

1. Assemble ligation reactions as follows: 2 μL of purified PCR product, 7.5 μL of H_2O, 2 μL of 10× ligase buffer, 1 μL of (0.5 m*M* final) 10 m*M* ATP (if not in buffer), 2.5% (5% final) 40% PEG 8000 (optional), 2 μL of T-linkers (5 μ*M* in 500 m*M* NaCl), and 1 μL of ligase (1 U/μL dilution).
2. Incubate at room temperature for 20 min to overnight.

3.3. Reamplification with T-Linker Primer

1. After ligation of the T-linker, reamplfy the products as follows: 5 μL of 10× PCR buffer, 7.5 μL of 10 m*M* $MgCl_2$, 5 μL of dNTPs, 10 μL of Red sucrose, 16.5 μL of H2O, 5 μL of TL-A (or HisTL-A) primer (10 μ*M*), ligation mixture, and 0.35 μL of *Taq* polymerase.
2. Cover the sample with mineral oil and amplify with 40 two-step cycles of 94°C for 0.5 min and 72°C for 2.5 min. (*see* **Note 5**).

3.4. Alternative Method: One-Tube Ligation/Reamplification

The whole T-linker reaction can be set up in one tube to make a "kit-like" product. This is done using a meltable barrier, as for hot-start PCR (we use AmpliGrease; **ref. 7**). The "top mix" contains the ligation reaction, to which the PCR-amplified band is added. After a suitable incubation to allow ligation, the reaction is heated to melt the barrier, and the second PCR is begun:

1. Set up a 100-μL PCR mix (*see* **Subheading 3.1.**) in a tube. Add only TL-A (or HisTL-A) as a primer, and no template.
2. Dispense approx 35 μL of petroleum jelly (AmpliGrease) onto the side of the tube with a syringe.
3. Heat the tube so that the grease melts to cover the bottom mix.
4. Allow to cool so that the grease resolidifies.
5. Add approx 30 μL of mineral oil on top of the grease.
6. Add 4 μL of ligation mix through the oil so the droplet rests above the grease barrier. The ligation mix is made as a master mix containing the following: 2 μL of T-linkers (5 μ*M* each in 50 m*M* NaCl), 2 μL of 10× ligase buffer, 1 μL of 10 m*M* ATP (if not in buffer), 0.5 μL of ligase (8 U/μL), and 10.5 μL of H_2O.
7. Add 1 μL of the PCR product to be cloned into the droplet of top ligation mix.
8. Use the following reaction conditions: 25°C for 1 h (ligation); 94°C for 15 s, 55°C for 15 s, and 72°C for 45 s for 30 amplification cycles (*see* **Note 6**).

3.5. Digestion and Cloning

The reamplified PCR product now has the restriction sites from the T-linker at its ends, and may be cloned using standard procedures. The protocol we usually use is as follows:

1. Run a small portion of the reamplified product on a checking gel to make sure you have enough of the correctly sized product with which to work. About 5 µL should give a clear band. With care, you should be able to use this same gel to check your DNA at several stages of the process. Do not let it dry out!
2. Add ~1 µg of uncut (supercoiled) plasmid vector to the tube containing the reamplified material. This will both act as a carrier during the purification process and help to make the effects of the restriction digests more obvious. You will probably not be able to see the slight size differences resulting from cutting off the ends of the T-linkers, but you should be able to see the difference as the vector is cut, and goes from being supercoiled to being linear.
3. Extract the DNA twice with phenol:chloroform and once with chloroform.
4. Precipitate the DNA by adding 2 vol of 95% ethanol containing 2% potassium acetate to the aqueous supernatant and spin on high speed in a microcentrifuge at 4°C for 30 min. Discard the supernatant and carefully rinse the pellet with cold 75% ethanol.
5. Resuspend the DNA pellet in distilled water and add the appropriate 10× restriction enzyme buffer. Add restriction enzymes and incubate until the vector is completely in the linear form on a checking gel. Directional cloning will require cutting with two enzymes. Manufacturers generally provide charts that indicate which buffer works reasonably well with both enzymes. Run a lane with uncut vector on the checking gel. Uncut vector should have three bands representing supercoiled, nicked circular, and (sometimes) linear forms. Cut vector should only have the linear form. Even small amounts of uncut vector will lead to high backgrounds on nonrecombinants.
6. Run the digested material on a preparative agarose gel and cut out the vector and insert bands. Minimize exposure to ultraviolet light.
7. Recover the DNA separately from each band. Many protocols are available for this: We usually use GeneClean for inserts larger than 250 bp.
8. Run about one-tenth to one-fifth of the DNA extracted from the band on a checking gel to roughly estimate the amount of DNA recovered. The relative amounts of DNA in each band are crudely estimated by visually comparing the brightness of the bands to those with a known amount of DNA. The vector bands should contain more or less known quantities of DNA, if you know how much you started with, and assume about 70% was recovered from the preparative gel.
9. Mix vector and insert in at least a 2:1 molar ratio and ligate. Use about 100 ng of vector per ligation. The ligation should resemble the following (for a larger insert, more DNA is needed for the same molar ratio): 100 ng vector (2.5 kb), 20 ng insert (250 bp), 2 µL of 10× ligation buffer, 1 µL of 10 m*M* ATP (if not in buffer), 0.5 µL of T4 DNA ligase, and up to 16.5 µL of H_2O.
10. Dilute the ligation 1:5, and use 1 µL to transform 20 µL of competent *E. coli*. Plate on appropriate antibiotic medium.
11. Recombinant colonies can be screened using PCR. Depending on which restriction site was used, you may be able to screen with a T-linker primer. For example, with inserts cloned nondirectionally using the *Eco*RI site of the NdeT-linker, NdeTL-A can be used for screening. Make a master mix for as many reactions as you need, in which a 10-µL reaction contains: 1 µL of 10× PCR buffer, 1 µL of dNTPs (2 m*M* each), 1.5 µL of $MgCl_2$ (10 m*M*), 2 µL of red sucrose dye, 1 µL of NdeTL-A (10 µ*M*), and 0.25 µL of *Taq* polymerase. Cover these small reactions with oil while picking colonies to prevent evaporation.
12. Touch a recombinant colony with the tip of a sterile toothpick, dip the end of the toothpick through the oil into the reaction, and swirl. Do not add enough bacteria to the reaction to make it cloudy. Just a few bacteria are sufficient. Too much bacterial matter, or

small amounts of agar, will inhibit the reaction. Be sure to include one reaction of a nonrecombinant ("blue") colony as a negative control.

13. Heat to 94°C for 1 min.
14. Amplify for 30 cycles using the following parameters: 94°C for 15 s, 55°C for 15 s, and 72°C for 45 s.

4. Notes

1. Why not use a T-vector? The method of choice for routine cloning of PCR products is probably a T-vector. One of the more important considerations is that with a T-vector you do not need to digest the insert with a restriction enzyme, so you do not need to worry about whether or not the insert contains that site. Also, reamplification with the T-linker provides another set of opportunities for the polymerase to introduce errors. However, T-linkers have several differences that can provide advantage in certain circumstances.
   a. The efficiency of ligation can theoretically be increased because much higher concentrations of linkers can be achieved.
   b. The efficiency of ligation does not need to be as high because the ligated product can be reamplified with one of the linker oligonucleotides to give a product that has added restriction sites at the ends.
   c. Oligonucleotides, such as those used to construct the T-linker, are quite stable, and remain usable for many years. Stability has been a problem with some of the commercially available T-vectors.
   d. Because the ends of a PCR product can be precisely defined and/or modified, and because T-linkers can be custom-made, it is possible to "split" a DNA sequence, such as a restriction site, between the PCR product and the T-linker so that a complete site is formed only at one end of the final product, without having to include the entire site in the primer made for amplifying a specific gene. This can save on the cost of oligonucleotides but still allow directional cloning.
   e. T-linkers should be more flexible in terms of using a variety of restriction sites in a variety of vectors.
2. The directional cloning of the coding sequence for the α3 acetylcholine receptor subunit illustrated in **Fig. 2** was accomplished by selecting the locations of the primers so that a *Hind*III site was created at only one end. However, by random chance, in one case of four, a PCR product made with primers not designed to complete the *Nde*I site will have a g at a given end, and will thus end up with an *Nde*I site. Similarly, one of 16 randomly chosen products will have *Nde*I sites at both ends. Thus, 3 of 16 randomly chosen PCR products will have a g at only one end, and could be cloned directionally using this T linker. A set of such linkers, with each depending on the presence of a different single nucleotide at the end of the PCR product, could therefore be used to directionally clone 75% of randomly chosen PCR products. Potential restriction enzymes for such complementable sites in T-linkers would be:

Site ends in	Restriction enzymes
. . . t(g)	*Nde*I, *Pvu*II (blunt), *Pm*lI (blunt)
. . . t(c)	*Eco*RI, *Eco*RV (blunt), *Aat*II, *Sac*I
. . . t(a)	*Sna*BI (blunt)
. . . t(t)	*Hind*III, *Ssp*I (blunt)

Three linkers would make up a complete set, that is, you do not need an a linker because any product that only has a at one end automatically has one of the other three bases at the other end. Together with the T-linkers using split *Nde*I and *Hind*III sites described here, a T-linker that introduces a split *Sac*I site, for example, would complete a set.

3. Potentially, other DNA sequences, such as a promoter for in vitro transcription, could be split so that a functional site is completed at only one end of the product. If the majority of the DNA sequence of such sites can be added by ligating a T-linker, this could provide significant cost savings compared to synthesizing target-specific primers containing such sequences.
4. Blunt-ended ligation of linkers has been used to add primer sequences to DNA fragments ***(9,10)***, but the T-linker application presented here is novel as far as we can tell. The T-linkers are more suitable for use with DNA fragments generated by PCR than blunt-ended linkers, because of the nontemplate derived as added by the polymerase. Because the sequences at the ends of PCR products can be easily manipulated by incorporating changes in the primers, DNA sequences, such as restriction sites, can be split between the T-linker and the PCR product. In general, this sort of site splitting is not practical except with PCR-generated fragments. One exception is fragments tailed with a known homopolymer, such as the poly A tail on eukaryotic mRNAs ***(11)***; such a linker is commercially available (Novagen, Madison, WI). Because the T-linker is added as an inverted repeat at the ends of the fragment, a single primer (TL-A) is used to reamplify after ligation. Single-primer amplification systems ***(12)*** have an advantage in that a "primer-dimer" cannot be formed from one primer, because this would produce an unamplifiable hairpin.
5. Because each product being cloned is subjected to an extra amplification, the overall frequency of errors should be increased, although our experience shows that the increase is not dramatic. In many cases, the risk of PCR errors is quite acceptable. For example, if one is producing an enzyme or other protein with a measurable function, the clones can be screened for activity. Deleterious mutations will thus be weeded out, and nondeleterious mutations may not matter (the $\alpha 3$ subunit in our example will be used as a potential ligand-binding protein, and clones will be screened on this basis). Similarly, if a sequence is to be used as a hybridization probe, a low frequency of base substitutions is frequently acceptable. In situations where no mutations are acceptable, clones must be screened by careful sequencing.
6. The pH and the magnesium concentration of the PCR are different from those of the ligation. Using a small ligation volume and a large PCR volume helps to correct the conditions for the PCR. Alternatively, the pH of the bottom mix can be increased, and the added magnesium reduced, so that the final pH and (Mg^{2+}) are correct after mixing. This allows use of smaller volumes. The ability to make a quick, automated one-tube ligation/reamplification reaction makes this setup intriguing. However, it is easier to repeat parts of the experiment (such as the reamplification) if you have leftovers from a separate ligation reaction.

Acknowledgments

This work was supported by research grants from the Muscular Dystrophy Association of America and from the Council for Tobacco Research, by the NIH grant N52319, and the NIDA Program Project grant DA05695. R. M. H. was the recipient of the Robert G. Sampson Neuromuscular Disease Research Fellowship from the Muscular Dystrophy Association.

References

1. Clark, J. M. (1988) Novel nontemplated nucleotide addition reactions catalyzed by procaryotic and eucaryotic DNA polymerases. *Nucleic Acids Res.* **16,** 9677–9686.

2. Mole, S. E., Iggo, R. D., and Lane, D. P. (1989) Using the polymerase chain reaction to modify expression plasmids for epitope mapping. *Nucleic Acids Res.* **17,** 3319.
3. Mead, D. A., Pey, N. K., Herrnstadt, C., Marcil, R. A., and Smith, L. M. (1991) A universal method for the direct cloning of PCR amplified nucleic acid. *Bio/Technology* **9,** 657–663.
4. Marchuk, D., Drumm, M., Saulino, A., and Collins, F. S. (1991) Construction of T-vectors, a rapid and general system for direct cloning of unmodified PCR products. *Nucleic Acids Res.* **19,** 1154.
5. Holton, T. A. and Graham, M. W. (1991) A simple and efficient method for direct cloning of PCR products using ddT-tailed vectors. *Nucleic Acids Res.* **19,** 1156.
6. Hoppe, B. L., Conti-Tronconi, B. M., and Horton, R. H. (1992) Gel loading dyes compatible with PCR. *BioTechniques* **12,** 679–680.
7. Horton, R. M., Hoppe, B. L., and Conti-Tronconi, B. M. (1994) AmpliGrease, “Hot Start” PCR using petroleum jelly. *BioTechniques* **16,** 42–43.
8. Tabor, S. and Richardson, C. C. (1985) A bacteriophage T7 RNA polymerase/promoter system for controlled exclusive expression of specific genes. *Proc. Natl. Acad. Sci. USA* **82,** 1074–1078.
9. Ko, M. S. H., Takahashi, N., Nishiguchi, K., and Abe, K. (1990) Unbiased amplification of a highly complex mixture of DNA fragments by ‘lone-linker’-tagged PCR. *Nucleic Acids Res.* **18,** 4293.
10. Bhat, G. J., Lodes, M. J., Myler, P. J., and Stuart, K. D. (1991) A simple method for cloning blunt ended DNA fragments. *Nucleic Acids Res.* **19,** 398.
11. Meissner, P. S. Sisk, W. P., and Berman, M. L. (1987) Bacteriophage lambda cloning system for the construction of directional cDNA libraries. *Proc. Natl. Acad. Sci. USA* **84,** 4171–4175.
12. Rich, J. J. and Willis, D. K. (1990) A single oligonucleotide can be used to rapidly isolate DNA sequences flanking a transposon Tn5 insertion by the polymerase chain reaction. *Nucleic Acids Res.* **18,** 6673–6676.

67

Cloning Gene Family Members Using PCR with Degenerate Oligonucleotide Primers

Gregory M. Preston

1. Introduction

1.1. What Are Gene Families?

As more and more genes are cloned and sequenced, it is apparent that nearly all genes are related to other genes. Similar genes are grouped into families, such as the collagen and globin gene families. There are also gene superfamilies. Gene superfamilies are composed of genes that have areas of high homology and areas of high divergence. Examples of gene superfamilies include the oncogenes, homeotic genes, and myosin genes. In most cases, the different members of a gene family carry out related functions. A detailed protocol for the cloning by degenerate oligonucleotide polymerase chain reaction (PCR) of members of the Aquaporin family of membrane water channels *(1,2)* is discussed here.

1.2. Advantages of PCR Cloning of Gene Family Members

There are several considerations that must be taken into account when determining the advantages of using PCR to identify members of a gene family over conventional cloning methods of screening a library with a related cDNA, a degenerate primer, or an antibody. It is recommended that after a clone is obtained by PCR, one uses this template to isolate the corresponding cDNA from a library because mutations can often be introduced in PCR cloning. Alternatively, sequencing two or more PCR clones from independent reactions will also meet this objective. The following is a list of some of the advantages of cloning gene family members by PCR.

1. Either one or two degenerate primers can be used in PCR cloning. When only one of the primers is degenerate, the other primer must be homologous to sequences in the phage or bacteriophage cloning vector *(3,4)* or to a synthetic linker sequence, as with RACE PCR. The advantage to using only one degenerate primer is that the resulting clones contain all of the genetic sequence downstream from the primer (be it 5′ or 3′ sequence). The disadvantage to this anchor PCR approach is that one of the primers is recognized by every gene in the starting material, resulting in single-strand amplification of all sequences. This is particularly notable when attempting to clone genes that are not abundant in the starting material. This disadvantage can often be ameliorated in part by using a nested

From: *Methods in Molecular Biology, Vol. 226: PCR Protocols, Second Edition*
Edited by: J. M. S. Bartlett and D. Stirling © Humana Press Inc., Totowa, NJ

amplification approach with two degenerate primers to preferentially amplify desired sequences.

2. It is possible to perform a PCR on first-strand cDNAs made from a small amount of RNA and, in theory, from a single cell. Several single-stranded "minilibraries" can be rapidly prepared and analyzed by PCR from a number of tissues at different stages of development or cell cultures under different hormonal conditions. Therefore, PCR cloning can potentially provide information about the timing of expression of an extremely rare gene family member, or messenger RNA splicing variants, that may not be present in a recombinant library.
3. Finally, the time and expense required to clone a gene should be considered. Relative to conventional cloning methods, PCR cloning can be more rapid, less expensive, and in some cases, the only feasible cloning strategy. It takes at least 4 d to screen 300,000 plaques from a λgt10 library. With PCR, an entire library containing 10^8 independent recombinants (~5.4 ng DNA) can be screened in one reaction. Again, to ensure authenticity of your PCR clones, you should either use the initial PCR clone to isolate a cDNA clone from a library or sequence at least two clones from independent PCRs.

1.3. Degenerate Oligonucleotide Theory and Codon Usage

Because the genetic code is degenerate, primers targeted to particular amino acid sequences must also be degenerate to encode the possible permutations in that sequence. Thus, a primer to a six- amino acid sequence that has 64 possible permutations can potentially recognize 64 different nucleotide sequences, one of which is to the target gene. If two such primers are used in a PCR, then there are 64 × 64 or 4096 possible permutations. The target DNA will be recognized by a small fraction (1/64) of both primers, and the amplification product from that gene will increase exponentially. However, some of the other 4095 possible permutations may recognize other gene products. This disadvantage can be ameliorated by performing nested amplifications and by using "guessmer" primers. A guessmer primer is made by considering the preferential codon usage exhibited by many species and tissues (*see* **Subheading 3.1.**). For instance, the four codons for alanine begin with GC. In the third position of this codon, G is rarely used in humans (~10.3% of the time) or rats (~8.0%), but often used in Escherichia coli (~35%) ***(5)***. This characteristic of codon usage may be advantageously used when designing degenerate oligonucleotide primers.

1.4. Strategy for Cloning Aquaporin Gene Family Members

In a related methods chapter ***(3)***, I described the cloning by degenerate primer PCR of Aquaporin-1 (formerly CHIP28) from a human fetal liver λgt11 cDNA library starting with the first 35 amino acids from the N-terminus of the purified protein. A full-length cDNA was subsequently isolated from an adult human bone marrow cDNA library ***(4)*** and following expression in Xenopus shown to encode a water selective channel ***(6)***. We now know that the Aquaporin family of molecular water channels includes genes expressed in diverse species, including bacteria, yeast, plants, insects, amphibians, and mammals ***(1,2,7)***. We have used degenerate oligonucleotide primers designed to highly conserved amino acids between the different members of the Aquaporin family to clone novel Aquaporin gene family cDNAs from rat brain (AQP4) and salivary gland (AQP5) libraries ***(8,9)***. In **Subheading 3.**, I will describe the creation of a new set of degenerate primers that we have used to clone, by degenerate primer PCR, Aquaporin homologs from a number of different tissues and species. **Subheading 3.**

Table 1
The Degenerate Nucleotide Alphabet

Letter	Specification
A	Adenosine
C	Cytidine
G	Guanosine
T	Thymidine
R	puRine (A or G)
Y	pYrimidine (C or T)
K	Keto (G or T)
M	aMino (A or C)
S	Strong (G or C)
W	Weak (A or T)
B	Not A (G, C, or T)
D	Not C (A, G, or T)
H	Not G (A, C, or T)
V	Not T (A, C, or G)
N	aNy (A, G, C, or T)
I	Inosine[a]

[a]Although inosine is not a true nucleotide, it is included in this degenerate nucleotide list because many researchers have employed inosine-containing oligonucleotide primers in cloning gene family members.

has been broken up into three parts: **Subheading 3.1.** describes the designing of the degenerate primers; **Subheading 3.2.** describes the PCR amplification with degenerate primers; and 3. **Subheading 3.3.** describes the subcloning and DNA sequencing of the specific PCR-amplified products.

2. Materials

2.1. Design of Degenerate Oligonucleotide Primers

No special materials are required here, except the amino acid sequence to which the degenerate primers will be designed and a codon usage table *(5)*. If the degenerate primers are going to be designed according to a family of related amino acid sequences, these sequences should be aligned using a multiple sequence alignment program. A degenerate nucleotide alphabet (**Table 1**) provides a single-letter designation for any combination of nucleotides. Some investigators have successfully used mixed primers containing inosine where degeneracy was maximal, assuming inosine is neutral with respect to base pairing, to amplify rare cDNAs by PCR *(10,11)*.

2.2. PCR Amplification with Degenerate Primers

For all buffers and reagents, distilled deionized water should be used. All buffers and reagents for PCR should be made up in distilled deionized 0.2-μ filtered water that has been autoclaved (PCR water) using sterile tubes and aerosol blocking pipet tips to prevent DNA contamination (*see* **Note 1**). All plastic supplies (microfuge tubes, pipet tips, and so on) should be sterilized by autoclaving or purchased sterile.

1. 10× PCR buffer: 100 m*M* Tris-HCl, pH 8.3; at 25°C, 500 m*M* KCl, 15 m*M* $MgCl_2$; and 0.1% w/v gelatin. Incubate at 50°C to melt the gelatin, filter sterilize, and store at –20°C (*see* **Note 2**).
2. dNTP stock solution (1.25 m*M* each of dATP, dGTP, dCTP, and dTTP) made by diluting commercially available deoxynucleotides with PCR water.
3. Thermostable DNA polymerase, such as Amplitaq DNA Polymerase (Perkin–Elmer Cetus, Norwalk, CT) supplied at 5 U/µL.
4. Mineral oil.
5. A programmable thermal cycler machine, available from a number of manufacturers, including Perkin–Elmer Cetus, MJ Research, and Stratagene.
6. Degenerate oligonucleotide primers should be purified by reverse-phase high-performance liquid chromatography or elution from acrylamide gels, dried down, resuspended at 20 pmol/µL in PCR-water, and stored at –20°C, preferably in aliquots.
7. The DNA template can be almost any DNA sample, including a single-stranded cDNA from a reverse transcription reaction, DNA from a phage library, and genomic DNA. The DNA is heat denatured at 99°C for 10 min and stored at 4 or –20°C.
8. Chloroform (*see* **Note 3**).
9. Tris-saturated phenol (*see* **Note 3**), prepared using ultra-pure redistilled crystalline phenol as recommended by the supplier (Gibco-BRL [product #5509], Gaithersburg, MD). Use polypropylene or glass tubes for preparation and storage.
10. PC9 (*see* **Note 3**): Mix equal volumes of buffer-saturated phenol, pH >7.2, and chloroform, extract twice with an equal volume of 100 m*M* Tris-HCl, pH 9.0, separate phases by centrifugation at room temperature for 5 min at 2000*g*, and store at 4 to –20°C for up to 1 mo.
11. AmAc (7.5 *M*) for precipitation of DNA. Ammonium acetate is preferred over sodium acetate because nucleotides and primers generally do not precipitate with it. Dissolve in water, filter through 0.2-µm membrane, and store at room temperature.
12. 100% ethanol stored at –20°C.
13. 70% ethanol stored at –20°C.
14. TE: 10 m*M* Tris, 0.2 m*M* EDTA, pH 8.0. Dissolve in water, filter through 0.2-µm membrane, and store at room temperature.
15. 50× TAE: 242 g of Tris-HCl base, 57.1 mL of acetic acid, 18.6 g of $Na_2(H_2O)_2EDTA$. Dissolve in water, adjust volume to 1 L, and filter through 0.2-µm membrane. Store at room temperature.
16. *Hae*III-digested φX174 DNA markers. Other DNA molecular weight markers can be used depending on availability and the size of the expected PCR-amplified products.
17. 6× gel loading buffer (GLOB): 0.25% bromophenol blue, 0.25% xylene cyanol FF, 1 m*M* EDTA, and 30% glycerol in water. Store up to 4 mo at 4°C.
18. Agarose gel electrophoresis apparatus and electrophoresis grade agarose. For the optimal resolution of DNA products >500 bp in length, NuSieve GTG agarose (FMC BioProducts) is recommended.
19. Ethidium bromide (EtBr; *see* **Note 3**). Ethidium bromide (10 mg/mL stock) prepared in water and stored at 4°C in a brown or foil-wrapped bottle. Use at 0.5 to 2.0 µg/mL in water for staining nucleic acids in agarose or acrylamide gels.
20. For the elution of specific PCR-amplified DNA products from agarose gels, several methods are available, including electroelution and electrophoresis onto DEAE-cellulose membranes ***(12,13)***. Several commercially available kits will also accomplish this task. I have had some success with GeneClean II (Bio 101, La Jolla, CA) for PCR products >500 bp in length, and with QIAEX (Qiagen, Chatsworth, CA) for products from 50 to 5000 bp. If you do not know the approximate size of the PCR-amplified products and wish to clone

all of them, the QIAquick-spin PCR purification kit is recommended (Qiagen) because this will remove all nucleotides and primers before attempting to clone. This kit is also recommended for purification of PCR products for secondary PCR-amplification reactions.

2.3. Cloning and DNA Sequencing of PCR-Amplified Products

1. From **Subheading 2.2., items 8–14** and **20**.
2. pBluescript II phagemid vector (Stratagene). A number of comparable bacterial expression vectors are available from several companies.
3. Restriction enzymes: *Eco*RV (for blunt-end ligation).
4. Calf intestinal alkaline phosphatase (CIP; New England Biolabs, Beverly, MA).
5. Klenow fragment of *Escherichia coli* DNA polymerase I (sequencing grade preferred) and 10 m*M* dNTP solution (dilute PCR or sequencing grade dNTPs).
6. T4 DNA ligase (1 or 5 U/μL) and 5× T4 DNA ligase buffer (Gibco-BRL).
7. Competent DH5α bacteria. Can be prepared ***(12,13)*** or purchased. Other bacterial strains can be substituted.
8. Ampicillin: 50 mg/mL stock in water, 0.2-μ filtered, and stored in aliquots at –20°C (*see* **Note 4**).
9. LB media: 10 g of bacto-tryptone, 5 g of bacto-yeast extract, and 10 g of NaCl dissolved in 1 L of water. Adjust pH to 7.0. Sterilize by autoclaving for 20 min on liquid cycle.
10. LB-Amp plates: Add 15 g of bacto-agar to 1000 mL of LB media before autoclaving for 20 min on the liquid cycle. Gently swirl the media on removing it from the autoclave to distribute the melted agar. Be careful: The fluid may be superheated and may boil over when swirled. Place the media in a 50°C water bath to cool. Add 1 mL of ampicillin, swirl to distribute, and pour 25 to 35 mL/90-mm plate. Carefully flame the surface of the media with a Bunsen burner to remove air bubbles before the agar hardens. Store inverted overnight at room temperature, then wrapped at 4°C for up to 6 mo.
11. IPTG: Dissolve 1 g of isopropylthiogalactoside in 4 mL of water, filter through 0.2-μm membrane, and store in aliquots at –20°C.
12. X-Gal: Dissolve 100 mg of 5-bromo-4-chloro-3-indolyl-β-D-galactopyranoside in 5 mL of dimethylformamide and stored at –20°C in a foil wrapped tube (light sensitive).
13. Plasmid DNA isolation equipment and supplies ***(12,13)*** or plasmid DNA isolation kits, available from many manufacturers.
14. Double-stranded DNA sequencing equipment and supplies ***(12,13)*** or access to a DNA sequencing core facility.

3. Methods

3.1. Design of Degenerate Oligonucleotide Primers

1. The first step in designing a degenerate primer is to select a conserved amino acid sequence and then determine the potential nucleotide sequence (or the complement of this sequence for a downstream primer), considering all possible permutations. If the amino acid sequence is relatively long, you can potentially design two or more degenerate primers. If only one is made, make it to sequences with a high (50–65%) GC content because these primers can be annealed under more stringent conditions (e.g., higher temperatures). **Figure 1** shows an alignment of the amino acid sequences for several members of the Aquaporin gene family in the two most highly conserved regions. Also shown is the consensus amino acid sequence, the degenerate nucleotide sequence, and the sequence of the primers we used to isolate Aquaporin gene family members. Interestingly, not only are these two regions highly conserved among the different members of this gene

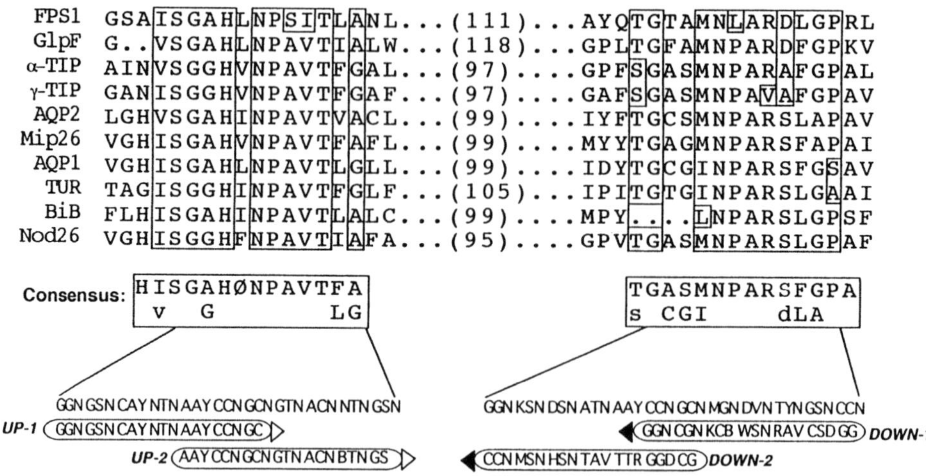

Fig. 1. Design of degenerate primers to amplify Aquaporin gene family members. Top, the amino acid sequences of 10 MIP family proteins, including the S. cerevisiae FPS1 *(23)*, *E. coli* GlpF *(24)*, α- and γ-tonoplast intrinsic proteins (TIP) of Arabidopsis thaliana *(25)*, the vasopressin-responsive water channel of rat renal collecting tubules (AQP2) *(26)*, the major intrinsic protein (MIP) of bovine lens fiber membranes *(27)*, human Aquaporin-1 *(4)*, turgor responsive gene (TUR) 7a from Pisum stivum *(28)*, the Drosophila neurogenic big brain protein *(29)*, and the Rhyzodium root Nodulin-26 peribacteroid membrane protein *(30)* were aligned by the PILEUP program of progressive alignments *(31)* using a gap weight of 3.0 and a gap length of 0.1 running on a VAX computer system. The two most highly conserved regions are shown, separated by the number of intervening amino acids. The most highly conserved amino acids are enclosed. Middle, below the aligned sequences, the consensus amino acid sequences are shown. Bottom, from part of the consensus amino acid sequences, the degenerate nucleotide sequences were determined (using the degenerate nucleotide alphabet from **Table 1**) followed by the sequences for the degenerate oligonucleotide primers.

family, but they are also highly related to each other, with the conserved motif being (T/S)GxxxNPAxx(F/L)G, that has been speculated to have resulted from an ancient internal duplication in a primordial bacterial organism, because this repeat has persisted in Aquaporin homologs from bacteria through plants and mammals *(1,6,14)*. These two regions are functionally related, both contributing to the formation of the water pore in Aquaporin-1 *(15)*.

2. The next step is to determine the number of permutations in the nucleotide sequence. There are 192 permutations ([2 × 4] × 3 × 4 × 2) in the sequence 5′-YTN-ATH-GGN-GAR-3′, which encodes the hypothetical amino acid sequence Leu-Ile-Gly-Glu. We can reduce the degeneracy by making educated guesses in the nucleotide sequence, that is, by making a guessmer. The 3′-end of a primer should contain all possible permutations in the amino acid sequence because *Taq* DNA polymerase will not extend a prime with a mismatch at the extending (3′) end. If the above primer was to a human gene, a potential guessmer would be 5′-CTB-ATY-GGN-GAR-3′, which only contains 64 permutations. This guessmer is proposed by taking into account the preferential codon usage for leucine and isoleucine in humans *(5)*.
3. The degeneracy of a primer can be reduced further by incorporating inosine residues in the place of N. The advantages of using inosine-containing primers is that they have a reduced number of permutations and that the inosine reportedly base pairs equally well with all

four nucleotides, creating a single bond in all cases ***(10)***. The disadvantage is that inosines reduce the annealing temperature of the primer. I have not used inosine-containing primers in my studies.

4. It is often convenient to incorporate restriction endonuclease sites at the 5′-ends of a primer to facilitate cloning into plasmid vectors ***(4,8,9)***. Different restriction sites can be added to the 5′-ends of different primers so the products can be cloned directionally. However, not all restriction enzymes can recognize cognate sites at the ends of a double-stranded DNA molecule equally well. This difficulty can often be reduced by adding a two to four nucleotide 5′-overhang before the beginning of the restriction enzyme site (*see* **Note 5**). Some of the best restriction enzymes sites to use are *Eco*RI, *Bam*HI, and *Xba*I. Catalogs from New England Biolabs have a list of the ability of different restriction enzymes to recognize short base-pair sequences. A potential pitfall of this approach would be the occurrence of the same restriction site within the amplified product as used on the end of one of the primers. Therefore, only part of the amplified product would be cloned.
5. The final consideration you should make is the identity of the 3′ most nucleotide. The nucleotide on the 3′-end of a primer should preferably be G or C and not N, I, or T. The reason for this is that thymidine (and supposedly inosine) can nonspecifically prime on any sequence. Guanosines and cytidine are preferred since they form three H-bonds at the end of the primer, a degree stronger than an A:T base pair.

3.2. PCR Amplification and DNA Purification

The template for these reactions can be the DNA in a phage library or the first-strand cDNA from a reverse transcription reaction on RNA. A phage library with a titer of 5×10^9 pfu/mL would contain, in a 5-µL aliquot, 2.5×10^7 pfu (~1.5 ng of DNA). Before PCR amplification, the DNA is heat denatured at 99°C for 10 min.

3.2.1. PCR (see **Notes 1** and **6**)

In all cases, the DNA template should also be PCR amplified with the individual degenerate primers to determine whether any of the bands amplified are derived from one of the degenerate primer pools. A DNA-free control is required to assess if there is contaminating DNA in any of the other reagents.

1. Pipet into 0.5-mL microcentrifuge tubes in the following order: 58.5 µL of PCR-water that has been autoclaved; 10 µL of 10× PCR buffer (*see* **Note 2**); 16 µL of 1.25 m*M* dNTP stock solution; 5.0 µL of primer up-1; 5.0 µL of primer down-1; and 5.0 µL of heat-denatured library or cDNA (1–100 ng). If several reactions are being set up concurrently, a master reaction mix can be made up consisting of all the reagents used in all of the reactions, such as the PCR water, reaction buffer, and dNTPs.
2. Briefly vortex each sample and spin for 10 s in a microfuge. Overlay each sample with 2 to 3 drops of mineral oil.
3. Amplify by hot-start PCR using the following cycle parameters. Pause the thermocycler in step 4-cycle 1 and add 0.5 µL of Amplitaq DNA polymerase to each tube. Then, run for 95°C, 5 min (initial denaturation); 94°C, 60 s (denaturation); 50°C, 90 s (annealing; *see* **Note 7**); 72°C, 60 s (extension); cycle 29 times to **step 2**; 72°C, 4 min; and 10°C hold.

3.2.2. DNA Isolation and Gel Electrophoresis Analysis

1. Remove the reaction tubes from the thermal cycler and add 200 µL of chloroform. Spin for 10 s in a microfuge to separate the oil-chloroform layer from the aqueous layer. Carefully transfer the aqueous layer to a clean microfuge tube.

2. Remove the AmpliTaq DNA polymerase by extracting the aqueous phase twice with 100 µL of PC9 (*see* **Note 3**). Spin for 2 min in a microfuge to separate the lower organic layer from the upper aqueous layer and transfer the aqueous layer to clean microfuge tube. This step is essential before digesting the DNA with restriction enzymes for directional cloning (*see* **Subheading 3.3.**) because the polymerase can precipitate, and in the presence of nucleotides, fill in recessed 3′ termini on DNA.
3. AmAc-EtOH precipitation: To a 100 µL of DNA sample add 50 µL of 7.5 *M* AmAc (50% vol). Vortex briefly to mix. Precipitate the DNA with 350 µL of 100% ethanol (2–2.5 vol). Vortex the samples for 15 s and ice for 15 min. Spin down the DNA at 12,000*g* for 15 min at 4°C in a microfuge. Decant the aqueous waste. Add 250 µL of 70% ethanol. Vortex briefly and spin another 5 min at 4°C. Decant the ethanol and allow the pellets to dry inverted at room temperature, or dry in a Speed-Vac for 2 to 10 min.
4. Resuspend in 20 µL of PCR water.
5. The next step is to resolve an aliquot (2–10 µL) of the PCR fragments by gel electrophoresis. Small DNA products (>300 bp) can be resolved at high resolution on 5 to 10% polyacrylamide gels ***(12,13)***. Moderate-sized PCR products (150–1000 bp) should be resolved on 2 to 4% NuSieve agarose gels (in 1× TAE buffer). Larger PCR products (>500 bp) can be resolved on 0.8 to 2% agarose gels (1× TAE buffer).
6. After the bromophenol blue dye has reached the end of the gel, soak the gel for 5 to 30 min in about 10 vol of water containing 1 µg/mL EtBr (*see* **Note 3**). Then view and photograph the gel under ultraviolet light. As shown in **Fig. 1**, there is little variability in the distance between the NPA motifs with the known members of the Aquaporin gene family. PCR amplification of the known Aquaporins cDNAs using the internal degenerate primers would generate products from 345 to 415 bp. A typical result is shown in **Fig. 2**.

3.2.3. Secondary PCR Amplifications and DNA Purification

Based on the results from gel electrophoresis of the PCR-amplified DNA products, a decision must be made on what to do next. The options are the following.

1. Amplify by PCR from the initial DNA sample under different conditions.
2. Amplify by PCR from a different DNA sample under the same conditions. (Different $MgCl_2$ concentration, annealing temperature, or primers, *see* **Notes 2**, **6**, and **7**).
3. Gel purify a band(s) of DNA from the gel for cloning or to reamplify by PCR.
4. Purify all PCR-amplified DNA fragments for cloning or to reamplify by PCR.
5. Reamplify by PCR with the same or an internal pair of degenerate primers.

Options 1 and 2 are self explanatory. If you want to gel purify a particular band or group of bands from an agarose gel, a number of procedures and kits are available (*see* **Subheading 2.2.**). If you plan on immediately cloning a PCR band(s), you may want to run the rest of the initial PCR on another gel to increase the recovery of DNA. It is also possible to recover specific DNA fragments from an acrylamide gel ***(3,12,13)***. To purify all PCR-amplified DNA fragments from the remaining sample, a number of methods are available, including the QIAquick-spin PCR purification kit, which can be used instead of **steps 1–3** in **Subheading 3.2.2.** (Qiagen). Finally, aliquots of DNA purified from a gel or from the initial PCR (1–10%) can be reamplified by PCR with either the same or an internal pair of degenerate oligonucleotide primers (*see* **Note 1**).

When attempting to identify a gene family homolog from a tissue that is known to express a homolog(s), a number of tricks can be tried to enrich the final PCR sample

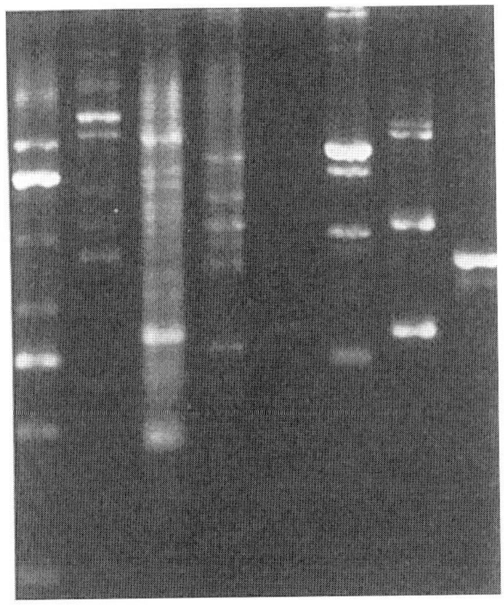

Fig. 2. Gel electrophoresis analysis of PCR-amplified DNA. DNA isolated from a human kidney cDNA library in bacteriophage λgt10 was amplified with degenerate primers up-1 (lanes 1, 5, and 6), up-2 (lanes 2, 7, and 8), down-1 (lanes 3, 5, and 7), and down 2 (lanes 4, 6, and 8). Reactions containing 5×10^6 pfu of heat-denatured phage DNA, 100 pmol of degenerate primers, and 1.5 m*M* $MgCl_2$ in a 100-μL volume were subject to 40 cycles of PCR amplification under the following parameters: 94°C for 60 s, 48°C for 90 s, and 72°C for 60 s. After chloroform extraction and ethanol precipitation, the DNA was resuspended in 20 μL of water, and 5 μL was electrophoresed into a 4% NuSieve agarose gel in 1× TAE. The gel was stained with ethidium bromide and photographed. The relative mobility of *Hae*III digested ϕX174 DNA markers is shown on the right. The bracket shows the size range of known members of this gene family from the primers used.

for new homologs. Because the degenerate oligonucleotide primers are designed from the sequence of the known gene family members, these primers will likely be biased for those homologs. Aquaporin-1 is abundant in the capillaries around the salivary glands and throughout the body, but absent in the salivary gland ***(16)***. To identify a salivary homology of the Aquaporin gene family, we used a rat salivary gland cDNA library that also contained Aquaporin-1 cDNAs, presumably from the surrounding capillaries. We first amplified the cDNA library with an external set of degenerate primers, digested the PCR-amplified DNAs with the restriction enzyme *Pst*I (which cuts between the NPA motifs of rat AQP1), and reamplified with an internal pair of primers. We again digested with *Pst*I to digest the rat AQP1 DNAs, then cloned and sequenced the DNA fragments between 350 and 450 bp ***(9)***. This strategy would not work if the resulting cDNA (AQP5) also contained a *Pst*I site. By trying different restriction enzymes that cut DNA infrequently (6–8 bp-recognition sites), a number of new homologs will preferentially be identified. Alternatively, after cloning the DNA products into bacterial expression vectors, bacterial colony lift hybridization can be used to identify colonies containing inserts for known gene family members ***(3,12,13)***.

3.3. Cloning and DNA Sequencing of PCR-Amplified Products

3.3.1. Preparation of Vector for Ligation

1. For blunt-end ligations, digest 1 μg of pBluescript II KS phagemid DNA (Stratagene) with 10 U *Eco*RV in a 50-μL vol. Incubate at 37°C for 2 h. For cohesive-end ligations, similarly digest the vector with the appropriate restriction enzyme(s).
2. For both blunt-end ligations and cohesive-end ligations where the vector has been digested with only one restriction enzyme, it is necessary to remove the 5′-phosphate from the vector to inhibit the vector from self ligating. This is accomplished by treating the vector with CIP according to the manufacturer's recommendations. Note that 1 μg of a 3-kbp linear DNA molecule contains 1 pmol of 5′ overhangs (*Bam*HI), blunt-ends (*Eco*RV), or 3′-overhangs (*Pst*I), depending on the enzyme that digested it. Afterward, add EDTA to 5 m*M* and heat-kill the enzyme at 65°C for 1 h. Adjust the volume to 50 to 100 μL with TE and extract once with Tris-saturated phenol, twice with PC9, and twice with chloroform. Back extract each organic layer with 50 μL of TE and pool with the final sample. AmAc-EtOH precipitate (*see* **Subheading 3.2.2.**) and resuspend in 10 μL of water.
3. If the insert is going to be directionally cloned into the vector, just extract once with 50 μL of PC9, AmAc-EtOH precipitate (*see* **Subheading 3.2.2.**) and resuspend in 10 μL of water.

3.3.2. Preparation of Inserts for Ligation

AmpliTaq and other thermostable DNA polymerases often fail to completely fill in the ends of the double-stranded DNA products, thus leaving recessed 3′ termini that can be filled in with the Klenow fragment of *E. coli* DNA polymerase I. This should be done whether or not the DNA is going to be digested with restriction enzymes added to the ends of the primers for directional cloning (*see* **Subheading 3.1.**).

1. AmAc-EtOH precipitate the DNA (*see* **Subheading 3.2.2.**) and resuspend in 15 μL of water.
2. Add 2 μL of 10× restriction enzyme reaction buffer. Klenow DNA polymerase works well in most restriction enzyme digestion buffers (10× REact 2 or 3 from Gibco-BRL). If the DNA is going to be subsequently digested with a restriction enzyme(s), use the buffer for that enzyme.
3. Add 2 μL of 10 m*M* dNTP solution. Then, add Klenow DNA polymerase (1 U/μg DNA) and incubate at room temperature for 15 min.
4. Heat-inactivate the enzyme at 75°C for 10 min. If the DNA is going to be directly used in ligation reactions, it is not necessary to purify the DNA from the unincorporated dNTPs because they will not inhibit T4 DNA ligase. To concentrate the DNA sample, proceed with **step 6**.
5. PCR products containing restriction sites on their ends should now be digested with the restriction enzymes. Incubate in the appropriate buffer, using 20 U of enzyme/μg of DNA and incubating for 2 to 4 h at the proper temperature.
6. Extract the DNA once or twice with PC9 and precipitate with AmAc-EtOH as described above (*see* **Subheading 3.2.2.**). Resuspend the final pellet in 5 to 10 μL of water.

3.3.3. DNA Ligation and Bacterial Transformation

1. At this point, it is often advantageous to run a small aliquot of the different DNA fragments on a gel to assess their approximate concentrations and purity. Ideally, you want at least a 2:1 molar ratio of insert to vector in the ligation reactions. If necessary return to the above procedures to isolate more DNA for the ligation reaction.

2. Set up the ligation reactions with the vector and insert similar to the following:
   a. Reaction 1: 1 μL of vector (10 ng; vector control);
   b. Reaction 2: 1 μL of vector + 1 μL of insert (~10 ng insert);
   c. Reaction 3: 1 μL of vector + 4 μL of insert.

   Then add 2 μL of 5× T4 DNA ligase buffer (Gibco-BRL) and water to 9.5 μL. If the buffer is more than 4-mo-old, the ATP may be depleted. Therefore, add fresh ATP to a final concentration of 1 m*M*.
3. For cohesive-end ligations add 0.5 μL of T4 DNA ligase (1 U/μL), gently mix, spin 5 s in a microfuge, and incubate at 15°C for 10 to 20 h. For blunt-end ligations add 1 μL of T4 DNA ligase (5 U/μL), gently mix, spin 5 s in a microfuge, and incubate at 25°C (or room temperature) for 1 to 12 h. Stop the reaction by heating at 75°C for 10 min and store the samples at –20°C.
4. Set up a bacterial transformation with competent DH5α bacteria or a comparable strain of bacteria. Be sure to include a positive control (10 ng of undigested vector DNA) and a negative control (water). To 1.5-mL microfuge tubes, add half of the ligation mix (5 μL) or 5 μL of control DNA or water and 50 μL of competent bacteria (thawed slowly on ice); incubate on ice for 30 min. Heat-shock at 42°C for 2 min. Return to ice for 1 min. Add 200 μL of LB media containing 10% glycerol. Mix gently and allow bacteria to recover and express the ampicillin resistance gene by incubating at 37°C for 1 h.
5. Prewarm LB-Amp plates at 37°C for 45 min. About 30 min before plating the bacteria on the plates, add 40 μL of X-Gal and 4 μL of IPTG and quickly spread over the entire surface of the plate using a sterile glass spreader. Spread 20 to 200 μL of the transformation reactions on these plates. Allow the inoculum to absorb into the agar and incubate the plates inverted at 37°C for 12 to 24 h (*see* **Note 4**). Afterward, placing the plates at 4°C for 2 to 4 h will help enhance the blue color development.

3.3.4. Plasmid DNA Minipreps and DNA Sequencing

1. Colonies that contain active β-galactosidase will appear blue, whereas those containing a disrupted LacZ gene will be white. Set up minicultures by inoculating individual white colonies into 2 mL of LB media containing ampicillin. After growing at 37°C overnight, isolate the plasmid DNA. Resuspend the DNA in 20 to 50 μL of water or TE.
2. Digest 5 to 20 μL of the DNA with the appropriate restriction enzymes and analyze by agarose gel electrophoresis (*see* **Subheading 3.2.3.**).
3. Perform double-stranded DNA sequencing on recombinants containing inserts in the expected size range.

4. Notes

1. All PCRs should be set up in sterile laminar flow hoods using pipet tips containing filters (aerosol-resistant tips) to prevent the contamination of samples, primers, nucleotides, and reaction buffers by DNA. If the PCR is going to be reamplified by PCR, all possible intervening steps should also be performed in a sterile hood with the same precautions to prevent DNA contamination. These precautions should also be extended to all extractions and reactions on the nucleic acid (RNA or DNA) through the last PCR. Likewise, all primers, nucleotides, and reaction buffers for PCR should be made up and aliquoted using similar precautions. All buffers for PCR should be made with great care using sterile disposable plastic or baked glass, and restricted for use with aerosol-resistant pipet tips.
2. Standard PCR buffers contain 15 m*M* $MgCl_2$ (1.5 m*M* final concentration). In many cases, changes in the $MgCl_2$ concentration will have significant consequences on the amplification of specific bands. In PCR-amplifying the four exons of the AQP1 gene,

$MgCl_2$ concentrations between 0.7 and 1.0 m*M* gave the best results ***(17,18)***; however, $MgCl_2$ concentrations between 0.5 and 5.0 m*M* have been reported.

3. Organic solvents and ethidium bromide are hazardous materials. Always handle with tremendous caution, wearing gloves and eye protection. Contact your hazardous waste department for proper disposal procedures in your area.
4. Ampicillin-resistant bacteria secrete β-lactamase into the media, which rapidly inactivates the antibiotic in regions surrounding the growing bacterial colony. Thus, when bacteria are growing at a high density or for long periods (>16 h), ampicillin-sensitive satellite colonies will appear around the primary colonies (which are white in blue-white selections). This problem can be ameliorated (but not eliminated) by substitution of carbenicillin for ampicillin on agar plates.
5. When designing primers with restriction enzyme sites and 5′-overhangs, note that the 5′-overhang should not contain sequences complementary to the sequence just 3′ of the restriction site because this would facilitate the production of primer-dimers. Consider the primer 5′-ggg.agatct.CCCAGCTAGCTAGCT-3′, which has a *Xba*I site proceeded by a 5′-ggg and followed by a CCC-3′. These 12 nucleotides on the 5′-end are palindromic and can therefore easily dimerize with another like primer. A better 5′-overhang would be 5′-cac.
6. When cloning a gene from a recombinant library by PCR, remember that not all genes are created equally. Genes with high G:C contents have proven more difficult to clone than most. Several researchers have made contributions in a search for factors to enhance the specificity of PCRs. Nonionic detergents, such a Nonident P-40, can be incorporated in rapid sample preparations for PCR analysis without significantly affecting *Taq* polymerase activity ***(19)***. In some cases, such detergents are absolutely required to reproducibly detect a specific product ***(20)*** presumably because of inter- and intrastrand secondary structure. Tetramethylammonium chloride has been shown to enhance the specificity of PCRs by reducing nonspecific priming events ***(21)***. Commercially available PCR enhancers are also available.
7. A critical parameter when attempting to clone by PCR is the selection of a primer annealing temperature. This is especially true when using degenerate primers. The primer melting temperature (T_m) is calculated by adding 2° for A:T base pairs, 3° for G:C base pairs; 2° for N:N base pairs, and 1° for I:N base pairs. Most PCR chapters suggest you calculate the T_m and set the primer annealing temperature to 5 to 10°C below the lowest T_m. Distantly related gene superfamily members have been cloned using this rationale ***(22)***. However, I have found that higher annealing temperatures are helpful in reducing nonspecific priming, which can significantly affect reactions containing degenerate primers.

Acknowledgments

I thank my colleagues, especially Peter Agre and William B. Guggino, for their support and helpful discussions. This work was supported in part by NIH grants HL33991 and HL48268 to Peter Agre.

References

1. Reizer, J., Reizer, A., and Saier, M. H., Jr. (1993) The MIP family of integral membrane channel proteins: sequence comparisons, evolutionary relationships, reconstructed pathways of evolution, and proposed functional differentiation of the two repeated halves of the proteins. *Crit. Rev. Biochem. Mol. Biol.* **28,** 235–257.
2. Knepper, M. A. (1994) The aquaporin family of molecular water channels. *Proc. Natl. Acad. Sci. USA* **91,** 6255–6258.

3. Preston, G. M. (1993) Use of degenerate oligonucleotide primers and the polymerase chain reaction to clone gene family members, in *Methods in Molecular Biology*, vol. 15: PCR Protocols: Current Methods and Applications. (White, B. A., ed.) Humana, Totowa, NJ, pp. 317–337.
4. Preston, G. M. and Agre, P. (1991) Isolation of the cDNA for erythrocyte integral membrane protein of 28 kilodaltons: Member of an ancient channel family. *Proc. Natl. Acad. Sci. USA* **88,** 11,110–11,114.
5. Wada, K.-N., Aota, S.-I., Tsuchiya, R., Ishibashi, F., Gojobori, T., and Ikemura, T. (1990) Codon usage tabulated from the GenBank genetic sequence data. *Nucleic Acids Res.* **18,** 2367–2411.
6. Preston, G. M., Carroll, T. P., Guggino, W. B., and Agre, P. (1992) Appearance of water channels in Xenopus oocytes expressing red cell CHIP28 protein. *Science* **256,** 385–387.
7. Chrispeels, M. J. and Agre, P. (1994) Aquaporins: Water channel proteins of plant and animal cells. *TIBS* **19,** 421–425.
8. Jung, J. S., Bhat, B. V., Preston, G. M., Guggino, W. B., Baraban, J. M., and Agre P. (1994) Molecular characterization of an aquaporin cDNA from brain: candidate osmoreceptor and regulator of water balance. *Proc. Natl. Acad. Sci. USA* **91,** 13,052–13,056.
9. Raina, S., Preston, G. M., Guggino, W. B., and Agre, P. (1995) Molecular cloning and characterization of an aquaporin cDNA from salivary, lacrimal and respiratory tissues. *J. Biol. Chem.* **270,** 1908–1912.
10. Knoth, K., Roberds, S., Poteet, C., and Tamkun, M. (1988) Highly degenerate, inosine-containing primers specifically amplify rare cDNA using the polymerase chain reaction. *Nucleic Acids Res.* **16,** 10,932.
11. Chérif-Zahar, B., Bloy, C., Kim, C. L. V., Blanchard, D., Bailly, P., Hermand, P., et al. (1990) Molecular cloning and protein structure of a human blood group Rh polypeptide. *Proc. Natl. Acad. Sci. USA* **87,** 6243–6247.
12. Sambrook, J., Fritsch, E. F., and Maniatis, T., eds. (1989) *Molecular Cloning: A Laboratory Manual.* Cold Spring Harbor Laboratory, Cold Spring Harbor, NY.
13. Ausubel, F. M., Brent, R., Kingston, R. E., Moore, D. D., Seidman, J. G., Smith, J. A., et al., eds. (1994) *Current Protocols in Molecular Biology.* Greene Publishing/Wiley-Interscience, New York.
14. Wistow, G. J., Pisano, M. M., and Chepelinsky, A. B. (1991) Tandem sequence repeats in transmembrane channel proteins. *TIBS* **16,** 170–171.
15. Jung, J. S., Preston, G. M., Smith, B. L., Guggino, W. B., and Agre, P. (1994) Molecular structure of the water channel through aquaporin CHIP: The hourglass model. *J. Biol. Chem.* **269,** 14,648–14,654.
16. Nielsen, S., Smith, B. L., Christensen, E. I., and Agre, P. (1993) Distribution of the aquaporin CHIP in secretory and resorptive epithelia and capillary endothelia. *Proc. Natl. Acad. Sci. USA* **90,** 7275–7279.
17. Smith, B. L., Preston, G. M., Spring, F. A., Anstee, D. J., and Agre, P. (1994) Human red cell Aquaporin CHIP, I. molecular characterization of ABH and Colton blood group antigens. *J. Clin. Invest.* **94,** 1043–1049.
18. Preston, G. M., Smith, B. L., Zeidel, M. L., Moulds, J. J., and Agre, P. (1994) Mutations in aquaporin-1 in phenotypically normal humans without functional CHIP water channels. *Science* **265,** 1585–1587.
19. Weyant, R. S., Edmonds, P., and Swaminathan, B. (1990) Effects of ionic and nonionic detergents on the *Taq* polymerase. *Biotechnology* **9,** 308–309.
20. Bookstein, R., Lai, C.-C., To, H., and Lee, W.-H. (1990) PCR-based detection of a polymorphic BamHI site in intron 1 of the human retinoblastoma (RB) gene. *Nucleic Acids Res.* **18,** 1666.

21. Hung, T., Mak, K., and Fong, K. (1990) A specificity enhancer for polymerase chain reaction. *Nucleic Acids Res.* **18,** 4953.
22. Zhao, Z.-Y. and Joho, R. H. (1990) Isolation of distantly related members in a multigene family using the polymerase chain reaction technique. *Biochem. Biophys. Res. Commun.* **167,** 174–182.
23. Aelst, L. V., Hohmann, S., Zimmermann, F. K., Jans, A. W. H., and Thevelein, J. M. (1991) A yeast homologue of the bovine lens fiber MIP gene family complements the growth defect of a Saccharomyces cerevisiae mutant on fermentable sugars but not its defect in glucose-induced RAS-mediated cAMP signalling. *EMBO J.* **10,** 2095–2104.
25. Muramatsu, S. and Mizuno, T. (1989) Nucleotide sequence of the region encompassing the glpKF operon and its upstream region containing a bent DNA sequence of Escherichia coli. *Nucleic Acids Res.* **17,** 4378.
25. Höfte, H., Hubbard, L., Reizer, J., Ludevid, D., Herman, E. M., and Chrispeels, M. J. (1992) Vegetative and seed-specific forms of Tonoplast Intrinsic Protein in the vacuolar membrane of Arabidopsis thaliana. *Plant Physiol.* **99,** 561–570.
26. Fushimi, K., Uchida, S., Hara, Y., Hirata, Y., Marumo, F., and Sasaki, S. (1993) Cloning and expression of apical membrane water channel of rat kidney collecting tubule. *Nature* **361,** 549–552.
27. Gorin, M. B., Yancey, S. B., Cline, J., Revel, J.-R., and Horwitz, J. (1984) The major intrinsic protein (MIP) of the bovine lens fiber membrane: characterization and structure based on cDNA cloning. *Cell* **39,** 49–59.
28. Guerrero, F. D., Jones, J. T., and Mullet, J. E. (1990) Turgor-responsive gene transcription and RNA levels increase rapidly when pea shoots are wilted: sequence and expression of three induced genes. *Plant Mol. Biol.* **15,** 11–26.
29. Rao, Y., Jan, L. Y., and Jan, Y. N. (1990) Similarity of the product of the Drosophila neurogenic gene big brain to transmembrane channel proteins. *Nature* **345,** 163–167.
30. Fortin, M. G., Morrison, N. A., and Verma, D. P. S. (1987) Nodulin-26, a peribacteroid membrane nodulin is expressed independently of the development of the peribacteroid compartment. *Nucleic Acids Res.* **15,** 813–824.
31. Feng, D.-F. and Doolittle, R. F. (1990) Progressive alignment and phylogenetic tree construction of protein sequences, in *Methods in Enzymology*, vol. 183: Molecular Evolution: Computer Analysis of Protein and Nucleic Acid Sequences (Doolittle, R. F., ed.), Academic, New York, pp. 375–387.

68

cDNA Libraries from a Low Amount of Cells

Philippe Ravassard, Christine Icard-Liepkalns, Jacques Mallet, and Jean Baptiste Dumas Milne Edwards

1. Introduction

1.1. Application of the SLIC Strategy to cDNA Libraries: General Considerations

Conventional cDNA library construction often requires a minimum available amount of material (typically 1 or 2 μg of polyA$^+$ RNA). For complex organs, such as brain, or certain species, such as humans, as well as subsets of cell types, this condition is often difficult to fulfill. Amplification by polymerase chain reaction (PCR) can be used to circumvent this limitation because it is a powerful method to obtain working quantities of low-abundance DNAs. To effectively apply this method, known sequences need to be attached to the ends of the single-stranded cDNA (ss-cDNA). One at the 5′ end of the ss-cDNA is added during the priming of the synthesis; the other, at the 3′ end, is covalently attached by ligation using the SLIC strategy. With known DNA sequences attached to both ends of the synthetized cDNA, minute quantities can be amplified with sequence-specific primers to provide sufficient material to successfully generate and screen cDNA libraries. The overall scheme is illustrated in **Fig. 1**.

Obviously, the goal in constructing such a library is to maintain the representation of every mRNA in the total RNA population, even low-abundant ones. Because the reverse transcription and the single-strand ligation do not modify the overall proportion of each molecule, only the PCR step may introduce an important bias in cDNA representation. Thus, for cDNA library construction using this methodology, it is important to optimize the synthesis of the sequence-tagged cDNA and to take steps to limit any amplification bias.

1.2. Simultaneous PCR Amplification of a Complex DNA Mixture Generates an Important Size Bias

The sequence-tagged cDNAs correspond to a large population of molecules with indentical ends. The only difference between these molecules is their relative size and sequence. Therefore, keeping the representativity of the cDNAs after PCR requires a constant amplification yield regardless of the size and sequence of the original cDNA population. This, unfortunately cannot be achieved by PCR. In fact, coamplification of

From: *Methods in Molecular Biology, Vol. 226: PCR Protocols, Second Edition*
Edited by: J. M. S. Bartlett and D. Stirling © Humana Press Inc., Totowa, NJ

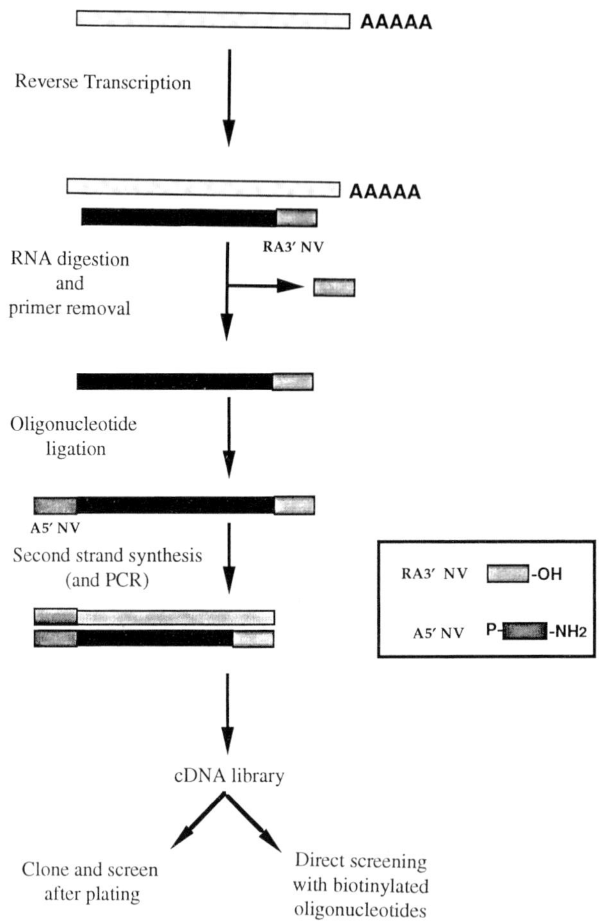

Fig. 1. Constructing cDNA libraries using the SLIC strategy. Reverse transcription is performed with a random primer RA3′ NV. RNAs are digested and primer is removed to avoid its ligation to the A5′ NV oligonucleotide. A5' NV bears a phosphate group at its 5′ extremity to allow ligation to the 3′ end of the ss-cDNA. To avoid formation of concatemers A5′ NV bears an amino group at its 3′ end. Two rounds of nested PCR are performed to generate a ready-to-use cDNA library.

short and long molecules with the same primers always leads to a selective amplification of the shorter ones, even though the longer ones were originally more abundant *(1)*.

Attempts to enlarge the average size of the PCR products have been made with the help of long-range PCR procedures, such as Pfu dilutions *(2)*, Taq extender (Stratagene, La Jolla, CA), and Expand PCR system (Boehringer, Indianapolis, IN). The results were not significantly different if compared with the classic PCR techniques (data not shown).

In conclusion, the size bias represents the major limitation in constructing PCR cDNA libraries. Thus, to avoid bias during the amplification step the average size of the cDNA library must be between 0.8 and 1 kbp, which is far below the usual average length of a ss-cDNA.

Name	5' Sequence 3'
A3'NV	ATCGTTGAGACTCGTACCAGCAGAGTCACGAGAGAGACTACACGGTACTGG(T)14
RA 3'NV	ATCGTTGAGACTCGTACCAGCAGAGTCACGAGAGAGACTACACGGTACTGG(N)9
A3'_1	ATCGTTGAGACTCGTACCAGCAGAG
A3'_2	TCGTACCAGCAGAGTCACGAGAGAG
A3'_3	CACGAGAGAGACTACACGGTACTGG

Fig. 2. Oligonucleotides used for 3′ anchored PCR. The three oligonucleotides A3′_1,2,3 are designed from A3′ NV and RA3′ NV to be used in PCR experiments. Note that these three primers have to be used with A5′_1,2,3.

1.3. ss-cDNA Synthesis

The constraint caused by the size limitation of the ss-cDNA forbids the use of oligo-dT to prime the reverse transcription. If such a priming strategy is chosen, this will lead to a 3′-UTR-cDNA library. To obtain a cDNA library representative of the sequences of all messenger RNA, priming with random primer RA3′ NV must be performed.

Experimentally, the ss-cDNA is synthesized with a random primer RA3′ NV (**Fig. 2**) and a radiolabled nucleotide. The average size is determined by alkaline gel electrophoresis and autoradiography. The incubation time with the reverse transcriptase is calibrated in order to generate an ss-cDNA whose average length is between 0.8 and 1 kbp. With those conditions the representation of any mRNA will be optimal in the library.

1.4. Amplification of the cDNA Library Based on the SLIC Strategy

We have used this strategy to generate a cDNA library from newborn rat cervical superior ganglia (CSG). Total RNA was prepared from one single CSG, and polyA$^+$ RNA was prepared with oligo dT-coated magnetic beads (Dynabeads mRNA purification kit). Half of the material, which corresponds to polyA$^+$ RNA of about 5000 cells, was used to synthesize ss-cDNA primed with RA3′ NV. An incubation time of 30 min was optimum to generate an average size of 1 kbp. After removal of the primer, the ss-cDNA was ligated to A5′ NV (note that after each step the different primers are removed). These ss-cDNA were amplified by two rounds of nested PCR. To increase the specificity of the PCR reaction, we have used the touchdown PCR protocol *(3)*. The primers used for the nested PCR were A5′_1 ∞ A3′_1 and A5′_2 ∞ A3′_2, respectively. One-twentieth of the reaction was cloned in a blunt-end vector, yielding 2 ∞ 10^5 colonies. Analysis by direct PCR on colonies of 96 randomly chosen clones indicated that the average size of the library was about 900 bases. The striking result is that the size dispersion around the mean is extremely low when compared with a conventional library. Finally, 5000 primary clones were screened with a TH oligonucleotide, yielding two positive TH clones (0.04%). Thus, no major distortion had been introduced in the abundance of TH clones, since TH represents 0.05% of CSG mRNA.

1.5. Direct Screening of the PCR Library with Biotinylated Oligonucleotides

One of the major difficulties in making a cDNA library is the cloning of the double-stranded cDNA (ds-cDNA). We tested a direct screening protocol of uncloned ds-cDNA. After the second nested PCR, the amount of ds-cDNA was about 2 to 5 μg. We directly hybridized the denatured ds-cDNA with a biotinylated TH specific primer. After the hybridization reaction, probe-cDNA hybrids were separated from unhybridized DNA using streptavidin-coated magnetic beads (Dynabeads M-280). After various washing steps, the captured cDNA was amplified using the third nested PCR primers (A5′3 and A3′3) directly onto the beads. The PCR product was cloned, and more than 85% of them were TH clones as analyzed by partial sequencing. Biotinylated cDNA probes, instead of oligonucleotides, can also be used to screen a PCR library *(4)*.

1.6. Application of the SLIC Method to Subtractive Libraries

Subtraction cloning strategies could be modified and certainly improved taking advantage of the generation of cDNA molecules exhibiting two defined extremities. Basically, tracer cDNA synthesized on mRNA from source A is hybridized to sequences of driver mRNA, which is isolated from a different but usually related source B. The tracer cDNAs that do not become hybridized with driver mRNA represent an enriched population of sequences expressed only in A cells. These are used for constructing an A-cell specific cDNA library.

The SLIC strategy provides DNA molecules with two defined ends. This offers the opportunity to work on cDNA from populations A and B with two different sets of SLIC primers, A (A5′ NV and RA3′ NV) and B (B5′ NV and RB3′ NV; **Fig. 3**). During the PCR amplification of the tracer population B, a pair of biotinylated primers is used. Thus, this population can be captured and pure single-stranded molecules immobilized on magnetic beads. Then, after hybridization with the amplified A population, the unhybridized population corresponds to the A-cell specific sequences and can be used easily to generate substracted libraries or probes. This strategy gives for the first time the opportunity to realize such subtractive libraries with a very small amount of input material.

1.7. Conclusion

The SLIC method is a powerful and unique tool to synthesize cDNA and substractive libraries from a limited number of cells. It is unfortunately extremely difficult to generate full-length libraries. Nevertheless, cloning 5′ or 3′ ends of a cDNA is no longer a limiting step because anchored PCR can be easily performed. In this case, the same ss-cDNA that was used to generate the library can also be used as a matrix to isolate both ends of the incomplete clone.

Name	5' Sequence 3'
B3'NV	AGCATCAGCCAAGACCCAGCGAAGCACCCTCGAGCTACCGCG(T)14
RB3'NV	AGCATCAGCCAAGACCCAGCGAAGCACCCTCGAGCTACCGCG(N)9
B3'_1	AGCATCAGCCAAGACCCAGCGAAGC
B3'_2	AGCGAAGCACCCTCGAGCTACCGCG
B5'NV	CTGACGACGAATTCCTTAGCGATTGGTGAGGGCATTCGGGTC
B5'_1	GACCCGAATGCCCTCACCAATCGCT
B5'_2	CAATCGCTAAGGAATTCGTCGTCAG

Fig. 3. Alternative oligonucleotides used for subtractive cDNA library construction. For the B oligonucleotides the same modifications (5′ phosphate and 3′ amino) as the A primers have to be used.

2. Materials

2.1. Oligonucleotides

1. PCR library primers: For the 5′ end of the ss-cDNA, use RA3′ NV and the related oligonucleotides (A3_1,2,3). For the 3′ end use A5′NV, A5′_1,2,3. All oligonucleotides must bear 5′ and 3′ hydroxyl group. Only A5′NV must have a phosphate group to its 5′ end to allow ligation to the 3′ end of the ss-cDNA. To avoid selfligation of A5′ NV, its 3′ end must be protected with an amino group. Those modifications are performed by any oligonucleotide suppliers. Apply the same rules when using B5′ NV and RB3′ NV. Note that only two related oligonucleotides are available for both ends.
2. Biotinylated screening primers: Ask your oligonucleotide supplier for 5′ biotinylated primers with a seven carbon spacer. Do not use an 11 carbon spacer because this will dramatically decrease the capture yield. If this primer is designed to hybridize in the middle of a molecule, add 4 to 6 random nucleotides at the 5′ end to facilitate the interaction with the streptavidin magnetic beads.

2.2. RNA Extraction (see Note 1)

1. Starting material: Fresh pelleted cells, store at –80°C on collection. Fresh pieces of organs, or tissue frozen in liquid nitrogen and stored at –80°C.
2. PolytronR TP1200 (if organs are used).
3. RNAZol reagent (BIOPROBE).
4. DT40 (Pharmacia, [Piscataway, NJ]; #17-0270-01) Prepare a 5 mg/mL stock solution in ddH_2O. Aliquot and store at –20°C.
5. $CHCl_3$.
6. Isopropanol.
7. 100% Ethanol.
8. 70% Ethanol.

9. ddH_2O.
10. Dynabeads mRNA purification kit (Dynal, Lake Success, NY).

2.3. Synthesis of the ss-cDNA

1. Water bath at 70 and 42°C.
2. Dry-ice powder and ice.
3. 10× FSB buffer: 1 *M* Tris-HCl, pH 8.4 (at 42°C), 1.2 *M* KCl, 100 m*M* $MgCl_2$. Aliquot and store at –20°C.
4. DTT (100 m*M*).
5. Acetylated bovine serum albumin, 5 mg/mL (RNase-free; Life Technologies/BRL, Gaithersburg, MD).
6. RNasin, 36 U/μL (Promega Biotech, Madison, WI).
7. dNTPs, 10 m*M* each (Use lithium-free dNTPs).
8. Sodium pyrophosphate (20 m*M*; PPi).
9. [α^{32}P] dATP, 3000 Ci/mmol (Amersham, Arlington Heights, IL).
10. AMV reverse transcriptase, 10 U/μL (Promega Biotech, Madison, WI).
11. RA3′ NV (50 ng/μL).

2.4. Synthesis Yield Determination

1. DE81 Whatman paper.
2. Na_2HPO_4 (0.5 *M*).
3. 100% Ethanol.
4. Aqueous scintillation cocktail.

2.5. Removal of Primers

1. Prep-A-Gene DNA purification Kit (Bio-Rad [Richmond, CA]; #732-6010). The silica matrix used in this kit does not bind RNA or small DNA molecules (cut off around 100 nucleotides) under oxidizing conditions.
2. Water baths at 90 and 65°C.

2.6. Ligation of the ss-cDNA to the Modified Oligonucleotide

Modified oligonucleotide: 5′ phosphate and 3′ NH_2 A5′NV (or B5′NV).

2.7. PCR Amplification

PCR reagents, including primers.

2.8. Direct Screening of the PCR Library with Biotinylated Oligonucleotides

1. Biotinylated primer (*see* **Subheading 2.1.2.**).
2. Streptavidin Dynabeads (10 mg/mL Dynabeads M-280 Streptavidin, DYNAL, Lake Success, NY).
3. Magnetic concentrator (Dynal MPC).
4. Rotating wheel.
5. 20× SSPE solution: 200 m*M* NaH_2PO_4, pH 7.4, 3.6 *M* NaCl, 20 m*M* EDTA *(5)*.
6. 20× SSC: 300 m*M* sodium citrate, pH 7.0, 3 *M* NaCl *(5)*.
7. 50× Denhardt's solution: 1% each of BSA, Ficoll 400, and PVP *(5)*.
8. 20% SDS.
9. Sonicated salmon sperm DNA.

10. 10× PCR buffer (*see* **Chapter 1**).
11. 42°C incubator.

2.9. Blunt-End Cloning of PCR Products

1. Unpurified PCR products.
2. T4 DNA polymerase (4 U/µL, Amersham).
3. Agarose gel (choose agarose concentration according to the PCR product length).
4. QIAEX II purification kit (Qiagen, Inc., Chatsworth, CA).
5. ATP (3 and 8 m*M*).
6. T4 polynucleotide kinase 5 U/µL (Amersham).
7. 10× T4 polynucleotide kinase buffer (Amersham).
8. Dephosphorylated SmaI pUC19 vector (Appligene).
9. T4 DNA ligase, 4 U/µL.
10. Electrocompetent XLI blue cells.
11. *Escherichia coli* electroporation equipment.

2.10. Direct PCR on Colonies

1. Inoculating needles.
2. PCR reagent, including primers. With pUC vectors, use M13 universal sequencing primer and M13 reverse sequence primer.

3. Methods

3.1. RNA Extraction

All manipulations must be performed in an RNase-free environment and with PCR anticontamination material. In our hands, the best extraction yield for low amount of material is obtained with the RNAZol kit. When working with tissues, use a polytronR TP1200 to homogenize in the RNAzol solution.

Follow the supplier's instructions with two important modifications.

1. Add 1 µL of DT40 (5 mg/mL) in the RNAZol solution prior to the homogenization step. This will increase the extraction yield.
2. At the end, resuspend the pellet in 20 µL of ddH_2O and store at –80°C.

If the extraction of $polyA^+$ RNA is required, use the Dynabeads mRNA purification kit (Dynal); do not use the Dynabeads mRNA DIRECT kit. Elute from the magnetic beads with 20 µL of ddH_2O instead of elution buffer and store at –80°C.

3.2. Synthesis of the ss-cDNA (see Note 2)

The final volume for the reverse transcription is 50 µL.

3.2.1. AMV RTase Preincubation Mix

1. Add, on ice, in a sequential manner the following reagents: 1.5 µL of H_2O, 3 µL of 10× FSB, 2.5 µL of 0.1 *M* DTT, 3 µL of 10 m*M* each dNTP (lithium free), 1 µL of 5 mg/mL BSA, 5 µL [$\alpha^{32}P$] dATP (100 U/µL) (3000 Ci/mmol), 1 µL of RNasin (36 U/µL), 10 µL of 20 m*M* PPi, and 1 µL of AMV RTase.
2. Preincubate on ice for 30 min.

3.2.2. RNA Mix Preparation

Prepare this mix during the preincubation of the AMV reverse transcriptase.

1. Dilute the RNA in 17 µL of ddH_2O (*see* **Note 3**).
2. Add 1 µL of 50 ng/µL RA3′NV (or any anchored random primer).
3. Add 2 µL of 10× FSB.
4. Heat the tubes at 70°C for 15 min.
5. Spin and freeze in dry-ice powder.
6. Let it thaw on ice.

3.2.3. Reverse Transcription

1. Assemble the preincubation mix and the RNA mix and incubate at 42°C for 20 min to 1 h.
2. To stop the reaction, add 1 µL of 0.5 *M* EDTA. To determine the optimal incubation time, perform five cDNA synthesis in 10 µL final reaction volume and stop the incubation every 10 min after 20 min initial reaction time. Load 5 to 10 µL on an alkaline agarose gel *(5)* and measure the average length. The optimal reaction time will be the one that gives an average size of 0.8 to 1 kbp.

3.3. Synthesis Yield Determination

1. Take 1 µL of the cDNA and dilute it to 10 µL in ddH_2O.
2. Spot 5 µL of the dilution on two pieces of DE81 Whatman filter.
3. Wash one filter only in 0.5 *M* Na_2HPO_4 for 10 min.
4. Repeat **step 3** two more times and dry this filter in 100% ethanol.
5. Dry both filters in air for 15 min.
6. Add 10 mL of aqueous scintillation cocktail to the washed and unwashed filters.
7. Count the ^{32}P activity. The washed filter activity corresponds to the incorporated activity (I), whereas the unwashed one corresponds to the total activity (T).
8. The synthesized ss-cDNA mass (M) is given by the formula:

$$M = (I/T) \times (\text{total dATP mass during reaction}) \times 4 \tag{1}$$

In the conditions used, M(ng) = (I/T) × 792. The overall yield should be between 25 and 30% of the starting mass of RNAs.

3.4. Removal of Primers

Use Prep-A-Gene DNA purification kit. Follow the supplier's instructions with the following important modifications.

1. To remove primer after the ss-cDNA synthesis, heat for 5 min at 90 to 95°C to denature the RNA:DNA heteroduplexes. Add 150 µL of binding buffer, mix, then add 5 µL of resuspended matrix. Mix well. Incubate for 10 min at room temperature.
2. To remove primer after ligation or PCR, do not perform this denaturation step.
3. For purification do not use a starting volume smaller than 50 µL.
4. Carefully remove all the wash buffer after the last wash. Traces of ethanol can be removed by drying the tubes for 3 min in a SpeedVac or equivalent rotary vacuum desiccator.
5. Elute with 5 to 10 µL of ddH_2O for 5 min at 65°C, then spin for 30 s and collect supernatant.
6. Alternative procedures can be followed.

3.5. Ligation of the ss-cDNAs to the Modified Oligonucleotide (see Note 4)

1. Remove primer before the PCR amplification step.

3.6. PCR Amplification (see Notes 5 and 6)

The general conditions used for both PCR amplifications are: 50 µL of reaction volume, hot start, and touchdown PCR.

1. First PCR: Use A5′_1 and A3′_1 and half of the purified ligation mixture.
2. Second PCR: Use A5′_2 and A3′_2 and one tenth of the purified first PCR.
3. For both PCRs:
   a. Use as final concentration 200 µ*M* of dNTPs, 0.8 µ*M* of each primer, and 1.5 m*M* of $MgCl_2$.
   b. Hot start: Add 0.5 µL of Taq DNA polymerase (5 U/µL) below the mineral oil when the reaction mixture reaches 80°C.
   c. Perform the following touch down PCR cycles: denaturation 93°C for 3 min; two cycles of 94°C, 30 s/70°C, 45 s/72°C, 1.5 min; two cycles of 94°C, 30 s/69°C, 45 s/72°C, 1.5 min; two cycles of 94°C, 30 s/68°C, 45 s/72°C, 1.5 min; two cycles of 94°C, 30 s/67°C, 45 s/72°C, 1.5 min; two cycles of 94°C, 30 s/66°C, 45 s/72°C, 1.5 min; 25 cycles of 94°C, 30 s/65°C, 45 s/72°C, 1.5 min; and cool down to 4°C.
   d. Remove primer after each PCR.

   After the second nested PCR the amount of ds-cDNA is about 2 to 5 µg.

3.7. Direct Sceening of the PCR Library with Biotinylated Oligonucleotides

3.7.1. Hybridization with the Biotinylated Oligonucleotide (see Note 7)

1. Use 500 ng of purified ds-cDNA. The volume should not exceed 8 µL.
2. Add 2 µL of the 100 ng/µL biotinylated oligonucleotide.
3. Adjust volume to 10 µL.
4. Heat denature for 5 min at 95°C.
5. Immediately add 100 µL of hybridization buffer: 5× SSPE, 5× Denhardt's solution, 1% SDS.
6. Incubate overnight at 42°C.

3.7.2. Separation of Probe-cDNA Hybrids

After the hybridization reaction, probe-cDNA hybrids are separated from unhybridized DNA using Streptavidin-coated magnetic beads.

1. Prehybridize Dynabeads with salmon sperm DNA by washing 20 µL of 10 mg/mL Dynabeads twice with 50 µL of hybridization buffer containing 250 µg/mL salmon sperm DNA. Incubate for 2 h at room temperature on a rotating wheel.
2. Mix the Dynabeads with the probe-cDNA solution. Incubate 15 to 30 min at room temperature on a rotating wheel.
3. The hybrids captured by the beads are washed twice with 1× SSC, 1% SDS, then twice with 0.1× SSC, 1% SDS. Washes are performed at 42°C for 20 min each.

4. Wash twice with 1× PCR buffer, 5% SDS, for 5 min at room temperature.
5. Wash with 1× PCR buffer until SDS is completely removed. To do this, change the microtube after every wash.

3.7.3. PCR Amplification of the Captured cDNA

1. Transfer one fourth of the beads with the captured cDNA into a PCR tube. Make sure that all traces of SDS are removed.
2. Perform PCR amplification with A5′_3 and A3′_3. Use the same protocol as described in **Subheading 3.6.**

3.8. Blunt-End Cloning of PCR Products

For blunt-end cloning, the 3′ overhanging extremities of the PCR product are removed with T4 DNA Polymerase (3′–5′ exonuclease activity). Oligonucleotides usually have 5′ hydroxyl ends. To allow ligation of the PCR product those extremities have to be phosphorylated by T4 Polynucleotide kinase (T4 PNK).

1. At the end of the amplification reaction, add to the PCR mixture 1 µL of T4 DNA polymerase (4 U/µL). Incubate for 20 min at 16°C. Do not allow the temperature to rise above 16°C.
2. Load the PCR product on a preparative agarose gel.
3. Cut the desired bands and purify the DNA with the QIAEX II purification kit. Follow the supplier's recommendations.
4. Elute DNA from the silica matrix with 10 µL of ddH_2O.
5. Add 1.5 µL of 10× T4 PNK buffer, 1 µL of 3 m*M* ATP, 1.5 µL of ddH_2O, and 1 µL of T4 PNK (5 U/µL).
6. Incubate at 37°C for 30 min.
7. Heat inactivate the enzyme at 75°C for 20 min.
8. The PCR product is ready for ligation. Use the same buffer as the phosphorylation reaction, a dephosphorylated blunt-end vector (e.g., pUC19 SmaI), and a final concentration of ATP of 0.8 m*M*.
9. Transform by electroporation and plate on the appropriate selection medium.

3.9. Direct PCR on Colonies (see Note 8)

1. Pick a colony with an inoculating needle.
2. Touch the bottom of a 0.5-mL microtube (or a well in a microtiter plate) with the needle.
3. Inoculate with the same needle, 3 mL of liquid bacterial growth medium in a 15-mL tube, and incubate at 37°C for 18 h.
4. Repeat **steps 1** to **3** for all the colonies to be analyzed.
5. Prepare the PCR mixture on ice as follows. The volumes given are sufficient for one reaction: 2.5 µL of 10× PCR buffer, 2 µL of 2.5 m*M* dNTP, 2 µL of 50 m*M* $MgCl_2$, 2 µL of primer 1 (50 ng/µL), 1 µL of primer 2 (50 ng/µL), 0.1 U of Taq DNA polymerase, and ddH_2O to 25 µL.
6. Distribute the PCR mixture into every tube on ice and add 100 µL of mineral oil.
7. Place the tubes or the microtiter plate in the thermal cycler and run the following program: 3 min denaturation at 93°C; 35 cycles of 94°C for 30 s, 55°C for 45 s, and 72°C for 1 min/kb.
8. Analyze the PCR products on an agarose gel.
9. Prepare plasmid DNA of the positive clones from the cultures (prepared in **step 3**). Use this plasmid DNA for sequencing.

4. Notes

1. All the material used in this manipulation must be very clean and at least sterilized. Wear gloves throughout the manipulation to avoid RNase contamination. The same precautions should be followed in the synthesis of the ss-cDNA. These precautions represent the lower level of protection against RNase and it is advisable to read **ref. *6*** carefully. Because PCR has to be performed later on this material, use anticontamination tips and aliquot every solution.
2. During this manipulation, prepare controls that will be used in the ligation and PCR experiments. Prepare samples without AMV RTase and samples without RNA. This will lead to three different controls.
3. We have successfully used as little as 10 ng of polyA$^+$ RNA. Do not exceed 1 µg of total RNA.
4. Prepare nonligated samples composed of the same mixture as described in **Subheading 3.5.** without the T4 RNA ligase. Include each control of the cDNA synthesis.
5. After removal of the primers, perform PCR amplification. Do not forget to include a PCR control without DNA for both amplifications.
6. To analyze each PCR amplification, load 5 µL of (one tenth) of the PCR product on an agarose gel. After ethidium bromide staining, a signal could be observed after the first PCR but it is generally obtained after the second nested PCR.
7. We have used degenerated primers to screen the PCR library. The amount of primer and Dynabeads should be at least five times above the amount described in **Subheading 3.7.**
8. We have developped a direct PCR analysis to determine the sizes of inserted fragments and to rule out false positive recombinant clones.

References

1. Boularand, S., Darmon, M. C., and Mallet, J. (1995) The human tryptophan hydroxylase gene: An unusual complexity in the 5′ untranslated region. *J. Biol. Chem.* **270,** 3748–3756.
2. Barnes, W. M. (1994) PCR amplification of up to 35-kb DNA with high fidelity and high yield from λ bacteriophage templates. *Proc. Natl. Acad. Sci. USA* **91,** 2216–2220.
3. Don, R. H., Cox, P. T., Wainwright, B. J., Baker, K., and Mattick, J. S. (1991) "Touch down" PCR to circumvent spurious priming during gene amplification. *Nucleic Acids Res.* **19,** 4008.
4. Abe, K. (1992) Rapid isolation of desired sequences from lone linker PCR amplified cDNA mixtures: Application to identification and recovery of expressed sequences in cloned genomic DNA. *Mammalian Genome* **2,** 252–259.
5. Sambrook, J., Fritsch, E. F., and Maniatis, T. (1989) *Molecular Cloning: A Laboratory Manual,* 2nd ed., Cold Spring Harbor Laboratory, Cold Spring Harbor, NY [SSPE, p. B.13, SSC, p. B.13, Denhart's solution, p. B.15, Alkaline gel electrophoresis, p. B.23].
6. Blumberg, D. D. (1987) Creating a ribonuclease free environment. *Methods Enzymol.* **152,** 20–24.

69

Creation of Chimeric Junctions, Deletions, and Insertions by PCR

Genevieve Pont-Kingdon

1. Introduction

Recombinant polymerase chain reaction (PCR) *(1)* is the method of choice if one wants to modify a cloned DNA. It is a versatile technique that allows operations as different as creation of deletions, addition of small insertions, site-directed mutagenesis, and construction of chimeric molecules at any chosen location in the molecule of interest (*see* **Note 1**). This chapter describes in detail a simplification of the original recombinant PCR method. This fast and efficient method has been successful in fusing two different sequences with precision *(2–4)*. It can also be used to create deletions or insert small fragments of DNA.

The method (*see* **Fig. 1**) relies on a "chimeric primer" (C) and two outside primers (A and B). The final product can be obtained in one or two rounds of PCR. The figure illustrates the construction of a chimeric molecule in which two different templates are joined. The creation of a deletion, or the introduction of a small insertion within a given template, would follow the same pathway (*see* **Note 1**). In all cases, the new junction is designed in the chimeric primer; the 3′ half of the chimeric primer pairs with one of the templates (or one side of the deletion/insertion point), and its 5′ half has homology with the other template (or the other side of the deletion/insertion point).

Both templates and the three primers (A–C) are placed in a reaction tube (**step 1**) and one PCR is performed. During the first cycles, only the primers A and C can prime exponential amplification (**step 2**). This amplification reaction gives rise to an "intermediate fragment" (**step 3**), that can itself act as a primer. One of its 3′-ending strands anneals to the second template and is extended (**step 4**). This extension provides a template (**step 5**) for exponential amplification of the final product using the primers A and B (**step 6**). To limit the amplification of the intermediate fragment to the first cycles of PCR, the chimeric primer is used at a lower concentration than the two outside primers *(3,4)*.

2. Materials

1. Primers (*see* **Subheading 3.1.**).
2. DNA templates, linearized at a site outside the region to be amplified.

From: *Methods in Molecular Biology, Vol. 226: PCR Protocols, Second Edition*
Edited by: J. M. S. Bartlett and D. Stirling © Humana Press Inc., Totowa, NJ

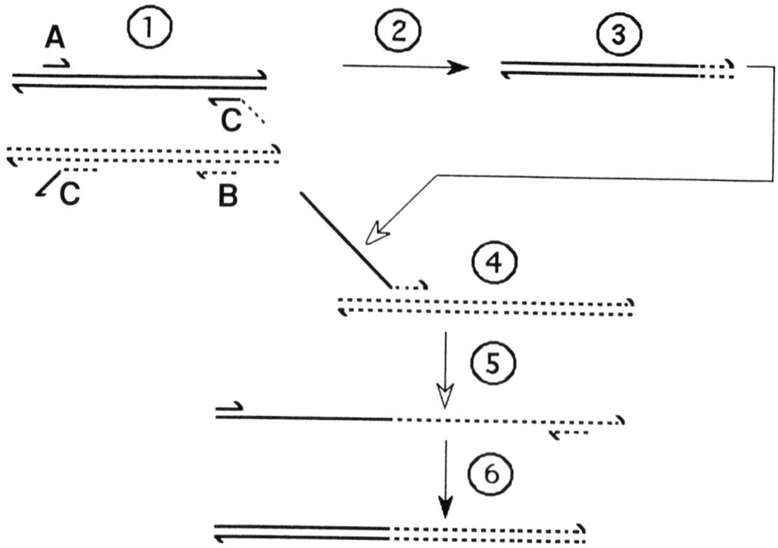

Fig. 1. Construction of a chimeric product. See text for explanation of **steps 1–6**. The dsDNA templates are the plain and dotted double lines. The outside primers are short, plain **(A)**, or dotted **(B)** single lines. The drawing of the chimeric primer (half plain, half dotted) reflects its homologies to the templates. Half arrowheads indicate 3′ ends. Closed arrows indicate amplification **steps** (**2** and **6**).

3. GeneAmp PCR Core Reagent Kit (Perkin–Elmer, Foster City, CA) containing: AmpliTaq DNA polymerase (5 U/μL), Gene Amp dNTPs (10 m*M* solutions of dATP, dGTP, dTTP, and dCTP), GeneAMP 10× PCR buffer II (100 m*M* Tris-HCl, pH 8.3; 500 m*M* KCl), and 25 m*M* $MgCl_2$ solution.
4. Mineral oil.
5. Restriction endonucleases and buffers.
6. Agarose gel electrophoresis reagents and equipment.
7. Phenol:$CHCl_3$:isoamyl alcohol (25:24:1, v:v:v).
8. Ammonium acetate (7.5 *M*).
9. 100% Ethanol.
10. 70% Ethanol.

3. Methods

3.1. Design of Chimeric Primer

The chimeric primer is crucial because it contains the new junction and because sequences on each side of the junction serve as primers in different **steps** (i.e., **2** and **4** in **Fig. 1**) of the reaction. To allow priming by the nucleotides found on each side of the junction, the new junction should be placed in the middle of an oligonucleotide of sufficient length. Otherwise, the design of a chimeric primer should follow the classical rules of primer-design ***(5)***. Our chimeric primer was a 34-mer, with 17 bases homologous to one template and 17 bases homologous to the other. Some slightly longer chimeric primers (36- and 40-mer) have been used ***(3,4)***. *See* **Note 2** for more information on the design of the chimeric primer.

3.2. Design of Outside Primers

Fewer constraints apply to the design of the two outside primers, and therefore following the general rules of primer design should be adequate. In principle, the outside primers can be kilobase pairs away (within the limit of PCR feasibility) from the new junction. However, because their location determines the size of the product that will be cloned, the choice of their position is important. For several reasons, it is advantageous to plan the cloning of a fragment of few hundred nucleotides instead of a longer one: First, the smaller the fragment is, the less chance there is to find PCR-induced mutations in the final clone. Second, a smaller piece of DNA has to be sequenced to verify the integrity of the newly cloned DNA. This can be achieved by using primers that anneal close to the junction and have "built in" restriction sites at their 5′ ends. Another approach is to choose primers that anneal further from the junction and clone with restriction sites that closely surround the chimeric junction in the final product. The choice between these two possibilities depends on the cloning strategy and the availability of cloning sites in the amplification product. If the chimeric fragment has to be cloned into a new vector, restriction sites unique in both vector and chimeric fragment can be engineered at the 5′ ends of the two outside primers. If the chimeric junction has to be replaced by cloning in one of the original templates, restriction sites that exist in this template and in the chimeric junction have to be used.

The size of the outside primers can be different than the size of the chimeric primer. We have been successful with a 34-mer chimeric primer, and two outside primers of 23 and 20 nucleotides, respectively. If the size of the two outside primers is very different than half the size of the chimeric primer, series of cycles with different annealing temperatures can be performed (*see* **Note 3**).

3.3. Procedure

1. Assemble the components in one 50-µL reaction containing: both templates (10 fmol each), both outside primers (25 pmol each), chimeric primer (1 pmol), 50 µ*M* each dNTP (*see* **Note 4**), 1× AmpliTaq DNA polymerase buffer, 1.5 m*M* $MgCl_2$ (*see* **Note 5**), and 2.5 U of AmpliTaq DNA polymerase.
2. Top PCR mix with 50 µL of mineral oil.
3. Perform PCR as follows (*see* **Note 3**):
   a. Three to five initial cycles to allow initiation and amplification (**steps 2–4** in **Fig. 1**) using the chimeric primer: 95°C for 30 s, T1 (*see* **Note 3**) for 30 s, and 72°C for 1 min for each kilobase of intermediate fragment.
   b. Twenty to thirty cycles to amplify the final chimeric product: 95°C for 30 s, T2 (*see* **Note 3**) for 30 s and 72°C for 1 min for each kilobase of chimeric product.
4. Directly analyze 10 µL of the amplified DNA by standard procedures (restriction enzyme digests and electrophoresis, *see* **Note 6**).
5. To clone the new junction, clean the PCR product by extraction with an equal volume of phenol:chloroform:isoamyl alcohol (25:24:1).
6. Transfer the aqueous phase to a sterile microcentrifuge tube. Remove the excess dNTP (it can inhibit T4 ligase) by ethanol precipitation. Add 1/2 vol of 7.5 *M* ammonium acetate and 2.5 vol of 100% ethanol.
7. Mix well and incubate at room temperature for 10 min.
8. Spin in a microfuge at maximum speed for 5 min.

9. Invert tube and allow to drain.
10. Wash pellet with 70% ethanol.
11. Resuspend the DNA in 20–50 μL of TE. Quantify the DNA by UV spectrophotometry and use for cloning.

4. Notes

1. The applications of this method are diverse; it allows the creation of:
   a. Chimeric molecules: This method is well suited to cases in which the two molecules to be joined are unrelated. In the case where the two templates are fused in a region of homology, another PCR technique *(6)* might be preferred.
   b. Deletions: A set of different deletions can be easily obtained from the same template by using a set of different chimeric primers and only one set of outside primers.
   c. Insertions: The size of potential insertion using this technique is limited to the size of the chimeric primer. A restriction site sequence, sandwiched into a chimeric primer, could be introduced at will into any DNA.
2. As in the Megaprimer method *(7)*, the technique described here uses a PCR product as a primer. It has been mentioned for the Megaprimer method *(8)* that mutations can be found in the final product because of the tendency of *Taq* polymerase to add nontemplated nucleotides at the 3′ end of newly synthesized DNA strands. The frequency of these mutations should be low since 3′-ending DNA strands carrying nontemplated nucleotides should not prime well for the synthesis of the final product. Although we did not observe such mutations in the two final clones that we obtained and sequenced, this phenomenon could apply here, and we encourage the reader to refer to **Note 1** in **ref.** *8* for a complete discussion.
3. In this technique, each half of the chimeric primer must anneal with its target. We have limited the size of our chimeric primer to 34 nucleotides, giving us 17 nucleotides for each half. Because our outside primers are 20 and 23 nucleotides in length, we felt that it was necessary to plan a first set of cycles with a lower annealing temperature to allow the stable annealing of each half of the chimeric primer in the **steps 2** and **4**. This precaution is not necessary if all the sequences with a "priming" role (each half of the chimeric primer and both outside primers) are close in length and in G+C content.

   The temperature T1 is the approximate "annealing temperature" ([number of G + C × 4°C] + [number of A + T × 2°C] – 10°C) of the less stable half of the chimeric primer. The temperature T2 is the approximate annealing temperature of the less stable of the two outside primers.
4. To limit the number of PCR-induced mutations in the final chimeric product, a low concentration of each dNTP (50 μ*M*) is used.
5. The optimal $MgCl_2$ concentration can vary among different pairs of primers. It is wise to define the best $MgCl_2$ concentration in a test experiment *(9)*. In fact, it is possible that the two consecutive PCRs that occur in the tube have incompatible requirements for $MgCl_2$. If this is the case, the chimeric product can be obtained by a two-step method *(2)*. In this alternative, the intermediate fragment obtained during the first reaction is extracted with phenol:chloroform:isoamyl alcohol and purified from excess primers by precipitation with isopropanol from 2 *M* NH_4OAc. The purified DNA is then used as a primer in a second reaction that contains the second template and the two outside primers.
6. It is often stated that the oil that tops the PCR has to be removed to properly load the DNA into the well of an electrophoresis gel. We found that this step is not needed if the pipet tip is cleaned with tissue paper (Kimwipe) just after it has been filled with a DNA sample. In fact we found that the oil left in the tube allows for longer conservation of the sample at 4°C.

Acknowledgment

I want to thank Janet E. Lindsley and Dana Carroll for helpful comments.

References

1. Higuchi, R. (1989) Using PCR to engineer DNA, in *PCR Technology: Principles and Applications for DNA Amplification*, Stockton, New York, pp. 61–70.
2. Pont-Kingdon, G. (1994) Construction of chimeric molecules by a two-step recombinant PCR method. *BioTechniques* **16,** 1010–1011.
3. Cao, Y. (1990) Direct cloning of a chimeric gene fused by the polymerase chain reaction. *Technique* **2,** 109–111.
4. Yon, J. and Fried, M. (1989) Precise gene fusion by PCR. *Nucleic Acids Res.* **17,** 4895.
5. Sharrocks, A. D. (1994) *The design of primers for PCR, in PCR Technology: Current Innovations* (Griffin, G. G. and Griffin, A. M., eds.), CRC, Boca Raton, FL, pp. 5–11.
6. Klug, J., Wolf, M., and Beato, M. (1991) Creating chimeric molecules by PCR directed homologous DNA recombination. *Nucleic Acids Res.* **19,** 2793.
7. Sarkar, G. and Sommer, S. S. (1990) The "megaprimer" method of site-directed mutagenesis. *Biotechniques* **8,** 404–407.
8. Barik, S. (1993) Site-directed mutagenesis by double polymerase chain reaction, in *Methods in Molecular Biology*, vol. 15: PCR Protocols: Current Methods and Applications (White, B. A., ed.), Humana, Totowa, NJ, pp. 277–286.
9. Saiki, R. K. (1989) The design and optimization of the PCR, in *PCR Technology: Principles and Applications for DNA Amplification* (Erlich, H. A., ed.), Stockton, New York, pp. 7–16.

70

Recombination and Site-Directed Mutagenesis Using Recombination PCR

Douglas H. Jones and Stanley C. Winistorfer

1. Introduction

The polymerase chain reaction (PCR) *(1)* provides a rapid means for the recombination and site-directed mutagenesis of DNA *(2)*. DNA modification can occur during PCR because the primers are incorporated into the ends of the PCR product. The simplest PCR-based method for site-directed mutagenesis and DNA recombination is recombination PCR.

Recombination PCR uses in vivo recombination in *Escherichia coli* to generate site-directed mutants and recombinant constructs *(3,4)*. In the recombination PCR method, PCR adds homologous ends to DNA. These homologous ends mediate recombination between these linear PCR products in *E. coli*, resulting in the formation of DNA joints in vivo. If two PCR products have homologous ends that can recombine to form a circle, and if this circle constitutes a selectable plasmid, *E. coli* can be readily transformed by the linear PCR products. Recombination PCR has almost no steps apart from PCR amplification and transformation of *E. coli*, and this method works well in Rec A minus *E. coli* strains used routinely in cloning. Since the introduction of this method in 1991, it has been used by numerous investigators *(5–9)*. One example of DNA recombination using recombination PCR is illustrated in **Fig. 1**, which shows a protocol for amplifying a portion of a donor plasmid and inserting it in a recipient plasmid at a defined position and orientation. The donor plasmid is shown on the left side and the recipient plasmid is on the right side. The steps corresponding to this figure are briefly outlined below:

1. The DNA segment that is to be inserted into the recipient plasmid is amplified from the donor plasmid using primers 1 and 2. In a separate PCR amplification, the recipient plasmid is amplified with primers 3 and 4. The 5′ regions of primers 1 and 2 that do not anneal to the donor plasmid are complementary to primers 4 and 3, respectively (or 3 and 4, depending on the orientation of the insert desired in the recombinant construct). Frequently, a plasmid template can be linearized outside the region to be amplified by restriction endonuclease digestion before PCR amplification. When this can be done, the PCR product does not need to be purified, because linearized plasmids transform *E. coli* inefficiently. If a plasmid template cannot be linearized by restriction endonuclease digestion outside the region to be amplified prior to PCR amplification, the PCR product must

From: *Methods in Molecular Biology, Vol. 226: PCR Protocols, Second Edition*
Edited by: J. M. S. Bartlett and D. Stirling © Humana Press Inc., Totowa, NJ

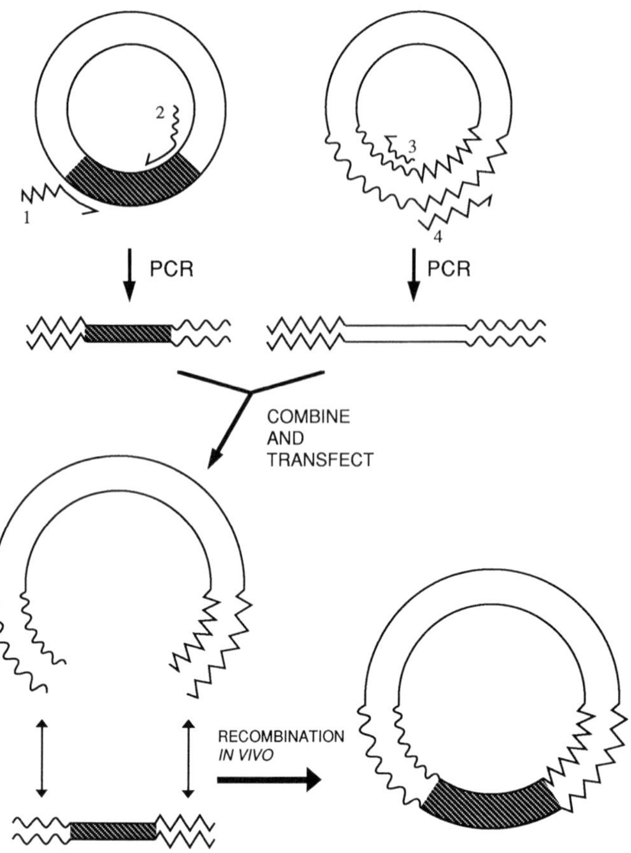

Fig. 1. Diagram illustrating DNA recombination using recombination PCR. The primers are numbered hemiarrows. The insert is the cross-hatched region. Smooth circles represent the DNA strands of the donor plasmid. Circles with wavy and jagged portions represent DNA strands of the recipient plasmid. Reprinted by permission from *Biotechniques* **10,** 62–66.

be removed from the plasmid before transformation to prevent background transformants arising from the supercoiled plasmid template. PCR product purification is accomplished either by agarose gel purification followed by glass bead extraction or by adding the restriction endonuclease *Dpn*I to the PCR mixture. *Dpn*I is a restriction endonuclease that digests methylated GATC sites. These sites are methylated in the plasmid by strains of *E. coli* used routinely in cloning (by dam methylase), but are not methylated in the PCR products, permitting *Dpn*I to digest the plasmid without cutting the PCR product ***(10)***.

2. The two PCR products are combined and used to transform MAX efficiency competent *E. coli* (BRL, Life Technologies, Gaithersburg, MD). If each plasmid template is restriction endonuclease digested outside the region to be amplified prior to PCR amplification, the two crude PCR products can simply be combined, and the resulting mixture used to transform *E. coli*.

In a simple variation of this recombination PCR strategy, the inserted segment can be an unmodified PCR product. In that case, primers 3 and 4 have 5′ ends that are homologous to the ends of the PCR fragment to be inserted, and the recipient plasmid is linearized by restriction endonuclease digestion prior to PCR amplification. We routinely use this approach to clone any PCR product ***(4)***.

In recombination PCR, the sum goal of the two amplifications is to yield two PCR products where each end of one product is homologous to a distinct end of the other PCR product. Because the amplifying primer sequences are incorporated into the ends of a PCR product, so long as primers 1 and 2 contain regions that are complementary to regions of primers 3 and 4 (or 4 and 3), the PCR products will contain ends that are homologous to each other, and these primer-determined DNA ends do not need to be determined by the original donor or recipient templates. The only requirement of this recombination PCR strategy is that primers 1 and 2 must have regions of complementarity to primers 3 and 4. Therefore, recombination PCR can be used not only to generate recombinant constructs, such as gene chimeras, but also for the site-directed mutagenesis of two distal sites concurrently (**Fig. 2**) or for the rapid site-directed mutagenesis of single sites (**Fig. 3**) *(11)*. In the point mutagenesis protocol illustrated in **Fig. 3**, the plasmid is linearized by restriction endonuclease digestion before each PCR amplification. In each of the two amplifications, the mutating primers (primers 1 and 3) mutate the identical base pair so that the mutated ends of each product are homologous to each other and the nonmutating primers (primers 2 and 4) are also designed to produce ends that are homologous to each other. Both unpurified PCR products are combined to transform *E. coli*, generating clones with the mutation of interest.

2. Materials

1. *Taq* DNA polymerase (AmpliTaq 5 U/mL; Perkin–Elmer, Norwalk CT) (*see* **Note1**).
2. 10× PCR buffer II: 500 m*M* KCl, 100 m*M* Tris-HCl, pH 8.3.
3. $MgCl_2$ solution (25 m*M*).
4. Stocks (10 m*M*) of each dATP, dCTP, dTTP, and dGTP, neutralized to pH 7.0 with NaOH.
5. Restriction endonucleases (New England BioLabs, Beverly, MA).
6. PCR primers. In **Fig. 1**, PCR amplification with primers 1 and 2 results in a product with 24 to 30 bp of homology with the products of primers 3 and 4. For PCR primers that introduce mutations, *see* **Note 2**.
7. Agarose.
8. Ethidium bromide.
9. TAE buffer: 40 m*M* Tris-acetate, 2 m*M* EDTA, pH 8.5 *(12)*.
10. Geneclean (Bio 101, La Jolla, CA).
11. TE buffer: 10 m*M* Tris-HCl, pH 8.0, 1 m*M* EDTA.
12. MAX Efficiency DH5α Competent *E. coli* (BRL, Life Technologies). Once a tube is thawed it should not be reused (*see* **Note 3**).
13. SOC Media *(13)*.
14. LB plates with 100 μg/mL ampicillin *(14)*.
15. Luria-Bertani medium (LB broth) *(15)*.

3. Methods

1. Linearize the plasmid template by restriction endonuclease digestion outside the region to be amplified, if possible. The plasmid digest does not need to be purified prior to its use as a PCR template (*see* **Notes 4** and **5**).
2. Assemble a PCR in a total volume of 50 μL containing the following: 2 ng of plasmid template, 25 pmol of each primer, 200 μ*M* each dNTP, 1X PCR buffer, 2.5 m*M* $MgCl_2$, and 1.25 U DNA *Taq* DNA polymerase. (*see* **Note 6**).
3. Perform PCR amplification using the following parameters (*see* **Note 6**): 94°C for 1 min (initial denaturation), 94°C for 30 s (denaturation), 50°C for 30 s (anneal), 72°C for 1 min/kb of PCR product (extension), 14–20 amplification cycles, and 72°C for 7 min (final extension step).

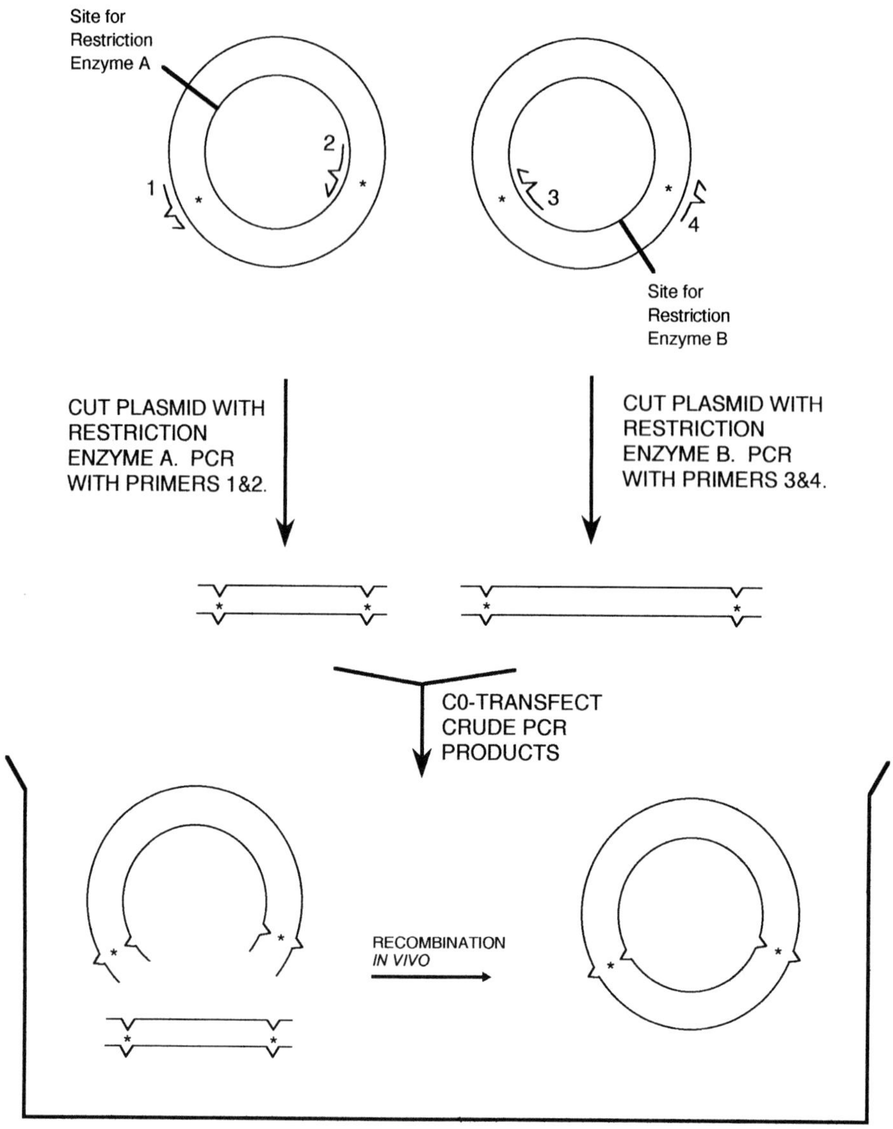

Fig. 2. Diagram illustrating site-directed mutagenesis of two distal sites using recombination PCR. The primers are numbered hemiarrows. Asterisks designate the mutagenesis sites. There is no purification of the PCR products. Notches designate point mismatches in the primers and resulting mutations in the PCR products. Reprinted by permission from *Technique* **2,** 273–278.

4. Visualize the PCR product on an agarose minigel. If 5 µL of the PCR product can be clearly seen following ethidium bromide staining (>15 ng/5 µL), there is enough product.
5. Withdraw 2.5 µL from each PCR tube (typically 10–60 ng/2.5 µL) and then combine the two samples. If the PCR template is linearized by restriction endonuclease digestion outside the region to be amplified, no purification of PCR products is necessary (*see* **Note 5**).
6. Transform MAX efficiency competent *E. coli* (BRL) with the 5-µL sample containing the two PCR products. Maintaining an even molar ratio of one product to another is not necessary. Transformation is carried out following the manufacturer's instructions with the following modifications:

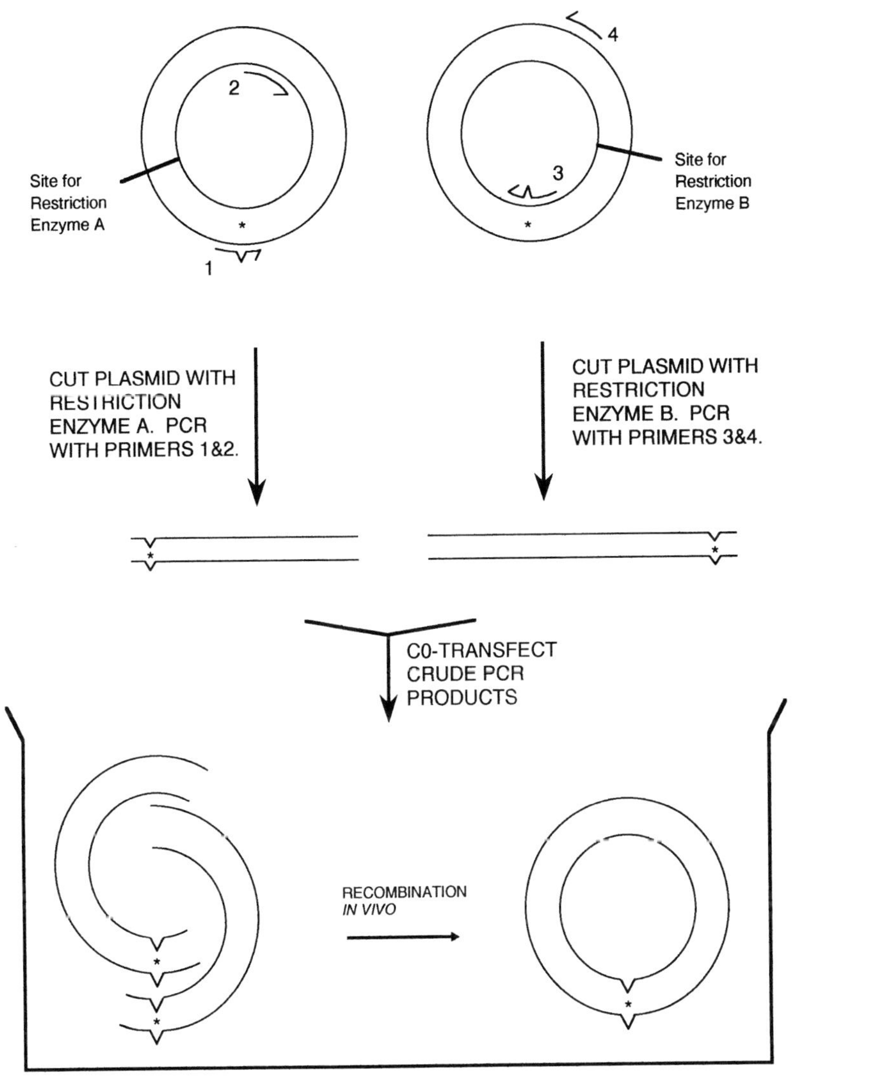

Fig. 3. Diagram illustrating site-directed mutagenesis of a single site using recombination PCR with 4 primers. The primers are numbered hemiarrows. The asterisk designates the mutagenesis site. Primer 2 is complementary to primer 4. Restriction endonuclease sites A and B bracket the insert. Notches designate point mismatches in the primers and resulting mutations in the PCR products. There is no purification of the PCR products. For each additional single site-directed mutagenesis reaction, only a new primer 1 and 3 need to be synthesized, and the same cut templates can be used. Reprinted by permission from *Technique* **2,** 273–278.

a. Use 50 µL of *E. coli* for each sample transformed, because this is effective and less expensive than the 100 µL recommended.
b. After incubation at 37°C in a shaker for 1 h, plate the entire sample onto an LB plate containing 100 µg/mL ampicillin.
c. Once an aliquot of bacteria is thawed, do not use it again.

Then, set up the following transformations: Plate A: 2.5 µL of PCR 1 + 2.5 µL of PCR 2; Plate B: 2.5 µL of PCR 1 + 2.5 µL of TE; Plate C: 2.5 µL of PCR 2 + 2.5 µL of TE; Plate D: 0.5 ng of a supercoiled template in 5 µL of TE; and Plate E: 5 µL of TE.

The yield of colonies from plate A is >2 times that from plate B + C, confirming a high percentage of recombinants in plate A. Plate D is a transformation control, and should yield a thick lawn of colonies. Plate E is an antibiotic control, and should yield no colonies since the bacteria that have not been transformed are sensitive to ampicillin. Only 25 µL of cells are used for the control plates D and plate E, so that only one BRL tube, which contains 200 µL of bacteria, needs to be used.

7. Screen plasmids by placing individual colonies in 2 mL of LB broth containing 100 µg/mL of ampicillin. Grow the colonies at 37°C for 6–24 h.
8. Screen the plasmids by removing 2 µL of the LB broth, place it directly in a PCR tube, and amplify for 25 cycles (see **steps 2–4**) using primers that flank the mutated site or insert (e.g., M13 primers) ***(16)***.
9. In a mutagenesis protocol, a base pair can be mutated to either create or remove a restriction endonuclease site. In particular, the degenerate amino acid code allows one to create or eliminate a restriction endonuclease recognition site without altering the amino acid encoded at that site. Screen for the mutation by adding 3 U of the restriction endonuclease and 1 µL of the appropriate 10× restriction buffer directly to 5 µL of the unpurified PCR product in a total volume of 10 µL.
10. Assess cutting by minigel analysis. Typically, 50 to 100% of the clones contain the recombinant of interest. Then, purify the plasmid and sequence the mutated region (*see* **Note 7**).

4. Notes

1. Other investigators have used Pfu DNA polymerase instead of *Taq* DNA polymerase in recombination PCR ***(6)***. Pfu DNA polymerase has better fidelity than *Taq* DNA polymerase ***(17)***.
2. The primers that introduce point mutations (primers 1–4 in **Fig. 2** and primers 1 and 3 in **Fig. 3**) are designed to generate 15to 45 bp of homology between each end of one PCR product relative to the other PCR product. In all recombination PCR protocols, 24 bp of homology works very well, and alterations that generate long regions of homology do not work noticeably better. Decreasing the length of homology from 25 to 12 bp in an early protocol did decrease the transformation efficiency four- to five-fold. Single point mismatches lie no closer than six nucleotides from the 3′ end of a primer and are frequently placed toward the middle. Placing point mutations near the 5′ end of each mutating primer will generate two PCR products whose mutated ends have >24-bp of homology. Multiple point mismatches should be placed in the middle or toward the 5′ end of a primer, with primer lengths long enough to create 24-bp of homology between the mutated ends of the two PCR products. Primers that are nonmutating are generally 24 to 30 nucleotides long. These nonmutating primers can be designed to anneal to the β-lactamase gene so that they can be used with a variety of different plasmids. Frequently, the mutating and nonmutating primers are designed to be perfect complements to each other.

   For site-directed mutagenesis, since unique restriction endonuclease recognition sites almost always bracket the insert, the same linearized templates can be used for the mutagenesis of any single site in the insert. Primers 2 and 4 are nonmutating (see **Fig. 3**), and are conserved for each new site targeted for mutagenesis, so that only two new primers need to be generated for each site targeted for mutagenesis (via primers 1 and 3). Furthermore, only approx one half of the length of the entire template needs to be amplified in each of the two PCR amplifications, facilitating the mutagenesis of large constructs and permitting considerable flexibility in the primer design and sequence. Recombination PCR has been used to mutate constructs up to 7.1 kb ***(18)***.

3. Because the transformation efficiency is low, highly competent bacteria (transfection efficiency >1 × 10^9/µg of monomer pUC19) should be used. Using restriction endonuclease digested templates, the transformation efficiency is about 10 colonies with the mutation/ng total DNA used to transform *E. coli*.
4. After 14 amplification cycles, the PCR product yield is much higher when using a linear template than when using a supercoiled template.
5. If a plasmid template cannot be linearized outside the region to be amplified before PCR amplification, the PCR product must be removed from the supercoiled plasmid template. This can be accomplished either by agarose gel electrophoresis and extraction using GeneClean or by digestion with the restriction endonuclease *Dpn*I. When agarose gel resolution and GeneClean extraction are used, the entire PCR product should be gel purified and reconstituted in 25 to 30 µL of TE, and 2.5 µL is combined with the 2.5 µL of the other PCR product before transformation. If *Dpn*I is used, add 20 U of *Dpn*I directly to 25 µL of the PCR sample using the recommended 10× *Dpn*I buffer in a final total volume of 30 µL, and incubate the mixture at 37°C for 1 h. No further purification of the PCR product is necessary.
6. The exact buffer components and conditions for PCR vary with different primers and template.
7. There is always the possibility of a sequence error in a single clone after PCR amplification. The altered region should be sequenced, and one may choose to clone a restriction fragment containing the mutated or recombined region of interest into a construct that has not undergone PCR amplification.

Acknowledgments

This work was supported by the Roy J. Carver Charitable Trust, the University of Iowa through funds generated by the Childrens Miracle Network Telethon, and by National Institutes of Health grant R01 HG00569. We thank Jim Hartley for suggesting use of the restriction endonuclease *Dpn*I to remove supercoiled template from the PCR mixture.

References

1. Mullis, K., Faloona, F., Scharf, S., Saiki, R., Horn, G., and Erlich, H. (1986) Specific enzymatic amplification of DNA in vitro: the polymerase chain reaction, in *Cold Spring Harbor Symposia on Quantitative Biology*, vol. LI, Cold Spring Harbor Laboratory, Cold Spring Harbor, NY, pp. 263–273.
2. White, B. (1993) *Methods in Molecular Biology, vol. 15, PCR Protocols: Current Methods and Applications*, Humana, Totowa, NJ.
3. Jones, D. H. and Howard, B. H. (1991) A rapid method for recombination and site-specific mutagenesis by placing homologous ends on DNA using polymerase chain reaction. *Biotechniques* **10,** 62–66.
4. Jones, D. H. (1994) PCR mutagenesis and recombination in vivo. *PCR Methods Appl.* **3,** S141–S148.
5. Coco, W. M., Rothmel, R. K., Henikoff, S., and Chakrabarty, A. M. (1993) Nucleotide sequence and initial functional characterization of the clcR gene encoding a LysR family activator of the clcABD chlorocatechol operon in Pseudomonas putida. *J. Bacteriol.* **175,** 417–427.
6. Fridovich-Keil, J. L. and Jinks-Robertson, S. (1993) A yeast expression system for human galactose-1-phosphate uridylyltransferase. *Proc. Natl. Acad. Sci. USA* **90,** 398–402.

7. Goulden, M. G., Kohm, B. A., Santa Cruz, S., Kavanagh, T. A., and Baulcombe, D. C. (1993) A feature of the coat protein of potato virus X affects both induced virus resistance in potato and viral fitness. *Virology* **197,** 293–302.
8. Gibbs, J. S., Regier, D. A., and Desrosiers, R. C. (1994) Construction and in vitro properties of HIV-1 mutants with deletions in "nonessential" genes. *AIDS Res. Hum. Retrovir.* **10,** 343–350.
9. Singh, K. K., Small, G. M., and Lewin, A. S. (1992) Alternative topogenic signals in peroxisomal citrate synthase of Saccharomyces cerevisiae. *Mol. Cell. Biol.* **12,** 5593–5599.
10. Weiner, M. P., Costa, G. L., Schoettlin, W., Cline, J., Mathur, E., and Bauer, J. C. (1994) Site-directed mutagenesis of double-stranded DNA by the polymerase chain reaction. *Gene* **151,** 119–123.
11. Jones, D. H. and Winistorfer, S. C. (1992) Recombinant circle PCR and recombination PCR for site-specific mutagenesis without PCR product purification. *Biotechniques* **12,** 528–534.
12. Sambrook, J., Fritsch, E. F., and Maniatis, T. (1989) *Molecular Cloning: A Laboratory Manual*, 2nd ed., Cold Spring Harbor Laboratory, Cold Spring Harbor, NY, p. B23.
13. Sambrook, J., Fritsch, E. F., and Maniatis, T. (1989) *Molecular Cloning: A Laboratory Manual*, 2nd ed., Cold Spring Harbor Laboratory, Cold Spring Harbor, NY, p. A2.
14. Sambrook, J., Fritsch, E. F., and Maniatis, T. (1989) *Molecular Cloning: A Laboratory Manual*, 2nd ed., Cold Spring Harbor Laboratory, Cold Spring Harbor, NY, p. A4.
15. Sambrook, J., Fritsch, E. F., and Maniatis, T. (1989) *Molecular Cloning: A Laboratory Manual*, 2nd ed., Cold Spring Harbor Laboratory, Cold Spring Harbor, NY, p. A1.
16. Liang, W. and Johnson, J. P. (1988) Rapid plasmid insert amplification with polymerase chain reaction. *Nucleic Acids Res.* **16,** 3579.
17. Lundberg, K. S., Shoemaker, D. D., Adams, M. W. W., Short, J. M., Sorge, J. A., and Mathur, E. J. (1991) High-fidelity amplification using a thermostable DNA polymerase isolated from Pyrococcus furiosus. *Gene* **108,** 1–6.
18. Yao, Z., Jones, D. H., and Grose, C. (1992) Site-directed mutagenesis of herpesvirus glycoprotein phosphorylation sites by recombination polymerase chain reaction. *PCR Methods Appl.* **1,** 205–207.

71

Megaprimer PCR

Application in Mutagenesis and Gene Fusion

Emily Burke and Sailen Barik

1. Introduction

Since the advent of the polymerase chain reaction (PCR), a variety of PCR-based procedures of mutagenesis have been developed through the use of synthetic primers encoding the mutation. Among these, the megaprimer method and related ones *(1–5)* remain some of the simplest and most versatile. Variations and improvements of the basic technique have been suggested over the past few years; these include a combination of magapriming and overlap extension, improvement of yield, use of single-stranded DNA, avoidance of unwanted mutations arising from nontemplated insertions by *Taq* polymerase, and the inclusion of various kinds of mutations, including multiple, nonadjacent ones *(2–11)*. The basic method (**Fig. 1**) requires three oligonucleotide primers and two PCRs (termed PCR-1 and -2 here) using the wild-type DNA as templatc *(1,2,8,10)*. The "mutant" primer is represented by M and the two "outside" primers by A and B. The M primer may encode a substitution, a deletion, an insertion, or a combination of these mutations, thus providing versatility while using the same basic strategy *(10)*. The first PCR (PCR-1) is performed using the mutant primer M and one of the outside primers, such as A (**Fig. 1**). The double-stranded product A-M is purified and used as a primer (hence the name megaprimer; **ref. *1***) in the second PCR (PCR-2) together with the other outside primer, B. Although both strands of the megaprimer may prime on the respective complementary strands of the template, the fundamental principles of PCR amplification ensure that only the one that extends to the other primer, that is, B in Fig. 1, will be exponentially amplified into the double stranded product in PCR-2. As mentioned, the wild-type DNA is used as template in both PCRs. This article describes the most optimized megaprimer method in our experience and has drawn freely on the improvements described by various authors *(3–28)*.

1.1. Improving the Yield of PCR-2

Poor yields from PCR-2 have sometimes been reported even when proper primer design (see above) was followed, especially when the megaprimer is large (0.8 kb and above). Although the exact reasons remain unclear, the most likely reasons are

From: *Methods in Molecular Biology, Vol. 226: PCR Protocols, Second Edition*
Edited by: J. M. S. Bartlett and D. Stirling © Humana Press Inc., Totowa, NJ

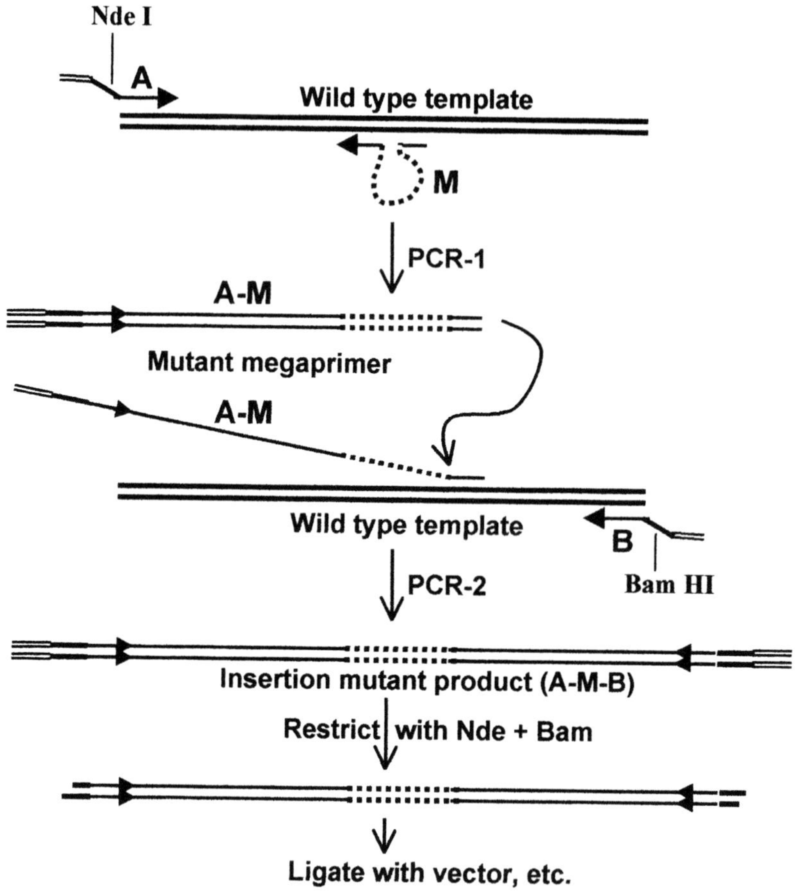

Fig. 1. The basic megaprimer method. Primers A, B, M, and the priming strand of the megaprimer AM are indicated by thinner lines with arrowhead, while the thicker double lines represent the wild type template (usually part of a plasmid clone, not shown). Primers A and B contain restriction sites (e.g., *Nde*I and *Bam*HI) indicated as thicker regions, and extra "clamp" sequence at the 5′ end indicated by double lines. The sequence to be inserted is shown as the dotted region in primer M and the subsequent PCR products. The final product containing the insertion is restricted and cloned.

the unique features of the megaprimer, viz., its double-stranded nature and large size. Strand separation of the double-stranded megaprimer is essentially achieved in the denaturation steps of the PCR cycle. Under some conditions, however, self-annealing of the megaprimer apparently tends to reduce the yield of the product *(4)*.

Various solutions to this problem have been suggested. In one approach, a biotin tag is added to the 5′ end of primer A, which would generate a biotin-labeled megaprimer in PCR-1. After denaturation, the biotinylated strand of the megaprimer is purified on avidin attached to magnetic beads *(26)*. In another method *(27)*, the use of two parallel templates allowed the inclusion of two outside primers as well as the megaprimer in PCR-2, resulting in a direct amplification of the final product. Use of a "one-tube" method (described above), when properly optimized, should eliminate loss of megaprimer during the purification step. Other strategies for increasing the yield of PCR-2 involve optimizing the concentrations of both template and megaprimer. In some instances, the use of higher amounts of template (in the microgram range, as opposed to nanogram quantities used in standard PCR) in PCR-2 has been shown to dramatically increase the product yield ***(4)***. Unfortunately, higher concentrations of template also tend to increase mispriming by a megaprimer with a mismatched 3′ end (our unpublished results). Thus, a more effective strategy may be to increase the amount of the megaprimer. A method that we have found useful is to perform the first several cycles of PCR-2 with the megaprimer only. After this initial asymmetric PCR, the small primer is added ***(11)***. In an optimization of this strategy ***(28)***, the starting concentration of megaprimer is increased to 6 μg (from 25 ng) per 100 μL of PCR-2. We have adopted a combination of the last two approaches in this article.

2. Materials

2.1. Template

About 100 ng of DNA template to be mutated (e.g., a gene cloned in a plasmid).

2.2. Primers

100 pg of oligonucleotide primers A and B, and 50 ng of mutant primer M; one primer, say A, in the opposite sense, and other primer, B, in the same sense as the mutant primer M (**Fig. 1**). Include restriction sites, preferably unique, in these primers so that the final product can be efficiently digested with restriction enzymes and cloned. The mutant primer may be designed to contain a point mutation, or insertion, or deletion, as desired (*see* **Notes 1** and **2**).

2.3. PCR Buffer

10× PCR buffer for *Pfu* polymerase (Stratagene Cloning Systems, La Jolla, CA) is: 200 m*M* Tris-HCl (pH 8.0–8.3); 100 m*M* KCl; 20 m*M* $MgCl_2$; 60 m*M* ammonium sulfate; 1% Triton X-100 100 μg/mL nuclease free BSA; the buffer is usually supplied with the enzyme by most manufacturers.

2.4. Deoxyribonucleotides

The final dNTP concentration is generally 200 μ*M* for each nucleotide. Make a stock dNTP mix containing 2 m*M* of each dNTP (dATP, dCTP, dGTP, dTTP); we make it by adding 50 μL of 10 m*M* stock solutions of each nucleotide, available commercially, into 50 μL H_2O, to produce 250 μL of the mix.

2.5. Analysis and Purification of DNA

A system for purifying the PCR products, such as gel electrophoresis, followed by recovery of the appropriate DNA band in the excised agarose fragment ***(8)***.

Wherever needed in this procedure, use deionized (e.g., Millipore) autoclaved water.

3. Method

3.1. PCR-1: Synthesis of the Megaprimer

1. It is assumed that the reader is familiar with standard PCR protocols. Use the following recipe for the first PCR. Make the following 100-μL reaction mix in an appropriate microcentrifuge tube (0.5 or 1.7 mL, dictated by the heating block of your thermal cycler): H_2O (75 μL); 10× PCR buffer (10 μL); 2 m*M* each of dNTP mix (10 UL; the final

concentration of each nucleotide is 200 μ*M*); Primer A (50 pmol); Primer M (50 pmol) (**Note 3**); DNA template (10–100 ng); and 2.5 U *Pfu* polymerase (0.5 μL);or 2.5 U *Taq* plus 0.1 U *Pfu* polymerase) for a total of 100 μL.
2. Vortex well to mix, then spin briefly in a microfuge. If the thermal cycler has a heated lid, then proceed to do PCR; otherwise, reopen the tube, overlay the reaction mixture with enough mineral oil to cover the reaction (~100 μL for a 0.5-mL microfuge tube), then close cap. The tube is now ready for thermal cycling.
3. Perform PCR-1 using the following cycle profiles. Initial denaturation: 94°C, 3 min; 30 to 35 main cycles: 94°C, 1 min (denaturation): T° (depending on the T_m of the primers), 2 min (annealing); 72°C, appropriate time, depending on product length (extension); and final extension 72°C, 1.5 × *N* min.

   After synthesis, the samples are maintained at 4°C (called "soak" file in older Perkin–Elmer programs) for a specified time. Some instruments lack an active cooling mechanism and keep samples at an ambient temperature of about 20°C by circulating tap water around the heat block, which appears to be adequate for overnight runs; others just shut off at the end of the final extension.
4. After PCR, proceed directly to the next step if there is no oil overlay. Otherwise, first remove the oil as follows. (If oil is not removed completely, the sample will float up when loaded in horizontal agarose gels!). Add 200 μL of chloroform to each tube. The mineral oil and chloroform will mix to form a single phase and sink to the bottom of the tube. Spin for 30 s in a microfuge. Carefully collect ~80 μL of top aqueous layer and transfer to a fresh Eppendorf tube.
5. Purify the megaprimer using any standard procedures such as gel purification (*see* **Chapter 18**) and use it in PCR-2 below.

3.2. PCR-2: Synthesis of the Mutant Using the Megaprimer

1. Reconstitute 100-μL PCR as follows: 10× PCR buffer (10 μL); 2 m*M* each of dNTP mix (10 μL; final concentration of each nucleotide is 200 μ*M*); All of the recovered megaprimer (A-M) from the previous step (20–50 μL); DNA template (0.2 μg); Make up volume to 100 μL with H_2O; and mix well.
2. Start reaction essentially as described for PCR-1, except that a "hot-start" is preferred (*see* **Note 4**) and is performed as follows. When the reaction is in the annealing step of the first cycle, open the cap briefly, quickly add 0.5 μL of *Pfu* polymerase (2.5 U, or 2.5 U *Taq* plus 0.1 U *Pfu* polymerase), and mix by pipetting. Close the cap and let PCR continue.
3. After five cycles, when the reaction is again at an annealing step, promptly add 50 pmol of primer B, mix well, and let PCR continue another 30 cycles. (The small amounts of primer B and *Pfu* polymerase do not contribute significantly to the total reaction volume and, therefore, have been ignored in the volume calculations).
4. Do another PCR in parallel, using primers A and B (but no megaprimer) and the same wild-type template; use an aliquot (5 μL) of this PCR as a size marker when analyzing PCR-2 by gel electrophoresis. This will also help in identifying the real product (in PCR-2) among the wrong ones that sometimes result from mispriming.
5. Gel purify the final mutant PCR product essentially as described earlier for the purification of the megaprimer (*see* **Notes 5** and **6**).

4. Notes

1. Design of the mutant primer. Perhaps the most unique feature of the megaprimer method is that the product of one PCR becomes a primer in the next, which creates the following potential problem. *Taq* polymerase, as a result of its lack of proofreading activity, tends

to extend the product DNA beyond the template by adding one or two non-templated residues, predominantly As *(12)*. When the product is used as a primer in the next round of PCR (PCR-2), these nontemplated A residues may not match with the template and, therefore, will either abrogate amplification ***(13–15)*** or produce an undesired A-substitution. A variety of solutions to this problem have been recommended ***(5,8,10,16)***. The first is to design the mutant primer such that there is at least one T residue beyond the 5′ end of the primer sequence in the template. Thus, when the complementary strand incorporates a non-templated A at the 3′ end, it will still be complementary to the other strand. If the template sequence does not permit this, a second solution is to use a mixture of *Taq* and *Pfu* DNA polymerases in 20:1 ratio in PCR-2 ***(3)*** or to use *Pfu* exclusively. This is what we have recommended in this chapter. The 3′ exonuclease activity of *Pfu* should remove any mismatch at the 3′ end of the megaprimer; however, this proofreading ability also necessitates the addition of at least 10 perfectly matched bases on both the 5′ and 3′ ends of the mutagenic primer ***(8,10,13,17)***. Finally, one can use enzymes, such as mung bean nuclease, that will remove nontemplated nucleotides from the megaprimer ***(16)***. In addition to these unique considerations, the general rules of primer design described below, should be followed.

2. Length of the megaprimer. Try to avoid making megaprimers (A-M) that approach the size of the final, full-length product (gene) A-B (*see* **Fig. 1**). Briefly, if M is too close to B, it will make separation of AB and AM (unincorporated, left-over megaprimer) difficult after PCR-2. When the mutation is to be created near B, one should make an M primer of the opposite polarity, and synthesize BM megaprimer (rather than AM), and then do PCR-2 with BM megaprimer and A primer. When the mutation is at or very near the 5′ or 3′ end of the gene (within 1–50 nucleotides), there is no need to use the megaprimer method; one can simply incorporate the mutation in either A or B primer and do a straightforward PCR using A and B primers! For borderline situations, such as when the mutation is, for example, 120 nucleotides away from the 5′ end of the gene, incorporation of the mutation in primer A may make the primer too big to synthesize; or else, it will make the megaprimer AM too short to purify away from primer B. In such a case, simply back up primer A to a few hundred bases further upstream to make the AM megaprimer longer. In general, realize that primers A and B can be located virtually anywhere on either side of the mutant primer M, and therefore, try to utilize this flexibility as an advantage when designing these primers.
3. Molar amount of megaprimer. Because the megaprimer is large, one needs to use a greater quantity of it to achieve the same number of moles as a smaller primer. Example: 50 pmol of a 20 nt-long single-stranded primer will equal 0.3 µg; however, 50 pmol of a 500 nt-long double-stranded megaprimer will equal 6 µg. A good yield and recovery of megaprimer is, therefore, important. If needed, do 2× 100 µL PCRs to generate the megaprimer. There is no need to remove the template DNA after PCR-1 because the same DNA will be used as template in PCR-2.
4. "Hot start" PCR-2. The hot-start technique used in PCR-2 works just as well as the more expensive commercial methods. Hot start tends to reduce false and nonspecific priming in PCR in general ***(29)*** and is particularly useful in PCR-2 of the megaprimer method (our unpublished observation).
5. Poor yield of mutant. If the final yield is poor, the surest strategy is to amplify a portion of the gel-purified mutant product in a third PCR (PCR-3) using primers A and B and hot start. This may also be necessary if PCR-2 produces nonspecific products in addition to the specific one. Before PCR-3 is conducted, however, it is very important to ensure that the mutant product of PCR-2 is well separated from the wild type template in the gel purification; otherwise, PCR-3 will amplify the wild-type DNA as well. The final gel-

purified mutant DNA (from either PCR-2 or PCR-3) is ready for a variety of applications, such as sequencing ***(30–36)*** or cloning ***(33,34)***.

6. Single-tube methods. Recently, various investigators have reported successful modifications of the megaprimer method in which the purification step is not required. One involves cleavage of the template, coupled with enzymatic removal of PCR-1 primers, to ensure amplification of the correct product in PCR-2 ***(24)***. A second possibility is to exploit the unusually high T_m of the megaprimer by designing a short, low Tm flanking primer for PCR-1, and a long flanking primer for PCR-2. This enables the use of a higher Tm for PCR-2 such that it will only allow annealing of the appropriate flanking primer ***(25)***. A third method uses a limiting amount of the first flanking primer, such that when the second flanking primer is added, the principle product will be the mutant DNA ***(17)***. Since we have not tested any of these modifications, the interested reader is advised to consult the original papers.

Acknowledgments

Research in the author's laboratory was supported in part by NIH Grant AI45803. S. B. is also a recipient of a Burroughs Wellcome New Initiatives in Malaria Research Award.

References

1. Sarkar, G. and Sommer, S. S. (1990) The "megaprimer" method of site-directed mutagenesis. *Biotechniques* **8,** 404–407.
2. Landt, O., Grunart, H.-P., and Hahn, U. (1990) A general method for rapid mutagenesis using the polymerase chain reaction. *Gene* **96,** 125–128.
3. Perrin, S. and Gilliland, G. (1990) Site-specific mutagenesis using asymmetric polymerase chain reaction and a single mutant primer. *Nucleic Acids Res.* **18,** 7433–7438.
4. Barik, S. and Galinski, M. (1991) "Megaprimer" method of PCR: Increased template concentration improves yield. *Biotechniques* **10,** 489–490.
5. Kuipers, O. P., Boot, H. J., and de Vos, W. M. (1991) Improved site-directed mutagenesis method using PCR. *Nucleic Acids Res.* **19,** 4558.
6. Merino, E., Osuna, J., Bolivat, F., and Soberon, X. (1992) A general PCR-based method for single or combinatorial oligonucleotide-directed mutagenesis on pUC/M13 vectors. *BioTechniques* **12,** 508–510.
7. Aiyar, A. and Leis, J. (1993) Modification of the megaprimer method of PCR mutagenesis: Improved amplification of the final product. *Biotechniques* **14,** 366–369.
8. Barik, S. (1993) Site-directed mutagenesis by double polymerase chain reaction. *Methods Mol. Biol.* **15,** 277–286.
9. Pont-Kingdon, G. (1994) Construction of chimeric molecules by a two-step recombinant PCR method. *BioTechniques* **16,** 1010–1011.
10. Barik, S. (1995) Site-directed mutagenesis by PCR: Substitution, insertion, deletion, and gene fusion. *Methods Neurosci.* **26,** 309–323.
11. Datta, A. K. (1995) Efficient amplificaton using 'megaprimer' by asymmetric polymerase chain reaction. *Nucleic Acids Res.* **23,** 4530–4531.
12. Clark, J. M. (1988) Novel non-templated nucleotide addition reactions catalyzed by procaryotic and eucaryotic DNA polymerases. *Nucleic Acids Res.* **16,** 9677–9686.
13. Barnes, W. M. (1994) PCR amplification of up to 35-kb DNA with high fidelity and high yield from lambda bacteriophage templates. *Proc. Natl. Acad. Sci. USA* **91,** 2216–2220
14. Sarkar, G., Cassady, J., Bottema, C. D. K., and Sommer, S. S. (1990) Characterization of polymerase chain reaction amplification of specific alleles. *Anal. Biochem.* **186,** 64–68.

15. Kwok, S., Kellogg, D. E., McKinney, N., Spasic, D., Goda, L., Levenson, C., et al. (1990) Effects of primer-template mismatches on the polymerase chain reaction: Human immunodeficiency virus type I model studies. *Nucleic Acids Res.* **18,** 999–1005.
16. Chattopadhyay D. and Raha, T. (1997) PCR mutagenesis: treatment of the megaprimer with mung bean nuclease improves yield. *Biotechniques* **6,** 1054–1056.
17. Picard, V., Ersdal-Badju, E., Lu, A., and Bock, S. C. (1994) A rapid and efficient one-tube PCR-based mutagenesis technique using Pfu DNA polymerase. *Nucleic Acids Res.* **22,** 2587–2591.
18. Rychlik, W. (1995) Selection of primers for polymerase chain reaction. *Mol. Biotechnol.* **3,** 129–134.
19. New England Biolabs Catalog (1995) pp. 208–209.
20. Innis, M. A., Myambo, K. B., Gelfand, D. H., and Brow, M. A. D. (1988) DNA sequencing with *Thermus aquaticus* DNA polymerase and direct sequencing of polymerase chain reaction-amplified DNA. *Proc. Natl. Acad. Sci. USA* **85,** 9436–9440.
21. Suggs, S. V., Hirose, T., Miyake, E. H., Kawashima, M. J., Johnson, K. I., and Wallace, R. B. (1981) Use of synthetic oligodeoxyribonucleotides for the isolation of specific cloned DNA sequences, in *ICN-UCLA Symposium on Developmental Biology Using Purified Genes*, Vol. 23 (Brown, D. D., ed.),. Academic Press, New York, pp. 683–693.
22. Tautz, D. and Renz, M. (1983) An optimized freeze-squeeze method for recovering long DNA from agarose gels. *Anal. Biochem.* **132,** 14–19.
23. Zintz, C. B. and Beebe, D. C. (1991) Rapid re-amplification of PCR products purified in low melting point agarose gels. *Biotechniques* **11,** 158–162.
24. Seraphin, B. and Kandels-Lewis, S. 1996. An efficient PCR mutagenesis strategy without gel purification step that is amenable to automation. *Nucleic Acids Res.* **24,** 3276–3277.
25. Ke, S. H. and Madison, E. L. (1997) Rapid and efficient site-directed mutagenesis by single-tube 'megaprimer' PCR method. *Nucleic Acids Res.* **25,** 3371–3372.
26. Fieschi, J., Niccoli, P., Camilla, C., Chames, P., Chartier, M., and Baty, D. (1996) Polymerase chain reaction-based site-directed mutagenesis using magnetic beads. *Anal. Biochem.* **234,** 210–214.
27. Upender, M., Raj, L., and Weir, M. (1995) Megaprimer method for in vitro mutagenesis using parallel templates. *Biotechniques* **18,** 29–32.
28. Smith, A. M. and Klugman, K. P. (1997) "Megaprimer" method of PCR-based mutagenesis: the concentration of megaprimer is a critical factor. *Biotechniques* **22,** 438–442.
29. Mullis, K. B. (1991) The polymerase chain reaction in an anemic mode: how to avoid cold oligodeoxyribonuclear fusion. *PCR Methods Appl.* **1,** 1–4.
30. Meltzer, S. J. (1993) Direct sequencing of polymerase chain reaction products. *Methods Mol. Biol.* **15,** 137–142.
31. Sarkar, G. and Bolander, M. E. (1995) Semi exponential cycle sequencing. *Nucleic Acids Res.* **23,** 1269–1270.
32. Maniatis, T., Fritsch, E. F., and Sambrook, J. (1982) *Molecular Cloning: A Laboratory Manual.* Cold Spring Harbor Laboratory Press, Cold Spring Harbor, NY, pp. 113–119.
33. Barik, S. (1993) Expression and biochemical properties of a protein Ser/Thr phosphatase encoded by bacteriophage λ. *Proc. Natl. Acad. Sci. USA* **15,** 10,633–10,637.
34. Mazumder, B., Adhikary, G., and Barik, S. (1994) Bacterial expression of human respiratory syncytial viral phosphoprotein P and identification of Ser237 as the site of phosphorylation by cellular casein kinase II. *Virology* **205,** 93–103.
35. Krowczynska, A. M. and Henderson, M. B. (1992) Efficient purification of PCR products using ultrafiltration. *Biotechniques* **13,** 286–289.
36. Stoflet, E. S., Koeberl, D. D., Sarkar, G., and Sommer, S. S. (1988) Genomic amplification with transcript sequencing. *Science* **239,** 491–494.

Index

Q

R